Resilience in Ecology and Health

Resilience in Ecology and Health

Edited by

Gerard Magill and James Benedict

Cambridge Scholars Publishing

Resilience in Ecology and Health

Edited by Gerard Magill and James Benedict

This book first published 2023

Cambridge Scholars Publishing

Lady Stephenson Library, Newcastle upon Tyne, NE6 2PA, UK

British Library Cataloguing in Publication Data
A catalogue record for this book is available from the British Library

ISBN (10): 1-5275-3657-2
ISBN (13): 978-1-5275-3657-9

We dedicate this book to Duquesne University in tribute to the Endowed Annual Conference Series on the Integrity of Creation.

TABLE OF CONTENTS

INTRODUCTION

GERARD MAGILL

In 2015, the President of Duquesne University in Pittsburgh in the United States (Charles J. Dougherty) commissioned an endowed annual academic conference series on the *Integrity of Creation* to celebrate the organization's Spiritan mission. The University is Catholic, being founded by members of the Congregation of the Holy Spirit, the Spiritans.[1] The current University President, Ken Gormley, continues to provide outstanding support for the conference, inspiring excellence as the series develops.

This conference series addresses emerging concerns and pivotal problems about our planet's environment and ecology, including this study, *Resilience in Ecology and Health*. Each conference is an interdisciplinary endeavor in the sense that presenters and participants from different disciplines are invited to engage each other in civil discourse on the selected topic. The conference has three goals: to provide a scholarly opportunity to engage with established and emerging research; to foster interdisciplinary discourse; and, to enlighten public awareness and discussion on the selected issues.

In Fall 2015, the topic of the inaugural conference was *Climate Change* as an urgent concern regarding the Integrity of Creation.[2] Before the conference occurred, there was the publication in May 2015 of the environmental encyclical of Pope Francis, *Laudato Si': Praise Be To You. On Care for Our Common Home*.[3] (As this book goes to press, Pope Francis is about to promulgate a new Apostolic Exhortation called *Laudate Deum*. Subsequent books in this series will discuss its significance for global climate, sustainability and resilience.) The insights of *Laudato Si'* continue to enlighten the discussion in this chapter. The Pope invited "every person living on this planet," "all people of good will," "to enter into dialogue with all people about our common home" as "a shared inheritance."[4] The call of Pope Francis to foster a "respect for the Integrity of Creation" is very similar to the focus of the Spiritan mission.[5] The second annual conference and its proceedings focused on the topic of *Integral Ecology: Protecting our Common Home*,[6] the conference title being inspired by a chapter in the encyclical of Pope Francis.[7] The third annual conference and its proceedings focused on the topic of *The Global Water Crisis*,[8] again inspired by the encyclical of Pope Francis.[9] The fourth annual conference and its proceedings focused on the topic, *The Global Sustainability Challenge*, reflecting the recognition of the Earth Charter by Pope Francis in his encyclical.[10] The fifth annual conference and its proceedings focused on the topic, *Toward a Healthy Planet*.[11] The sixth annual conference and its proceedings focused on the topic, *Strands of Sustainability*.[12] The seventh annual conference and its proceedings in this volume discuss *Resilience in Ecology and Health*,[13] connecting sustainability with resilience, ecology, climate, health and related issues.

As this conference series evolves, many other topics will be discussed to shed light on the *Integrity of Creation* from multiple perspectives. To safeguard our planet, we must be attentive to ecological and environmental concerns with air pollution, problems that arise from toxicity in the land and ocean regarding food sources and biodiversity, and many other crises, not least of which is how to anticipate the movement of vast populations from coastal regions that may become permanently flooded. These topics all highlight the basic problem of global sustainability and resilience.

The presentations at the conferences result from a peer-reviewed process for inclusion at the conference and in these published proceedings. The book chapters reflect the conference presentations and have been written to appeal to a general audience with rigorous scholarship, depicting the interdisciplinary focus of the discussions. The chapters are organized into several disciplines and categories that relate together in an integral manner. Each section is designed to present a variety of perspectives regarding resilience. The organization of chapters adopts this layout: resilience and ecology, resilience and health, resilience and biotechnology, and generational challenges regarding ecology and the environment.

The first section focuses on resilience and ecology to explain that sustainability and resilience in higher education depend on inclusivity, to present a Hegelian reflection on rewilding, and to explain the significance of ecological resilience for the future of humanity. The next section focuses on resilience and health to engage a variety of perspectives: assessing practical ethical challenges for resilience in healthcare in multicultural societies; considering the resilient tension between autonomy/rights and population/common good; highlighting the resilience of justice in public health in dealing with a pandemic; articulating the resilience of the principle of double effect reasoning in healthcare; exploring resilience in the diagnosis and treatment of psychiatric conditions; discussing how substance abuse treatment and resilience can decrease recidivism among incarcerated patients; and exploring the resilience of clinical spiritual care in a pluralistic care environment. The subsequent section focuses on resilience and biotechnology to address these emerging perspectives: ecological perspectives in a biotechnological age; the hermeneutics of nudging to enlighten the resilient reciprocity between transhumanism and nudging; resilience in embryo intervention to resolve serious genetic disease; resilience in genetic testing in the pursuit of sustainable healthcare; the resilient use of hyperspectral imaging in the research of cancer imaging; and the resilient use of photoacoustic flow cytometry for analyzing *Pseudomonas Aeruginosa* with bacteriophage tags. Finally, there is a section on generational challenges on resilience where undergraduate students present pivotal concerns around two sets of topics. The first set of topics deals with ecology and climate, discussing the following: the papal encyclical *Laudato Si'* and the ecology of creation; mitigation and adaptation regarding climate change through a just transition; anxiety and climate change; democracy's impact on climate change; climate change and democracy more generally. The second set of topics deals with the environment and conservation, discussing the following: technical progress and society; animal experimentation and the environment; agricultural pesticides and marine wildlife; primary causes and impacts of coral bleaching and future conservation outlooks; and recent California wildfires. The book's brief conclusion considers the ethical imperative to foster the reciprocity between resilience, ecology, and health within the general context of the climate challenge.

A few words of acknowledgment are appropriate to recognize the contribution of many in planning the annual conference series that has led to this collection of conference proceedings. Above all, the establishment of an endowment by President Charles J. Dougherty at Duquesne University to support this annual academic conference series presents a lovely legacy. Also, there is sincere appreciation for the continuing support of Ken Gormley as our current University President and for the ongoing support of the Spiritan Congregation that sponsors the University. The editors are very grateful for the advice and oversight of a highly dedicated Conference Planning Committee, an outstanding University librarian, Ted Bergfelt, and a superb support staff, including a very gifted group of international graduate students who provide outstanding support for the annual conference that leads to our published proceedings. Finally, we extend heartfelt gratitude to Ercan Avci, Ph.D., the conference coordinator. His detailed planning assured a seamless scholarly event with its multitude of posters and presentations and his meticulous formatting of the chapters in this work facilitated a timely submission of the manuscript to the publisher.

Notes

[1] See, http://www.duq.edu/about/mission-and-identity; also see, http://www.spiritans.org.
[2] See the conference proceedings, Gerard Magill, Kia Aramesh, eds., *The Urgency of Climate Change* (Newcastle Upon Tyne, UK: Cambridge Scholars Publishing, 2017), Introduction, xi-xiv. Permission has been provided to reiterate in the Introduction of this book materials from the Introduction to the first book in this series, *The Urgency of Climate Change*.
[3] Pope Francis, *Laudato Si': Praise Be To You. Encyclical Letter of the Holy Father Francis on Care for Our Common Home* (Vatican City: Libreria Editrice Vaticana, 2015).
[4] Pope Francis, *Laudato Si'*, no. 3, 28, 93.
[5] Pope Francis, *Laudato Si'*, no. 130.
[6] See the conference proceedings, Gerard Magill, Jordan Potter, eds., *Integral Ecology: Protecting Our Common Home* (Newcastle Upon Tyne, UK: Cambridge Scholars Publishing, 2018).
[7] Pope Francis, *Laudato Si'*, no. 124 (in chapter three), and chapter four.
[8] See the conference proceedings, Gerard Magill, James Benedict, eds., *Cascading Challenges in the Global Water Crisis* (Newcastle Upon Tyne, UK: Cambridge Scholars Publishing, 2019).
[9] Pope Francis, *Laudato Si'*, chapter 1, section II, "The Issue of Water," no. 27-31.
[10] See the conference proceedings, Gerard Magill, James Benedict, eds., *The Global Sustainability Challenges* (Newcastle Upon Tyne, UK: Cambridge Scholars Publishing, 2020). Also, see, Pope Francis, *Laudato Si'*, no. 207, refers to the *Earth Charter* for a sustainable, global society that emphasizes justice and peace (The Hague, 29 June 2000, at https://earthcharter.org/discover/what-is-the-earth-charter).
[11] See the conference proceedings, Gerard Magill, James Benedict, eds., *Toward a Healthy Planet* (Newcastle Upon Tyne, UK: Cambridge Scholars Publishing, 2021).
[12] See the conference proceedings, Magill, Gerard, James Benedict, eds., *Strands of Sustainability* (Newcastle Upon Tyne, UK: Cambridge Scholars Publishing, 2023).
[13] See the conference proceedings, Magill, Gerard, James Benedict, eds., *Resilience, Ecology and Health* (Newcastle Upon Tyne, UK: Cambridge Scholars Publishing, forthcoming).

Literature

Earth Charter. 2000. The Hague. At, https://earthcharter.org/discover /what-is-the-earth-charter.

Pope Francis. 2015. *Laudato Si': Praise Be To You. Encyclical Letter of the Holy Father Francis on Care for Our Common Home*. Vatican City: Libreria Editrice Vaticana.

Magill, Gerard, James Benedict, eds., forthcoming. *Resilience, Ecology and Health*. Newcastle Upon Tyne, UK: Cambridge Scholars Publishing.
Magill, Gerard, James Benedict, eds. 2023. *Strands of Sustainability*. Newcastle Upon Tyne, UK: Cambridge Scholars Publishing.
Magill, Gerard, James Benedict, eds. 2021. *Toward a Healthy Planet*. Newcastle Upon Tyne, UK: Cambridge Scholars Publishing.
Magill, Gerard, James Benedict, eds. 2020. *The Global Sustainability Challenge*. Newcastle Upon Tyne, UK: Cambridge Scholars Publishing.
Magill, Gerard, James Benedict, eds. 2019. *Cascading Challenges in the Global Water Crisis*. Newcastle Upon Tyne, UK: Cambridge Scholars Publishing.
Magill, Gerard, Jordan Potter, eds. 2018. *Integral Ecology: Protecting Our Common Home*. Newcastle Upon Tyne, UK: Cambridge Scholars Publishing.
Magill, Gerard, Kia Aramesh, eds. 2017. *The Urgency of Climate Change*. Newcastle Upon Tyne, UK: Cambridge Scholars Publishing.

CHAPTER ONE

PIVOTAL PERSPECTIVES REGARDING RESILIENCE IN ECOLOGY AND HEALTH

GERARD MAGILL

Introduction

This book continues a series of works that address the urgency of the Sustainable Development Goals adopted by the United Nations Development Programme in 2015,[1] such as slashing carbon dioxide emissions to net-zero if we are to prevent the average global temperature from increasing beyond 1.5 celsius above preindustrial levels.[2]

This book presents the proceedings of the 7th annual endowed conference on the Integrity of Creation, fostering the wholeness of creation from interdisciplinary perspectives. This 7th annual conference selected the topic, *Climate Resilience: Collaboration, Adaptation, and Action*. The perspectives in the study represent a variety of disciplines that engage each other in an integrative way. The chapters have been organized to let the dialogue unfold in an overlapping manner, with points of view developing from different angles.[3] The sections below summarize crucial standpoints as we seek to protect our common home. As mentioned in the book's Introduction, there are several sections to bring coherence to the contributions. Discourse on resilience can inspire engagement from many fields, including ecology and health, hence the title of the book, *Resilience in Ecology and Health*. The first section engages resilience and ecology, the second section addresses resilience and health, the third section discusses resilience and biotechnology, and the fourth section considers generational challenges, including ecology and climate as well as the environment and conservation. This opening chapter introduces many pivotal perspectives that appear throughout the book (using the abstracts submitted by the various contributors).[4] This overview is designed to assist readers in keeping an eye on the big picture as they explore the various topics. As the chapters develop, there is a cumulative sense of building resilience to advance the integrity of creation as a function of the common good, belonging to all and meant for all (adopting a phrase from the Papal encyclical, *Laudato Si'*).[5] What follows is a brief review of the chapters in each of the main sections of the book.

Resilience and Ecology

This first section of the book engages resilience and ecology to engage several fascinating topics. The first topic examines how the future of resilience and sustainability in higher education depends on inclusivity. Institutions of higher education can become more resilient in the face of climate change and contribute positively to a sustainable future. Here, resilience and sustainability foster a field of study and practice to create a relationship with the Earth within the limits of the planet's ecological systems that benefits all living beings. We need change in our institutional structures and current relationship with one another and the planet, developing a framework of truth, courage, and inclusivity.

The second topic presents a philosophical analysis by offering a Hegelian reflection on rewilding. This occurs by a detailed analysis of part of the last chapter "Absolute Knowing" in Hegel's *Phenomenology of Spirit*. This involves an attitude that can be called "Speculative Moral Receptivity," which is a moment in the becoming of Hegel's "Preserving Idea." Crucially, this Idea engages biodiversity. This interpretation is unorthodox: it involves moving beyond anthropocentrism, presenting the culmination of Hegel's phenomenology as a dialectical becoming that develops from receptivity to preservation, via biodiversity. This fascinating stance is consistent with current scientific efforts to rewild, as a kind of dialectical thinking that is itself a rewilding. The discussion gives intriguing examples of scientific, rewilding views of the earth Hegel would endorse today.

The third topic then advances the focus to consider ecological resilience as crucial for the future of humanity, specifically considering climate change as a critical issue in bioethics discourse because of its serious threat to humanity and creation. The fundamental causes of climate change include population growth, urbanization, and industrialization as co-dependent variables that lead to global warming. Because climate change leads to the alteration of ecosystem functioning, it exerts a negative effect on plants, animals, and human beings. To mitigate the effects of climate change, we need to alter fundamental assumption towards creation and to develop skills to improve climate resiliency via collaboration, adaptation, and strategic action.

Resilience and Health

This section of the book considers several pivotal perspectives regarding the reciprocity between resilience and health. The first topic examines resilience in healthcare to assessing practical and ethical challenges in multicultural societies. Ethical and practical challenges in multicultural societies prompt crucial risks to the sustainability of healthcare services and require exploring specific resolutions for governing these challenges. However, the prevalence of cultural and religious diversity and its impact on healthcare services lead to practical problems, such as language barriers, and can complicate distinct ethical standpoints. It is important to underline the primary practical and ethical challenges in multicultural societies and to make suggestions on how to alleviate these challenges. Cultural and religious diversity cause both practical and ethical difficulties, and practical challenges inevitably trigger ethical problems. There is a need to strengthen cultural competence, providing patients with language assistance services, and educating healthcare professionals to reduce practical problems. Regarding ethical issues, it is crucial to emphasize similarities rather than unduly stress differences among different cultures and religions. Additionally, it is necessary to support enhancing universal agreements on common ethical problems and to avoid absolute positions on ethical issues to manage different arguments and facilitate the sustainability of healthcare services.

The second topic presents the resilient tension between emphases on autonomy/rights and population/common good. Healthcare electronic data is everywhere–on computers, phones, tablets, and virtually any device that can access the internet. Patient electronic health records, medication information, genetic testing results, and so many more are out there. What is being done with all this information, and did patients consent to its use in public health interests? It is crucial to address whether it is ethically justifiable for big data banks to use personal health information without consent for public health surveillance. The answer to this question requires a deeper look at the tension between personal autonomy/human rights and population health/common good. These concepts need to be applied in the real world by describing big data, what it is, and how it can help healthcare, in addition to its ethical considerations and limitations. It is critical to understand data usage and to provide an explanation of the ethical basis for public health surveillance, including the hazards and obligations related to the issue of personal privacy vs public health surveillance.

The third topic further explores the public health perspective by considering the resilience of justice in public health when encountering a pandemic. Public health, being critical for global health, concerns the entire population rather than an individual. Hence, a focus on public health calls for a collective effort to maintain general well-being. The main concern of public health is prevention. When a health crisis occurs, public health authorities step forward to help mitigate the problem. Since ethics is required in general health issues, the critical priority of ethics in public health includes three general perspectives. Generally, the rapport between health and ethics needs to be recognized. More specifically, the principle of justice is indispensable for the work of public health. And practically, the principle of justice in public health is especially significant during a pandemic, such as Covid-19.

The fourth topic explains how a broad ethical approach is needed for resilience in health by explaining the complexities involved in double effect reasoning in healthcare. Written on the human heart is the desire to do good and to avoid evil. This desire is a primary consideration while making complicated decisions in healthcare. Circumstances can arise requiring an action that results in both a good and bad result. An overview of the history of this ethical principle is provided to consider practical applications that can foster resilience in health.

The fifth topic further explores practical aspect of resilience in health by examining the diagnosis and treatment of psychiatric conditions. As research has continued to develop, enhanced treatment methods have been used for patients with psychiatric conditions. The current status of psychiatric care can be reviewed through categories of diagnostic techniques and treatments. Diagnostic procedures include specialized physical examinations, lab tests, and psychiatric evaluations. Treatment methods include psychosocial methods, which include psychotherapy, art and music therapy, and psychoeducation, or medication interventions such as antipsychotics, benzodiazepines, and antidepressants. Though the diagnostic and treatment methods have progressed and served as a foundation for psychiatric care, the current methods lead to gaps in long-term care that must be further addressed in research.

The sixth topic continues this practical focus on resilience and health by explaining how substance abuse treatment and resilience can decrease recidivism among incarcerated patients. More than half of all prisoners meet the diagnostic criteria for drug abuse or dependence. For example, a national survey of New Mexico's jails by the Bureau of Justice Statistics showed that 70% of inmates met criteria for substance abuse, as opposed to the 9% of the United States general population. Since the beginning of data collection in 2001, no manner of death has increased more than drug overdoses and alcohol intoxication. Treatment of substance abuse disorders and fostering of resilience together lead to a decrease in crime rates and recidivism.

The seventh topic continues the practical focus on resilience and health by considering the significance of clinical spiritual care in a pluralistic health care environment. There are numerous codes of professional ethics for different health care related fields; clinical chaplains, as a progressively professionalizing field, have likewise created and continue to develop professional ethics expectation to foster ethics resilience in the field. A crucial contribution deals with a popular approach in ethics theory (called Principlism) that focuses on the four widely acknowledged primary ethical principles of autonomy, nonmaleficence, beneficence, and justice. The discussion explains how clinical chaplain care engages these primary ethical principles in four arenas: the patient's autonomy and the chaplain's faith, nonmaleficence and proselytization, beneficence and clinical chaplain care, and finally, justice and patients of other faiths.

Resilience and Biotechnology

The next section of the book engages several emerging perspectives on resilience regarding biotechnology. The first topic examines ecological resilience in a biotechnological era regarding xenotransplantation. Researchers for a long time have utilized animals to find appropriate medical interventions for human suffering. Hence, an ethical assessment is needed regarding the use of animals employed in xenotransplantation. Ecological resilience focuses on interventions that enhance both nature's and humanity's benefits, adopting the approach of common morality as a principal source of moral justification. When humans utilize other species for their ends, such as via xenotransplantation, ecological resilience requires caution when balancing risks and benefits in such endeavors.

The second topic adopts a futuristic stance to focus on the resilient reciprocity between transhumanism and nudging in order to clarify the significance of nudges in these biotechnological processes. The discussion considers pivotal issues regarding of integrating technology into human existence. Transhumanism is a movement that is presented as the avenue to overcome our physical and mental limits. It is firmly rooted in Enlightenment humanism that supplanted religious dogma with human reason. Transhumanism presents new opportunities to merge technology with our biology. The movement benefits from using nudge theory. Nudge theory seeks to assist people in making better decisions by using choice architecture. Issues arise when the assumptions about nudging someone toward an outcome are questioned, such as what is the ideal body that we are nudging people toward.

The third topic deals with breakthroughs regarding resilience in embryo intervention to resolve serious genetic disease. Embryo gene intervention has been debated for many years. The discussion here highlights the pivotal distinction between somatic and germline gene therapy to consider a variety of practical issues including preimplantation genetic testing, ex vivo gene therapy, and gene therapy with emphasis on SCID and spinal muscular atrophy.

The fourth topic continues the focus on genetics to consider resilience in genetic testing. Preimplantation genetic diagnosis (PGD) involves testing embryos for genetic variations and predisposition to disease. This fast developing technology can help humanity eradicate disease while reckoning with just how far society is willing to go regarding embryo science. For example, may PGD be permitted to enable parents to choose the sex of their offspring? Perhaps even more controversially, PGD allows prospective parents to choose to purposely give their potential children genes for disabilities like deafness or dwarfism. An ethical rationalization for genetic testing can benefit from adopting four standard principles in bioethics (autonomy, nonmaleficence, beneficence, and justice) as well as the philosophical ideas of utilitarianism and deontology. Because genetic testing can enhance the futures of individuals and humanity, it is crucial for a resilient approach to health care.

The fifth topic engages another very practical focus in biomedical engineering to consider resilience in research on cancer imaging. Medical imaging began with the discovery of X-rays in 1895 by German Professor of physics, Wilhelm Rontgen. Over the years, many different imaging modalities have been introduced into the medical field to diagnose diseases and make progress in the medical industry. Scientists try to push the boundaries of medicine as far as possible to ensure current and future generations are privileged with the best technology. Each modality, from CT scans to multispectral and fluorescence imaging have been engineered to address a specific issue. Each modality uses different technological advances and methods to see into the human body allowing us to create images to address issues within human anatomy correctly and accurately. More specifically, hyperspectral imaging can be especially helpful for cancer imaging A sound bioethical analysis is indispensable for the adoption of hyperspectral imaging in the medical field of cancer research.

The sixth topic continues this science focus on biomedical engineering to consider the resilient use of photoacoustic flow cytometry for analyzing *Pseudomonas Aeruginosa* with bacteriophage tags. The number of daily bacterial infections is climbing, and the CDC explains that this is due to the antibiotic-resistant threat. Hence, it is important to find a faster way of bacterial identification. It currently takes 1-4 days for a medical lab to get a successful culture diagnosis to identify bacterial infection. Photoacoustic flow cytometry can be used as an alternative method resulting in swift identification within an hour. This research may provide evidence of a more resilient method to identify bacteria in the medical setting.

Generational Challenges

This section of the book considers generational challenges that students encounter. These challenges are organized into two over-lapping areas: generational challenges on ecology and climate and generational challenges on the environment and conservation.

On the generational challenges on ecology and climate there are five perspectives. First, there is a fascinating philosophical account of resilience in the ecology of creation. In 2015 Pope Francis wrote his Encyclical Letter *Laudato Sí* on the environment. He outlines a Christian view of the natural world as created and sustained by God for our use and preservation for our lives and the lives of all creatures. Following the inspiration of St. Francis of Assisi, the Pope warns against the narrowing of our engagement with the natural world only to our anthropocentric interests. He attributes many environmental harms specific to this widespread attitude.

Second, the approach of Pope Francis is applied to resilience in climate change to discuss mitigation and adaptation through a just transition. Climate change is the largest threat to humanity, but it doesn't impact everyone equally. It is often socioeconomically sensitive populations who suffer the most; for example, Native Americans and the urban poor are directly affected by rising temperatures while being the least likely to contribute to climate change. In combating climate change, it is imperative that adaptation and resilience policies serve to mitigate the historic inequities already present in society. Doing so can benefit from the application of the philosophy of justice as fairness.

Third, the need for resilience regarding climate inevitable causes anxiety. In psychological studies, results have shown an increase in anxiety levels connected to the current climate change crisis. Environmental dangers, including overpopulation, limited resources, destruction of ecosystems, and a variety of natural disasters, are contributing to changes in the mental health field. There are different types of anxiety responses and treatment options for individuals experiencing symptoms of climate change-induced anxiety. Climate change is a global issue that affects every living being. To better address this anxious response, fostering resilience can be helpful.

Fourth, another perspective on climate is the resilient impact of democracy on climate change. For democracy to be a successful mitigation tool for climate change, it has to follow a course of action, adaptation, and collaboration in order to be successful. The word democracy means "rule by the people" and therefore people are the reason why democracy can focus itself on climate change. Because people are the heart of democracy, change cannot happen fully without their consent. It is essential that resilient democracy is a factor towards making a difference in the mitigation efforts against climate change.

Fifth, when considering climate change and resilient democracy, it is crucial to foster the willingness of governments for both short term and long-term solutions to help the environment.

On the generational challenges on the environment and conservation there are another five perspectives. First, it is necessary to appreciate the negative impact of technical progress on the resilience of society. Two things can be noted about modern society: industrialization has negatively affected the natural environment and that many citizens point to feelings of loneliness and isolation. As society has progressed into a more industrial state, communities and individuals increasingly became separated from their environment. As society distanced itself from the environment, the environment was seen as a mere "resource" to be used for capitalistic and industrial means. By separating the environment from society, individuals and communities lost their natural connection to the environmental world. Likewise, as the environment and nature were alienated and commodified, communities suffered alienation as more traditional means of socialization became weaker. Further, as industrial societies grew, the individual became more and more isolated, leading the individual to be seen as a commodity. Ultimately, individuals became objectified for capitalistic exploitation and consumerism, further isolating us from each other, their communities, and the environment.

Second, another critical way to foster a resilient environment is to address the ethical problem of animal experimentation. Many organizations and people such as People for the Ethical Treatment of Animals are against testing animals for any means, such as cosmetic, industrial, or medicinal. Peter Singer is a well-known philosopher who does not support animal experimentation. Singer explains three main arguments to defend his views in his article "Animal Liberation." Other than the harm done to animals, the process of animal experimentation has negative effects on the environment, such as human health and the amount of waste being produced. Since animal experimentation has proven to be so harmful, researchers are examining alternative methods that will benefit animals, humans, and our environment.

Third, resilience in the environment also requires being attentive to agricultural pesticides and marine wildlife. The pesticides and nutrients used in agricultural farming are running into our major bodies of water, allowing for the depletion of oxygen, and the harming and killing of marine wildlife. This directly affects our oceans and other water sources, but also the wildlife that resides within them and the different forms of life that may use them as a resource, including humans. A philosophical lens can aid in guiding our beliefs about what is right and wrong when it comes to looking at the environment, especially when analyzing what is causing damage to it, including water sources and their wildlife counterparts.

Fourth, resilience in the environment includes critical consideration of the causes and impacts of coral bleaching. The issue of mass coral bleaching has threatened coral reefs in every region for the past 10 decades. These events occur when the algae endosymbionts of coral are lost, resulting in a rapid paling of color. Studies indicate factors of climate change, such as increased temperatures and variations in precipitation, as the causes of coral bleaching events. Other human-driven components such as reckless diving practices and pollution runoff play a role in coral bleaching. This decline in coral health, directly and indirectly, affects a variety of organisms and ecosystems due to the importance of coral within aquatic communities. Coral reefs provide habitats, shelter, and spawning grounds for marine life, as well as protection of coasts from storms, waves, and floods. We rely upon reefs for food, resources, and income from recreational activities. Due to our extensive significance, we must reduce destruction by conducting research and implementing stricter laws.

Fifth, an increasingly problematic challenge to a resilient environment is the increasing range of wildfires, as evidenced so prominently in California. The climate in California has always been fire-prone, with low levels of rainfall and high heat in the summertime that dries out vegetation. Often, all that is needed is a spark to turn the dry brush on the ground into fuel. California continues to experience dramatic wildfire seasons, with records being broken regularly regarding both number and size of the fires. While these conditions are not new to the area, the fires are worsening every season. The increasing severity of the fires can in part be attributed to the changing global climate, but also to the history of fire suppression in the area. There were some unique features of the 2020 season, such as the new records it set, and the COVID-19 pandemic that may have effects on future seasons.

Conclusion

In the concluding chapter, the analysis identifies some features of resilience in ecology and health (different from those discussed in the book's chapters) that have emerged with prominence in recent literature. This discussion emphasizes the ongoing ethical imperative to foster ecology and health in order to build a more resilient world.

Notes

[1] See, United Nations Development Programme, *Sustainable Development Goals* (2015), at, https://www.undp.org/content/undp/en/home/sustainable-development-goals.html.

[2] International Energy Agency, *Net Zero by 2050* (May 2021), https://www.iea.org/reports/net-zero-by-2050.

[3] The organization of this chapter adopts the same approach as previous books in this annual series of conference proceedings.

[4] This discussion reflects the chapters in this book that constitute proceedings of the conference on *Climate Resilience: Collaboration, Adaptation, and Action* in Spring 2022.

[5] Pope Francis, *Laudato Si': Praise Be To You. Encyclical Letter of the Holy Father Francis on Care for Our Common Home* (Vatican City: Libreria Editrice Vaticana, 2015), §23.

Literature

International Energy Agency, *Net Zero by 2050*. At, https://www.iea.org/reports/net-zero-by-2050.

Pope Francis. 2015. *Laudato Si': Praise Be To You. Encyclical Letter of the Holy Father Francis on Care for Our Common Home*. Vatican City: Libreria Editrice Vaticana.

United Nations. 2019. *Climate Change*, at https://unfccc.int/event/cop-25.

United Nations Development Programme. 2015. *Sustainable Development Goals*. At, https://www.undp.org/content/undp/en/home/sustainable-development-goals.html.

I. Resilience and Ecology

CHAPTER TWO

THE FUTURE OF SUSTAINABILITY AND RESILIENCE IN HIGHER EDUCATION DEPENDS ON INCLUSIVITY

MEGHAN CHAPPLE

Introduction

This chapter is intended to support the Spiritan mission of Duquesne University celebrated at the Integrity of Creation Conference. Both the presentation and chapter are based on my professional experience in creating change for sustainability.

Through academics and practice, Catholic institutions of higher education can become more resilient in the face of climate change and contribute positively to a sustainable future. I define sustainability as a field of study and practice to create a relationship with the Earth within the limits of the planet's ecological systems that benefits all living beings. To make the change towards a more sustainable university, we need change in our institutional structures and current relationship with one another and the planet. Here I share my approach to making change for sustainability in higher education with institutions and individuals. I reference how we have applied an approach I developed during my time in the Office of Sustainability at Georgetown, a Jesuit university. Based on two decades of work in the professional field of sustainability, I believe the resilience of our universities and colleges and the future of sustainability in higher education can be realized through a framework of truth, courage, and inclusivity.

A Framework of Truth, Courage, and Inclusivity for Making Change for Sustainability

Truth

To make effective changes in higher education for a sustainable future, we must be willing to face both the scientific and social truth of the situation in the world and at our institutions. Only by acknowledging the situation with an understanding of the planetary limits, the social impacts, and economic models, can we begin to understand the current system and look to new paradigms and new solutions.

Globally society is facing daunting challenges as pertains to our relationship with the Earth. Taking into account research-driven, evidence-based facts, we see that the choices we have been making as a global human society toward unlimited economic growth have negative consequences related to levels of consumption, biodiversity loss, and disproportionate negative impacts on marginalized human populations.

Fortunately, we have many options and choices to address these challenges. For example, there exists a personal will by consumers in the global community to change their habits for the better. Another promising insight is that the more justice informed and inclusive our solution development processes are, the more effective they can be.

Actions that curb global warming to 1.5°C will significantly reduce damages. Long-term costs to manage losses will escalate if we do not take action today. Given the urgency of the issues, swift action is necessary to mitigate the worst potential outcomes. We can use the hard truths and the hopeful truths to inform how and when we want to make a change.

Here are some basic facts about the current situation in the world:

- Consumption—The richest 1% of the global population emits two times more carbon emissions than the poorest 50%.[1]
- Biodiversity—Extinction rates are tens to hundreds of times higher than they have been in the last 10 million years.[2]
- Racial Disparities—In the United States Black and African Americans and Latinos are 30 to 50 percent more likely than whites to work or live in areas with the highest projected increases in climate-related joblessness, illness, and death.[3]
- Personal Will—Fifty percent of global consumers are willing to change purchasing habits to help reduce environmental impact.[4]

- Inclusivity—When women participate equally with men, climate policy interventions are more effective.[5]
- Time Sensitivity—We need to reduce climate emissions by 45% by 2030 to stay below catastrophic levels of a 1.5°C increase in the average global temperature.[6]

When we face these truths, we can choose to see that we are all part of a system. The good news is we are interconnected and can find levers for change in the system. With more than a thousand Catholic institutions of higher education globally,[7] the potential role of this sector in addressing the sustainability of human and ecological life on our planet Earth is noteworthy. Catholic colleges and universities provide education to future leaders on sustainability, contribute knowledge and insight from faculty researchers, and play critical roles as anchor institutions in local and global communities that practice what we teach and preach about sustainability. And as declared by Pope Francis in referencing Pope John Paul II in the *Laudato Si'* encyclical on climate change, Catholic institutions have a responsibility to improve their impact on the sustainability of the Earth.[8]

Courage

An important tool for making change is the courage to process the emotional aspect of this work both in terms of facing the truth of our current situation and in terms of collectively developing and envisioning an alternative future.

In facing the truth of the situation, we see that we have created a world where wildfires and extreme weather are becoming the norm, sea levels are rising and eroding coastlines, land is becoming devastatingly arid for agriculture, oceans are swamped with plastic waste, species are going extinct at astronomical rates, disease and famine are spreading due to ecological disruption, and climate migration is stressing governments and economies. Climate anxiety is a researched and documented phenomenon. In Generation Dread Britt Wray states that we are experiencing an existential crisis and people are enduring sadness, stress, and fear about the future of the planet.[9]

You may want to dismiss or ignore any unpleasant feelings you have about the state of our planet that we call home. Ironically, these feelings may be your best tool for making change in yourself and in the system of which you are a part.

Imagining a better future can seem pointless. Identifying scalable solutions can be daunting. Building support for global fixes can feel overwhelming. It takes courage to imagine something different, to dare to dream. Creating new visions is essential to making change.

Courage requires vulnerability which according to Brené Brown involves uncertainty, risk, and emotional exposure.[10] Making change requires courage to face uncomfortable personal feelings and to dare to create something different. We can use curiosity, inquiry, and self-reflection to transform our discomfort into a tool for change.

Processing our emotions can help us shift our personal views on the way the world can be and our role in it. This is one way to transcend paradigms, which Donella Meadows points out is the most foundational way to create change in a system.[11]

Inclusivity

If we assume that university decision-making is the proverbial table, inclusivity in sustainability planning and action means creating opportunities for and then committing to the process of having people *claim* their seats at the table, as well as building genuine relationships with and *creating* seats at the table for marginalized voices in meaningful and reciprocal ways. This includes marginalized voices from human populations as well as flora and fauna populations. Pope Francis refers to this as "responding to the cry of the Earth and the cry of the poor."

"We are faced not with two separate crises, one environmental and the other social, but rather with one complex crisis which is both social and environmental… Today, however, we have to realize that a true ecological approach always becomes a social approach; it must integrate questions of justice in debates on the environment, to hear both the cry of the earth and the cry of the poor."[12]

There are well-documented benefits to diversity and inclusion in work settings.[13] The potential advantages of inclusivity in the sustainability process include diverse perspectives, momentum for change, and equity in practice. First, bringing in a diversity of voices provides an opportunity for different perspectives to influence the university's approach to creating a more sustainable future. This may lead to different approaches to how the institution relates to the planet and even assesses its relationship to the planet. New perspectives can help fertilize the ground for paradigm shifts that will lead to long-term change.

Secondly, inclusivity in the process stands to build a shared collective vision and momentum at an institutional level. The more parties from across the institution, as well as those who fall between institutional structures, who are involved in creating a new vision for the future, the more relevant that vision is to them, and thus the more momentum there is towards that vision. Peter Senge has been teaching for decades that it is not just what the vision is, but what the vision does.[14] And as adrienne maree brown points out that what we focus on grows.[15] Collectively envisioning a more sustainable future shifts our frame of mind and focuses our attention toward a new way of being.

Thirdly, inclusivity in the sustainability process increases the opportunity for equity both in terms of addressing and even eliminating inequities of environmental harms, and in terms of creating a just transition to a better future. These are two sides of the same coin – minimizing harm and generating a new way that aspires to avoid harm in the first place.

Applying the Framework of Truth, Courage, and Inclusivity for Sustainability at Georgetown University

Sustainability at Georgetown

The situation on sustainability at Georgetown is rich in its complexity. As a responsible member of the global community, the university has made voluntary public commitments to address sustainability. The commitments are part of collective efforts across the higher education sector to make change that is at a larger scale than one university acting alone. They include:

- Laudato Si' Seven Year Action Plan
- U7+ Climate Action Plan
- International Sustainable Campus Network
- Ivy+ Sustainability Consortium
- Washington, D.C. College & University Sustainability Pledge

The university has made progress in taking action to address its commitments to create a sustainable future. Starting in 1984 Georgetown constructed one of the largest rooftop solar arrays at the time on the roof of the Intercultural Center (ICC) on campus. New research and academic initiatives have emerged and thrived including the Climate Center that launched in 2009 and most recently the Earth Commons Institute in 2021. Along the way in 2020, the university signed a large-scale offsite renewable energy power purchase agreement for 100,000 megawatt-hours of electricity to support renewable energy development in the Mid-Atlantic region. That same year the university enacted a policy to divest the endowment from public securities of fossil fuel companies within five years and to divest from existing private investments in those companies over 10 years. Perhaps the most impactful internal cultural step, in 2021 the university instituted "Care for Our Common Home," as part of the Spirit of Georgetown values[16] inspired by Pope Francis' encyclical letter Laudato Si' to care for the Earth.

At Georgetown, the Office of Sustainability has launched a planning process for sustainability. We are exploring the role each functional unit can play in addressing the practice of sustainability at the university as described in Table 1. Identifying how to integrate sustainability across the university is essential to making collective change. By addressing sustainability in practice across administrative functions, we provide the foundation for academic impact. The Office of Sustainability is collaborating with internal partners in these units to generate creativity in the implementation of our sustainability goals and targets.

Table 1. Potential Areas for Integration of Sustainability into University Functions at Georgetown

The Academy	Sustainability in teaching and research to educate future leaders, generate thought leadership, host dialogue, and offer policy and scientific solutions
Mission & Ministry	Ecological spirituality and lifestyles
Community Relations	Engagement with local communities, collaborative projects, experiential learning, community-engaged research
Athletics	Sustainability in athletic fields, facilities, concessions, travel, food, apparel, merchandise
Human Resources & Student Services	Green benefits, cultural change, diversity, inclusion, mental health, wellbeing
Information Systems	Smart green performance, transparency, servers
Campus Infrastructure	Sustainability in facilities, grounds, energy, water, zero waste, transportation
Procurement	Circularity, plastics, food, embedded carbon, healthy materials
Investment	Sustainability in the endowment, retirement plans, treasury

Truth

A key component of facing the truth at an institution of higher education is assessing progress. At Georgetown, we have more truths to uncover to make continued strides toward a sustainable future. A large part of our work entails setting goals and targets for action and then collecting data to indicate how far along we are in meeting those targets. We are finding that there is always more to be done to deliver a sustainable future.

We are also committed to providing a transparent process. We share data on the university's past performance in areas of interest including the baseline impact, benchmark with others in the sector, targets that we aim to reach, and the rate of progress the university has made.

Courage and Inclusivity

The sustainability planning process at Georgetown is designed to gather input from a range of students, faculty, and staff, and is intentionally engaging representatives of typically marginalized groups. The Office of Sustainability's intention is to generate momentum with members of the university community to courageously create a collective vision. We are working through the difficult emotions that arise in the members during this change process, and endeavor to acknowledge the emotions as a normal part of the change process.

University leaders are part of three groups who provide input through meetings facilitated by the Office of Sustainability, as listed in Table 2. The Office also intends to gather input and feedback from external partners in the private, public, and non-profit sectors.

Through meetings and workshops facilitated by the Office of Sustainability, members of the aforementioned groups at the university provided input to a collective vision for sustainability at Georgetown to include the areas listed in Table 3. Themes that we will carry throughout these focus areas include equity, economics, and climate change.

Table 2. Georgetown University Engagement Structure for Sustainability Plan

Executive Steering Committee	Senior university leadership who provide ambition level and endorsement of the plan.
Sustainability Leadership Council	Leaders with working knowledge of university functions to authorize a plan that strikes the right balance between ambitious and achievable and to charge their teams on implementation.
Sustainability Advisory Group	Representative student, faculty, and alumni leaders in the Georgetown community to shape sustainability priorities and socialize the plan rollout such that it resonates with the entire university community.
External Collaboration Network	Panel of external collaborators with a broad perspective to ensure the plan sets up Georgetown to be a leader in sustainability.

Table 3. Scope of Georgetown Sustainability Plan: Focus Areas

Climate Change	Buildings, energy, adaptation & resilience, and investments
Materials	Waste management, plastics, reuse
Water	Stormwater, water efficiency, water procurement
Food	Procurement, service, urban agriculture, composting
Nature	Biodiversity, landscape, plantings, habitats
Transportation	Commuting, EVs, accessibility, air travel

During the sustainability planning process at Georgetown, the Office of Sustainability provided training for university leaders on bringing emotional courage to the effort. During an exercise with university leaders, we reviewed the dire state of the planet and the urgency for change, we asked these questions of university leaders:

1. Recognize your feelings about the state of the planet. Ask, “What do I feel?”
2. Feel your feelings. Ask, “How do I feel?” such as the sensation in your body.
3. Listen to your feelings. Ask, “What power do I have to do something in response to these feelings?”

Several members of the group courageously tapped into their curiosity and used self-reflection to uncover the motivations and actions behind their emotional discomfort. One person stated that they felt guilty and then found that they were inspired to take responsibility on campus for the problems. Another person expressed a feeling of being overwhelmed and was able to transition that into a commitment to take a first step with their team towards making change. The leaders then broke into small groups and began making suggestions on how to address the sustainability issues at Georgetown through their units including operations, procurement, facilities, student affairs, human resources, public affairs, and others. Their feelings led them to solutions and ideas. The process allowed them to share ideas for a different future, and acknowledge their power in making change.

Environmental equity and transparency are integral components of the sustainability plan and will be woven into the scope of the sustainability plan. As the Office leads the university through the planning and implementation of goals and targets in each of the sustainability focus areas, we will take into consideration the disparities created by the environmental impacts of the university's actions, and work to develop solutions that are inclusive and create a more just transition to a sustainable future.

Whose Truth: Sustainability & Indigenous Philosophy
As part of the inclusive sustainability planning process, the Georgetown Office of Sustainability is incorporating insights that are shifting paradigms. For example, the Office of Sustainability co-hosted a workshop with faculty expert Dr. Shelbi Nahwilet Meissner.[17] The purpose of the workshop was to help faculty, students, and staff begin to explore tensions and possibilities for convergence between sustainability and indigenous philosophy. Participants left with a better understanding of the limits of the sustainability planning process within the current Western model and the opportunity for more divergent thinking. The Western model promotes goals and targets, while indigenous philosophy rests on tenets of relationality and reciprocity. We explored the implications for sourcing food at the university, land ownership, and our relationship with nature. While there were countless moments of discomfort, the group acknowledged that participants were coming from a place of caring, which could emerge into a coalition of caring. The workshop helped to build inclusivity in the sustainability planning process. It also uncovered the truth from different perspectives and built understanding and trust.

Conclusion

The professional field of putting sustainability into practice at institutions of higher education has matured over the last three decades. At the primary level, the practice of creating a more sustainable university requires management methods such as: data analysis of the current state, setting goals and targets, administrative structures and incentives for implementation, financial analysis of proposed solutions, the latitude to pilot technological and policy innovations, and even cooperation with partners outside of the university.

While that may be the case, through my personal experience I have learned that making change for a more sustainable future through higher education also requires additional virtues-based resources of truth, courage, and inclusivity. Acknowledging and understanding the power that comes in the practice of these virtues is an important part of making change, because it may help us navigate challenges, tensions, and resistance that are sure to arise.

Key to creating more resilience in higher education in the change process for sustainability is ensuring the process is inclusive. I believe that when different perspectives are included in a collaborative way, where power is shared, then a shared vision for sustainability emerges that provides a positive focus to work towards together. A shared vision is more resilient in the face of tensions because it takes into account potentially differing perspectives that might otherwise put the vision for sustainability at odds. Moreover, the process of creating a shared vision that is inclusive is likely to build relationships and trust. Change is more probable when there are meaningful relationships, and when there is trust between stakeholders, trust in leadership, and trust in the shared vision.

I invite you to draw on your individual and collective commitment and skills to create more truth, courage, and inclusivity to make change for a more sustainable future.

Notes

[1] United Nations Environment Programme, UNEP Copenhagen Climate Centre, *Emissions Gap Report 2020*, (Nairobi: UNEP 2020), xxv.

[2] S. Díaz, J. Settele, E. S. Brondízio, H. T. Ngo, M. Guèze, J. Agard, A. Arneth, P. Balvanera, K. A. Brauman, S. H. M. Butchart, K. M. A. Chan, L. A. Garibaldi, K. Ichii, J. Liu, S. M. Subramanian, G. F. Midgley, P. Miloslavich, Z. Molnár, D. Obura, A. Pfaff, S. Polasky, A. Purvis, J. Razzaque, B. Reyers, R. Roy Chowdhury, Y. J. Shin, I. J. Visseren-Hamakers, K. J. Willis, and C. N. Zayas (eds.), *Summary for policymakers of the global assessment report on biodiversity*

and ecosystem services of the Intergovernmental Science-Policy Platform on Biodiversity and Ecosystem Services (Bonn: IPBES Secretariat, 2019), 26.

[3] U.S. Environmental Protection Agency, *Climate Change and Social Vulnerability in the United States: A Focus on Six Impacts* (Washington: U.S. Environmental Protection Agency, 2021), 6.

[4] Cindy Anderson, Dave Zaharchuk, Jane Cheung, Janet Mertens, and Steven Peterson, *Home for the holidays* (Armonk, NY: IBM Corporation, 2020), 4. www.ibm.com/thought-leadership/institute-business-value/report/ 2020-holiday-consumer-trends#. Accessed 27 March 2022.

[5] Nathan J. Cook, Tara Grillos, and Krister P. Andersson, "Gender Quotas Increase the Equality and Effectiveness of Climate Policy Interventions," *Nature Climate Change* 9 (2019): 330–34.

[6] V. Masson-Delmotte, P. Zhai, H.-O. Pörtner, D. Roberts, J. Skea, P.R. Shukla, A. Pirani, W. Moufouma-Okia, C. Péan, R. Pidcock, S. Connors, J.B.R. Matthews, Y. Chen, X. Zhou, M.I. Gomis, E. Lonnoy, T. Maycock, M. Tignor, and T. Waterfield (eds.), "Summary for Policymakers." In, *Global Warming of 1.5°C. An IPCC Special Report on the impacts of global warming of 1.5°C above pre-industrial levels and related global greenhouse gas emission pathways, in the context of strengthening the global response to the threat of climate change, sustainable development, and efforts to eradicate poverty*. (Cambridge and New York: Cambridge University Press, 2018), 3-24.

[7] "2 out of 3 Catholic Universities in 3rd World," *Collegium Newsletter*, May 2003, 1. *Collegium*, www.collegium.org/sites/default/files/resources/fall04.pdf. Accessed 16 Feb. 2022.

[8] Pope Francis, *Encyclical Letter "Laudato Si'" (On Care for Our Common Home): Of the Holy Father Francis* (Washington: United States Conference of Catholic Bishops, 2015), 31.

[9] Britt Wray, *Generation Dread: Finding purpose in an age of climate crisis*. (Toronto: Penguin Random House, 2022), 8-9.

[10] Brené Brown, *Daring Greatly: How the Courage to Be Vulnerable Transforms the Way We Live, Love, Parent and Lead*. (London: Portfolio Penguin, 2013), 32.

[11] Donella Meadows, *Leverage Points: Places to Intervene in a System*. (Academy for Systems Change, 1996-2022) *The Donella Meadows Project*, https://donellameadows.org/archives/leverage-points-places-to-intervene-in-a-system/. Accessed 15 March 2022.

[12] Pope Francis, 23.

[13] Scott E. Page, *The Diversity Bonus: How great teams pay off in the knowledge economy*, (Princeton and Oxford: Princeton University Press, 2017), 6.

[14] Peter Senge, *The Fifth Discipline: The art and practice of the learning organization* (New York: Random House Books, 2006), 249-250.

[15] adrienne maree brown, "gifting my attention," WordPress, 9 August 2012, https://adriennemareebrown.net/2012/08/09/giftingmyattention/. Accessed 11 February 2022.

[16] "New Georgetown Value Highlights University Commitment to Planet," *University News*, 21 April 2021, Georgetown University,

www.georgetown.edu/news/new-georgetown-value-highlights-university-commitment-to-planet/. Accessed 29 April 2022.

[17] "Indigenous Philosophy and Creating a More Sustainable Future for Georgetown," 25 April 2022, Georgetown College, https://college.georgetown.edu/news-story/indigenous-philosophy-sustainability/. Accessed 30 April 2022.

Literature

Anderson, Cindy, Zaharchuk, Dave, Cheung, Jane, Mertens, Janet, and Peterson, Steven. 2022. *Home for the holidays.* Armonk, NY: IBM Corporation. www.ibm.com/thought-leadership/institute-business-value/report/2020-holiday-consumer-trends#.

brown, adrienne maree. 2012 "gifting my attention." WordPress, 9 August 2, https://adriennemareebrown.net/2012/08/09/giftingmyattention/.

Brown, Brené. 2013. *Daring Greatly: How the Courage to Be Vulnerable Transforms the Way We Live, Love, Parent and Lead.* London: Portfolio Penguin.

Cook, Nathan J., Grillos, Tara and Andersson, Krister P., 2019. "Gender Quotas Increase the Equality and Effectiveness of Climate Policy Interventions." Nature Climate Change 9: 330–34.

Díaz, S., J. Settele, E. S. Brondízio, H. T. Ngo, M. Guèze, J. Agard, A. Arneth, P. Balvanera, K. A. Brauman, S. H. M. Butchart, K. M. A. Chan, L. A. Garibaldi, K. Ichii, J. Liu, S. M. Subramanian, G. F. Midgley, P. Miloslavich, Z. Molnár, D. Obura, A. Pfaff, S. Polasky, A. Purvis, J. Razzaque, B. Reyers, R. Roy Chowdhury, Y. J. Shin, I. J. Visseren-Hamakers, K. J. Willis, and C. N. Zayas (eds.). 2019. Summary for policymakers of the global assessment report on biodiversity and ecosystem services of the Intergovernmental Science-Policy Platform on Biodiversity and Ecosystem Services. Bonn: IPBES Secretariat.

"Indigenous Philosophy and Creating a More Sustainable Future for Georgetown." 2022. 25 April. Georgetown College, https://college.georgetown.edu/news-story/indigenous-philosophy-sustainability/.

Masson-Delmotte, V., P. Zhai, H.-O. Pörtner, D. Roberts, J. Skea, P.R. Shukla, A. Pirani, W. Moufouma-Okia, C. Péan, R. Pidcock, S. Connors, J.B.R. Matthews, Y. Chen, X. Zhou, M.I. Gomis, E. Lonnoy, T. Maycock, M. Tignor, and T. Waterfield (eds.). 2018. "Summary for Policymakers." In, Global Warming of 1.5°C. An IPCC Special Report on the impacts of global warming of 1.5°C above pre-industrial levels and related global greenhouse gas emission pathways, in the context of strengthening the global response to the threat of climate change, sustainable development, and efforts to eradicate poverty. Cambridge and New York: Cambridge University Press.

Meadows, Donella. N.d. *Leverage Points: Places to Intervene in a System.* (Academy for Systems Change, 1996-2022) The Donella Meadows Project, https://donellameadows.org/archives/leverage-points-places-to-intervene-in-a-system/.

"New Georgetown Value Highlights University Commitment to Planet." 2021. University News, 21 April 2021. Georgetown University. www.georgetown.edu/news/new-georgetown-value-highlights-university-commitment-to-planet/.

Page, Scott E. 2017. *The Diversity Bonus: How great teams pay off in the knowledge economy.* Princeton and Oxford: Princeton University Press.

Pope Francis. 2015. Encyclical Letter "*Laudato Si'*" (On Care for Our Common Home): Of the Holy Father Francis (Washington, D.C.: United States Conference of Catholic Bishops.

Senge, Peter. 2006. *The Fifth Discipline: The art and practice of the learning organization*. (New York: Random House Books. "2 out of 3 Catholic Universities in 3rd World." 2003. Collegium Newsletter, May, 1. Collegium, www.collegium.org/sites/default/files/resources/fall04.pdf.

United Nations Environment Programme. 2020 UNEP Copenhagen Climate Centre. Emissions Gap Report 2020. Nairobi: UNEP.

U.S. Environmental Protection Agency. 2021. Climate Change and Social Vulnerability in the United States: A Focus on Six Impacts. Washington: U.S. Environmental Protection Agency.

Wray, Britt. 2022.*Generation Dread: Finding purpose in an age of climate crisis. Toronto: Penguin Random House.*

Chapter Three

Rewilding: A Hegelian Reflection

Jennifer Ann Bates

Only a major shift in moral reasoning, with greater commitment given to the rest of life, can meet this greatest challenge of the century. (Edward O. Wilson)[1]

...peace will be found only in a thorough critical examination of our reason. (Immanuel Kant)[2]

They [mosses] earn their freedom from the pain of change by total surrender to the ways of rain. (Robin Wall Kimmerer)[3]

[T]he context of the earth in the universe is something that needs profundity...it requires poetry. (William Shatner in a CNN interview with Kate Bolduan, after he had traveled to space)[4]

Introduction

Part I introduces rewilding (E. O. Wilson's idea of it, among others) and the call to change our moral attitude toward the environment. As German Idealists, we can answer this by espousing virtues only if we also involve critique and freedom. The philosopher G. W. F. Hegel is an exemplar of how to best involve these. I introduce what I call "environmentalism" and "Speculative Moral Receptivity".

Part II concerns Hegelian phenomenological reflection. Any "view from above" must be concretely actual and complex. I briefly contrast Hegel's *Natural Law* and *Science of Logic* with the Socratic view from above, to fend off closed versions of Hegelian dialectic and eco-fascisms. I discuss Hegel's "Absolute Knowing" by doing a line-by-line analysis of a passage at the end of Hegel's *Phenomenology of Spirit*. This analysis shows that Hegel's protagonist—"natural consciousness"—attains "Absolute Knowing" as much *as* nature, as he/she/they do *as subject.* This is consistent with Hegel's dictum that "everything turns on grasping and expressing the True, not only as *Substance* but equally as *Subject*" (*Phenomenology of Spirit* paragraph 17). In "Absolute Knowing", Subject and Substance equally, reciprocally, and in dialectical relation are Hegel's Nature as rewilding. I call this Hegel's "Preserving Idea". To reflect complexly (rather than abstractly) on Hegel's Preserving Idea, I contrast that Idea with other metaphysical and scientific "views from above": I explain the method of Hegel's Preserving Idea in the *Science of Logic*, the arc of the whole movement of the Preserving Idea in his *Encyclopedia of Philosophical Sciences*, and compare this method and movement with the role of the Preserving Idea in *Phenomenology of Spirit.*

In Part III, I show that Hegel's dialectic is consistent with current scientific rewilding practices. I name views from above that Hegel would endorse today. Hegel would also endorse many of Wilson's hopeful conclusions, albeit with critique, deepening, and complexification. I raise objections to Wilson and the objection that my reading of Hegel is anthropocentric. I reply that it is not anthropocentric, because biodiversity is the central principle between Speculative Moral Receptivity and the Preserving Idea. Hegel's dialectic is a reflective experience of the whole that is itself a rewilding. This reflection is consistent with my individual experience of rewilding my backyard; I briefly describe my experience of Speculative Moral Receptivity in that case of rewilding.

Part I: Introduction to Rewilding, to the Call for a New Moral Perspective, and to Hegel

A) General Definition of Rewilding:

Rewilding can occur through restoration, for example, "the planned reintroduction of a plant or animal… to increase biodiversity and restore the health of an ecosystem".[5] It can also occur through the preservation and conservation of existing areas of dense biodiversity by letting them develop without human intervention. Rewilding also has complex and deep meanings in ways I hope to show.

One of today's most avid proponents of rewilding is Edward O. Wilson. He writes:

> The global conservation movement has temporarily mitigated but hardly stopped the ongoing extinction of species. The rate of loss is instead accelerating. If biodiversity is to be returned to the baseline level of

> extinction that existed before the spread of humanity, and thus saved for future generations, the conservation effort must be raised to a new level. The only solution to the "Sixth Extinction" is to increase the area of inviolable natural reserves to half the surface of the Earth or greater.[6]

By "half the earth" Wilson does not mean drawing a line mid-way around the earth, he means focusing on areas of high biodiversity. His view has its detractors. I mention some later. But I am convinced his idea of rewilding is a good one.[7]

A snapshot of Wilson's Half-Earth Project website (see Figure 1)[8] offers us one of those famous "views from above" of the earth that we are now able to have thanks to satellites and space flight. I include this picture here because later in this chapter, I discuss the philosophical nature of the views from above.

Figure 1:

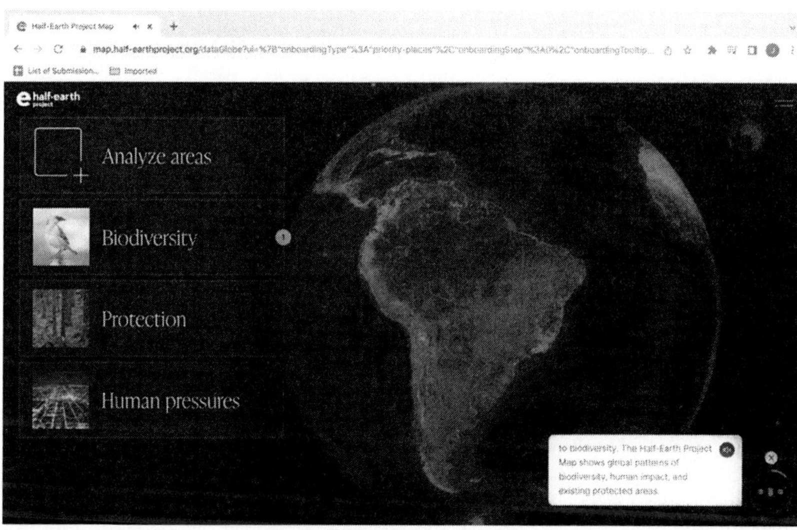

Another example of rewilding organizations is the Rewilding Institute, whose "Rewilding Earth Mission" is "[t]o develop and promote the ideas and strategies to advance continental-scale conservation in North America and beyond, particularly the need for large carnivores and a permeable landscape for their movement, and to offer a bold, scientifically-credible, practically achievable, and hopeful vision for the future of wild Nature and human civilization."[9] There are many more rewilding organizations.

My interest in rewilding lies in my commitment to moving beyond anthropocentric ethics (that is, beyond human-centered ethics). I teach a course at Duquesne University on "Philosophy of the Environment" in which we read and evaluate articles about extending moral standing beyond humans to animals, ecosystems, the land, and even the planet.[10]

News of efforts to move beyond anthropocentric ethics is easily found on the web today, for example, this website: "Do salmon have rights? A Legal Battle Between Seattle and the Sauk-Suiattle Indian Tribe Could Test The Rights of Nature Movement";[11] or this one: "The Rights of Nature – Can an Ecosystem Bear Legal Rights?"[12] Another example provides a legal case in which moral standing is extended to whole ecosystems: "Colorado River Ecosystem vs. State of Colorado" in 2017.[13]

B) But I am a philosopher, not a scientist or a lawyer. Why should philosophers discuss rewilding? We should because environmentalists call for a change in our *moral* reflections.[14] Wilson writes:

> Only a major shift in moral reasoning, with greater commitment given to the rest of life, can meet this greatest challenge of the century. Wildlands are our birthplace….Millions of years of evolution are indelibly encoded in our genes. History without the wildlands is no history at all.[15]
>
> …I believe we've learned enough to adopt a transcendent moral precept concerning the rest of life. It is simple and easy to say: Do no further harm to the biosphere.[16]

Also, religious leaders, like Pope Francis, call for a moral change, and for everyone to engage: "I urgently appeal…for a new dialogue about how we are shaping the future of our planet. We need a conversation which includes everyone, since the environmental challenge we are undergoing, and its human roots, concern and affect us all."[17]

To accomplish this, philosophers and theologians often propose that we adopt a set of virtues. For example, I recently I heard Dr. Jame Schaefer, Professor Emerita, Department of Theology at Marquette University, discuss a set of Catholic virtues.[18] In a secular vein, the philosopher Jan Zwicky proposes a Platonic set of virtues.[19]

Catholic virtues:	**Platonic virtues:**
Prudence	Awareness/humility
Justice	Courage
Temperance	Justice
Fortitude	Contemplative practice
Compassion	

And there are Buddhist versions, too, such as this one about asking ourselves "3 questions to build resilience – and change the world," presented by the Zen Buddhist nun Sister True Dedication, at the Countdown Summit in October 2021.[20]

German Idealists, however, argue that virtues require critique and freedom. As Hegel's predecessor Kant claimed, virtues are excellent, but without a good will, without our free and good intentions, virtues can be misguided.[21]

C) G. W. F. Hegel:

In my view, G. W. F. Hegel is the one who digs deepest into the relationship between our freedom, critique, and virtue-ethical contexts. I first discussed this in 2003 in "An Inquiry Into The Nature of Environmentally Sound Thinking."[22] In that article, I surveyed a variety of environmental ethical attitudes and discussed how for Hegel, consciousness is necessarily economic; I drew mostly on his *earlier* work, the *Lectures on Psychology* (1803-4). In this chapter today, I focus instead on Hegel's account of the final form of this relationship between freedom and ethics, the *absolute* dialectic between them. This involves onto-epistemology, or what I like to call an "environ-mentalism." For this, we need to look at Hegel's *later* works (1807 and 1832).

In doing this, I am answering Deep Ecology's challenge. The philosopher and ecologist Arne Naess is the author of the movement called "Deep Ecology".[23] Naess takes extending morality as far as one can go: he asks us to delve into our deepest ontological, epistemic, and religious attitudes and commitments, to see how they are affecting our view of nature, and then he asks us to question these attitudes and commitments. Let us do this now with Hegel.

G. W. F. Hegel (1770-1831) was a German Idealist philosopher; a post-Kantian. He was brought up in the Protestant tradition. For the deepest Hegelian reflection, we have to look at what he calls "Absolute Knowing" at the end of his book the *Phenomenology of Spirit* (1807). Hegel referred to the *Phenomenology of Spirit* as a "science of experience".[24] In it thought and being are revealed to be dialectically identical in the sense of the identity of identity and difference.

It takes at least seven years to read the *Phenomenology of Spirit* properly. So I am going to focus on only three lines at the end of it. My conclusion will be that a Hegelian today must assert that we are not stewards of the earth: there are lots of biodiverse gardeners and we are all members of a *complex, dialectically evolving, phenomenological garden.* I am going to argue an unorthodox reading of Hegel: that the culmination of his phenomenology presents experience as a dialectical becoming that develops from receptivity to preservation, via biodiversity.

This chapter is nonetheless for both environmentalists and Hegelians. For environmentalists, I offer the moral attitude of dialectical "environ-mentalism" that I just described and explain in what follows. (I know there is a television series about cars called "Environ-mental",[25] but that is not what I mean here!). For Hegelians: I offer a new reading of Hegel's *Phenomenology of Spirit*, by suggesting that we can read the end of the book as consistent with rewilding practices, as well as itself a rewilding of thinking. In the process, I read the classic Hegelian transition from logic to nature to spirit, in a new light.

This reading of Hegel is different from the Hegel scholar Nigel Tubbs' "Green Metaphysics."[26] Tubbs argues that Liberal Arts and education in general (and education in Hegelianism in particular) are essentially green: human minds have great potential for education; thinking is a renewable resource, does not pollute, and costs almost nothing. Tubbs' ideas are great: I cite him in a chapter I wrote for a previous IOC conference, in which I argue that we must preserve the Liberal Arts alongside the environment.

But my use of Hegel today involves moving beyond anthropocentrism, thus beyond our concern with education and sustainability; I advocate that we look at how Hegel's protagonist-consciousness in the *Phenomenology of Spirit* attains "Absolute Knowing" as much *as* nature, as he/she/they do *as subject.* This is consistent with Hegel's dictum that "everything turns on grasping and expressing the True, not only as *Substance*, but equally as *Subject*."[27]

Specifically, I show that Hegelian "Absolute Knowing" is itself a rewilding because in "Absolute Knowing" the dialectic of subject and substance is Hegel's Absolute Spirit as Nature; I show that the phenomenological *experience* of this is speculative moral receptivity opening onto and into (bio)diversity; this is our experience of what I call Hegel's "Preserving Idea."

There are problems with my reading of Hegel that I can only briefly touch upon. But I hope that contemporary Hegelians will have something new to consider about his formidable and important 1807 work.

Part II: Hegel's Phenomenological "Absolute Knowing"

A) Any Absolute Knowing "view from above" must be concretely actual ("complex"):

Let me start by dispensing with erroneous ways to conceive of Hegel's "Absolute Knowing" as views from above of the whole that are abstract and "immediate" (rather than mediated and therefore complex). Examples of such could be a view in the history of philosophy such as Socrates' description of the pure soul's view of the world at the end of Plato's *Phaedo*,[28] or today, non-complex satellite images of the planet. These views can be metaphysical or technological ones of the immediately perceived whole.

In contrast, in Hegel's philosophy, any view that claims to be universal must go from being an abstract immediacy to being mediated. So any "view from above", including Hegel's Absolute Knowing, is only (as) true as the real and recollected *becoming* of the content of that view. This is the concretely actual, complex truth of the thing at hand (*die Sache selbst*).

An example of moving from abstraction to complexity is found in Hegel's account of *Natural Law*. H. B. Acton explains Hegel's distinction in that work between inadequate "scientific empiricism" and a "pure empiricism" which is more holistic.[29] In the former, "some single quality or function such as self-preservation or deterrence is selected from a complex social whole and held to be the prime moving force in it. But the selection is arbitrary and cannot do justice to the complex unity of the whole."[30] By contrast, the latter, holistic view, is more complex since it deals with relations as well. Acton explains, "[c]omplexity may engender a certain amount of confusion, but naively complex confusion is less misleading than simplified abstraction."[31] I return to this issue of complexity later.

This movement from abstraction to salutary complexity is clear in Hegel's *Science of Logic* of 1832. Here, Hegel develops his method as an advance which is also a "getting back nearer", moving from abstraction into concrete universality:

> It is in this manner that each step of the advance in the process of further determination, while getting further away from the indeterminate beginning is also getting back nearer to it, and that therefore, what at first sight may appear to be different, the retrogressive grounding of the beginning, and the progressive further determining of it, coincide and are the same.[32]

To understand this logical method in our experience and why it is not just an autarchic return of the same or some sort of romantic conservatism (let alone a basis for eco-fascism), and why it is *instead* consistent with rewilding science, as well as a kind of rewilding itself, we must look at a passage from the end of Hegel's *Phenomenology of Spirit.*

B) Hegel's "Absolute Knowing": a passage from paragraph 807 of *Phenomenology of Spirit*:

> The self-knowing Spirit knows not only itself but also the negative of itself, or its limit: to know one's limit is to know how to sacrifice oneself. This sacrifice is the externalization in which Spirit displays the process of its becoming Spirit in the form of *free contingent happening*, intuiting its pure Self as Time outside of it, and equally its Being as Space. This last becoming of Spirit, *Nature*, is its living immediate Becoming; Nature, the externalized Spirit, is in its existence nothing but this eternal externalization of its *continuing existence* and the movement which reinstates the *Subject.*[33]

Let us do a line-by-line analysis of this passage, starting with the last, line 3: "Nature, the externalized Spirit, is in its existence nothing but this eternal externalization of its *continuing existence* and the movement which reinstates the *Subject.*" There is one concrete reading of this reinstating of the subject by nature, this circular return, that we do not want. We can see it in E. O. Wilson's account of what could happen to us:

> If our species, self-appointed Lords of Earth, accidentally burned the planet's surface to a crisp or an outsized asteroid were to wipe out all organisms on the surface, life could go on in the planet's deep biosphere. There, its microbes and invertebrate predators would persist, shielded and unconcerned in their dark refugia, drawing their energy and substance from rocks, resistant to heat, perhaps evolving over hundreds of millions of years, eventually to reach the surface and generate a diversity of multiple-celled organisms, thence against heavy odds a human-grade metazoan. Whereby the great cosmic cycle might give intelligence on Earth a second chance.[34]

Clearly, we do not want this kind of reinstatement of subjectivity. Therefore, let me propose instead that Hegel's Absolute Knowing reinstates the subject because it is *a kind of existential rewilding, here and now,* as expressed in line 1 of the passage: "The self-knowing Spirit knows not only itself but also the negative of itself, or its limit: to know one's limit is to know how to sacrifice oneself."

Let us analyze this line. First, Hegel expects that we, his readers, have, by this endpoint of his book, come to realize ourselves as makers and selfexpressers in and through otherness. This means that we realize that there is a negative moment (a limit) in our experience that is alienating to us.

But that alienation does not remain untouched by the process, because there is a second negation. The process is a double negation, a "negation of the negation" through which we move and reflect on what has happened. Thus, on the one hand, we sacrifice ourselves to the necessity of that alienating moment but also realize that in that very movement, we are born anew into a new moment, a moment that preserves in recollection the moment that has passed and opens us to the next moment with this past lesson in mind. (Tubbs celebrates this about Hegelian education.)

At its deepest and most comprehensive, this is Hegel's *philosophical* comprehension of the incarnation, death, and resurrection of the divine. Importantly, as such, it is not just a human movement, it is the movement of Absolute Being as Absolute Spirit.

The second and third lines of the paragraph are the hardest to interpret. They reveal Hegel's unorthodox Christian view that God and Nature are not separate entities and that Absolute Spirit is the dialectical life of that whole. It is a sacrifice but also externalization and preservation. What does this mean? Let us look at line 2: "This sacrifice is the externalization in which Spirit displays the process of its becoming Spirit in the form of *free contingent happening*, intuiting its pure Self as Time outside of it, and equally its Being as Space. This last becoming of Spirit, *Nature*, is its living immediate Becoming."

In my reading, line 2 reveals that the sacrifice is what I call Speculative Moral Receptivity. Speculative Moral Receptivity is a person's receptivity to becoming as Nature. Rather than reinstating the subject in some post-extinction time, this is an existential setting up of subjectivity via the unity of thought and being as Nature.[35] Nature is nothing other than this externalization "in which Spirit displays the process of its becoming Spirit in the form of free contingent happening…."

This receptivity and becoming as free contingent happening is not an invitation to chaos, it is an invitation into biodiversity. It is not only a moral attitude *toward* nature, it is also Spirit/nature's active reinstatement of subjectivity, because, as today's science amply shows (and as I discuss more below), receptivity to biodiversity is how we preserve subjectivity.

At this point, we are right to ask: which subjectivity is hereby reinstated? In an earlier article, "Hegel and the Concept of Extinction," I argued that Hegel's phenomenological dialectic is a hermeneutic that includes other animal subjectivities. In my analysis today of Hegel's Absolute Knowing, I think the question goes even deeper. Let me finish the line-by-line analysis and then explain this.

Let us analyze line 2 in more detail: "becoming Spirit in the form of *free contingent happening*, intuiting its pure Self as Time outside of it, and equally its Being as Space." This phrase is the most difficult. What could this mean for you and me, sitting here now? Speculative Moral Receptivity means that we allow ourselves to become as free contingent happening. We allow ourselves to be, like exhaled breath, just becoming as time and as space. To experience what is. But what *is* this externalized Nature that Spirit has become?

We start to see an answer by looking at the next and last paragraph in the *Phenomenology of Spirit*. Our receptivity is paired, in that paragraph (808), with the idea of the self being reborn out of this receptivity *into* what was preserved from the dialectical development thus far (historically, materially, productively, phenomenologically), i.e., what has been preserved up to this moment of sacrifice. (We recall from the logical method discussed above, that we advance in a way that is a return into: "each step of the *advance* in the process of further determination, while getting further away from the indeterminate beginning is also *getting back nearer* to it".[36])

Sacrifice into free contingent being is not a mere openness, nor is the return into what is preserved (i.e. being reborn) merely a return to prior identities. The dialectic of receptivity and preservation is the becoming of differences, a diversification, a complexification. These two moments--receptivity and preservation--together constitute what Hegel calls the "Calvary of Absolute Spirit."[37]

But this is not an anthropocentric reading of Absolute Spirit. An anthropocentric attitude toward this development would be "one-sided", not dialectical; it would be an act merely from the side of the subject. Here, instead, subject and substance are Spirit/Nature's becoming, a becoming that sacrifices itself into contingency, and in doing so, reinstates (preserves) subjectivity.

My (as yet abstract) proposal about the meaning of the end of Hegel's book is that Spirit's becoming as Nature, Absolute Spirit, *is* rewilding. Phenomenologically--for us as experiencers, as Absolute Knowers of this Absolute Spirit--this involves sacrifice (Speculative Moral Receptivity) and reinstantiation (preservation), *via* the middle term of free contingency as biodiversity.

I propose, therefore, that rewilding is, alongside the three *traditional* Hegelian forms of Absolute Spirit (art, religion, and philosophy), itself a form of Absolute Spirit's becoming.

Here ends my line-by-line analysis. Now I need to explain all this—to make it less abstract, and more complex--by looking at Hegel's method in more detail, and by showing how free contingent happening, as biodiversity, connects receptivity with the Preserving Idea.

C) Let us return to the Socratic view from above and compare it with Hegel's method:

According to Plato's Socrates, immortal souls see the Forms of things prior to those Forms becoming instantiated in material earthly life. The basic Platonic view is that no instantiation of a triangle, cat or tree, or human, or anything, is the Form itself, or as perfect as the Form; to know the Form, the soul must separate itself from the distractions of bodily senses.[38]

By contrast, for Hegel, knowledge moves from immediate universality of what is thought or experienced, *through* time and space, to *concrete* universality. Concrete universality is the completion, the complexly comprehended whole of the first immediate, abstract universality.

Thus for Hegel, instantiation (or reinstating) is the *truth* of the Form, rather than the other way around: "each individual member as ensouled by the method is reflected into itself, so that in returning into the beginning it is at the same time the beginning of a new member."[39] Hegel explains that, for us, phenomenologically: "The True is the Whole. But the whole is nothing other than the essence consummating itself through its development."[40] Spirit must ensoul itself because it must "realize and reveal what is at first only *inward*…to vindicate it for Spirit's certainty of itself."[41] Let me explain this relation between methodological ensoulment and phenomenological ensoulment.

The *form* of the whole *method* can be grasped by Spirit. In other words, because phenomenology is structured by the dialectical method moving in time, we can arrive, experientially, phenomenologically, *via* that method, at the knowledge *of* the method. Hegel describes this arrival at the end of the *Phenomenology of Spirit*. It is the complete moment when the Self "sets aside its Time-form, comprehends this intuiting, and is a comprehended and comprehending intuiting."[42] This logical comprehension gives us the pure method of Absolute Knowing.

However, in the *Phenomenology of Spirit*, Hegel, consistent with this very method, explains that this formal, logical grasp of the method is not enough for consciousness, since consciousness wants to be certain of its truth.[43] To achieve that certainty, consciousness needs to experience this absolute content in time and space. That is, we need to experience the absolute *externalized substantial* side, not just a subjective, ideational, self-consciously grasped absolute method. One side without the other becomes an abstraction. As experiencers, we need the conscious existential side of this method of Absolute Knowing. It is following this realization of the logical method, therefore, that we read the passage about Spirit externalizing itself as Nature, the line we analyzed above.

Our task as Absolute Knowers is therefore also not only historical comprehension. It is recollection and existence: it is existentially certain *being*. This is important because Spirit does not *exist* anywhere else: "As Spirit that knows what it is, it does not exist before, and nowhere at all, till after the completion of its work of compelling its imperfect 'shape' to procure for its consciousness the 'shape' of its essence, and in this way to equate its *self-consciousness* with its *consciousness*."[44]

In the final paragraphs of the *Phenomenology of Spirit*, Hegel explains that now that we have absolute knowledge of the method as well as the certainty of self-conscious experience of it, we can go on to study the pure method in its fullness (i.e., in all its determinations, developments and movements) in Hegel's *Science of Logic,* or, we go back to the beginning of the *Phenomenology* and do all the moves of conscious experience again, this time knowing and recognizing them as we read and experience them.[45]

But based on the passage we read above (which follows this claim about the two directions we can now take), we can and ought to return to the "beginning" in an existential sense as well. In other words, the passage we studied, along with the rest of the last two paragraphs of the book, indicate that we should go back to the "beginning" in the sense of phenomenologically reading our own lives, going back to the "beginning" of each moment of our own "books", and "read" what is happening, existentially. (This is thus the method practiced in phenomenological experience: an "advance" which is also a "getting back nearer", moving from abstraction backward into concrete universality.) That, I believe, is the daily practice and experience of what Hegel calls, in the last words of the book, the foaming "chalice of this realm of spirits", without which we are "lifeless and alone."[46] I have argued above that a central part of this practice and experience is Speculative Moral Receptivity to biodiversity, and that as such, it involves a sacrificing of oneself into pure contingent happening as Nature, which is part of the recollecting/recycling process of the Preserving Idea.

However, because we are (in) a dialectic between subject and substance in time, we have a propensity to become one-sidedly abstract, on one side or the other, in a moment. (This can happen with either the too-specific experience or the too-general idea or method.) To avoid this propensity to put our ideas or our experience ahead of our actual being, let me now apply the method existentially to my own theorizing about Hegel today. That is, let me now do the required critical work of making an "advance" which is also a "getting back nearer," with regard to my understanding of what I have been calling Hegel's "Preserving Idea".

In what follows, I begin with the abstract idea of the method, then make it concrete by instantiating it in scientific, and then personal, rewilding experience.

D) The Preserving Idea: explained abstractly as *method* in the *Science of Logic*, and then as *movements* (from logic to nature to spirit) in the *Encyclopedia of the Philosophical Sciences*; the method and movements as compared with the Preserving Idea in the *Phenomenology of Spirit*.

i) The *Science of Logic*:

In the *Science of Logic*, we are dealing with the pure logical method. It is Hegel's *speculative* logical form of an immediate universal becoming its own whole through its own logical development. Hegel refers to the *Science of Logic* as God's essence before creation.[47] This does not put essence before *existence* because for Hegel God's essence is the *self-moving* logic of Being. (While this theological language deserves more discussion, I am going to focus here on the logic as speculative method.)

The *Science of Logic* tells us that the method is to circle back and into any content. It is never about grasping only the abstraction; it is about allowing the abstraction to fill out and develop. We mentioned earlier that the form of the immediate becoming its own whole is an "*advance*" which is also a '*getting back nearer* to it'. This expresses a speculative logical "circle."[48] This circling does not express a final separation (as when Socrates imagines his disembodied soul knowing the Forms). Rather, the method, even while expressing itself *formally* as a method, explains itself as emersion and fulfillment; even though the logical method is not explained as a comprehending in time, any comprehension in time follows this logic. Therefore, we always start with abstraction and then develop with it through its mediations. Anyone who asserts that an abstract stands on its own, is making a proposition that itself needs to be justified by filling that proposition out and letting it develop. Hegel explains that otherwise, such propositions

> have an untrue content, for they convert what we know to be finite and untrue into something incontestable and absolute, namely, a *limited* cognition determined as *form* and *instrument relatively to its content*; this untrue cognition is itself also the form, the process of seeking grounds, that is retrogressive. The method of truth, too, knows the beginning to be incomplete, because it is a beginning; but at the same time it knows this incompleteness to be a necessity, because the truth only comes to be itself through the negativity of immediacy.[49]

To provide examples, let us go back to the view from above. Whether we begin with a satellite view with data collection, or abstract conceptions of the whole of the planet or its ultimate purpose or meaning, Hegel's method critically evaluates the starting viewpoint and allows its content to develop. Therefore, we critically evaluate the satellite viewpoint as well as the questions we ask about the data it collects, and likewise any presuppositions within our universal ideas of the purpose and meaning of our view and its content. But it is not just that we *do* this critique as a willed action: the nature of the dialectic necessarily impels any one-sidedness to become through its othernesses. When we ignore this process, we get tripped up in the necessary contradictions that occur.

In other words, reflection cannot be merely abstract surveillance, nor can its content be mobilized by a single goal. No one can truly say *only* that they care about "the planet" or are "holists", for whether surveillance is technological or metaphysical, whether it is a view of the planet, of a part of it, of any ontological "whole," or of a commitment to any of these, *all reflection becomes*. Each abstraction becomes a concrete universal (which itself is then subject to critique and develops through its differences); the path of each abstraction as it becomes more concrete, expresses the method. The method is a reflective, preserving, recuperative, complex economy. That complex, developing economy of the whole is the method of the Preserving Idea.

Now, let us move from method to the arc of the Idea externalizing itself in nature, and its/our subsequent comprehension of this self-externalization in thinking.

ii) The *Encyclopedia of the Philosophical Sciences*:

Over the course of the three books of his *Encyclopedia of the Philosophical Sciences*, Hegel explains how thought and being reinstate each other in reality, their movements following the circular method that we saw outlined above in the *Logic*. The arc of the Preserving Idea's self-actualization is as follows: at the end of the *Encyclopedia Logic* (the first book), the Idea externalizes itself into nature; then, the idea as nature (the second book) develops from simple physics to complex biology, to humans with their reflective inwardness and their collectivity as Spirit; and then, Spirit (the third book) comes to know all its own phenomenological, psychological and social shapes, up to and including Absolute Spirit's *expression* in politics, art, religion, and philosophy (Spirit finally also *seeing* in these the logic of its developing being and thinking). The method throughout is thus a spiraling return, from Idea to Nature to Spirit to cognized method to externalization, and so forth.

In this whole transition, we see the becoming of what I have been calling "The Preserving Idea": it is a reflective logical method that externalizes itself as Nature (substance) and develops (into) itself as subject.

iii) The Preserving Idea's logical method and movements compared with its role in the *Phenomenology of Spirit*:

In the *Phenomenology of Spirit*, instead of dealing with the moving "essence of God" as we did in the *Science of Logic*, we are dealing with conscious experience. This experience is the *phenomenologized* method. It consists of raising conscious experience from being unaware, or only partially aware, of the dialectical, subject-substance method by which it is conscious, to full self-consciousness. In other words, it consists in developing what conscious experience is by filling out and developing its moments over time, more and more self-consciously, thereby revealing the dialectical method as well as the content by which consciousness is what it is. The endpoint is the phenomenological form of the whole as a "process of Spirits" in the world. As I explained earlier, this endpoint in the *Phenomenology of Spirit* is the beginning point of the *Science of Logic* (the ability to do pure logic) but also the standpoint from which we can go back to the beginning of any experience and grasp the becoming of that experience as the becoming of *der Begriff* (in my environmentalist reading, the "Preserving Idea"). We can experience in consciousness the phenomenological richness of the Preserving Idea's becoming as experience (the foaming chalice). I argued above in my line-by-line analysis that this experience moves beyond the merely human, into Spirit's becoming as Nature, and that that is what joins Absolute Knowing with Absolute Spirit in the arc of the Preserving Idea's becoming.

The *Science of Logic* and *Phenomenology* are therefore different "takes" on the same thing: the Logic is the *logos* of creation, and the *Phenomenology of Spirit* is our minds as creation. In other words: in Hegel's "Absolute Knowing," we existentially enact in consciousness, what the *Logic* spells out as logically occurring to the Absolute Idea: it externalizes itself as a nature that then re-instantiates subjectivity. What the *Logic* gives us is the insight into the absolute dialectical method of what for us in time and space feels like negation and recuperation. Preservation is this circular self-creation in and through nature.

The self in the *Logic* is the negating and Preserving Idea, whereas, in the *Phenomenology*, it is Absolute Spirit, our multiple consciousnesses, moving through circles of limiting, sacrifice, and reinstantiation of subjectivity-in-objectivity and objectivity-in-subjectivity.

iv) My unorthodox reading is, therefore, that the role of nature is more pivotal in this arc of Hegel's system than traditionally understood: nature is the process of subjective and substantive differentiation through externalities, without which consciousness is not possible.

However, my explanation thus far is *still* too abstract. Therefore, once again following Hegel's method, we must ask: how can we reflect on Hegel's Preserving Idea without being abstract about it? The answer lies in applying the method further and filling it in with experience. Fortunately, by now, we know that abstractions, including my reading above of Hegel's Preserving Idea, are not a problem, since for Hegel, "illegitimate abstraction sets the dialectical method in motion";[50] and, according to Hegel, the "truth itself resides only in the extended course of the process and in the conclusion."[51] Therefore let us proceed with the dialectical account of the Preserving Idea, in order to move from abstractions about it to some of its concrete complexities.

Part III: Hegelian Rewilding: a Preserving Idea

Experiencing Hegel's Preserving Idea as Rewilding:

A) Summary thus far of our abstraction-to-be-complexified:

Let me begin by recalling what we have discussed thus far and our goal. We ruled out an undesirable view of the whole as an evolutionary "reinstating [of] the subject" in terms of mass extinction and eventual regrowth of beings like us over millions of years. Instead, I proposed a speculative moral attitude, a dialectical Hegelian reflection, that can handle the wildernesses of difference, *now*, in a way that forecloses extinctions. To this end, we analyzed several lines from the end of the *Phenomenology of Spirit*, which revealed that Absolute Knowing is a Spirit that externalizes itself as Nature. We compared this with the formal method at work in those lines, as described in Hegel's *Science of Logic*. We saw that there is no mere form of the good of surveillance: a view of the whole must be allowed to develop; we must keep an eye, many eyes, on any form of surveillance. The contents of any surveillance will develop in their complex relations: whether it is a philosophical view or a satellite view from above, all views from above fall into contradictions as they become more complexly mediated, so our best method is to self-consciously engage in that (self-)criticism, rather than taking any view as "something incontestable and absolute."[52] We saw that the method Hegel develops is one of working backward, an "advance" which is also a "getting back nearer", moving from abstraction backward into concrete universality.

In other words, from our analysis of the *Phenomenology of Spirit*, we know that our reflective attitude requires critical, double-negation: a dialectical evaluation and development, not just of the substance of our view, but of the formal methods by which we take views of the whole in experience (i.e., dialectical development not just of the whole's content but of any form we think it has and of the "eyes" with which we are viewing it). We adopt a critical understanding of surveillance, of the fact that we are doing that surveillance, and why we are doing it. Our goal is therefore not able to remain confined in an abstraction, even when, in the spirit of my Chapter, the abstraction is called "Hegel's Preserving Idea."

We also can see now that we cannot *know* this Preserving Idea in its immediacy, for we only know the time of the mediations of it that we experience. The time of that mediation reveals the whole to us only in and as those moments, their histories, and potential futures. Outcomes presumed as givens must be thought through, and therefore, it is not a given that there is a sustainable idea in which we survive, or, that we will experience total extinction. To fill in the abstraction of the dialectical moment we find ourselves in *now*, we must, as Hegel writes, let ourselves go into the free contingency of experience, and thereby develop the concrete universal that is the complex and open-ended truth of the Preserving Idea.

We have seen how this method works logically, but we also need to do it in time, consciously, phenomenologically, and experientially. This involves sacrifice and allowing ourselves to experience the whole (ourselves included) as free contingent happening, as nature.

In this spirit, therefore, let me give some content to Hegel's preserving method by looking at our rewilding experience of the whole *today*.

B) Hegel's dialectic is consistent with *scientific* rewilding today:

Hegelian reflection of the whole (earth), with its goal of concrete universality, is consistent with the whole complex range of what E. O Wilson calls for. On the one hand, Wilson calls for taxonomist boots on the ground: analyzing nature in its details, and figuring out the histories and relations of biological entities to their pasts and each other. On the other, Wilson writes that we should harness all of the technologies available to us, from restoration to geoengineering, including links between "biology, nanotechnology, and robotics", as well as "synthetic biology", AI, and Whole Brain Emulation Science.[53] In other words: whether a Socratic view of the pure soul from above or a satellite view of the earth, it is better to have a Hegelian-kind of reflection of the whole, knowing that we have only just begun to understand, rather than an abstract whole with which our spirits are, in some disembodied way, aesthetically contented merely because we see it as the form of the beautiful or the good.

This is not a return to some romantic naturalism (Hegel was deeply critical of the romantics). It is aligned with science and is appropriately hopeful.

i) Scientific views from above that Hegel would therefore endorse include the gathering of data through satellite technologies, such as:

GEOBON's data-rich view from above with its biodiversity projects (Figure 2)[54]:

Figure 2:

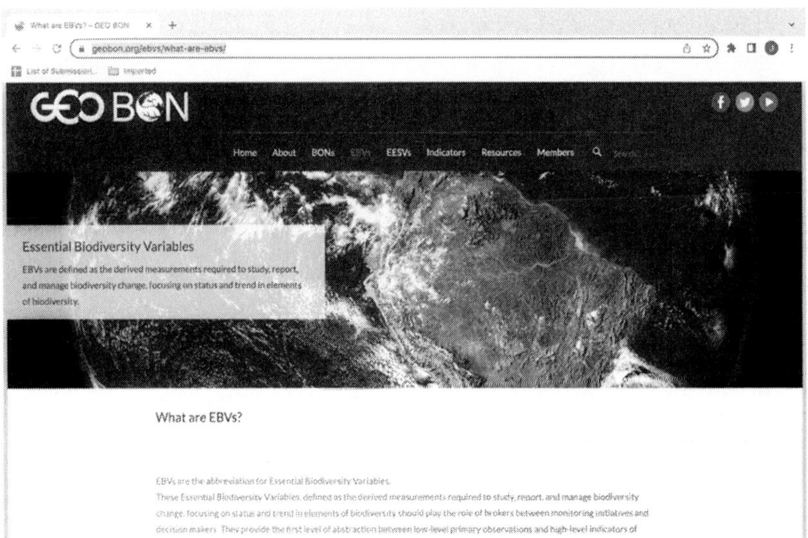

"GEO BON has facilitated the development or enhancement of at least 25 national biodiversity observation systems, representing most of the Earth's major biomes, that are coordinated and can contribute to regional and global biodiversity assessments."[55]

Other views from above that Hegel would endorse are Al Gore's DSCOVR,[56] and The Map Of Life.[57]

Hegel would no doubt also endorse the Half-Earth project that I mentioned at the start (see Figure 1).[58]

All of these views from above are effectively retracing the path back toward the biodiversity needed for the planet to be a healthy biological whole, and thus to systems that keep reinstating subjectivities. This is a receptive awakening to biodiversity and preservation of it that reinstates subjectivity. Wilson writes:

> We will come awake. Meanwhile, digital technologies have serendipitously made it possible to complete the census of global biodiversity, and with it to determine the status of each of the millions of species that compose the Earth's fauna and flora. That process is already under way .... If biodiversity is given space and security, most of the large fraction of species now endangered will regain sustainability on their own.[59]

In sum, Hegelians today should agree with the scientists who assert that our very subjectivity depends upon the rewilding preservation of biodiversity.[60]

ii) Hegel would also endorse many of Wilson's hopeful and urgent conclusions in *Half-Earth*. For example, Wilson writes:

> Furthermore, advances made in synthetic biology, artificial intelligence, whole brain emulation, and other, similar mathematically based disciplines can be imported to create an authentic, predictive science of ecology. In it, the interrelations of species will be explored as fervently as we now search through our own bodies for health and longevity.[61]

Hegel's dialectic teaches us to take Wilson's sometimes abstract optimism critically, deepening it into its diversities, and making it more concrete by reflecting it through its differences.

iii) One might object that there are problems with Wilson and this view from above:

First, Wilson presents the questionable idea that there is in evolution a kind of hidden hand similar to what Adam Smith argued was present in capitalist markets (we might dub this "the invisible hand abstraction"). Wilson writes:

> Just as natural selection drives organic evolution by competition among genes to produce more copies of themselves per unit cost in the next generation, raising benefit-to-cost of production drives the evolution of the economy. Almost all of the competition in a free market, other than in military technology, raises the average quality of life.[62]

This view is rightly critiqued by Bram Büscher and Robert Fletcher.[63] Paul Hawken's arguments in "Natural Capitalism" can also be harnessed in a critical reply to Wilson's citation here.[64] Hawken discusses the work that nature contributes to the functioning of life on earth (life as a whole, complex biosphere), work so complex and wide-ranging that humans could not possibly accomplish it *even* (I would add) *with the help of AI.* But my use here of Hawken's argument serves only to critique Wilson's presumption about a hidden hand: otherwise, Hawken's call to let nature do its work, and to recognize this work, aligns completely with Wilson's call for rewilding.

It might be objected, too, that what I am calling "Hegel's Preserving Idea" is subject to just this kind of criticism, i.e., that the Preserving Idea is a kind of "hidden hand". I hope that I have shown otherwise. Taking the Preserving Idea in the sense of a hidden hand would make it into an abstraction. I have argued instead that the truth of any of Hegel's concrete universalities, especially that of what I am calling his Preserving Idea, involves knowing limits, sacrificing standpoints in becoming pure contingency, and preservation of complex multiplicities in open-ended development. In contrast to the abstraction of a hidden hand, Hegel's dialectical method, which develops through the movement of experience, is a kind of rewilding. In consciously practicing Hegelian rewilding reflection, we consciously become in and through a substantial-subjective dialectical development, returning to subjectivity at each moment, enriched, different, becoming in and through otherness.

A second objection might be that these Hegelian and scientific "views from above" are anthropocentric. My first reply is that they are not anthropocentric views when thought through dialectically. Hegel's dialectic shows that the truth about humans is that we are part of a living phenomenological garden in which we are not only gardeners but necessarily also receptive to and preserved by other biotic gardeners.

My second reply is that biodiversity lies between Speculative Moral Receptivity and the Preserving Idea. Rewilding is essentially about preserving biodiversity. We exist in the midst of and are inextricably part of, a complex biodiverse world, and we participate experientially in that when we practice moral receptivity. When we attend to this, we see and experience what science today has already proven: biodiversity reduces disease and increases flourishing; it is a *sine qua non* of our survival.

Jason Rohr provides evidence of this. His careful scientific work shows that the *loss* of biodiversity increases disease transmission in wildlife and humans and that the opposite occurs when we preserve biodiversity.[65] This scientific discovery connects well with my argument that Hegel's dialectical Absolute Knowing (of the whole) is a kind of rewilding. In case you are not yet convinced, let me elaborate.

C) Hegel's dialectic is a reflective experience of the whole as a kind of rewilding:

We recall the phrase "becoming Spirit in the form of *free contingent happening*." This is not inciting chaos, it is thinking properly. The difference between *wild* and *chaos* is captured by Deep Ecology's distinction between Complex and Complicated: an old-growth forest is a rich *Gestalt*, a bio-system of complex, inter-determining life forms; by contrast, garbage by the roadside, or a can thrown into a rewilding patch, is complicated chaos in which the things are not related as a biological self-sustaining system.[66]

The wild as complex prevents the system from closing in on itself or being a hegemonic return to the same: the view of the whole advocated at the end of Hegel's *Phenomenology of Spirit* is complex, and as such it exceeds the boundaries that any Anthropos could impose upon it.

While this might feel like engendering chaos, it is rather a critical, complex attitude of the whole whose truth can only be got at by our continuous, careful study of what is, by going backward – not to a previous ideal wilderness—but back down into the on-going wilderness, especially that wilderness that is our human interfacings with biodiverse life and the bio-systems that sustain all lives on earth.[67]

D) Finally, Hegel's dialectic as a rewilding is consistent with the individual experience of rewilding a portion of land. An anecdote, and some suggestions:

First, let me offer a pictured experiment of what it means to rewild thinking. Imagine the famous statue by Rodin called "The Thinker"; now imagine the following. Instead of an immobile statue inside a completely walled-in museum, imagine a real-life person in the same pose as The Thinker, out of doors and looking at a rewilded patch of land. (My husband took such a photograph of me looking at my rewilded backyard.) With this image in mind, let me describe my own experience of rewilding.

i) Rewilding my backyard: a case of Speculative Moral Receptivity as rewilding:

The hardest thing about rewilding my yard was to be *receptive* because to do that I had to do nothing – to do nothing about the "weeds", to do nothing about the neighbors and mowing companies who hated the weeds, to do nothing about the thousands of bees (who as it turned out were lovely, not at all dangerous – indeed one day I reached out and pet a bee, and it just kept on as if nothing unusual were happening); to do nothing until I could gradually see the multiverse of lives being lived in the buzzing, active goings-on.[68] At least in this one spot, nature was coming back to life, flowers, birds, millions of bugs, and yes, pesty deer would make their way through my tiny yard, no doubt leaving tics, and also eating some of my favorite flowers…. Again, I did nothing.

It was an experience of sacrifice and the ongoing challenge of the Preserving Idea: my Hegelian reflection is that part of rewilding is actively not-knowing and not-doing, which for a philosopher and a gardener, is a sacrifice. This is, in part, an awareness akin to what the philosopher Jan Zwicky and others write about: it is a contemplative virtue; it also shares in the frequently expressed view that only experiencing the wild will give us the love of the wild necessary to want to save it.[69] Awareness of wilderness helps us get outside our anthropocentric attitudes. I share the view that unless we can do this, we cannot experience properly. What Hegel contributes alongside this view, is his idea of dialectical mediation: the unavoidable economy of concrete universality in the relation of thinking and being; receptivity to biodiverse nature as the developing, Preserving Idea.

Hegel's philosophy contributes the realization that our recuperation is into what has been made (by many beings), that we make things; that even as I sat there in my yard, I was not separate from building a space, that there is no outside to the mediation, that if there is a wild, it was always already something we thought, and that if there is a wild, it is always already something in which we are entangled, and that we cannot get out of that mediating dialectic.

There is no mere view from above or mere immersion within. Any true view from above is in reality a complex ensoulment. Today, when we recall Hegel's famous phrase "What is rational is actual and what is actual is rational"[70] we should think: environ-mentalism.

ii) Suggestions:

Duquesne University (indeed all College campuses) should fence off a patch of campus and let it go wild, and put a bench facing it (and remove the garbage that people will inevitably throw into it, but otherwise let it be) so that we all (humans and other biotas) can directly experience it.

Also, rewilding is preventative medicine. I propose, therefore, that the new program in Osteopathy at Duquesne University should include a program of study harnessing the Liberal Arts, to promote different kinds of 'environ-mentalism.'

In conclusion: **Rewild Thinking!**

Notes

[1] Edward O. Wilson, *Half-Earth: Our Planet's Fight for Life* (New York/London: Liveright Publishing Corporation, 2016), 211.

[2] Immanuel Kant, *Grounding for the Metaphysics of Morals*, Third Edition, translated by James W. Ellington (Indianapolis/Cambridge: Hackett Publishing Company, INC., 1981), 17.

[3] Robin Wall Kimmerer, *Gathering Moss: A Natural and Cultural History of Mosses* (Corvallis: Oregon State University Press, 2003), 35.

[4] William Shatner, on CNN's "At This Hour With Kate Bolduan," December 9, 2021:
https://archive.org/details/CNNW_20211209_160000_At_This_Hour_With_Kate_Bolduan/start/3540/end/3600-3660

[5] https://www.merriam-webster.com/dictionaly/rewilding.

[6] Wilson, *Half-Earth*, 167.

[7] Why half the earth is explained at: Half-earthproject.org/discover-half-earth/#why-half.

[8] HALF-EARTH PROJECT is a trademark of E.O. Wilson Biodiversity Foundation, Inc. See https://map.half-earthproject.org/dataGlobe. (Copyright permission granted by the E.O. Wilson Biodiversity Foundation, August 22, 2022).

[9] Rewilding.org/about-tri/our-mission.

[10] In 1861, the philosopher John Stuart Mill extended his moral principle to all sentient beings:

> "According to the greatest happiness principle…the ultimate end, with reference to and for the sake of which all other things are desirable…is an existence exempt as far as possible from pain, and as rich as possible in enjoyments, both in point of quantity and quality….secured to all mankind; and not to them only, but, so far as the nature of things admits, to the whole sentient creation." (Stuart Mill, John, *Utilitarianism*, 2nd edition, edited by George Sher, [Indianapolis/Cambridge: Hackett Publishing Company, Inc. 2001], 12).

Today, Mill's utilitarianism presents just one of many different kinds of principled positions for extending moral standing to living things other than humans (for other examples, see Sections III and IV of *The Environmental Ethics & Policiy Book*, eds. Donald VanDeVeer and Christine Pierce, Third Edition [Belmont, CA: Wadsworth/Thomson Learning, 2003] 114-310).

[11] Joseph Lee, Grist.org/indigenous/do-salmon-have-rights/ March 01, 2022.

[12] Tiffany Challe, News.climate.columbia.edu/2021/004/22/rights-of-nature-lawsuits, April 22, 2021.

[13] Ibid.

[14] This point was made recently by the theologian Jame Schaefer in her response to Jason Rohr's presentation "Is Biodiversity Loss Increasing Diseases of Wildlife and Humans, And if so, What Can We Do?", Environmental Justice and Sustainability Lecture Series, Duquesne University (organized by Siena's Franciscan Center for Catholic Studies, Duquesne University's Theology Department, the

Notre Dame Club of Northeastern New York, and Siena's Departments of Religious Studies and Environmental Studies and Sciences), March 16, 2022, (Virtual).
[15] Wilson, *Half-Earth*, 211.
[16] Ibid., 212.
[17] Pope Francis, *Laudato Si': On Care for our Common Home*, Encyclical Letter (Huntington, IN: Our Sunday Visitor Publishing Division, Our Sunday Visitor, Inc. and Vatican City: Libreria Editrice Vaticana, 2015), 14.
[18] Jame Schaefer, Respondent. (See endnote 14.)
[19] Jan Zwicky, "A Ship From Delos" in, Robert Bringhurst and Jan Zwicky, *Learning to Die: Wisdom in the Age of Climate Crisis* (Saskatchewan: University of Regina Press, 2018), 41-71.
[20] Ted.com/talks/sister_true_dedication_3_questions_to_build_resilience_and_change_the_world
[21] Kant, *Grounding For the Metaphysics of Morals*, 7.
[22] Jennifer Bates, "An Inquiry Into The Nature of Environmentally Sound Thinking" *Environmental Ethics,* 25(2) (2003): 193-197.
[23] Arne Naess, "The Shallow and The Deep, Long-Range Ecology Movement", *Inquiry* 16 (1973): 95-100.
[24] G. W. F. Hegel, *Phenomenology of Spirit,* translated by A. V. Miller (Oxford: Oxford University Press. 1977), 21, paragraph 36.
[25] https://thegrandtour.fandom.com/wiki/Enviro-mental.
[26] Nigel Tubbs, "Green metaphysics: A sustainable and renewable liberal arts education," *Educational Philosophy and Theory* 50, no. 11 (2018): 1469-5812.
[27] Hegel, *Phenomenology of Spirit* 9-10, paragraph 17.
[28] Plato, "Phaedo," in, *The Collected Dialogues of Plato Including the Letters,* edited by Edith Hamilton and Huntington Cairns, Bollingen Series LXXI (Princeton, NJ: Princeton University Press, 1961), 40-98. 108c-115b, 90-95.
[29] H. B. Acton, "Introduction," in, Hegel, G. W. F., *Natural Law*, translated by T. M. Knox, Introduction by H. B. Acton, (Pennsylvania: University of Pennsylvania Press, 1975), 9-47.
18-21.
[30] Ibid., 20-21.
[31] Ibid., 19.
[32] G. W. F. Hegel, *Hegel's Science of Logic*, edited by Lewis, H. D., translated by Miller, A. V., Humanities Paperback Library (Atlantic Highlands, NJ: Humanities Press International, Inc., 1969), 841-42.
[33] Hegel, *Phenomenology of Spirit* (second-to-last paragraph), 492, paragraph 807.
[34] Wilson, *Half-Earth*, 130.
[35] As the Canadian Hegelian H. S. Harris translates it, the last line expresses the dialectical unity of the Divine and Nature: "As alienated Spirit, Nature is just its eternal uttering, and the movement that sets up the subject." (Harris, H. S., *Hegel's Ladder*, Volumes 1& 2 [Indianapolis, Cambridge: Hackett Publishing Company, 1997], Vol 2, 745).
[36] Hegel, *Science of Logic*, 841.
[37] Hegel, *Phenomenology of Spirit*, 493, paragraph 808.
[38] See for example Plato's "Phaedo", 65b-67b, 48-49, and 74a-76d., 57-59.

[39] Hegel, *Science of Logic*, 842.
[40] Hegel, *Phenomenology of Spirit*, 11, paragraph 20. At the end of the *Phenomenology*, Hegel uses the words "concealment" and "disclosure", "revelation" and "destiny and necessity". These words express the *process of Spirit in Time*, "enriching" itself to the point where it has grasped the "entire substance and has absorbed into itself the entire structure of the essentialities of substance." (Hegel, *Phenomenology of Spirit*, 487, paragraph 801).
[41] Hegel, *Phenomenology of Spirit*, 487, paragraph 801.
[42] Hegel, *Phenomenology of Spirit*, 487, paragraph 801.
[43] Hegel, *Phenomenology of Spirit*, 491-492, paragraph 806-7.
[44] Hegel, *Phenomenology of Spirit*, 486, paragraph 799.
[45] Hegel, *Phenomenology of Spirit*, 490-91, paragraphs 805-806.
[46] Hegel, *Phenomenology of Spirit*, 493, paragraph 808.
[47] "It can therefore be said that this content is the exposition of God as he is in his eternal essence before the creation of nature and a finite mind." (Hegel, *Science of Logic*, 50).
[48] Hegel, *Science of Logic*, 841and 842.
[49] Ibid., 481.
[50] Acton, "Introduction", 21.
[51] Hegel, *Science of Logic*, 842.
[52] Hegel, *Science of Logic*, 841.
[53] Wilson, *Half-Earth*, 204-06.
[54] https://geobon.org/ebvs/what-are-ebvs/. (Copyright permission granted by email by Adriana Radulovici, PhD, Executive Secretary, GEO BON, on July 18, 2022.)
[55] https://geobon.org/about/vision-goals/
[56] algore.com/project/the-deep-space-climate-observatory
[57] mol.org
[58] https://map.half-earthproject.org/
[59] Wilson, *Half-Earth*, 205-6.
[60] Anyone skeptical about AI helping should see Graeme Green's "Numbers Game: Can AI Save the Elephant?" *The Guardian Weekly*, March 4, 2022, 24-25.
[61] Wilson, *Half-Earth*, 206.
[62] Ibid., 191.
[63] https://aeon.co/ideas/why-e-o-wilson-is-wrong-about-how-to-save-the-earth
[64] Paul Hawken, "excerpts from Natural Capitalism," *Mother Jones* March/April (1997): https://www.motherjones.com/politics/1997/03/natural-capitalism/
[65] Jason Rohr, Galla Professor and Chair of the Department of Biological Sciences at Notre Dame, "Is Biodiversity Loss Increasing Diseases of Wildlife and Humans, And if so, What Can We Do?", Environmental Justice and Sustainability Lecture Series, Duquesne University (organized by Siena's Franciscan Center for Catholic Studies, Duquesne University's Theology Department, the Notre Dame Club of Northeastern New York, and Siena's Departments of Religious Studies and Environmental Studies and Sciences), March 16, 2022, (Virtual).
[66] Naess, "The Shallow and the Deep", 97.
[67] See Hegel's *Science of Logic*, 841 and *Phenomenology*, 11, paragraphs 20-21. At this point, I would like to make my *apologia*. To *fully* explain all of this, I

would need to explain more about Hegel's *Phenomenology of Spirit,* in particular how Absolute Knowing offers a solution to unhappy consciousness idealisms (including the idealism of a rewilded biosphere); I would have to deal more precisely politically and environmentally with the fact that his dialectical mediation is unavoidably always an *economy* of universality; I would have to analyze the end of Hegel's *Greater Logic* in which the Absolute Conceptual Idea externalizes itself into Nature, and discuss what Reflection means in the *Greater Logic*; I would have to explain how my reading does not turn Hegel into his contemporary Schelling, and clarify further that Hegel was not a nature-lover in any romantic sense; I would have to explain how my reading of Hegel handles the antiquated science Hegel writes about in his *Philosophy of Nature*, and finally and most importantly, why I can hold this *non-anthropocentric* Hegelian view despite Hegel's clear preference in his works on art, religion and politics for expressly *human-mediated* objectivity. All of this is beyond the scope of this chapter. But in any case, I believe that my reading of the end of the *Phenomenology of Spirit* is a legitimate one for our time, and therefore, in a way, implied in the text; if Hegel were alive today, aware of the threat of extinction and of the role of biodiversity in the constitution and preservation of life on earth, I believe he would agree.

[68] I recently found a kindred spirit in the television documentary "My Garden of a Thousand Bees", filmed and narrated by Martin Dohrn, on, PBS Television Series *Nature*, directed by David Allen, produced by Gaby Bastyra, (Thirteen Productions LLC, Passion Planet LTD and Ammonite LTD., 2021), Season 40, Episode 1.

[69] By contrast, "The most dangerous worldview is the worldview of those who have not viewed the world" (Alexander von Humboldt, as cited in Wilson, *Half-Earth*, 79).

[70] G. W. F. Hegel, *Hegel's Philosophy of Right*, trans. T. M. Knox (London, Oxford, New York: Oxford University Press, 1952) 10.

Literature

Acton, H. B. 1975. "Introduction." In, G. W. F. Hegel, *Natural Law*. Translated by T. M. Knox, Introduction by H. B. Acton. (Pennsylvania: University of Pennsylvania Press): 9-47.

algore.com/project/the-deep-space-climate-observatory. N.d.

Bates, Jennifer. 2003. "An Inquiry Into The Nature of Environmentally Sound Thinking." *Environmental Ethics* 25(2):193-197.

Challe, Tiffany. 2021. News.climate.columbia.edu/2021/004/22/rights-of-nature-lawsuits. April 22.

Gorongosa.org/map-of-life/. N.d.

Green, Graeme. 2022. "Numbers Game: Can AI Save the Elephant?" *The Guardian Weekly* March 4, 24-25.

Half-earthproject.org/discover-half-earth/#why-half. N.d.

Harris, H. S. 1997. *Hegel's Ladder*. Volumes 1& 2. Indianapolis, Cambridge: Hackett Publishing Company.

Hawken, Paul. 1997. "Excerpts from Natural Capitalism." In, *Mother Jones* March/April: https://www.motherjones.com/politics/1997/03/natural-capitalism/.
Hegel, G. W. F. 1969. *Hegel's Science of Logic*. Editor Lewis, H. D. Translator Miller, A. V. Humanities Paperback Library. Atlantic Highlands, NJ: Humanities Press International, Inc.
Hegel, G. W. F. 1977. *Phenomenology of Spirit*. Translated by A. V. Miller. Oxford: Oxford University Press.
Hegel, G. W. F. 1952. *Hegel's Philosophy of Right*. Translated by T. M. Knox. London, Oxford, New York: Oxford University Press.
https://aeon.co/ideas/why-e-o-wilson-is-wrong-about-how-to-save-the-earth
https://geobon.org/about/vision-goals/
https://geobon.org/ebvs/what-are-ebvs/
https://map.half-earthproject.org/
https://www.merriam-webster.com/dictionaly/rewilding.
https://thegrandtour.fandom.com/wiki/Enviro-mental.
Kant, Immanuel. 1981. *Grounding for the Metaphysics of Morals*. Third Edition. Translator W. Ellington. James. Indianapolis/Cambridge: Hackett Publishing Company, Inc.
Kimmerer, Robin Wall. 2003. *Gathering Moss: A Natural and Cultural History of Mosses*. Corvallis: Oregon State University Press.
Lee, Joseph. 2022. Grist.org/indigenous/do-salmon-have-rights/. March 1.
Mill, John Stuart. 2001. *Utilitarianism*, 2nd Edition. Ed. Sher. George. Indianapolis/Cambridge: Hackett Publishing Company, Inc1.
mol.org. (Map of Life.) N.d.
"My Garden of a Thousand Bees." 2021.Filmed and narrated by Martin Dohrn. PBS Television Series *Nature*. Directed by David Allen, produced by Gaby Bastyra. Thirteen Productions LLC, Passion Planet LTD and Ammonite Ltd. Season 40, Episode 1.
Naess, Arne. 1973. "The Shallow and The Deep, Long-Range Ecology Movement". *Inquiry* 16: 95-100.
Plato. "Phaedo." 1961. In *The Collected Dialogues of Plato Including the Letters*. Eds. Hamilton, Edith. Cairns, Huntington. Bollingen Series LXXI. Princeton, NJ: Princeton University Press. 40-98.
Pope Francis. 2015. *Laudato Si': On Care for our Common Home*, Encyclical Letter. Huntington, IN: Our Sunday Visitor Publishing Division, Our Sunday Visitor, Inc. and Vatican City: Libreria Editrice Vaticana.
Rewilding.org/about-tri/our-mission. N.d.

Rohr, Jason. 2022. "Is Biodiversity Loss Increasing Diseases of Wildlife and Humans, And if so, What Can We Do?", Environmental Justice and Sustainability Lecture Series, Duquesne University. Organized by Siena's Franciscan Center for Catholic Studies, Duquesne University's Theology Department, the Notre Dame Club of Northeastern New York, and Siena's Departments of Religious Studies and Environmental Studies and Sciences, March 16. (Virtual).

Schaefer, Jame. 2022. Respondent to the Lecture by Rohr. Jason. "Is Biodiversity Loss Increasing Diseases of Wildlife and Humans, And if so, What Can We Do?", Environmental Justice and Sustainability Lecture Series, Duquesne University. Organized by Siena's Franciscan Center for Catholic Studies, Duquesne University's Theology Department, the Notre Dame Club of Northeastern New York, and Siena's Departments of Religious Studies and Environmental Studies and Sciences, March 16. (Virtual).

Shatner, William. 2021. Interviewed by Bolduan, Kate. CNN "At This Hour With Kate Bolduan." December 9: https://archive.org/details/CNNW_20211209_160000_At_This_Hour_With_Kate_Bolduan/start/3540/end/3600-3660

Ted.com/talks/sister_true_dedication_3_questions_to_build_resilience_and_change_the_world. N.d.

The Environmental Ethics & Policy Book. 2003.. Donald VanDeVeer and Christine Pierce, eds. Third Edition. Belmont, CA: Wadsworth/Thomson Learning. 114-310.

Tubbs, Nigel. 2008. *Education in Hegel*. London and New York: Continuum International Publication Group.

Tubbs, Nigel. 2018. "Green metaphysics: A sustainable and renewable liberal arts education." *Educational Philosophy and Theory* 50, no. 11: 1469-5812.

Wilson, Edward O. 2016. *Half-Earth: Our Planet's Fight for Life*. New York/London: Liveright Publishing Corporation.

Zwicky, Jan. 2018. "A Ship From Delos." In Robert Bringhurst and Jan Zwicky, *Learning to Die: Wisdom in the Age of Climate Crisis. Saskatchewan: University of Regina Press. 41-71.*

CHAPTER FOUR

ECOLOGICAL RESILIENCY AND THE FUTURE OF HUMANITY

FERDINAND OKAFOR

Introduction

Ecological change is a front-burner discussion in bioethics. Experience over the years and available studies reveal how drastically climate change is taken place. At the heart of climate change is the ecological crisis. Ecological crisis is an issue of great concern because of its threat to the immediate and future well-being of humanity and creation. The epic of ecological crisis is global warming and its allied aftermath. If the trend in global warming remains unchecked, the future of humanity will be in serious jeopardy. Hence, how the future will look is an issue of great concern.

Climatic change is an inherent consequence of population growth, urbanization, and industrialization. They are regarded as dependent variables that lead to global warming and other forms of ecological abuse by human beings. The argument in this chapter is that climate change as a response to macro and micro human activities is inevitable because the factors that are responsible for it are also inevitable. However, the effect can be attenuated. The cumulative effect of climate change on humanity is the emergence of various health hazards such as genetic disorders, neurological and developmental disorders, and psychiatric and personality disorders.

To mitigate the effects of climate change, there is a need to first alter the fundamental assumption towards creation by altering the attitude of everybody towards creation. This will consequently lead to alteration in the consequences of climate change. Secondly, there is a need to develop skills to improve climate resiliency since climate change is inevitable. This chapter further maintains that climate resiliency is a task for everyone. To achieve climate resiliency; creating awareness inter alia collaboration, adaptation, and strategic action plans are advanced as veritable means.

This chapter is divided into two parts. The first part examines the impact of climate change on humanity and how climate change is a response to macro and micro human activities. While the second part examines how climate resiliency can be promoted. It examines the role of each person in promoting climate resiliency.

Alteration of the Ecosystem as the Foundation of Climate Change

The term ecosystem refers to the interaction among living organisms, their environment, and abiotic factors in the environment. It includes plants, animals, species, air, water, sun, natural resources, biosphere, and biodiversity. Biosphere refers to the influence zone of organisms on chemical elements fluxes.[1] It is the layer of the earth that consists of a complex web of life.[2] While biodiversity refers to varieties of life.[3] Organisms and their environment constitute what is known as nature. Nature refers to the atmosphere, lithosphere, anthroposphere, astysphere, and agrosphere. Human beings are an integral part of nature.[4] But they belong to the part of nature known as the biosphere.[5] There is a symbiotic relationship between human beings and nature. Nature and human beings provide support for each other. The advancement of nature depends on human intervention. Similarly, the well-being of human beings is dependent on the care for nature.[6] The subsistence of human life on earth is dependent on the existence of plants, animals, other organisms, and natural resources. Organisms perform critical functions in the functionality of the ecosystem.[7] Every organism is relevant to the sustenance of the ecosystem. Without the presence of living and non-living organisms, human life will go into extinction. Human beings depend on plants and animals for food to be able to persist and thrive.[8] Through the interactions that occur among organisms and other abiotic elements in the environment, the exchange of energy and transportation of mineral nutrients and energy occurs among organisms themselves as well as across the different domains in the ecosystem.[9]

The relationship between human beings and nature is not just physical, it is also social and religious.[10] Nature provides an inroad into the spiritual dimension of life. The adaptive interplay between living organisms, the climate, and broader earth system processes has resulted in a resilient biosphere.[11] The organization of life in the ecosystem is mostly dependent on climate.[12] Unfortunately, there is a drastic alteration of the ecosystem due to Anthropocene activities. These activities include agriculture, urbanization, population growth, and industrialization. This alteration of the ecosystem exerts tremendous influence on its resiliency. Therefore, the obligation to conserve and preserve the integrity of nature is incumbent on all.

Ecological Crisis: Identifying the Problem

The fundamental assumption that governs the human relationship with nature is that nature is static whereas it is changing. Change and orderliness are fundamental and inherent principles in the ecosystem. The term change implies evolution. The different elements that constitute the ecosystem are evolving from one stage of development to another. Apart from change being a natural process, it can also be facilitated by human intervention. Human activities such as industrialization, urbanization, and population can accelerate the process of change in the ecosystem. Change is an integral dimension of nature; it is an irreversible process. Therefore, it is impossible for nature to be restored to its pristine form. The implications of the erroneous assumption about nature are undue manipulation of nature, degradation of nature, and unpreparedness towards the consequences of ecological abuse. Ecological abuse is a daily occurrence. It is constantly perpetuated by human beings through different kinds of pollution.

Pollution is the release of harmful substances and the deposit of residue into the ecosystem. It originates from macro and micro human activities. The term macro human activities refer to environmental degradation arising from the activities of multinational companies. While micro human activities refer to ecological abuse arising from individual abuse of the environment. We all generate small ecological damage.[13] Macro-human activities arise from industrialization and urbanization. Industrialization is a tremendous source of pollution. Pollution creates an imbalance in the ecosystem. In short, it leads to ecological crisis. Ecological crisis is an unfavorable reality that confronts humanity presently. It is a concept that underscores overpopulation, climate change, deforestation, and acid rain. Ecological crisis is a major problem in the world, and it constitutes a topic of discussion because of the environmental, social, economic, and political implications.[14] The epic of ecological crisis is global warming with climate change as its allied aftermath. Some people tend to conflate climate change with global warming. Whereas the two terms are different. Climate change refers to abiotic changes in the earth. While global warming refers to changes in temperature.[15]

Climate change is an ethical issue as well as a moral issue because it involves human action towards nature. The term climate change refers to alterations in temperature and weather conditions due to human activities. This alteration leads to the greenhouse effect or global warming effect. The aftermath of global warming is climate change and a rise in the sea level.[16] Global climate change has an impact on temperature. The first indicator of climate change is temperature change. Among all the effects of climate change, the thermal effect is regarded as the most direct.[17]

The evidence of climate change has been noticeable since the industrial revolution in 1880 to the present time.[18] Since the beginning of the industrial revolution, humanity has witnessed the disastrous effect of rising temperatures.[19] The disturbance of the balance of the earth's climate has resulted in extreme weather events since the 1950s. These include heat waves, drought, hurricanes, superstorms, and tornadoes.[20] The changes in the world have been described as *rapidification.*[21] It refers to the acceleration of changes affecting humanity and the planet as well as a more intensified pace of life.

The Root of the Problem: Population Growth, Industrialization, and Urbanization

Alteration in temperature and weather arises from either natural processes or anthropogenic activities. Most of the changes are dependent on human interference with the environment.[22] Ecological concern has been described as a "tragic consequence" of unchecked human activity.[23] The main factors that are responsible for climate change are population growth, industrialization, and urbanization. Climate change has a link with population growth, urbanization, and environmental pollution. It is a symptom of ecological abuse.[24]

Population is a key causative factor in climate change. The population of human beings in the world is growing at an alarming rate. Between 1900 and 2013 the population of human beings in the world grew from 1.5 billion to 7 billion.[25] And between 1970 and 2004 the human population in the world increased by almost 70 percent.[26] The world's population is growing at a rate of 1.1 percent.[27] And the population of human beings living on the earth is now about 8 billion. It is expected to stabilize between 9-11 billion by the end of the 21st century.[28] This has serious implications for climate change. Human population has an impact on the consumption and demand for food resources and natural resources.[29] Population growth directly contributes to the rise in greenhouse gas.[30] Population leads to more consumption, and consumption necessitates more production to meet human needs. The challenge is whether the earth will be able to cater for the growing population of humanity.

Population also influences industrialization. To meet human demands, industrialization becomes a way out, and consequently becomes a problem for humanity. The more population, the more the production to keep humanity alive. So, population growth leads to more emissions of greenhouse gases. Human population is, therefore, a determining factor in greenhouse gas emissions.[31]

The agricultural system is a major source of greenhouse gas emissions; it constitutes a contributor to climate change.[32] Apart from the greenhouse effect, agriculture leads to deforestation. The drive for massive agricultural production necessitates deforestation to provide land for agriculture. Agriculture is not limited to crop production, but it includes animal grazing, which is considered to have a more deleterious effect on the ecosystem.[33] Agriculture is linked to deforestation. Through agriculture, the natural environment has been taken over.[34] Deforestation has a negative impact on humanity. It leads to erosion and the destruction of the ecosystem. The problem of flooding which is a major problem in some parts of the world is not unconnected with deforestation. The presence of methane in the atmosphere has increased due to deforestation. Methane is regarded as a source of global warming. Methane emission is partly due to anthropogenic activities such as coal mining, oil, and natural gas, and agricultural activities such as livestock and rice farming. Methane can last as long as 12 years before it can disappear from the atmosphere.[35]

Another causative factor of the ecological crisis is urbanization. Urbanization is rooted in the human quest for development. Human beings are tending more toward urbanization. The percentage of the world population living in urban centers is about 55%. It is estimated that by mid twenty-first century 7 out of 10 persons will be living in urban centers.[36] Urban centers account for about 70 percent of Carbon IV Oxide emission.[37] Population growth and urbanization lead to encroachment of unoccupied terrain and deforestation. Most often, human beings engage in deforestation in the guise of development, without replacement of the trees that are harvested. The consequence of encroachment is alterations of the ecosystem and climate and biodiversity loss. While the population of human beings is increasing, other species are decreasing.

Human activities such as industrialization, and bush burning release greenhouse gases into the atmosphere. The increased presence of greenhouse gases such as Carbon IV Oxide (CO_2), Nitrous Oxide (N_2O), Water vapor (H_2O), Menthane (CH_4), Ozone (O_3), and Sulphur Hexafluoride (SF_6) in the atmosphere leads to alteration of the balance in the biosphere and warming effect in the planet, earth. The twenty-first century is regarded as the warmest period in the history of humanity with the persistent increase in the warming of the biosphere.[38] The use of fossil fuels is a dangerous endeavor for humanity. It generates and releases Carbon (IV) Oxide into the atmosphere. The presence of Carbon (IV) Oxide leads to a rise in temperature. It takes centuries for Carbon (IV) Oxide to disappear from the atmosphere. Continuous deforestation and global warming have adverse consequences for humanity.[39]

Climate Change: An Inevitable Reality

Climate change is an inevitable problem because the variables that propel it are inevitable. Heraclitus, the Greek Philosopher opined that change is the only permanent reality. Climate change is not new in the history of the earth, but the realization of climate change only came to human consciousness because of its effect.[40] Every age of human development ushers in a new problem. Population growth, urbanization, and industrialization are variables that necessitate climate change. They are necessary events in human well-being. The transition from industrialization to technology and new method of production have an inherent ecological crisis.

The cumulative effect of climate change on humanity is the emergence of various health hazards such as genetic disorders, neurological and developmental disorders, and psychiatric and personality disorders. The effects of climate change though inevitable, can however be attenuated.

Climate Change: Implications and Matters Arising

Global Climate change is a global ethical issue as it poses adverse effects on creation and a serious threat to human existence. Climate change can be identified as extreme weather events, such as too much rain or extreme drought. The consequences of undermining the integrity of creation will affect every human being, including future generations. Although, every human being is affected differently. People with underlining health issues, aged people, and infants may be more affected by some climatic changes.

The effect of climate change is evident in health hazards, impact on agriculture, extinction of some species, and flooding. Flooding is an indicator of climate change. The North Pole is warmer than it used to be. The rise in temperature melts the polar ice resulting in a serious rise in the sea level. It is conjectured that by 2050 the sea level could rise by one meter.[41] Miami Beach and other arcas near the coast are now flood zones. Many nations have experienced great flooding in recent years due to climate change.

Health Hazard

One of the consequences of ecological change is health hazards. Exposure to atmospheric pollutants can lead to various health hazards and spread of diseases. Some people develop health problems due to inhaling smoke from cooking, combustion from cars, or industrial fumes. Exposure to pesticides, fertilizers, fungicides, herbicides, and agrotoxins, can also lead to health hazards.[42] The commonest health hazard today are mental health, genetic, and psychiatric disorders. Mental health disorder is associated with the brain. The brain is the most sensitive part of the body. The brain, genes, and mind are directly affected by events and hazards in the environment. Hence, there is an upsurge in mental, genetic, and neurodevelopmental disorders. No age in the history of humanity has had the experience of mental health, genetic, and neurodevelopmental disorders that we have seen in the twenty-first century. Since the environment has an impact on the mind, genes, and brain; hazards and alteration in the ecosystem will lead to the alteration of the mind, gene, and brain, and indeed, the alteration of human life.

Autism spectrum disorder is a genetic and neuro disorder that has been discovered to have a link with environmental hazards. Studies reveal that autism spectrum disorder can be caused by epigenetic factors. The interaction between the environment and genes as well as exposure to pollutants, can cause epigenetic processes like attachment and detachment to occur rapidly or less rapidly.[43] Exposure to harmful hazards can increase the chances of autoimmune deficiencies in children. Autoimmune deficiency is a characteristic feature of autism spectrum disorders.[44] Apart from causing health vulnerabilities, climate change can intensify already existing health problems. This means that there is a synergy between health and climate change.[45] The implication is that humanity will be faced with high intensity of mental health, genetic, and neuro and psychiatric disorders if ecological crisis is unchecked. This shows how health has been diversified to include not only somatic problems but mental, neuro, psychiatric, and genetic disorders.

Agriculture

Climate change has a serious impact on agriculture. Plants need rain and sun to survive. The alteration of weather conditions has an impact on agricultural produce, especially in the sub-Saharan African continent, where agriculture is still at the primitive stage, irrigation is still less practiced, and agriculture is heavily dependent on rainfall and sunlight. Climate change leads to irregular weather conditions. It can affect crop water requirements.[46] Studies show that rainfall variability has a significant impact on agriculture productivity in Africa.[47] Apart from Africa, India is also home to many poor people. It is estimated that about 65% of the Indian population depends on agriculture. These are also adversely affected by climate change.[48] The implication is that in Africa and India, a shortage in food production is bound to occur as is already the case in most parts of Africa. This will consequently lead to a debt burden in developing countries.

In addition, climate change leads to the loss of biodiversity. Loss of biodiversity is regarded as the most critical problem in climate change. Many species of wildlife and pollinators have been driven into extinction because of deforestation, ecological abuse, and climate change.

Human beings, directly and indirectly, influence the constitution of species through pollution, ecological abuse, artificial selection, and controlled reproduction of animals, crops, and plants.[49] Many plants depend on pests and pollinators for cross-pollination to occur. Unfortunately, many of these agents of pollination are going to extinction.

The Poor People and Developing Countries

Climate change has a more negative impact on the poor people of the earth. The poor lack the necessities to thrive in life which already makes them unprotected against hazards in life. Many poor people especially in developing countries live in areas that are exposed to intense pollution, and they rely on natural resources and ecosystem services such as agriculture, fishing, and forestry as their means of livelihood. The persistent increase in pollution and climate change exerts a negative effect on natural resources, which is their means of livelihood. This makes their condition worse.[50] Besides, their daily exposure to environmental hazards and their inability to move along with the changes in the world have serious negative impacts on their health and well-being, especially as they have limited access to basic social services. The cumulative effect of these negative experiences is a reduced life span and growing inequality in the standard of living.

In a complementary advancement, climate change will impact more on developing countries of the world than developed countries. Economic growth, health indicators, water availability, production of food, and the ecosystem in developing countries are already being affected by climate change. The African continent has been spotted as a red zone of climate.[51] Some countries in Africa, especially Nigeria, are experiencing yearly flooding.[52] These countries lack the abilities and resources to meet these challenges.

Promoting Climate Resiliency: A Task for All

The use of the term climate resiliency implies the existence of climatic threats. It is concerned with the ability to cope with risk and recover from shocks, and it is intrinsically linked with the reduction of vulnerability, increased adaptive capacities, and sustainability. It includes the ability to adapt, persist, and develop with changing climatic conditions.[53]

The climate is a common possession. It belongs to all, and it is meant for the well-being of everyone.[54] Unfortunately, it is beset with crisis due to ecological abuse. Human beings have an obligation to care for nature for human well-being. Responsibilities towards nature are derivatives of the interconnectivity between human beings and nature. Therefore, it is important to evaluate human conduct toward nature.[55] Promoting the resiliency of nature demands a lifestyle change and requires collaboration, adaptation, and an action plan. The type of intervention response depends on the type of climatic threat and the population that is affected. Therefore, it is crucial to study the situation and provide a correct intervention to mitigate the challenge posed by ecological crises. To promote resiliency in health, for instance, more experts in behavioral interventions will be needed to promote resiliency especially as most health problems will be linked to genetic, mental health, and neurodevelopmental disorders or psychiatric disorders. The implication is that with climate change, not all health problems will require pharmacological intervention.

Collaboration

Humanity is related and originates from a common stock and has common problems. Whatever affects one, affects all. The Covid-19 pandemic has revealed how interrelated humanity is. It is either we are saved together, or we perish together. The universe is a common heritage, and climate is a common good. Climate change results from collective human activities, and it affects everyone. So, we are united by an issue of common concern.[56] Therefore, the effort of all is needed to promote climate resilience. A global problem requires collective responsibility.[57] Everyone has a shared responsibility towards a common good. It is imperative that developed nations and developing nations collaborate to develop a unified plan to salvage the ecosystem from abuse. Climate change as a global problem requires long-term collaboration among nations to resolve the consequences.

Ecological crisis offers opportunities for growth and retardation depending on the decision that is made to address the crisis. The future of humanity will be bountiful if human beings can understand their influence on the biosphere and make an appropriate decision.[58] A decision is crucial to the well-being of the human person. The collective decision to promote a balance in creation will advance the well-being of humanity and creation. Improvements in health and standard of living have enhanced longer life for humanity than before. Many people live longer now than before.[59] To promote climate resiliency, changes in lifestyle, production, and consumption are needed.[60] Lifestyle is governed by mindset or assumption. Therefore, to change lifestyle, the mindset and attitude towards the environment will need to be altered. This can be done through enlightenment or sensitization of everyone. The environment should be regarded as a home.[61] Collaboration in creating awareness of respect for ecology is a veritable means of promoting ecological resilience.

The core of ecological awareness should focus on how industrialization, urbanization, and population growth contribute to climate change. Everybody must become aware of the causes of climate change, how it affects everybody, especially the poor and future generations, and how each person can develop ecological resilience and avoid ecological abuse. Ecological abuse is a lifestyle that is governed by a mindset. People cannot behave differently unless they become aware of the consequences of their behavior. Awareness can lead to the development of adaptive skills to cope with climate change.

Adaptation

Climate change poses a great risk to humanity. No one is exempted from the effects of climate change. However, poor countries are most affected, and the most vulnerable populations are children, the aged, and people with disability and underlining ailment.[62] Thus far, adaptation remains a major strategy for promoting resiliency to climate change.

Adaptations refer to adjustment to actual or expected climate and its effect. The purpose of adaptation is to reduce the harm caused by climate change and increase the benefits.[63] This implies that humanity should develop strategies to cope with the challenges of climate change. Adaptation strategy is built on the post-cautionary principle. It is a realist view, which emphasizes adaptation to the inevitable consequences of climate change. Adaptation plays a dual role. It is based on the understanding that reducing greenhouse gases alone is insufficient to deal with the consequences of climate change. Adaptation entails acknowledging the impact of climate change, facilitating adequate adaptation skills, and cooperating in preparing for adaptation to the impacts of adaptation.[64]

Adaptation strategy is dependent on the vulnerable population. The population as well as the climatic problem that is affecting them needs to be studied and the intervention needs to be carefully planned. So, the nature of a problem and the affected population determine the nature of the response. The greatest climatic threat for people with disability is extreme weather events and the key adaptation strategy is inclusive emergency planning, response, and recovery. But for older people and those who experience chronic ailments, the threat is specifically from heat waves, and the adaptation intervention is the maintenance of cool places for shelter from the heat.[65]

Action Plan

To minimize climate change, humanity needs to engage in concrete actions that promote resilience. Whatever action that is undertaken should be a responsible and virtuous action with due consideration for the wellbeing of future generations. As human beings we are interdependent. The actions of the people in this generation have serious consequences on future generations, just as the actions of past generations have an impact on our generation. In utilizing the resources of the earth, intergenerational justice should be duly considered. That implies that people of future generations should be given due consideration.[66]

The first fundamental value to be promoted is respect for nature. Respect for nature is one of the United Nations Millennium Declarations. It is a fundamental international value.[67] Ecological abuse has been perpetuated for a long. Abuse emerges from a lack of respect. To avert the consequences of ecological abuse, global *ecological conversion* is inevitable. Ecological conversion entails global changes in lifestyle, models of production and consumption, and established structures of power that govern societies.[68] Lifestyle is regulated by values. A change of lifestyle implies a change in the values that necessitate a given lifestyle. Too much priority has been dedicated to economic values to the detriment of the values of health. There is a need to revert this scale of preference, by giving greater emphasis to the value of health.[69] In addition,

every nation should enact laws to protect living species and natural resources from abuse.

Secondly, planting trees is already a practice that some people are already propagating. The planting of more trees can help to reduce Carbon (IV) Oxide and increase Oxygen output in nature. Human life depends on Oxygen, while plants need Carbon (IV) Oxide for photosynthesis. Research studies have shown that Carbon (IV) Oxide concentration can stimulate photosynthesis and growth in plants.[70] However, whether industrially generated Carbon (IV) Oxide is safe for plants, remains unproven. Plants can also help to reduce flood effects. Plants can absorb water. Flooding occurs more in places without vegetation. Therefore, planting of trees will mitigate the effect of climate change.

Conclusion

This chapter is an effort to first discover the foundation of climate change. Human activities such as industrialization, urbanization, and population are identified in this chapter as the causes of climate change. The chapter maintains that climate change is a response to macro and micro human activities. Rising temperatures, health hazards, and low agricultural productivity due to drought or too much rain are identified as indicators of climate change. If climate change remains unchecked, the future of humanity will be jeopardized. Therefore, to mitigate the consequences of ecological abuse, the chapter argues for ecological resilience. Ecological resilience entails collaboration, adaptation, and action plans.

In conclusion, to resolve the problem of climate change, the effort of everyone is needed to promote resiliency and adaptation.

Notes

[1] Stefan Norra, "The Bisophere in Times of Global Urbanization," *Journal of Geochemical Exploration* 147, no. Part A (December 1, 2014): 52. https://doi.org/10.1016/j.gexplo.2014.06.004.

[2] Carl Folke, Stephen Polasky, Johan Rockstrom, Victor Galaz, Frances Westley, Michele Lamont, Marten Schefer, et. al., "Our Future in the Anthropocene

Biosphere," *Ambio: A Journal of Environment and Society* 50, no. 4 (April 1, 2021): 835. https://doi.org/10.1007/s13280-021-01544-8.

[3] Henk Ten Have, *Global Bioethics: An Introduction* (New York: Routeledge, 2016), 229.

[4] ten Have, *Global Bioethics: An Introduction*, 231.

[5] Norra, "The Bisophere in Times of Global Urbanization," 53.

[6] Have, *Global Bioethics: An Introduction*, 228.

[7] Folke, et. al., "Our Future in the Anthropocene Biosphere." 842.

[8] Kent E. Portney, *Sustainability.* MIT Press Essential Knowledge Series (The MIT Press, 2015), 4.
https://search-ebscohost-com.authenticate.library.duq.edu/login.aspx?direct=true&db=cat03556a&AN=duebr.EBC4397950&site=eds-live&scope=site.

[9] Eric Post, "Ecosystem Function and Dynamics," *In Ecology of Climate Change: The Importance of Biotic Interactions.* Monographs in Population Biology (Princeton University Press, 2013), 249.
https://search-ebscohost-com.authenticate.library.duq.edu/login.aspx?direct=true&db=edspmu&AN=edspmu.MUSE9781400846139.13&site=eds-live&scope=site.

[10] Karsten Paerregaard, "Communicating the Inevitable: Climate Awareness, Climate Discord, and Climate Research in Peru's Highland Communities," *Environmental Communication* 14, no. 1 (February 2020): 113.
https://doi.org/10.1080/17524032.2019.1626754.

[11] Folke, et. al. "Our Future in the Anthropocene Biosphere," 836.

[12] Marie-Antoinette Melieres, and Chloe Marechal. *Climate Change: Past, Present, and Future* (Chichester, West Sussex, UK: Wiley-Blackwell, 2015):12.
https://search-ebscohost-com.authenticate.library.duq.edu/login.aspx?direct=true&db=nlebk&AN=948372&site=eds-live&scope=site.

[13] Pope Francis, *Laudato Si,* n. 8.
https://www.vatican.va/content/dam/francesco/pdf/encyclicals/documents/papa-francesco 20150524 encyclica-laudato-si en.pdf

[14] Pope Francis, *Laudato Si,* n. 25.

[15] Eric S. Post, *Ecology of Climate Change, [Electronic Resource]: The Importance of Biotic* Interactions (Princeton University Press, 2013), 3.
https://search-ebscohost-com.authenticate.library.duq.edu/login.aspx?direct=true&db=cat03556a&AN=duebr.EBC1160068&site=eds-live&scope=site.

[16] Pope Francis, *Laudato Si,* n. 23.

[17] Shu Manqi, "The Human Health Impacts of Global Climate Change," *E3S Web of Conferences* 204 (January 1, 2020):01006.
https://doi.org/10.1051/e3sconf/202020401005.

[18] Kaufui Vincent Wong, *Climate Change,* Environmental Engineering Collection, Momentum Press, 2016, 41.
https://search-ebscohost-com.authenticate.library.duq.edu/login.aspx?direct=true&db=cat03556a&AN=duebr.EBC4307182&site=eds-live&scope=site.

[19] Paul H. Carr, "What is Climate Change Doing to Us and for Us?" *Zygon* 53, no. 2 (June 2018): 444.
https://search-ebscohost-com.authenticate.library.duq.edu/login.aspx?direct=true&db=rfh&AN=ATLAiE58180528001102&site=eds-live&scope=site.

[20] Wong, *Climate Change*, 43.
[21] Pope Francis, *Laudato Si,* n. 18.
[22] Norra, "The Bisophere in Times of Global Urbanization," 53.
[23] Pope Francis, *Laudato Si,* n. 4.
[24] Pope Francis, *Laudato Si,* n. 8.
[25] Melieres, and Marechal, *Climate Change: Past, Present, and Future,* 364.
[26] Melieres, and Marechal, *Climate Change: Past, Present, and Future,* 328.
[27] Melieres, and Marechal, *Climate Change: Past, Present, and Future,* 364.
[28] Folke, et. al., "Our Future in the Anthropocene Biosphere," 836.
[29] Melieres, and Marechal. *Climate Change: Past, Present, and Future,* 364.
[30] J. Perry Gustafson, Peter H. Raven, and Paul R. Ehrlich, *Population, Agriculture, and Biodiversity: Problems and* Prospects (Columbia: University of Missouri, 2020), 18. https://search-ebscohost-com.authenticate.library.duq.edu/login.aspx?direct=true&db=nlebk&AN=2462650&site=eds-live&scope=site.
[31] Melieres, and Marechal, *Climate Change: Past, Present, and Future,* 328.
[32] Gustafson, et.al., *Population, Agriculture, and Biodiversity: Problems and* Prospects, 15.
[33] Gustafson, et.al., *Population, Agriculture, and Biodiversity: Problems and* Prospects, 9.
[34] Carol Hand, *Climate Change: Our Warming Earth. History of* Science (Minneapolis: Essential Library, 2015), 28. https://search-ebscohost-com.authenticate.library.duq.edu/login.aspx?direct=true&db=nlebk&AN=878389&site=eds-live&scope=site.
[35] Wong, *Climate Change*, 41.
[36] Folke, et. al., "Our Future in the Anthropocene Biosphere," 837.
[37] Folke, et. al., "Our Future in the Anthropocene Biosphere," 840.
[38] Folke, et. al., "Our Future in the Anthropocene Biosphere," 839.
[39] Folke, et. al., "Our Future in the Anthropocene Biosphere," 842.
[40] Carol Hand, *Climate Change: Our Warming Earth. History of* Science, 10.
[41] Carr, "What is Climate Change Doing to Us and for Us?" 447.
[42] Pope Francis, *Laudato Si,* n. 20.
[43] Richard C. Francis, *Epigenetics: How Environment Shapes Our Genes* (New York: W.W Norton & Company, 2012), xi.
[44] L.A Sealey, B.W. Hughes, A. N. Sriskanda, et.al., "Environmental Factors in the Development of Autism Spectrum Disorders," *Environment International* 88 (March 1, 2016): 289. https://doi.org/10.1016/j.envint.2015.12.021
[45] Lauren Cullum, "An Analysis of the International Climate Change Adaptation Regime and Its Response to Global Public Health Concerns," *UCLA Journal of Environmental Law and Policy,* vol. 37, no. 2, (Jan. 2019): 194. https://search-ebscohost-com.authenticate.library.duq.edu/login.aspx?direct=true&db=edshol&AN=edshol.hein.journals.uclalp37.11&site=eds-live&scope=site.
[46] *Climate Change, Water and Agriculture: Towards Resilient* Systems (OECD Studies on Water. OECD, 2014), 17. https://search-ebscohost-com.authenticate.library.duq.edu/login.aspx?direct=true&db=cat05821a&AN=ecp.EBC3121257&site=eds-live&scope=site.

[47] Kolawole Ogundari and Raphael Onyeaghala, "The effects of Climate Change on African Agricultural Productivity Growth Revisited," *Environmental Science and Pollution Research International* 28, no. 23 (June 2021): 30040.
https://doi.org/10.1007/s11356-021-12684-5.
[48] Banshaikupar Lyngdoh Mawlong, *Climate Change and Developing* Countries (Cambridge Scholars Publishing, 2018), 102.
https://search-ebscohost-com.authenticate.library.duq.edu/login.aspx?direct=true&db=cat03556a&AN=duebr.EBC5549071&site=eds-live&scope=site.
[49] Folke, et. al., "Our Future in the Anthropocene Biosphere," 837.
[50] Pope Francis, *Laudato Si,* n. 25.
[51] Mawlong, *Climate Change and Developing* Countries, 147.
[52] Mawlong, *Climate Change and Developing* Countries, 5.
[53] *Climate Change, Water and Agriculture: Towards Resilient* Systems, 48.
[54] Pope Francis, *Laudato Si,* n. 23.
[55] ten Have, *Global Bioethics: An Introduction*, 231.
[56] Pope Francis, *Laudato Si,* n. 7.
[57] ten Have, *Global Bioethics: An Introduction*, 214.
[58] Norra, "The Bisophere in Times of Global Urbanization," 56.
[59] Folke, et. al., "Our Future in the Anthropocene Biosphere." 837.
[60] Pope Francis, *Laudato Si,* n. 23.
[61] Pope Francis, *Laudato Si,* n. 17.
[62] Rae Walker, Wendy Mason, and Karen De Heer, *Climate Change Adaptation: For Health and Social Services* (CSIRO Publishing, 2015), xix.
https://search-ebscohost-com.authenticate.library.duq.edu/login.aspx?direct=true&db=cat03556a&AN=duebr.EBC4010517&site=eds-live&scope=site.
[63] George Luber, and Jay Lemery, *Global Climate and Human Health: From Science to Practice.*
(Jossey-Bass, 2015), 547.
https://search-ebscohost-com.authenticate.library.duq.edu/login.aspx?direct=true&db=cat03556a&AN=duebr.EBC4187157&site=eds-live&scope=site.
[64] Cullum, "An Analysis of the International Climate Change Adaptation Regime and Its Response to Global Public Health Concerns," 207.
[65] Walker, et. al., *Climate Change Adaptation: For Health and Social Services*, xix.
[66] Mikael Stenmark, *Environmental Ethics and Policy-Making* (Florence: Taylor& Francis Group, 2016): 40, Accessed May 18, 2022. ProQuest Ebook Central.
[67] "United Nations Millennium Declaration." *United Nations Chronicle* 37, no 3 (January 1, 2000): 39.
https://search-ebscohost-com.authenticate.library.duq.edu/login.aspx?direct=true&db=edshol&AN=edshol.hein.unl.unchron0037.98&site=eds-live&scope=site.
[68] Pope Francis, *Laudato Si,* n. 5.
[69] ten Have, *Global Bioethics: An Introduction*, 231.
[70] Lewis H. Ziska, "Rising Carbon Dioxide and Global Nutrition: Evidence and Action Needed," *Plants (Basel, Switzerland)* 11, no. 7 (April 6, 2022):1
https://doi.org/10.3390/plants11071000.

Literature

Carr, Paul H. 2018. "What is Climate Change Doing to Us and for Us?" *Zygon* 53, no. 2 (June): 443-461. https://search-ebscohost-com.authenticate.library.duq.edu/login.aspx?direct=true&db=rfh&AN=ATLAiE5818052800l102&site=eds-live&scope=site.

Climate Change, Water and Agriculture: Towards Resilient Systems. 2014. (OECD Studies on Water. OECD), 17. https://search-ebscohost-com.authenticate.library.duq.edu/login.aspx?direct=true&db=cat05821a&AN=ecp.EBC3121257&site=eds-live&scope=site.

Cullum, Lauren. 2019. "An Analysis of the International Climate Change Adaptation Regime and Its Response to Global Public Health Concerns." *UCLA Journal of Environmental Law and Policy,* vol. 37, no. 2, (Jan.): 191-228. https://search-ebscohost-com.authenticate.library.duq.edu/login.aspx?direct=true&db=edshol&AN=edshol.hein.journals.uclalp37.11&site=eds-live&scope=site.

Folke, Carl, Stephen Polasky, Johan Rockstrom, Victor Galaz, Frances Westley, Michele Lamont, Marten Schefer, et. al. 2021. "Our Future in the Anthropocene Biosphere." *Ambio: A Journal of Environment and Society* 50, no. 4 (April 1): 834-869. https://doi.org/10.1007/s13280-021-01544-8.

Francis, Richard C. 2012. *Epigenetics: How Environment Shapes Our Genes*. New York: W.W Norton & Company.

Gustafson, J. Perry Peter H. Raven, and Paul R. Ehrlich. 2020. *Population, Agriculture, and Biodiversity: Problems and* Prospects. Columbia: University of Missouri. https://search-ebscohost-com.authenticate.library.duq.edu/login.aspx?direct=true&db=nlebk&AN=2462650&site=eds-live&scope=site.

Hand, Carol. *2015. Climate Change: Our Warming Earth. History of* Science. Minneapolis: Essential Library. https://search-ebscohost-com.authenticate.library.duq.edu/login.aspx?direct=true&db=nlebk&AN=878389&site=eds-live&scope=site.

ten Have, Henk. 2016. *Global Bioethics: An Introduction.* New York: Routledge.

Luber, George, and Jay Lemery. 2015. *Global Climate and Human Health: From Science to Practice.*Jossey-Bass. https://search-ebscohost-com.authenticate.library.duq.edu/login.aspx?direct=true&db=cat03556a&AN=duebr.EBC4187157&site=eds-live&scope=site.

Manqi, Shu. 2020. "The Human Health Impacts of Global Climate Change." *E3S Web of Conferences* 204 (January 1):01006. https://doi.org/10.1051/e3sconf/202020401005.

Mawlong, Banshaikupar Lyngdoh. 2018. *Climate Change and Developing* Countries. Cambridge Scholars Publishing. https://search-ebscohost-com.authenticate.library.duq.edu/login.aspx?direct=true&db=cat03556a&AN=duebr.EBC5549071&site=eds-live&scope=site.

Melieres, Marie-Antoinette and Chloe Marechal. 2015. *Climate Change: Past, Present, and Future.* Chichester, West Sussex, UK: Wiley-Blackwell. https://search-ebscohost-com.authenticate.library.duq.edu/login.aspx?direct=true&db=nlebk&AN=948372&site=eds-live&scope=site.

Norra, Stefan. 2014. "The Bisophere in Times of Global Urbanization." *Journal of Geochemical Exploration* 147, no. Part A (December 1): 52-57. https://doi.org/10.1016/j.gexplo.2014.06.004.

Ogundari, Kolawole, and Raphael Onyeaghala. 2021. "The effects of Climate Change on African Agricultural Productivity Growth Revisited." *Environmental Science and Pollution Research International* 28, no. 23 (June): 30035-30045. https://doi.org/10.1007/s11356-021-12684-5.

Paerregaard, Karsten. 2020. "Communicating the Inevitable: Climate Awareness, Climate Discord, and Climate Research in Peru's Highland Communities." *Environmental Communication* 14, no. 1 (February): 112-125. https://doi.org/10.1080/17524032.2019.1626754.

Pope Francis, 2015. *Laudato Si.* https://www.vatican.va/content/dam/francesco/pdf/encyclicals/documents/papa-francesco 20150524 encyclica-laudato-si en.pdf

Portney, Kent E. 2015. *Sustainability.* MIT Press Essential Knowledge Series. The MIT Press. https://search-ebscohost-com.authenticate.library.duq.edu/login.aspx?direct=true&db=cat03556a&AN=duebr.EBC4397950&site=eds-live&scope=site.

Post, Eric S. 2013. *Ecology of Climate Change. [Electronic Resource]: The Importance of Biotic* Interactions. Princeton University Press. https://search-ebscohost-com.authenticate.library.duq.edu/login.aspx?direct=true&db=cat03556a&AN=duebr.EBC1160068&site=eds-live&scope=site.

Post, Eric. 2013. "Ecosystem Function and Dynamics." *In Ecology of Climate Change: The Importance of Biotic Interactions.* Monographs in Population Biology. Princeton University Press. https://search-ebscohost-com.authenticate.library.duq.edu/login.aspx?direct=true&db=edspmu&AN=edspmu.MUSE9781400846139.13&site=eds-live&scope=site.

Sealey, L.A., B.W. Hughes, A. N. Sriskanda, J. R. Guest, A.D. Gibson, L. Johnson-Williams, D. G. Pace, O. Bagasra. 2016. "Environmental Factors in the Development of Autism Spectrum Disorders." *Environment International* 88 (March 1): 288-298. https://doi.org/10.1016/j.envint.2015.12.021

Stenmark, Mikael. 2016. *Environmental Ethics and Policy-Making*. Florence: Taylor & Francis Group. Accessed May 18, 2022. ProQuest Ebook Central.

"United Nations Millennium Declaration." 2000. *United Nations Chronicle* 37, no 3 (January 1): 38-49. https://search-ebscohost-com.authenticate.library.duq.edu/login.aspx?direct=true&db=edshol&AN=edshol.hein.unl.unchron0037.98&site=eds-live&scope=site.

Walker, Rae, Wendy Mason, and Karen De Heer. 2015. *Climate Change Adaptation: For Health and Social Services* (CSIRO Publishing), xix. https://search-ebscohost-com.authenticate.library.duq.edu/login.aspx?direct=true&db=cat03556a&AN=duebr.EBC4010517&site=eds-live&scope=site.

Wong, Kaufui Vincent. 2016. *Climate Change.* Environmental Engineering Collection. Momentum Press. 41. https://search-ebscohost-com.authenticate.library.duq.edu/login.aspx?direct=true&db=cat03556a&AN=duebr.EBC4307182&site=eds-live&scope=site.

Ziska, Lewis H., 2022. "Rising Carbon Dioxide and Global Nutrition: Evidence and Action Needed." *Plants* (Basel, Switzerland) 11, no 7 (April 6): 1-11. https://doi.org/10.3390/plants11071000.

II. Resilience and Health

CHAPTER FIVE

RESILIENCE IN HEALTHCARE: ASSESSING PRACTICAL AND ETHICAL CHALLENGES IN MULTICULTURAL SOCIETIES

ERCAN AVCI

Introduction

Facing various practical and ethical challenges in any healthcare system is not surprising.[1] However, governing these challenges in multicultural societies requires more effort due to social, political, cultural, and religious diversity.[2] Practical challenges refer to the problems experienced by people of a particular group in a similar way when receiving healthcare services, such as language barriers and healthcare disparities, whereas ethical challenges indicate the problems stemming from individuals' distinct moral evaluations of a specific issue, like accepting or rejecting the withdrawal of life-sustaining support or determination of brain death. Based on the World Health Organization (WHO)'s definition, for quality of care, health services should be effective, safe, people-centered, timely, equitable, integrated, and efficient.[3] As Amina Pereno and Daniel Eriksson highlight, healthcare needs to establish a socially, economically, and environmentally sustainable system through continuous cooperation among health industries, health providers, and patients.[4] However, the presence of practical and ethical problems in a health system demonstrates the lack of collaboration among the stakeholders and the quality of services and affects the relationship between patients and providers and healthcare sustainability in the long term. At that point, healthcare organizations need to show a sufficient level of resilience to respond to these problems adequately. Christopher Nemeth et al. define resilience as "the intrinsic ability of a system to adjust its functioning prior to, during, or following changes and disturbances so that it can sustain required operations, even after a major mishap or in the presence of continuous stress."[5] In this context, eliminating or reducing relevant practical and ethical problems would improve the resilience and continuity of healthcare organizations. For this reason, this chapter aims to address some practical and ethical challenges and propose particular suggestions to manage these problems in order to achieve the sustainability of healthcare services in multicultural societies.

Practical Problems

Due to the interaction between them, distinguishing practical challenges from ethical ones may be difficult. Additionally, a holistic perspective could engender a better understanding of healthcare services instead of a dualistic approach. However, the need for producing practical solutions when experiencing practical problems urges us to focus on practical and ethical challenges separately. For example, even though a language barrier-based difficulty can cause certain ethical challenges, such as making autonomous decisions, pinpointing the problem with its unique aspect can help us apply practical remedies. For instance, a healthcare provider may alleviate language barriers by employing multilingual or bilingual personnel or providing interpretation services.[6] On the other hand, ethical problems in multicultural societies are more stubborn because of cultural and religious distinctions among persons and groups, which often makes it burdensome to reconcile different moral standpoints.[7]

The numbers show that many countries' multicultural social structure has been increasing. For instance, in the United States, the minority population is anticipated to be 47 % by 2050, which means that roughly 50% of the people in the United States will consist of minorities in the next three decades.[8] Even in some states, such as California and New Mexico, minorities already form the majority.[9] This reality brings about different practical complications in healthcare, such as language-based difficulties and healthcare disparities. However, resilience requires healthcare organizations to explore proper tools and strategies to manage organizational sustainability, and pinpointing the problems is the first step to achieving it.

Communication between Patients and Caregivers

Providing and receiving healthcare services rely on communication between patients and caregivers. The information given by a patient plays a prominent role in the patient's clinical evaluation. Similarly, the information provided by healthcare professionals is crucial for patients' participation in the treatment and relevant decision-making procedures.

Studies reveal that adequate and appropriate communication is essential in the patient-physician relationship to avoid medical errors, unnecessary testing and screening, and breaching patient autonomy.[10] In other words, linguistic concordance and proper communication between patients and healthcare providers is the first and most significant requirement for better health outcomes, overall quality, and patient satisfaction.

The demographic data demonstrate that more than 329 spoken languages in the United States, and roughly 32 million people speak a non-English language at home.[11] This reality is the first and foremost cause of miscommunication between care providers and patients. The lack of language accordance between patients and healthcare professionals influences the number of physician visits, affects preventive care, reduces patient satisfaction, impacts physicians' prescription decisions, and decreases overall quality.[12] The Commonwealth Fund 2001 Health Care Quality Survey indicates that 33% of Hispanics, 27% of Asian Americans, and 23% of African Americans have problems in communication with physicians. It also shows that, compared to English-speaking Hispanics, Spanish-speaking Hispanics are more likely to face difficulties communicating with their physicians.[13]

Communication does not merely refer to language proficiency or linguistic barriers. Many African Americans, at least new generations, are native speakers of English, but they still have communication obstacles in healthcare. The lack of health literacy among African Americans is a challenge indicating these communication problems.[14] At that point, healthcare providers' attitudes and behaviors and health policies towards cultural competence and cultural disparity become essential to create a good relationship between patients and healthcare professionals. Nevertheless, research studies prove that some physicians, especially white physicians, are not even cognizant of the problem. They assume they treat all patients equally and in the same manner without biases or prejudices. A survey conducted in 2001 reveals that only 25% of white physicians accept the existence of the disparities, while 77% of black physicians think that the healthcare system contains disparities.[15]

Physicians may claim that patients' cultural, social, and economic values, status, and backgrounds do not affect their attitudes and behaviors toward patients. However, Ryn and Burke argue that physicians' assumption that they are unbiased toward their patients is not true; physicians are definitely impacted by patients' "socio-demographic characteristics."[16] Furthermore, Smedley and his colleagues underscore that time pressure, limited information, and prejudices or biases affect care providers' decisions, attitudes, and behaviors.[17] Therefore, besides individual characteristics and approaches, health professionals' working conditions also influence their communication with patients.

Lack of communication between care providers and patients causes certain undesirable consequences, such as diagnostic errors, the underuse or overuse of screening and testing, poor quality of services, and less patient satisfaction.[18] Additionally, financial burdens are some other unwelcome outcomes of deficiency in communication between two primary stakeholders of healthcare. According to the report, *Economic Burden of Health Inequalities in the United States*, the total financial burden of health inequalities in the United States was $1.24 trillion from 2003 to 2006. This unwanted and unnecessary cost represents an incredible magnitude of wasted money, which is "more than the gross domestic product of India, the world's 12th-largest economy in 2008."[19]

Health and Healthcare Disparities

The term 'health disparities' refers to the differences in the burden of diseases, injuries, and mortalities among distinct population groups, whereas 'healthcare disparities' bears on the distinctions in access to care, utilization of care, and quality of care among various population groups.[20] Even though as a single term 'health and healthcare disparities' indicates the differences among all population groups, including race, ethnicity, age, gender, sexual orientation, socioeconomic status, and disability status, it is mostly used to merely address racial and ethnic disparities.[21] For example, the *Committee on Understanding and Eliminating Racial and Ethnic Disparities in Health Care* "defines disparities in healthcare as racial or ethnic differences in the quality of healthcare that are not due to access-related factors or clinical needs, preferences, and appropriateness of intervention."[22] In this context, in this chapter, the term *health and healthcare disparities* demonstrates the race- and ethnicity-based practical challenges in healthcare.

The *Committee on Understanding and Eliminating Racial and Ethnic Disparities in Health Care* reviewed more than 100 studies to determine whether racial and ethnic disparities occur in clinical practices and after examining a wide range of medical procedures and services regarding several illnesses, including cancer, cardiovascular care, mental health services, and AIDS, reached convincing data proving the existence of disparities. Additionally, the committee underscored that these disparities affect minorities more.[23] Furthermore, the *2014 National Healthcare Quality and Disparities Report* published by the Agency for Healthcare

Research and Quality (AHRQ) accentuates that despite some improvements, disparities are still widespread across the United States.[24]

According to the Henry J. Kaiser Family Foundation, in the United States, health and healthcare disparities create certain problems. Firstly, they exacerbate social injustice because they exemplify inequalities that many minority groups face daily. Secondly, they affect the quality of healthcare services that minorities receive; minorities do not merely lack access to healthcare, but also they receive poorer quality healthcare.[25] According to the AHRQ, "Blacks received worse care than Whites for about one-third of quality measures, while Hispanics, American Indians and Alaska Natives, and Asians received worse care than Whites for some quality measures and better care for some measures."[26] Finally, they cause the economy to waste approximately 309 billion dollars each year due to the lack of care or inefficiency in services provided to minorities.[27] Although some regulations, such as the National Standards for Culturally and Linguistically Appropriate Services (CLAS) and the Affordable Care Act (ACA), have been developed and put into effect, studies reveal that limited improvements have been achieved in health and healthcare disparities.[28] The magnitude of the problem, its historical background and existing economic disadvantages, language barriers, and discriminations require continuous and comprehensive resolutions.[29] In short, it is possible to state that, the persistent disparities in the United States bring about less and unequal access to care, poor quality, worse health benefits, and more uninsured and vulnerable people.[30]

Strengthening Cultural Competence

As mentioned above, multicultural societies, like the United States, face various practical challenges in healthcare, such as health and healthcare disparities, which directly affect healthcare organizations' level of resilience. Even though being persistent, these problems are not unsolvable. For example, cultural competence can be a resolution to eliminate, or at least, decrease the problems. Cultural competence refers to beliefs, attitudes, behaviors, and policies that facilitate concordant togetherness in a multicultural society.[31] Gregory Juckett describes cultural competence as "possessing knowledge, awareness, and respect for other cultures."[32] Therefore, cultural competence is an all-inclusive perspective that requires being cognizant of cultural differences, respecting distinct lifestyles, and creating a reconcilable cross-cultural environment. However, strengthening cultural competence necessitates adequate and satisfactory communication among all relevant parties, including healthcare professionals and patients.

For this reason, eliminating language barriers is the first and foremost prerequisite for fostering cultural competence.

Eliminating Language Barriers

Language barriers are one of the most confronted obstacles in healthcare delivery in multicultural societies, and the lack of communication between patients and care providers may cause irreversible problems. For this reason, National Standards for CLAS in Health and Health Care encourages healthcare organizations to provide patients with limited English proficiency language assistance services.[33] Furthermore, Title IV of the 1964 Civil Rights Act, the regulations of the Centers for Medicare and Medicaid Services (CMS) and the Joint Commission on Accreditation of Health Care Organizations, and the National Committee on Quality Assurance's Health Plan Employer Data and Information Set (HEDIS) request certain actions to support people with limited English proficiency.[34]

In general, language assistance refers to interpreter services. Healthcare institutions provide several types of interpreter services, such as using family or friends as interpreters, having professional interpreters, and employing bilingual staff members as interpreters.[35] All interpreter services aim to have a third party who can speak both the language of the caregiver and the language of the patient to assist in the communication between healthcare professionals and patients. Even though each healthcare organization can choose any interpreter services, such as in-person or distance interpretations, using an interpreter is the only way to allow care providers to communicate with patients who cannot speak/understand caregivers' language. However, the efficiency of interpretation services depends on several factors. Interpreters' proficiency in both languages, their knowledge to understand the medical jargon, and their ability to transfer nonverbal discourses are essential to the efficiency of these services. In communication, gestures are mostly as crucial as verbal expressions. In the case of distance interpretation, the absence of observing body language may lead to certain misunderstandings. Additionally, using family members as interpreters can bring about some other problems. For instance, in a male-dominant society, a husband's assistance may put the wife (patient) under pressure to express herself and her wishes. Therefore, even though it could be more costly for healthcare organizations, staffing professionals for face-to-face interpreters might be the most efficient way to overcome language barriers.[36]

Education

Language barriers are not the only challenges prompting disparities in healthcare. According to Brian D. Smedley et al., limited information, time pressure, and prejudice or bias are other sources of health and healthcare disparities.[37] Except for time pressure, all the mentioned problems entail providing healthcare professionals with education in cultural competence. In clinical assessment, the information disclosed by patients is decisive in caregivers' medical evaluation. Any misinterpretations may cause inaccurate clinical assessments and undesirable consequences. For this reason, healthcare professionals should have sufficient and suitable orientation and training programs to understand patients with different cultures correctly. Lack of cultural competence carries the risk of miscommunication between caregivers and patients and the risk of prejudices and biases in clinical judgments. Therefore, providing healthcare professionals with adequate cross-cultural education is a prominent component of cultural competence.[38]

A research study assessing resident physicians' preparedness for cross-cultural issues shows that 96% of resident physicians believe that cultural competence in healthcare is essential, but an overwhelming majority of them also state that "they were not prepared to provide specific components of cross-cultural care."[39] Another study evaluating nurses' knowledge and skills in cultural competence indicates that nurses do not consider themselves well-educated and prepared to respond to multicultural needs.[40] In this context, without providing healthcare professionals with adequate education about multicultural groups, cultural differences, and cross-cultural challenges, expecting them to produce and deliver satisfactory healthcare services may not be reasonable. To address this need, the *National Standards for CLAS in Health and Health Care* require educating and training the workforce of healthcare organizations, even senior executives, in accordance with cross-cultural demands.[41]

On an organizational level, cross-cultural education should begin when a new staff member is hired, and training programs should continue periodically. However, in a broader view, this education should be provided in the first place at medical schools, nursing schools, and other relevant schools. This education may be offered as separate or integrated into ethics courses because, as Ercan Avci underscores, promoting cultural competence is a fundamental goal of ethics education.[42] However, two research studies examining the teaching of ethics education in medical schools in the United Kingdom, the United States, and Canada demonstrate that even though the curriculums contain ethics education, the shortages and inefficiencies in faculty members, teaching methods, teaching period, and content preclude the possibility of getting desired outcomes.[43]

The primary goal of education and training in cultural competence is to strengthen individuals' knowledge, skill, and ability to establish good cross-cultural communication and produce and deliver consistent healthcare services. The possible outcomes of such education and training are threefold: raising awareness, increasing knowledge, and developing behavioral skills. Firstly, these educational activities could raise awareness of the educated/trained person towards cultural differences. Secondly, education in cultural competence can allow learners/participants to learn particular knowledge regarding other cultures. For instance, learning the different reactions of patients from different cultures to certain situations, such as eye contact, touching, and personal space, would be very helpful in relationships with multicultural populations.[44] Finally, educational programs can help health professionals to learn how to navigate distinct cultural features, attitudes, and behaviors in the therapeutic relationship.[45]

Ethical Challenges

Ethical challenges are not merely an issue of multicultural societies; ethical concerns, conflicts, and disagreements may occur at any stage of healthcare services and negatively influence the resilience of all healthcare organizations. Nonetheless, religions and cultural customs, features, and perspectives significantly impact individual and social behaviors, attitudes, and policies. For this reason, examining these components could help to understand the influence of religious and cultural pluralism on ethical issues.

Religious Pluralism

The impact of religious perspectives on individuals, groups, and societies creates a direct connection between religions and ethics. Religions can generate definite control and authority about particular cases on their followers. As Robert M. Veatch underscores, religions possess "mechanisms of knowing divine truth or divine will."[46] For a person of faith, divine truth refers to absolute knowledge and necessitates absolute obedience. In this sense, a physician, person, or institution who/which observes a religion's practices is supposed to exhibit a specific standpoint toward these religious discourses. For example, Roman Catholic healthcare organizations do not carry out some medical procedures, such as abortion, direct sterilization, and heterologous fertilization, because these medical practices are deemed morally wrong and prohibited by the Church, and the employees, patients, and partnerships of the Catholic healthcare organizations need to respect and uphold those ethical and religious directives.[47]

However, secular individuals, physicians, and bioethicists may not attribute any value to religious norms and reject any relationship between religions and ethics.[48] Biases, prejudices, and stereotypes against a particular faith can prevent people, even philosophers, from making fair and impartial judgments about the faith's teachings and ethical codes. Danial Callahan points out this matter by saying, "one of my toughest problems during the Hastings Center's first twenty years was persuading the philosophers to sit down with the theologians and to take them seriously. The secular philosophers could not give a damn for what the theologians were saying."[49] Callahan witnessed this situation about 30 years ago, but it may be difficult to claim that such biases have entirely been eliminated throughout this period because many people and groups still tend to repudiate any religious grounds for moral norms.[50]

Andrew Fagan perceives the position of some religions and cultures as highly risky to bioethical norms, values, and principles. He claims that "the character of some people's relationship with their cultural or religious community significantly constrains the possibilities for acting autonomously."[51] According to Fagan's autonomy-based evaluation, in societies excessively dominated by religious or cultural rules, persons cannot make autonomous decisions under the pressure of their religions or cultures. Furthermore, Fagan is concerned about how fundamental religions or cultures influence multicultural societies to impact/shape public services over time.[52]

Max Charlesworth states another opposing argument about religious-oriented ethical views. He sees a liberal structure, which implicitly indicates a secular course of actions, the precondition of an ethically pluralist society. He argues that a social consensus cannot be established in a pluralist society due to religious practices that carry biases and prevent reaching an agreement on particular ethical issues.[53] Catholic Church's stance on abortion, sterilization, euthanasia, assisted suicide, and embryo experiments exemplifies religious conservatism. Charlesworth claims that "a large number of contemporary Catholic theologians also recognize the positive advantages of the Church being in a liberal and ethically pluralist situation" but also addresses the current differences between the Catholic standpoint and the feminist ethical position in the beginning-of-life issues to demonstrate why a consensus between religious and secular views is very difficult or impossible because the Church believes in the sacredness of human life, whereas the feminist ethics respect women's choice.[54]

However, contrary to the arguments that certain religions or cultures threaten bioethical principles, some scholars assert that secular tensions highly pressure religious individuals, groups, and institutions.[55] The pressure does not merely come from external sources, such as the activities of the proponent of abortion or euthanasia, but also stems from internal sources. For instance, even though some religious healthcare institutions officially uphold religious perspectives on certain medical procedures, their employees may have distinct positions on the same issues. Therefore, such conflicts are internal tensions in religious healthcare institutions' practices. Additionally, the pressure on these organizations is anticipated to increase because their ethical responses to changes and cross-cultural problems cannot compete with the pace of medical and technological developments.[56]

Cultural Pluralism

Culture is a broad concept defined in various ways by social scientists. In a general framework, culture might be described as norms and values acquired from the social environment.[57] Even though various definitions of cultural pluralism are available in the literature, in this chapter, cultural pluralism refers to distinct cultural features of people living in multicultural societies.[58] The effects of cultural differences on ethical issues primarily appear in comprehending and applying the principles of respect for autonomy and justice. For example, informed consent and advance directive-related matters demonstrate substantial differences among cultures. The principle of respect for autonomy has dominant importance in Western culture, and the requirements of this principle are met by providing correct and thorough information to patients. In Western culture, regardless of the content of information, either good news or bad news, physicians are expected to inform patients to allow them to make autonomous decisions about their treatment processes and procedures. However, research studies show that many other communities refuse to disclose information to patients with serious illnesses when the news is bad to protect the patient from psychological harm. Similarly, giving advance directives or making end-of-life decisions could be considered inappropriate to their cultural norms.[59]

Western culture and bioethics are generally described as individualistic and autonomy-oriented, while non-Western cultures are defined as familial and paternalism-oriented.[60] Nevertheless, according to Leigh Turner, certain studies attribute an excessive emphasis to the autonomy-oriented characteristics of Western bioethics, because as research studies indicate, not sharing bad news with patients and family members is widespread across the world; even in Canada and the United States, several people and communities do not support the disclosure of end-of-life information to patients.[61]

Cultural differences also bring about distinct perspectives in implementing the principle of justice. Sırkku K. Hellsten expresses a striking example of this situation by elaborating on the international aid to the Karamoja region in Uganda about supplying food to the people struggling with starvation. Based on Western values, the aid team prioritized the children and pregnant women during the distribution of the food. However, they later realized that the food given to the children and pregnant women had been served to the elders due to the local social values that attribute ultimate respect to them.[62] This story demonstrates that the essential and unanimous values of a particular culture can diverge from the values of another one. As the example shows, children and pregnant women might have priority due to their vulnerability in Western culture, but another culture may prioritize the elders based on social norms.

Therefore, individuals, groups, and communities with distinct cultural traits may interpret and implement ethical principles and norms differently. One view could recognize the disclosure of bad news as a requirement of respect for autonomy, whereas another one may refuse it to protect the patient from psychological harm. Such ethical cases can prompt irreconcilable conflicts, but it is also possible to encounter more serious matters, such as female circumcision, which makes the issue of whether to respect cultural differences and practices very controversial.[63] In such cases, honoring differences might breach fundamental human rights. On the other hand, not respecting differences or limiting this area could generate some other challenges. For example, imposing certain moral values, norms, and principles on another culture may be perceived as cultural imperialism.[64]

As a result, today's religious and cultural reality necessitates underlining the following points: firstly, it should be accepted that the modern world has a religiously and culturally diverse social structure; secondly, several countries contain many social, religious, and cultural differences within their boundaries; thirdly, unifying or eliminating the existing diversity may not be possible or moral; and finally, a peaceful and plausible way must be discovered and applied to protect individuals and communities from external and internal pressures against their fundamental rights while preserving their unique values and characteristics.

Seeking a Common Ethical Ground in a Pluralist Society

Managing the current religious and cultural pluralism requires applying common standards, moral norms, or principles to the problems at hand in order to improve ethical resilience in healthcare institutions. However, individuals or groups sometimes insist on following "their own cultures, religious traditions and unique moral systems" rather than looking for available common standards.[65] This situation would engender a wide spectrum of religious and cultural perspectives and may cause insurmountable conflicts and problems because compromising with others on certain ethical views, such as in the event of abortion and euthanasia, could almost be impossible. Even though such challenges may make the position of bioethics in the multicultural world questionable, bioethicists strive to find an answer concerning whether it is feasible to create a common ground for controversial ethical issues. In this sense, the suggestions from two approaches – common morality and convergence hypothesis – can be assessed to explore whether a basic ground for some reconcilable ethical issues in multicultural societies can be established.

Common Morality

Morality is a set of values, rules, and standards determining the rightness and wrongness of human actions.[66] Individuals start learning moral norms from their family and social environment. Some norms, values, and standards bind everyone, whereas others are valid merely for a particular group of people. Not killing, not harming, and not lying are examples of moral norms shared by all people with different cultures, religions, and moral values. However, professional standards and religious or cultural traditions merely bind people from these groups or communities. For instance, medical norms and standards need to be taken into account only by health professionals rather than by every individual in society.[67] Beauchamp and Childress name this distinction as common morality and particular morality. They claim that there are certain core norms, values, and principles that their acceptability does not depend on cultural, religious, or individual standpoints. For example, "not to lie, not to steal others' property, to keep promises, to respect the rights of others, and not to kill or cause harm to others" are basic tenets shared by all people.[68] These moral norms recognized and accepted universally point out common morality. On the other hand, particular morality refers to the moral norms and standards that are supposed to be applied only by specific persons, groups, or communities.[69] For instance, the Catholic Church's *Ethical and Religious Directives* contain mandatory provisions merely for Catholic healthcare institutions.[70]

John-Stewart Gordon uses Beauchamp and Childress' approach to prove the existence of certain universally valid moral norms in order to resolve bioethical challenges.[71] According to Gordon, common morality "offers a promising means of explaining and solving cross-cultural issues in ethics without denying the importance and legitimacy of cultural diversity."[72] Even though criticizing some points of Beauchamp and Childress' view concerning common morality, Bernard Gert et al. draw a similar common morality perspective by defending the presence of certain universal moral norms that are valid and operative everywhere in every culture, religion, and society.[73] However, regarding cultural and religious differences in healthcare ethics, Gert et al. accentuate two points: firstly, there are not as many contentious issues as it has been stated; many apparently controversial matters do not obstruct people from making a decision; secondly, many people, even physicians and philosophers, find it more attractive to concentrate on irreconcilable ethical conflicts, even though such matters are a small part of the whole. Additionally, they claim that even though it is possible to provide exact answers to certain moral challenges, we may be unable to offer unique correct answers to some other moral difficulties. According to Gert and his colleagues, the purpose of common morality is to prevent or reduce certain harms, and it carries agreeable solutions to some moral problems but cannot produce resolutions to every case and every person; it only exhibits a general structure for rational and impartial individuals to find proper answers to moral problems.[74]

Beauchamp and Childress embody their ethical perspective with the four principles, while Gert et al. point out the five rules and five ideals to elaborate on their ethical framework.[75] Despite a prolonged debate between the two parties on morality and its reflections, both sides believe that there are some shared, recognized, and universally applied moral norms that form common morality.[76] Besides this similarity, another affinity between the two parties is about the absoluteness of their principles or rules. According to Beauchamp and Childress, the principles, rules, and rights they mention are not absolute; they are just prima facie principles, rules, and rights that can be overridden under certain circumstances.[77] Similarly, Gert and his colleagues justify the violation of the rules they express under the conditions of impartiality, rationality, and knowability.[78] The agreement at that point on prima facie of ethical norms, standards, or principles is crucially important to recognize and accept different norms from distinct cultures and religions. In other words, absolutism does not have any room for distinct views, and the diversity in multicultural societies cannot be handled through such an approach, but prima facie-based interpretations encompass the potential to compromise with distinct norms, values, or standards.

Convergence Hypothesis

The question of whether an ethical common ground can be generated is also examined by Robert M. Veatch, who suggests a normative common morality approach named *convergence hypothesis*.[79] According to Veatch, Hippocratic and professional ethics are not operative anymore, and there are currently two applicable ethical perspectives. The first one encompasses religious aspects and indicates a collaboration of laypeople and health professionals who trust a divine revelation-based moral structure. Religious groups or communities might apply this model to form a religion-driven healthcare system. However, as Veatch emphasizes, such approaches cannot be implemented in societies with secular moral values and norms. The second perspective, which refers to Veatch's *convergence hypothesis*, is a model representing both religious and secular people. Veatch asserts that *convergence hypothesis* contains a normative ethical framework compounding moderate religious moral norms with modest secular moral norms under two conditions: firstly, converging religious and secular norms must be "knowable by natural means of reason and experience" instead of divine revelation (moral norms must be derived from natural means rather than natural theology); secondly, moral norms must not claim absolute certainties due to the changes in the knowledge of morality and science as well as the fallibility of the human being, even in the interpretation of religious moral issues.[80]

Additionally, according to Veatch, health professionals and laypeople from various religions, cultures, and professions should come together to build a consensus on ethical principles, rights, and rules. To substantiate his argument, Veatch gives the example of the *Universal Declaration on Bioethics and Human Rights*, which was signed by the member states of the United Nations Educational, Scientific, and Cultural Organization (UNESCO) in 2005, as a significant international attempt to create a universal ethical ground.[81] Besides Robert M. Veatch, Leigh Turner highlights the importance of international organizations' endeavors to create a global culture. According to Turner, even though it might be challenging to engender universal ethical norms, research studies and international health institutions need to pay more attention to the constitution of international culture to increase the contribution of bioethics to emerge global norms and values.[82] Furthermore, as also accentuated by Veatch, Turner proposes concentrating more on the normative theories and aspects of ethics instead of the empirical examples of ethical issues to constitute a common ground for cross-cultural ethical difficulties.[83]

Conclusion

Multicultural societies confront several practical and ethical difficulties in healthcare. Due to the interaction between practical and ethical challenges, the problems of one side can inevitably influence the other one. As practical challenges, language barriers, the shortcoming of cross-cultural communication, and disparities in health and healthcare affect access to care, the quality and cost of care, and patient satisfaction. Some of the problems in multicultural societies are insistent but not unsolvable. Strengthening cultural competence through providing patients with language assistance services and educating healthcare professionals could reduce the severity of the problems and improve the resilience of healthcare organizations to respond to the problems. Furthermore, practical challenges may also cause serious ethical problems. For instance, language barriers result in averting autonomous decisions and limiting patients to receiving healthcare services according to their own values and preferences. Cultural and religious diversity requires reconciling distinct views in order to peacefully live together. Handling cultural and religious-oriented ethical problems demands increasing ethical awareness and cultural competency among healthcare professionals and treating each patient/person as a unique individual with her/his own values. As Gert and his colleagues mention, instead of exaggerating the differences, concentrating on common moral values and norms might help to create an ethical ground in healthcare.[84] Furthermore, supporting and enhancing universal agreements and avoiding absolute positions would alleviate some apparently irreconcilable ethical arguments and facilitate a practically, ethically, and socially sustainable healthcare structure in multicultural societies.

Notes

[1] "10 Global Health Issues to Track in 2021," WHO, accessed December 1, 2022, https://www.who.int/news-room/spotlight/10-global-health-issues-to-track-in-2021.
- Institute of Medicine, *The Learning Healthcare System: Workshop Summary* (Washington, DC: The National Academies Press, 2007), https://doi.org/10.17226/11903, 1-2.
[2] Mesfin Samuel Mulatu and John W. Berry, "Health Care Practice in a Multicultural Context: Western and Non-Western Assumptions," in *Handbook of Cultural Health Psychology*, eds. Shané S. Kazarian and David R. Evans, pp. 45-61 (Academic Press, 2001), 45.
[3] "Quality of Care," WHO, accessed December 23, 2022, https://www.who.int/health-topics/quality-of-care#tab=tab_1.
[4] Amina Pereno and Daniel Eriksson, "A Multi-stakeholder Perspective on Sustainable Healthcare: From 2030 Onwards," *Futures*, 122 (2020): 1-21.
[5] Christopher Nemeth et al., "Minding the Gaps: Creating Resilience in Health Care," in *Advances in Patient Safety: New Directions and Alternative Approaches (Vol. 3: Performance and Tools)* (Rockville: Agency for Healthcare Research and Quality, 2011).
[6] Cindy Brach and Irene Fraser, "Can Cultural Competency Reduce Racial and Ethnic Health Disparities? A Review and Conceptual Model," *Med Care Res Rev* 57 Suppl 1 (2000): 185.
[7] David F. Kelly, Gerard Magill, and Henk Ten Have, *Contemporary Catholic Health Care Ethics*, 2nd edition (Washington, D.C: Georgetown University Press, 2013), 127-129.
[8] Gregory Juckett, "Cross-Cultural Medicine," *American Family Physician* 72, no. 11 (2005): 2267.
[9] Cindy Brach and Irene Fraser, "Reducing Disparities throught Culturally Competent Health Care: An Analysis of the Business Case," *Quality Management in Health Care* 10, no. 4 (2002): 18.
[10] Brian D. Smedley, Adrienne Y. Stith, and Alan R. Nelson, eds. *Unequal Treatment: Confronting Racial and Ethnic Disparities in Health Care* (Washington: The National Academies Press, 2002), 141.
[11] Laurie M. Anderson, et al., "Culturally Competent Healthcare Systems. A Systematic Review," *Am J Prev Med* 24, no. 3 Suppl (2003): 68.
[12] Donna L. Washington, et al., "Transforming Clinical Practice to Eliminate Racial–Ethnic Disparities in Healthcare," *Journal of General Internal Medicine* 23, no. 5 (2008): 685-91.
- Diana Jill Burgess, et al., "Patient Race and Physicians' Decisions to Prescribe Opioids for Chronic Low Back Pain," *Social Science & Medicine* 67, no. 11 (2008): 1852-60.
- Joel S. Weissman et al., "Resident Physicians' Preparedness to Provide Cross-Cultural Care," *JAMA* 294, no. 9 (2005): 1058-67.

- Kevin Fiscella, et al., "Disparities in Health Care by Race, Ethnicity, and Language among the Insured: Findings from a National Sample," *Med Care* 40, no. 1 (2002): 52-9.
[13] Karen Scott Collins et al., *Diverse Communities, Common Concerns: Assessing Health Care Quality for Minority Americans* (New York: Commonwealth Fund, 2002), 9-17.
[14] Baraka Muvuka, et al., "Health Literacy in African-American Communities: Barriers and Strategies," *Health Lit Res Pract.* 4, no. 3 (2020): e138-e143, doi: 10.3928/24748307-20200617-01.
[15] Donna L. Washington, et al., 689.
[16] Michelle van Ryn and Jane Burke, "The Effect of Patient Race and Socio-Economic Status on Physicians' Perceptions of Patients," *Soc Sci Med* 50, no. 6 (2000): 813-828.
[17] Brian D. Smedley, Adrienne Y. Stith, and Alan R. Nelson, eds. 184-185.
[18] Ibid, 141.
[19] Thomas A. LaVeist, Darrell J. Gaskin, and Patrick Richard, "The Economic Burden of Health Inequalities in the United States," Washington, D.C.: Joint Center for Political and Economic Studies (2009): 6.
[20] "Focus on Health Care Disparities Key Facts," The Henry J. Kaiser Family Foundation, accessed November 9, 2022, https://kaiserfamilyfoundation.files.wordpress.com/2012/11/8396-disparities-in-health-and-health-care-five-key-questions-and-answers.pdf, 2.
[21] Christine Dehlendorf et al., "Health Disparities: Definitions and Measurements," *Am J Obstet Gynecol* (2010): 212.
[22] Brian D. Smedley, Adrienne Y. Stith, and Alan R. Nelson, eds. 32.
[23] Ibid, 38-79.
[24] Agency for Healthcare Research and Quality, *2014 National Healthcare Quality and Disparities Report*, AHRQ pub no. 15-0007 (2015), 1-32
[25] "Focus on Health Care Disparities Key Facts," The Henry J. Kaiser Family Foundation, accessed November 9, 2022, https://kaiserfamilyfoundation.files.wordpress.com/2012/11/8396-disparities-in-health-and-health-care-five-key-questions-and-answers.pdf, 3-4.
[26] Agency for Healthcare Research and Quality, *2014 National Healthcare Quality and Disparities Report*, AHRQ pub no. 15-0007 (2015), 18.
[27] "Focus on Health Care Disparities Key Facts," The Henry J. Kaiser Family Foundation, accessed November 9, 2022, https://kaiserfamilyfoundation.files.wordpress.com/2012/11/8396-disparities-in-health-and-health-care-five-key-questions-and-answers.pdf, 3.
[28] Agency for Healthcare Research and Quality, *2014 National Healthcare Quality and Disparities Report*, AHRQ pub no. 15-0007 (2015).
- "Focus on Health Care Disparities Key Facts," The Henry J. Kaiser Family Foundation, accessed November 9, 2022, https://kaiserfamilyfoundation.files.wordpress.com/2012/11/8396-disparities-in-health-and-health-care-five-key-questions-and-answers.pdf.
[29] Brian D. Smedley, Adrienne Y. Stith, and Alan R. Nelson, eds. 80-124.

[30] "Focus on Health Care Disparities Key Facts," The Henry J. Kaiser Family Foundation, accessed November 9, 2022, https://kaiserfamilyfoundation.files.wordpress.com/2012/11/8396-disparities-in-health-and-health-care-five-key-questions-and-answers.pdf, 7.
[31] Jean Lau Chin, "Culturally Competent Health Care," *Public Health Report* 115, no:1 (2000): 25-26.
- Laurie M. Anderson, et al., 68.
[32] Gregory Juckett, 2267.
[33] United States Department of Health and Human Services Office of Minority Health, *National Standards for Culturally and Linguistically Appropriate Services in Health and Health Care: A Blueprint for Advancing and Sustaining Clas Policy and Practice* (Washington, 2013), 13.
[34] Cindy Brach and Irene Fraser, "Reducing Disparities" 18-20.
[35] Ibid, 185.
[36] Jean Lau Chin, "Culturally Competent Health Care," *Public Health Report* 115, no:1 (2000): 28-29.
[37] Brian D. Smedley, Adrienne Y. Stith, and Alan R. Nelson, eds. 160-175.
[38] Ibid.
[39] Joel S. Weissman, et al., "Resident Physicians' Preparedness to Provide Cross-Cultural Care," *JAMA* 294, no. 9 (2005): 1058.
[40] Madeline M. Maier-Lorentz, "Transcultural Nursing: Its Importance in Nursing Practice," *J Cult Divers* 15, no. 1 (2008): 38.
[41] United States Department of Health and Human Services Office of Minority Health, *National Standards for Culturally and Linguistically Appropriate Services in Health and Health Care: A Blueprint for Advancing and Sustaining Clas Policy and Practice* (Washington, 2013), 13.
[42] Ercan Avci, "A Normative Analysis to Determine the Goals of Ethics Education through Utilizing Three Approaches: Rational Moral Education, Ethical Acculturation, and Learning Throughout Life," *International Journal of Ethics Education* (2017): 125.
[43] Karen Mattick and J Bligh, "Teaching and assessing medical ethics: where are we now?," *J Med Ethics* 32 (2006): 181-185.
- Lisa S. Lehmann and others, "A Survey of Medical Ethics Education at U.S. and Canadian Medical Schools," *Academic Medicine* 79, no. 7 (2004): 682-689.
[44] Madeline M. Maier-Lorentz, 28-29.
[45] Cindy Brach and Irene Fraser, "Can Cultural Competency Reduce Racial and Ethnic Health Disparities?" 185-18
[46] Robert M. Veatch, *Hippocratic, Religious, and Secular Medical Ethics: The Points of Conflict* (Washington D.C.: Georgetown University Press, 2012), 104.
[47] United States Conference of Catholic Bishops. Ethical and Religious Directives for Catholic Health Care Services, (Fourth printing, 2013), 7-32.
[48] Robert M. Veatch, 104-105.
[49] Albert R. Jonsen, *The Birth of Bioethics* (New York: Oxford University Press, 1998), 83-84, cited in Daniel Callahan, personal communication, December 11, 1996.

[50] Robert M. Veatch, 1-9.
[51] Andrew Fagan, "Challenging the Bioethical Application of the Autonomy Principle within Multicultural Societies," *J Appl Philos* 21, no. 1 (2004): 16.
[52] Ibid., 28-29.
[53] Max Charlesworth, "Don't Blame the 'Bio'--Blame the 'Ethics': Varieties of (Bio) Ethics and the Challenge of Pluralism." *J Bioeth Inq* 2, no. 1 (2005): 16.
[54] Ibid, 16-17.
[55] Phillip Thompson, "Seeking Common Ground in a World of Ethical Pluralism: A Review Essay of Moral Acquaintances: Methodology in Bioethics by Kevin Wm. Wildes, S.J.," *HEC Forum* 16, no. 2 (2004): 125-126.
[56] Ibid.
[57] Leigh Turner, "From the Local to the Global: Bioethics and the Concept of Culture," *J Med Philos* 30, no. 3 (2005): 305-306.
[58] Catherine Demangeot and Kizhekepat Sankaran, "Cultural pluralism: Uncovering consumption patterns in a multicultural environment," *Journal of Marketing Management* 28 (2012): 762.
[59] Leigh Turner, "From the Local to the Global,"307-311.
[60] Leigh Turner, "Bioethics in a Multicultural World: Medicine and Morality in Pluralistic Settings," *Health Care Anal* 11, no. 2 (2003): 104-109.
[61] Ibid.
[62] Sırkku K. Hellsten, "Global Bioethics: Utopia or Reality?," *Dev World Bioeth* 8, no. 2 (2008): 72-73.
[63] Leigh Turner, "Bioethics in a Multicultural World," 114.
[64] Sırkku K. Hellsten, 80-81.
[65] Chris Durante, "Bioethics in a Pluralistic Society: Bioethical Methodology in Lieu of Moral Diversity," *Med Health Care Philos* 12, no. 1 (Mar 2009): 35
[66] Ercan Avci, 127.
[67] Tom L. Beauchamp and James F. Childress, P*rinciples of Biomedical Ethics*, seventh edition (New York: Oxford University Press, 2013), 2-8.
[68] Ibid, 3.
[69] Ibid, 3-4.
[70] United States Conference of Catholic Bishops, Ethical and Religious Directives for Catholic Health Care Services, fourth printing, 2013.
[71] John S. Gordon, "Global Ethics and Principlism," *Kennedy Inst Ethics* J 21, no. 3 (2011): 253.
[72] John S. Gordon, 252.
[73] Bernard Gert, Charles M. Culver, and K. Danner Clouser, *Bioethics: A Systematic Approach*, second ed. (New York: Oxford University Press, 2006), 22-23.
- Tom L. Beauchamp and James F. Childress, 393-394.
[74] Bernard Gert, Charles M. Culver and K. Danner Clouser, 21-22.
[75] Bernard Gert, Charles M. Culver and K. Danner Clouser.
- Tom L. Beauchamp and James F. Childress.
[76] Bernard Gert, Charles M. Culver and K. Danner Clouser, 26.
- Tom L. Beauchamp and James F. Childress, 393-394.
[77] Tom L. Beauchamp and James F. Childress, 15-16.

[78] Bernard Gert, Charles M. Culver and K. Danner Clouser, 38-39.
[79] Robert M. Veatch159-160.
[80] Ibid, 160-163.
[81] Ibid, 190.
[82] Leigh Turner, "From the Local to the Global," 316-318.
[83] Karori Mbugua, "Respect for Cultural Diversity and the Empirical Turn in Bioethics: A Plea for Caution," *J Med Ethics Hist Med* 5 (2012): 1.
- Robert M. Veatch, 160.
[84] Bernard Gert, Charles M. Culver and K. Danner Clouser, 21-22.

Literature

Agency for Healthcare Research and Quality. 2015. *2014 National Healthcare Quality and Disparities Report*. AHRQ pub no. 15-0007.

Anderson, Laurie M., Susan C. Scrimshaw, Mindy T. Fullilove, Jonathan E. Fielding, Jacques Normand, and Task Force on Community Preventive Services. 2003. "Culturally Competent Healthcare Systems: A Systematic Review." *American Journal of Preventive Medicine* 24, no. 3: 68-79.

Avci, Ercan. 2017. "A Normative Analysis to Determine the Goals of Ethics Education through Utilizing Three Approaches: Rational Moral Education, Ethical Acculturation, and Learning Throughout Life." *International Journal of Ethics Education* (2017): 125-145.

Beauchamp, Tom L. and James F. Childress. 2013. *Principles of Biomedical Ethics*, seventh edition. New York: Oxford University Press.

Brach, Cindy and Irene Fraser. 2000. "Can Cultural Competency Reduce Racial and Ethnic Health Disparities? A Review and Conceptual Model." *Med Care Res Rev* 57 Suppl 1: 181-217.

—. 2002. "Reducing Disparities throught Culturally Competent Health Care: An Analysis of the Business Case," Quality Management in Health Care 10, no. 4: 15-28.

Burgess, Diana Jill, Megan Crowley-Matoka, Sean Phelan, John F. Dovidio, Robert Kerns, Craig Roth, Somnath Saha, and Michelle van Ryn. 2008. "Patient Race and Physicians' Decisions to Prescribe Opioids for Chronic Low Back Pain." *Social Science & Medicine* 67, no. 11: 1852-1860.

Charlesworth, Max. 2005. "Don't Blame the 'Bio'—Blame the 'Ethics': Varieties of (Bio) Ethics and the Challenge of Pluralism." *Journal of Bioethical Inquiry* 2: 10-17.

Chin, Jean Lau. 2000. "Culturally Competent Health Care." *Public Health Report* 115, no:1: 25-26.

Collins, Karen Scott, Dora L. Hughes, Michelle M. Doty, Brett L. Ives, Jennifer N. Edwards, and Katie Tenney. 2002. *Diverse Communities, Common Concerns: Assessing Health Care Quality for Minority Americans*. New York: Commonwealth Fund.

Dehlendorf, Christine, Allison S. Bryant, Heather G. Huddleston, Vanessa L. Jacoby, and Victor Y. Fujimoto. 2010. "Health Disparities: Definitions and Measurements." *American Journal of Obstetrics & Gynecology* 202, no. 3: 212-213.

Demangeot, Catherine, and Kizhekepat Sankaran. 2012. "Cultural pluralism: Uncovering Consumption Patterns in a Multicultural Environment." *Journal of Marketing Management* 28, no. 7-8: 760-783.

Durante, Chris. 2009. "Bioethics in a pluralistic society: Bioethical Methodology in lieu of Moral Diversity." *Medicine, Health Care and Philosophy* 12: 35-47.

Fagan, Andrew. "Challenging the Bioethical Application of the Autonomy Principle within Multicultural Societies." *Journal of Applied Philosophy* 21, no. 1 (2004): 15-31.

Fiscella, Kevin, Peter Franks, Mark P. Doescher, and Barry G. Saver. 2002. "Disparities in Health Care by Race, Ethnicity, and Language Among the Insured: Findings from a National Sample." *Medical Care*: 52-59.

Gert, Bernard, Charles M. Culver, and K. Danner Clouser. 2006. *Bioethics: A Systematic Approach*, second edition. New York: Oxford University Press.

Gordon, John-Stewart. 2011. "Global Ethics and Principlism." *Kennedy Institute of Ethics Journal* 21, no. 3: 251-276.

Hellsten, Sirkku K. 2008. "Global Bioethics: Utopia or Reality?." *Developing World Bioethics* 8, no. 2: 70-81.

Institute of Medicine. 2007. *The Learning Healthcare System: Workshop Summary*. Washington, DC: The National Academies Press. https://doi.org/10.17226/11903.

Jonsen, Albert R. 1998. *The Birth of Bioethics*. New York: Oxford University Press, pp. 83-84, cited in Daniel Callahan, personal communication, December 11, 1996.

Juckett, Gregory. 2005. "Cross-Cultural Medicine." *American Family Physician* 72, no. 11: 2267-2274.

Kelly, David F., Gerard Magill, and Henk Ten Have. 2013. *Contemporary Catholic Health Care Ethics*, 2nd edition. Washington, D.C: Georgetown University Press.

LaVeist, Thomas A., Darrell J. Gaskin, and Patrick Richard. 2009. "The Economic Burden of Health Inequalities in the United States." Washington, D.C.: Joint Center for Political and Economic Studies.

Lehmann, Lisa Soleymani, Willard S. Kasoff, Phoebe Koch, and Daniel D. Federman. 2004. "A Survey of Medical Ethics Education at US And Canadian Medical Schools." *Academic Medicine* 79, no. 7: 682-689.

Maier-Lorentz, Madeline M. 2008. "Transcultural Nursing: Its Importance in Nursing Practice." *Journal of Cultural Diversity* 15, no. 1: 37-43.

Mattick, Karen, and John Bligh. 2006. "Teaching and Assessing Medical Ethics: Where are We Now?." *Journal of Medical Ethics* 32, no. 3: 181-185.

Mbugua, Karori. 2012. "Respect for Cultural Diversity and the Empirical Turn in Bioethics: A Plea for Caution." *Journal of Medical Ethics and History of Medicine* 5.

Mulatu, Mesfin Samuel and John W. Berry. 2001. "Health Care Practice in a Multicultural Context: Western and Non-Western Assumptions." In *Handbook of Cultural Health Psychology*. Eds. Shané S. Kazarian and David R. Evans, pp. 45-61. Academic Press.

Muvuka, Baraka, Ryan M. Combs, Suur D. Ayangeakaa, Nida M. Ali, Monica L. Wendel, and Trinidad Jackson. 2020. "Health literacy in African-American Communities: Barriers and Strategies." *HLRP: Health Literacy Research and Practice* 4, no. 3: e138-e143.

Nemeth, Christopher, Robert Wears, David Woods, Erik Hollnagel, and Richard Cook. 2011. "Minding the Gaps: Creating Resilience in Health Care." In *Advances in Patient Safety: New Directions and Alternative Approaches (Vol. 3: Performance and Tools)*. Rockville: Agency for Healthcare Research and Quality.

Pereno, Amina and Daniel Eriksson. 2020. "A Multi-stakeholder Perspective on Sustainable Healthcare: From 2030 Onwards." *Futures*, 122: 1-21.

Smedley, Brian D., Adrienne Y. Stith, and Alan R. Nelson, eds. 2002. *Unequal Treatment: Confronting Racial and Ethnic Disparities in Health Care*. Washington: The National Academies Press.

The Henry J. Kaiser Family Foundation. 2022. "Focus on Health Care Disparities Key Facts." Accessed November 9. https://kaiserfamilyfoundation.files.wordpress.com/2012/11/8396-disparities-in-health-and-health-care-five-key-questions-and-answers.pdf.

Thompson, Phillip. 2004. "Seeking Common Ground in a World of Ethical Pluralism: A Review Essay of Moral Acquaintances: Methodology in Bioethcs by Kevin wm. Wildes, Sj." *HEC* F. 16: 114.

Turner, Leigh. 2003. "Bioethics in a Multicultural World: Medicine and Morality in Pluralistic Settings." *Health Care Analysis* 11: 99-117.

Turner, Leigh. 2005. "From the Local to the Global: Bioethics and the Concept of Culture." *The Journal of Medicine and Philosophy* 30, no. 3: 305-320.

United States Conference of Catholic Bishops. 2013. *Ethical and Religious Directives for Catholic Health Care Services*. Fourth printing.

United States Department of Health and Human Services Office of Minority Health. 2013. *National Standards for Culturally and Linguistically Appropriate Services in Health and Health Care: A Blueprint for Advancing and Sustaining Clas Policy and Practice*. Washington.

Van Ryn, Michelle, and Jane Burke. 2000. "The effect of Patient Race and Socio-Economic Status on Physicians' Perceptions of Patients." *Social Science & Medicine* 50, no. 6: 813-828.

Veatch, Robert M. 2012. *Hippocratic, Religious, and Secular Medical Ethics: The Points of Conflict*. Washington D.C.: Georgetown University Press.

Washington, Donna L., Jacqueline Bowles, Somnath Saha, Carol R. Horowitz, Sandra Moody-Ayers, Arleen F. Brown, Valerie E. Stone, Lisa A. Cooper, and Writing Group for the Society of General Internal Medicine, Disparities in Health Task Force. 2008. "Transforming Clinical Practice to Eliminate Racial–Ethnic Disparities in Healthcare." *Journal of General Internal Medicine* 23: 685-691.

Weissman, Joel S., Joseph Betancourt, Eric G. Campbell, Elyse R. Park, Minah Kim, Brian Clarridge, David Blumenthal, Karen C. Lee, and Angela W. Maina. 2005. "Resident Physicians' Preparedness to Provide Cross-Cultural Care." *JAMA* 294, no. 9: 1058-1067.

WHO. 2021. "10 Global Health Issues to Track in 2021." Accessed December 1, 2022. https://www.who.int/news-room/spotlight/10-global-health-issues-to-track-in-2021.

WHO. N. d. "Quality of Care." Accessed December 23, 2022. https://www.who.int/health- topics/quality-of-care#tab=tab_1.

CHAPTER SIX

THE RESILIENT TENSION BETWEEN INDIVIDUAL AUTONOMY/RIGHTS AND POPULATION/COMMON GOOD

ANDREW HARRINGTON

Introduction

Healthcare electronic data is everywhere, on computers, phones, tablets, and virtually any device that can access the internet. Patient electronic health records, medication information, genetic testing results, and so much more are out there. What is being done with all this information, and do patients consent to its use in public health interests? This chapter's focus will address whether it is ethically justifiable for big data banks to use personal health information without consent for public health surveillance. The answer to this question requires a deeper look at personal autonomy, human rights and limitations, population health and an individual's societal obligations, all while connecting to the concept of the common good. This chapter will then address how these concepts are applied in the real world by describing big data, what it is, and how it can help healthcare, in addition to its ethical considerations and limitations. This chapter will next discuss some of the issues with data usage and explain the ethical basis for public health surveillance, including the hazards and obligations related to this profession, before addressing the issue of personal privacy vs. public health surveillance.

Individual Autonomy vs. Population Health

Bioethicists nowadays are often classified based on how they see the interaction between the person and society. Ethicists have been characterized by terms like "communitarian" and "liberal," among many others, in their critiques and refutations of opposing viewpoints.[1] For a long time, bioethics was dominated by liberal individualism, which was built on the philosophical pillars of liberty, dignity, and privacy. However, ideologies that promote giving the concepts of collaboration, unity, and common good a more substantial role have begun to pose an increasing threat to it.[2] It has been argued that privacy laws, informed consent laws, and individual autonomy laws significantly inhibit medical research that aims to advance society. Although acquiring and analyzing information technology has advanced quickly, stringent laws defending individual rights have made it difficult to use such information.[3]

Even while many proponents of communitarianism do not overtly reject individual rights, it is difficult to create a compromise between such rights and concerns for the greater good.[4] The autonomy of individuals is often weighed against the interest of the public or collective good when considering the recommended ethical frameworks for genetic and clinical database systems. It ultimately turns into an either-or situation. This split may have resulted from the inability to see the intimate connection between autonomy and the common good. One possibility is to draw influence from Aristotle's political theory, which seeks to link autonomy with the general good in a clear, alluring, and persuasive manner.[5] This statement alone should help to bridge the gap between communitarian and liberal bioethics since Aristotle neither supports nor reflects either position.[6]

Personal Autonomy & Human Rights

Bioethics has often been built in the West on concepts and language connected to human rights. This idea has often been centered on the rights of individual patients vs. some kind of authority, such as medical paternalism, constraints on stopping treatment under the law, or the dull operations of hospital practice.[7] There is an increasing push for bioethics to move beyond discourse centered on rights. However, this term often refers to a shift toward a patient's or family's deeper narrative or values framework. Cultural values are sometimes included, but they are oriented in a way that is more unique than multiple and universal.[8] The move from a solitary mindset to a plural one is much more inclusive when bioethics and human rights are discussed.[9]

Human Rights

The quest for a bioethics concept that encompasses human rights that is broader and fuller than one that is not already acknowledged in all civilizations and older cultures that value community above individuals, would likely be the biggest barrier to this universal approach. It is clear how much cultural values influence how we see morality in biomedicine from the emphasis on independence and self-expression in Western bioethics, especially from American bioethicists.[10] Our beliefs about uniqueness and autonomy affect all medical transactions, such as consent and privacy in the patient-clinician relationship. The patient is seen as the center of decision-making competence and complete consent in the United States and other Western countries, although many other people accept the community or familial decision-making model, notably in Eastern-Asian cultures and nations.[11] In civilizations where family and community bonds are prioritized above all else, individual informed consent is essentially meaningless. Decision-making ability in these situations is only possible on a social level.[12] In other words, medical decisions are relational or made by numerous family members or close friends instead of just the individual receiving medical care. Populations in patriarchal or matriarchal societies may also have medical decisions made by the family's leader.

In Western civilizations, it is also believed that one's confidentiality of information should be preserved. However, this idea does not necessarily imply a universal application. The importance of privacy varies widely, according to social research on attitudes and customs related to cultural conceptions of the self.[13] To further understand the ethical complexity associated with personhood, more empirical study is needed. Therefore, the importance of autonomy is not a solution for bioethics on a global scale but rather a symptom of a deeper issue for most societies.[14]

Limitations of Autonomy

Furthermore, the autonomy concept that guides bioethics and law has real challenges. Autonomy is an ideal term since it alludes to a person's complete self-control. However, an individual's capacity to regulate behaviors, desires, or both may have flaws.[15] Jacob Rendtorff states, "The individual's capacity of reasoning may be limited or nonexistent, and this is not only the case for children, senile, insane persons, etc. but also for normal, intelligent people who feel themselves weak and dependent on others or who simply do not understand the scientific project in which they are asked to participate."[16] People may also base their conclusions on unreliable or incorrect information they have learned from various sources. Additionally, the person's wants or wishes could be unclear. The individual's sense of self is not always consistent, and there are instances when they are unsure about their desires.[17]

This image of the free and independent person with the power to make their own decisions about their own life has drawn substantial criticism for being so far from the realities of the clinical environment and the specifics of bioethics and law. Reciprocal self-determination is allegedly insufficient as a conceptual framework for comprehending the relationship between patients and physicians. According to the objection, autonomy assumes a person with no social connections. Additionally, it makes an overly strong assumption about a person's ability to make their own judgments. Rendtorff continues, "In extreme situations of medical treatment, it is not likely that people can make independent and autonomous decisions. They are vulnerable, weak, and not in control of the situation."[18] In addition, it is unlikely that individuals make independent and logical judgments during their daily lives without the assistance of others. In actuality, the need for individual autonomy is seldom met when making judgments in severe circumstances.[19]

Furthermore, others claim that diversity and freedom are not the sole unalterable principles. Sometimes having a strong sense of community, based on shared values, is a need for self-determination. In this setting, autonomy assumes an institutional and cultural basis built around the shared values of tolerance for variety and individual freedom. Because of this, an explanation of autonomy cannot be wholly libertarian; instead, it must acknowledge that each person is embedded in various social customs, ties, and relationships with others.[20] One loses sight of the delicate and sensitive aspects of the human situation that call for respect and care if one just concentrates on autonomy.[21]

Population Health & Societal Obligations

As Dr. Henk ten Have states in his book *Global Bioethics: An Introduction,* "Focus on the concept of individual autonomy implies a relative neglect of notions such as the common good, public interest, and community. The individualism of bioethics discourse makes it harder to examine issues such as resource allocation, technology assessment, the aims of healthcare, and justice."[22] Renée Fox, a sociology professor, argues in favor of this criticism, saying that "bioethics is centered on value complex of individual rights, self-determination, and privacy, at the expense of social responsibility and social justice."[23] The flimsy ethics of individual autonomy have historically focused on the patient issues brought on by clinical medicine and medical technology. Still, they cannot effectively handle the social and institutional framework of medical decisions.[24] Perhaps it is time to adopt a different principle and change how the West interacts with the concept of autonomy in bioethics and law.

Changing Ethical Frameworks

The classic Hippocratic medical ethics were created within a framework of professional ethics based on virtue, emphasizing the obligations of beneficence and secrecy on the part of physicians.[25] The foundation of the doctor-patient relationship was the patient's confidence in and respect for the doctor's authority. However, the emancipation that was taking place in many other spheres of life led to criticism of this paternalistic attitude in the late 1960s. According to Alfred Tauber, "Medical ethics was caught up in rights movements and took the position that patients needed to be empowered against paternalistic doctors and an intimidating technology that threatened to dehumanize patients."[26] The movement in favor of the individualism approach may also be seen as a response to the repressive past of eugenics and the forced involvement of people in activities under the guise of the greater good, namely the Tuskegee case in the United States and Nazi physicians' experiments on POWs. Thus, bioethics pledged to treat each person's liberty, privacy, and equality with the utmost regard.[27]

It has been claimed more and more often since the 1990s that bioethics committed a grave error by elevating human autonomy to the status of a core ethical concept at the expense of other crucial values. A few arguments have proposed new ethical frameworks for greater communal values like reciprocity, mutuality, solidarity, citizenship, and universality.[28] As communitarian ethicists constantly point out, humans are social creatures, and "our nature is distorted if we think of ourselves only as co-existing social atoms."[29]

These objections and issues suggest that an ethic that is more contextualized is necessary. The gathering, storing, and use of personal data, including genomic and health records, cannot be controlled by the same ethical principles as clinical testing, treatment-focused clinical care, public health, or other areas.[30] The distinctions between these settings and their internal disparities are attracting increasing attention in the modern day. Invasive biomedical research involving people is where the so-called guidelines for informed consent were developed. In other circumstances, such as demographically based genomic repositories or population health, they may not be appropriate. However, some detractors are concerned about creating a risky precedent. They fear that if ethical guidelines are loosened to allow for a study on human genetic databases, this might lead to modifications in other research areas and, ultimately, result in research subjects losing protection.[31] One concept related to the push to focus new ethical frameworks on more collective values is the common good.

Common Good

The welfare of society was a priority for the ancient Greek polis. The reciprocity of such benefits is essential to the survival of the shared benefits that belong to everyone in society, including peace, stability, success, fairness, and solidarity. It was believed that the goals of society and the individual would be compatible under normal conditions. In a dispute, the common good would be seen as superior. The core of Aristotle's political theory is his claim that, although neither is sufficient on its own, "the state is by nature clearly prior to the family and the individual since the whole is of necessity prior to the part."[32] Every citizen's well-being is intertwined with the republic, and individuals are exposed to the republic's virtue daily. The common good, for Aristotle, relates to "a good proper to, and attainable only by the community, yet individually shared by its members."[33] For Aristotle, the common good must be both collective and individual. It spans the objectives and reasons individuals seek in their social and interpersonal interactions instead of correlating with the whole of individual commodities.[34]

Both the virtue of the people and the institutional framework of the republic are recognized by Aristotle as requirements for the common good. A shared existence oriented to the common good is necessary to have collaborative pursuits, thus citizens must act fairly.[35] Unfortunately, many people lack justice and are always in conflict for things like power, money, or honor. Since there is always a shortage of outside resources, their lousy behavior causes fierce rivalry and ongoing disputes. As a result, reorienting people's moral priorities toward virtue via education is a crucial need for achieving the common good. For example, overcoming human disparities in interests, abilities, and jobs is one of the challenges to achieving the common good.[36]

According to Margit Sutrop, society is more than just a collection of different people vying for the same few resources.[37] We all have many common values as members of the community. Society's role extends beyond just giving individuals the resources to live fulfilling lives. It is challenging to formulate a definition of the common good that fully encapsulates and takes into account our shared interests.[38]

Suppose one views the common good in Aristotle's definition as equivalent to the public interest. People would then realize that what is required of them is careful consideration of its constraints and circumstances in addition to a commitment to the common good.[39] The interesting question is, how do such limitations apply to the growing big data problem and the genetic databases that could be used for the public interest of health and wellbeing? This question will be further investigated as to whether it is ethically justified to conduct public health surveillance (via biobanks, genetic data repositories, or other means) without patient consent. This will be done by first explaining what big data is, how it can help healthcare, and the ethical challenges of privacy and security that go along with it in a modern-day application of the battle between individual autonomy and rights versus public health and the common good.

The Big Data Problem

From the beginning of recorded history until 2003, there has accumulated around 5 billion gigabytes of information, making now the most data-rich time ever.[40] For comparison, in 2011, 5 billion gigabytes of data were generated every two *days*.[41] In 2013, 5 billion gigabytes were generated every ten *minutes*.[42] In 2015, 5 billion gigabytes of data were generated every ten *seconds*.[43] That is a lot of data, more than the human mind can comprehend, and it will only continue to exponentially increase over time, hence 'big data.' This is only the beginning when it comes to the big data problem – while it offers exciting and new avenues to healthcare by reducing costs, personalizing medicine, and predicting future health problems – many ethical challenges need to be addressed in these new developments. These obstacles include, but are not exclusive to, concerns with informed consent, privacy and security, and epistemological questions. These challenges must be addressed as the healthcare world moves forward with big data; otherwise, the system will fail to respect some of the most basic bioethical concepts of autonomy, consent, and justice.

What is Big Data & Why Does It Matter?

The healthcare sector has generated large volumes of data for many years. They will continue to do so due to patient records, compliance, regulatory obligations, and patient care, to mention a few.[44] Although much of this data is still in paper form, digitizing this vast quantity of data has become more prevalent in recent years. This transition is driven by a few variables, the most noteworthy of which is its potential to lower costs while improving the quality of healthcare service. These enormous data volumes, or "big data," have the potential to serve a range of healthcare tasks that might be advantageous to the majority, if not all, healthcare sectors.[45]

The Four Vs of Big Data

According to academics, there is not a single, generally agreed definition of big data. Four essential characteristics—volume, variety, velocity, and veracity—are often used to characterize big data.[46] Regularly created and collected health-related data will amass to a reasonable volume of data. Raghupathi et al. describe, "The already daunting volume of existing healthcare data includes personal medical records, radiology images, clinical trial data, FDA submissions, human genetics and population data genomic sequences, etc. Newer forms of big data such as 3-D imaging, genomics, and biometric sensor readings are also fueling this exponential growth."[47] Huge quantities of data exceed the capacities and retrieval capabilities of traditional data processing, organization, and storage and retrieval, including data warehouses, making the volume a crucial component of big data.[48] Big data hence necessitates scalable and efficient information management and storage strategies.[49]

Healthcare data may be found in many different places, including electronic health records (EHRs), medical tests, smartphone apps, and social media. The application of this information to improve healthcare is, however, quite limited. Big data methodologies allow for the effective connecting and analysis of data in many formats to address specific operational, commercial, or research problems.[50] Raghupathi et al. continue, "Future applications of real-time data, such as detecting infections as early as possible, identifying them swiftly and applying the right treatments (not just broad-spectrum antibiotics) could reduce patient morbidity and mortality and even prevent hospital outbreaks."[51] For newborns in the NICU, real-time processing is already being examined to discover potentially fatal infections earlier.[52] If real-time data analysis could be applied to vast volumes of healthcare data from all specialties, adding the second "v" for variety, healthcare would experience a renaissance.[53]

Velocity is another defining characteristic of big data. Massive volumes of continuously updated, variously structured data cannot be processed and analyzed in real-time by conventional IT health infrastructures.[54] Data management may be done faster and with greater flexibility thanks to big data infrastructure. It is a blessing that improvements in data management, especially in virtualization and cloud technologies, have made it simpler to design systems for more efficient data collection, storage, and analysis.[55] Real-time and fast data accumulation is taking place. Unprecedented issues arise from the steady influx of new data accumulated. The amount, variety, and collection of data needed for output retrieval, evaluation, comparison, and decision-making have changed along with the speed at which data is generated.[56]

According to some professionals and academics, the fourth element is veracity, sometimes referred to as data assurance. In other words, the large data, analysis, and results are accurate and reliable. Veracity is the desired outcome, but it has yet to become a reality. Data quality issues are of the highest relevance in this sector because correct information is essential for making survival decisions and because the quality of medical files, particularly unstructured information, is extremely inconsistent and sometimes wrong.[57] The most notorious instance is probably incorrect translations of shoddy handwriting on prescriptions.[58]

Advantages to Healthcare

According to Roski et al., "Big data may have the potential to create approximately $300 billion annually in value in the health care sector, two-thirds of which would be generated by lowering health care expenditures."[59] Big data has so far proven useful in the therapeutic field in several circumstances, such as with the advent of tailored medicine (particular diagnostics and therapies predicated on a patient's thorough risk assessment) in the treatment and care of cancer patients.[60] Another example is to use of decision support systems to match medications to the risk categories of specific patients. These systems are improved by automated detection of diagnostic imaging pictures, and literature mining.[61] Another example of how patient-generated data is utilized is personalizing screening and therapeutic options and educational communications to support desired patient behaviors.[62] A fourth, though not the last, example of how big data has already established its utility is the fact that population health assessments powered by big data have shown patterns that could not have been identified had fewer quantities of regularly structured data been examined instead.[63] The Centers for Medicare and Medicaid Services' use of big data fraud prevention and detection methodologies has superseded older manual documentation techniques. Over $4 billion in expenses have been recovered thanks to these methods in 2011 alone.[64]

Big data is helpful in healthcare research and development, which is another field. The research and development pipeline for medications and devices may be made leaner, quicker, and more focused using big data and predictive modeling.[65] Clinical research designers and participant recruiters may more effectively match medications to specific patients with the use of statistical approaches and algorithms, which lower trial failure rates and hasten the development of cutting-edge cures. Finally, before pharmaceuticals are put up for sale, big data analysis must be used to go through health records and clinical studies to uncover follow-up implications and harmful impacts.[66] Raghupathi et al. conclude, "These advantages also help areas of public health in which big data can analyze disease patterns and track disease outbreaks in transmission to improve public health surveillance and speed response. Faster development of more accurately targeted vaccines, such as choosing the annual influenza strains and turning large amounts of data into actionable information that can be used to identify needs, provide services, and protect and prevent crises, especially for the benefit of populations."[67]

Ethical Challenges

It should be no surprise that with all this patient information and data streaming around, this infrastructure has some ethical concerns and challenges. Many ethical issues are covered in the literature on big data, including informed consent, security (encryption and data protection), stewardship, and epistemological questions (such as those of power or control and the digital divide).[68] The scope of this section will address the two most popular themes in the literature of privacy and security and informed consent concerns with big data.

Privacy & Security

Our present and past are becoming obvious to individuals with the right access and expertise as our actions are captured and replayed in digital reality. For instance, the Rapid Information Overlay Technology (RIOT) software, created by US defense contractor Raytheon, reviews a single person and tries to make their daily activities perfectly clear using data that is freely available from social network and information systems affiliated with an IP address, among other sources.[69] This information was made public by the British newspaper *The Guardian*.[70] Data analysts use big data to assess our internet consumption habits, friendships, sleep patterns, and shopping preferences. This information is personalized only in a few circumstances, usually in intelligence circles.[71]

However, de-individualization is just one facet of anonymization, which involves taking away components that enable data to be related to a particular person.[72] Andrej Zwitter notes, "Location, gender, age, and other information relevant for the belongingness to a group and thus valuable for statistical analysis relate to the issue of group privacy."[73] Whether or not a dataset has been anonymized depends on how many and which group characteristics are still there. This would entail stripping the data of its substance if it were to be stripped of all aspects about any group. As a result, groups are continuously becoming more open even if the data is anonymous because it is de-individualized.[74] Privacy and security will continue to pose an ethical challenge to big data until the anonymization problem can be tackled. Another major ethical challenge that is part of the big data problem is informed consent, as explained in the next section.

Informed Consent

The concept of consent denotes the ability and right of an individual to exert control over information about them, to limit how it is used and reused by organizations like research institutes and healthcare providers, and to revoke such use or reuse.[75] According to existing consent regulations, it is better for society to emphasize personal data ownership above the social advantages that might be gained by disclosing sensitive data for a variety of reasons.[76] In contrast to rigid rules for every single conceivable use, an equilibrium between control of the person and informed disclosure in the context of public safety, environmental preservation, or other aims may be necessary for approval in order to effectively exploit the possibilities of big data.[77]

For instance, the Human Genome Project has put in place permission procedures that allow data to be used and reused for a variety of scientific purposes.[78] The project has tight guidelines for data protection. Even while the details of that use are unclear when the study asks for their consent, it also warns individuals as to how their volunteered personal data may be used and repurposed for their benefit and that of others. To maximize participants' chances of benefiting from the study, its permission procedures explicitly specify how participants' data may be shared with other scientific research projects and to what degree data can be safeguarded.[79] According to Tene & Polonetsky, "In examining such an approach and the potential for its wider adoption, policymakers should balance the value of data uses and potential privacy risks, assess the practicability of obtaining true and informed consent, and understand the enforceability of restrictions on future data sharing and use."[80]

Sometimes, gene sequencing research is restricted to "information altruists," or people eager to offer their data (and occasionally their identities) in the hope that their social standing or financial resources will sufficiently protect them from future prejudice or negative effects.[81] 'Radical honesty' approaches, in which people volunteer to share their de-identified genetic information with the public, are comparable. Establishing "honest broker" and "stewardship" consent models is another strategy to safeguard the interests of data subjects by having unbiased third parties negotiate broad consent agreements.[82] By focusing on professionalism or enacting sanctions for data misuse, it is feasible to delegate the duty of researchers who have access to participants' personal information to instill emotions of accountability in them. Prohibiting unethical research practices like re-identifying anonymized data would lessen any possible harm to data subjects.[83]

By serving as a jumping-off point and a point of reference for future study and debate on the ethics of big data, the quick analysis outlined in this section aims to change widespread and excessively optimistic views. An overview of big data through the four v's (volume, variety, velocity, and veracity) and why it matters in its advantages to the healthcare system was reviewed. The ethical challenges that must be addressed with this new avenue of processing data were mentioned, specifically privacy and security issues and informed consent problems. However, these issues are just the tip of the iceberg – many other ethical and technological issues must be addressed before big data comes into full swing in healthcare, which is a rapidly approaching deadline. One of the key topics at the center of the big data discussion is the conflict between upholding individual data privacy versus deanonymizing personal data to assist in public health surveillance. The challenges in big data use and the ethical justifications for public health surveillance are detailed in the next section.

Individual Data Privacy vs. Public Health Surveillance

By necessity, public health monitoring takes place without express patient agreement. The continued use of personally identifiable information for reporting communicable illnesses and other forms of public health monitoring has legitimate legal and scholarly backing. The ethical justification for monitoring without a patient's express permission is examined in this section using current medical and population health ethics topics. Lee et al. stress, "Overriding individual autonomy must be justified in terms of the obligation of public health to improve population health, reduce inequities, attend to the health of vulnerable and systematically disadvantaged persons, and prevent harm."[84] Additionally, data components obtained without permission must only cause the absolute minimum disruption, support efficient public health action, and be kept in a secure environment.[85]

Challenges in Data Use

Big data presents several challenges with its use – some were highlighted briefly in the previous section, such as privacy, security, and informed consent. This section will take a closer look at electronic healthcare (e-healthcare), specifically at privacy and confidentiality issues that physicians face when debating whether to disclose information such as an infectious disease to the public. In addition, additional data utilization issues will be addressed, and how those should be handled within the legal and professional ethics worlds—subject to future change as time and data changes.

Privacy & Confidentiality

The quick development of e-healthcare has significantly impacted information security for patients. Furthermore, breaches of patient confidentiality and privacy may occur quickly, in large numbers, and without being noticed in the e-healthcare setting.[86] State and professional organizations have acknowledged that safeguarding the information entrusted to healthcare professionals is a fundamental ethical obligation for all clinical staff in light of these e-healthcare possibilities and the populace's concerns about security and confidentiality. A key element that gives information management actions by those in charge of regulating personal health information ethical and moral dimensions is the fact that patients provide information to healthcare professionals in confidence and out of need.[87] Preserving the confidentiality of patients' individually identifiable health data is essential for e-healthcare to be successful and to live up to its promises.[88]

To get therapy and enhance their health, people must provide information to healthcare professionals. This information is provided on a confidential basis. The patients are aware that their present and future medical treatment is the primary reason for revealing, collecting, and retaining personal healthcare information. It raises serious privacy and confidentiality concerns when such private medical information is exploited for other reasons unrelated to their treatment.[89] The government claimed the common good to support additional usage of individual e-healthcare data in socially beneficial initiatives. Nevertheless, it is debatable whether financial considerations and further uses of patient data are acceptable.[90]

Calculating crucial data is necessary when planning and allocating resources over the country.[91] The sharing of sensitive medical information outside of patient care is also closely scrutinized in the national control of contagious and epidemic illnesses. To advance democracy at the national level, personal privacy and secrecy must be legally protected.[92] How is e-healthcare information used? Does one use have purpose over another? What happens if the use is ambiguous? These questions and more are answered in the next section involving e-healthcare utilization.

Utilization

Chronic difficulties with safeguarding e-healthcare information have arisen because of its many uses. Personal e-healthcare data's primary goal is to help clinical care choices for a specific identifiable individual.[93] Therefore, it is very important to identify each subject of healthcare information correctly and accurately for main use objectives. Additionally, the law and professional medical ethics preserve the patient's privacy and confidentiality.[94] The term "secondary uses" refers to further applications of e-healthcare data. The veil of protection effectively prohibits the secondary purposes and uses of health data that support controlling disease outbreaks.[95]

The secondary applications of clinical information might be considered as a compromise between the privacy rights of a person and society's desire to lower healthcare expenses while enhancing the quality of treatment and effectiveness within the delivery system. To evaluate the quality of treatment, plan and manage healthcare services, and conduct clinical or epidemiological research, it is essential to employ EHRs.[96] A well-managed healthcare system has thereby benefited patients more because of the secondary applications of electronic healthcare data.[97]

Every secondary utilization of health information, whether it benefits the patient or the public or not, such as utilizing it to deny a job or health insurance application, raises concerns about privacy and confidentiality as well as legal and ethical difficulties.[98] Through the many healthcare professions, ethical issues are handled. In the US, legislation like HIPAA that safeguards personal information is handled within the legal departments of hospitals or, if escalated, the courtroom.[99] The complete erasure of the subject's personally identifying information is thus required for secondary use of e-healthcare data, as is informed consent. An essential issue for e-healthcare is the elimination of personally identifying information, which is being addressed through anonymizing and pseudonymizing the data.[100]

Ethical Justification of Public Health Surveillance

Lee & Thacker define public health surveillance as "the ongoing, systematic collection, analysis, and interpretation of health-related data with the a priori purpose of preventing or controlling disease or injury, or of identifying unusual events of public health importance, followed by the dissemination and use of information for public health action."[101] Public health monitoring differs from other forms of monitoring, such as security, in that its primary goals are the prevention or control of sickness or harm as well as the improvement of public health.[102] In certain circumstances, public health monitoring is used only to assess the severity of health issues, characterize the course of a disease, identify epidemics and outbreaks of well-known or newly discovered pathogens, and track the dispersion and progression of health events.[103] In addition to supporting epidemiological and clinical research, it also creates and tests theories, evaluates prevention and management strategies, monitors isolation activities and infectious agent changes, pinpoints alterations in medical procedures, plans population health interventions including via means of assets, and appropriates and distributes funds for treatment and control.[104]

Considerations

Public health monitoring techniques may be regarded as morally admissible and systems as ethically justifiable if they comply with the positive operational criteria and abstain from transgressing the negative ones. Building a system for monitoring public health is the first action in current best practices, which involves having explicit knowledge of the system's public health goals to guarantee that the relevant data are gathered to address the critical issues.[105] Prioritizing health outcomes depending on the scale of the impact, assessed by frequency, severity, cost, or preventability, is often necessary for determining what public health concern to address.[106] The operational principles of gathering accurate information and responding quickly to the healthcare requirements of a population are met by this prioritizing.

The requirement for adequate data to guide both the management of public health and the ability to protect the privacy of individual patients' medical records must be balanced when deciding which data categories to collect.[107] Lee et al. emphasize the "Current recommended approach is to collect the minimum number and simplest data elements necessary to meet the goals of the system to minimize risk to individuals, thus meeting the operating principle of imposing the least possible infringement. Early engagement of partners and affected communities is recommended in developing public health surveillance systems, especially when the data are sensitive, or populations are particularly vulnerable."[108] Choosing what kind of information should be gathered in a surveillance system often involves consulting the impacted groups. In the 1990s, community involvement was successfully employed to win over objections to name-based HIV reporting, even in places like New York City and San Francisco, California, where there had been vocal early opposition.[109]

The best practices that are currently being used for the data stages, such as data gathering, aggregation, processing, and evaluation, cascade straight into a well-planned system.[110] Given the system's emphasis on public health, gathering the fewest and least sensitive data pieces is required. This is consistent with the tenets of the system, which include rationality, minimal intrusion, and attention to the root causes of illness. Lee et al. state "Collecting information from all affected cases (or a representative sample) is the hallmark of population-based public health surveillance, leaving no subgroup excluded from data collection or, perhaps more importantly, from identifying the need for intervention."[111] The lack of restriction due to demographical, behavioral, or social factors operationalizes the goals of removing injustices, correcting structural inequities, and inclusion.[112]

Data gathering is crucial to the fair and equitable distribution of treatment and preventative measures in a sector like public health that is scientifically based. Consider the case when a demographic is often left out of public health monitoring, perhaps as a consequence of receiving treatment from a certain kind of practitioner who has less reporting responsibility. If so, structural discrepancies in the benefits gained from system participation will be revealed to this group, which is underrepresented in the statistics.[113] Groseclose et al. continue "Acquiring accurate data during collection and collation is a minimal requirement of a public health surveillance system. To ensure data accuracy, ongoing monitoring and periodic evaluation are recommended for all systems."[114] By guaranteeing the use of the data gathered and giving proof that the advantages of data collection exceed the drawbacks, this approach helps satisfy the requirements of two operational principles. A complete understanding of the surveillance system is necessary for using suitable analytical methods and correctly interpreting results.[115]

The operational pillars of responsible interpretation ensure the effectiveness of the intervention, under this case, surveillance, so that advantages outweigh violations. The last component of best practices for data security is defined processes for how data are collected, transported, accessed, kept, utilized, or shared, as well as specifications for technical and physical security safeguards of data.[116] Lee et al. says "Limiting the number of printed and digital copies of case data, anonymizing and encrypting data during transport, limiting the number of users to the smallest possible, annual security training, and incorporating specific sanctions into annual confidentiality agreements signed by personnel working with surveillance data are some of the recommended best practices for maintaining confidentiality and data security."[117] These processes demonstrate how to protect confidentiality and prevent personal injury by protecting data and reducing the danger of damage. Additionally, they support little intervention.[118]

The best available procedures for information use are deduced from the core principle that data gathering must not be an end in itself but rather a means to improve the condition of the area where the information was gathered. These practices include information exchange and conversation with those who need to know it as well as the implementation of the data to health promotion strategies and initiatives to achieve the best outcomes.[119] It needs excellent communication techniques and an understanding of the people and organizations that can utilize the data from a monitoring system to take action for public health. Furthermore, Lee et al. explain "Effective communication models require assessing the quality of data, the definition of the purpose of the communication and the audience, the development of the message, selection of the channel, marketing of the information, implementation of the plan, and evaluation of the process and outcome."[120]

During this stage of the public health surveillance action, operating principles such as intervening in the data to benefit society, proving that advantages outweigh violations, and minimizing disparities by highlighting inequities and recommending initiatives to counteract them—are all applied.[121] When spreading unfavorable information about socially underdeveloped minorities, caution must be taken to guarantee that the least amount of damage is done. By engaging affected groups and community leaders on the best communication strategies and populations, the harm and heightened marginalization that may come from direct mainstream media messaging may be minimized.[122]

Risks and Responsibilities of Public Health Ethics

The public health authority is now responsible for the reported data after a reportable condition has been identified and submitted in accordance with the legal duties imposed on the healthcare practitioner.[123] It is a significant obligation to collect and store personally identifiable information. Since public trust is the cornerstone of public health monitoring, treating patients with respect and dignity is essential for the continuous operation of public health management.[124]

The public health system is aware of the consequences that individuals and communities face as a result of public health practitioners publishing name-based data about private medical matters. Potential damages have been recommended to be protected by ethical guidelines.[125] Lee et al. suggest "These include the following: a collection of individual identifiers only when essential to achieve the public health goals of the system, acquisition of the minimum amount of information needed to meet the public health objective, and engagement of affected communities and stakeholders to consult regarding the most effective way to disclose and disseminate the findings of public health surveillance data, especially when they might substantially add to the stigmatization of an already marginalized group."[126] For classification initiatives that pose the greatest social concerns, such as HIV and other STDs, the acquisition, preservation, and use of monitoring data must adhere to strict programming criteria.[127]

The agreement between the public health sector and the general community must include the utilization of monitoring data to improve health. The fundamental ethical condition that the risk involved in obtaining and preserving the data is justified by the possibility that it will be utilized to accomplish the intended goals continues to apply to every public health dataset. In the nineteenth century, Herman Biggs, a physician from New York who helped establish surveillance, was certain that "data were collected to be used, not to keep adding machines busy."[128] This attitude remained into the twenty-first century when William Foege, the director of the Centers for Disease Control and Prevention from 1977 to 1983, indicated that gathering data was crucial for limiting sickness and that spending resources should not be permitted if no mission is completed.[129]

In fact, it is still morally required that we utilize the data we acquire to advance public health; any exceptions to this rule need to be supported.

Ethics was often used as a defense for switching to personal identification tracking in public health monitoring programs that had previously used anonymized systems to best use the data for health optimization. As an example, the United States ended its national anonymous HIV monitoring program in 1995 among expectant mothers in pursuit of routinely checking expectant mothers and infants and sending the results to their pre-existing identity surveillance system.[130] This made it possible to share the findings with expectant moms and new mothers so they could decide whether to get chemoprophylaxis or early therapy.[131]

Overall, the United States public health sector has a proven history of handling private information responsibly. The need for rules and codes for the ethical gathering, preservation, and use of healthcare data is growing as more and more electronic storage makes it possible for easy accessibility, verification, and distribution.[132] Due to the diversity of authorities, public health programs confront difficulties at the municipal, state, and federal levels, but these difficulties are surmountable. Scientific and legal justifications for public health monitoring in the absence of explicit and specific patient permission, contemporary public health ethical frameworks and their guiding principles, and, at least in part, a thorough monitoring system involving disadvantaged populations, gathers the minimal amount of data necessary, stores information safely, and utilizes information for public health promotion.[133]

Conclusion

The dominant concept and practice of bioethics have been called for to change from a more individual-centered to a more global-centered emphasis throughout the last several decades. With this shift comes a move from less stress on personal autonomy and privacy to more emphasis on the community, solidarity, and the common good. Applying this shift in thought to today in America, the problem of data sharing for public surveillance comes to the forefront. The mainly Western notion of anti-paternalism and advocacy for the concept of respect for autonomy takes precedence in the medical field within the physician-patient relationship. However, there is more to bioethics than this simple relationship. Humans are interconnected and social beings and therefore are connected to a community, so the individual good should be the same and connected to the community good. This way of being connected to the community good can come through the analysis and use of electronic health records for public surveillance.

Combined with the possibilities of algorithmic computing and data sorting, vast quantities of public health information are now available to be used in numerous ways. Of course, with these data use and sharing come ethical considerations such as privacy, security, informed consent, and confidentiality. These issues must be addressed to the best extent possible to use this data for research – and anonymization and pseudonymization allow these possibilities to be addressed. However, with some public health data, this is not possible. As discussed in the last section, it can be ethically and legally permissible to use such data without the individual's informed consent. Several federal statutes that regulate entry to and the potential utilization of specific categories of information, like genetic information or data on other biometrics, offer challenges since they need explicit, informed consent for data use. Some of these challenges, such as privacy, confidentiality, surveillance, and large-scale biometrics applications, require further consideration in their ethical use and implementation in society.

Notes

[1] Margit Sutrop, "Changing Ethical Frameworks: From Individual Rights to the Common Good?," *Cambridge Quarterly of Healthcare Ethics* 20, no. 4 (2011): 533.
[2] Ibid.
[3] Ibid.
[4] Ibid., 534.
[5] Ibid., 534.
[6] Ibid., 534.
[7] David C. Thomasma, "Evolving Bioethics and International Human Rights," in *Autonomy and Human Rights in Health Care: An International Perspective*, ed. David N. Weisstub and Guillermo Diaz Pintos (Dordrecht: Springer, 2008), 14.
[8] Ibid.
[9] Ibid., 15.
[10] Ibid., 15.
[11] Ibid., 15.
[12] Ibid., 15.
[13] Ibid., 15.
[14] Ibid., 15.
[15] Jacob D. Rendtorff, "The Limitations and Accomplishments of Autonomy as a Basic Principle in Bioethics and Biolaw," in *Autonomy and Human Rights in Health Care: An International Perspective*, ed. David N. Weisstub and Guillermo Diaz Pintos (Dordrecht: Springer, 2008), 80.
[16] Ibid.
[17] Ibid.
[18] Ibid., 81.

[19] Ibid., 81.
[20] Warren T. Reich, ed., *Encyclopedia of Bioethics* (Washington, DC: Simon & Schuster, 1995), 219.
[21] Rendtorff, 81.
[22] Henk ten Have, *Global Bioethics: An Introduction* (London; New York, NY: Routledge, Taylor & Francis Group, 2016), 39.
[23] Renée C. Fox, *The Sociology of Medicine : A Participant Observer's View*, Prentice Hall Foundations of Modern Sociology Series (Englewood Cliffs, N.J.: Prentice Hall, 1989), 72.
[24] ten Have, 39.
[25] Sutrop, 534.
[26] Alfred I. Tauber, *Patient Autonomy and the Ethics of Responsibility*, Basic Bioethics (Cambridge, MA: MIT Press, 2005), 68.
[27] Sutrop, 534.
[28] Bartha Maria Knoppers and Ruth Chadwick, "Human Genetic Research: Emerging Trends in Ethics," *Nature Reviews Genetics* 6, no. 1 (2005): 75-79.
[29] Daniel Callahan, "Individual Good and Common Good: A Communitarian Approach to Bioethics," *Perspectives in Biology and Medicine* 46, no. 4 (2003): 496.
[30] Sutrop, 542.
[31] Tuija Takala, "Setting a Dangerous Precedent: Ethical Issues in Human Genetic Database Research," *Medical Law International* 8, no. 2 (2007): 105-137.
[32] Aristotle, *The Politics*, Cambridge Texts in the History of Political Thought (Cambridge: Cambridge University Press, 1988), 1253.
[33] Louis Dupré, "The Common Good and the Open Society," *The Review of Politics* 55, no. 4 (1993): 687.
[34] Sutrop, 535-536.
[35] Thomas W. Smith, "Aristotle on the Conditions for and Limits of the Common Good," *The American Political Science Review* 93, no. 3 (1999): 625-636.
[36] Sutrop, 536.
[37] Ibid.
[38] Ibid.
[39] Ibid., 542.
[40] Andrej Zwitter, "Big Data Ethics," *Big Data & Society* 1, no. 2 (2014): 2.
[41] Ibid.
[42] Ibid.
[43] Ibid.
[44] Wullianallur Raghupathi and Viju Raghupathi, "Big Data Analytics in Healthcare: Promise and Potential," *Health Information Science and Systems* 2, no. 3 (2014): 1.
[45] Ibid.
[46] Ibid., 3.
[47] Ibid., 3.
[48] J. Roski, G. W. Bo-Linn, and T. A. Andrews, "Creating Value in Health Care through Big Data: Opportunities and Policy Implications," *Health Affairs* 33, no. 7 (2014): 1116.

[49] Ibid.
[50] Ibid.
[51] Raghupathi and Raghupathi, 3.
[52] Ibid.
[53] Ibid.
[54] Roski, Bo-Linn, and Andrews, 1116.
[55] Raghupathi and Raghupathi, 3.
[56] Ibid.
[57] Ibid., 4.
[58] Ibid., 4.
[59] Roski, Bo-Linn, and Andrews, 1116.
[60] Andrew McAfee and Erik Brynjolfsson, "Big Data: The Management Revolution; Exploiting Vast New Flows of Information Can Radically Improve Your Company's Performance. But First You'll Have to Change Your Decision-Making Culture," *Harvard Business Review* 90, no. 10 (2012): 61-68.
[61] Roski, Bo-Linn, and Andrews, 1116.
[62] Ibid.
[63] Ibid.
[64] Ibid.
[65] Raghupathi and Raghupathi, 3.
[66] Ibid.
[67] Ibid.
[68] Brent D. Mittelstadt and Luciano Floridi, "The Ethics of Big Data: Current and Foreseeable Issues in Biomedical Contexts," *Science and Engineering Ethics* 22, no. 2 (2016): 312.
[69] Ryan Gallagher, "Software That Tracks People on Social Media Created by Defense Firm," *The Guardian* 2013, 1.
[70] Ibid.
[71] Zwitter, 4.
[72] Ibid.
[73] Ibid.
[74] Ibid.
[75] Roski, Bo-Linn, and Andrews, 1119.
[76] Ibid.
[77] Ibid.
[78] Jeantine E. Lunshof et al., "From Genetic Privacy to Open Consent," *Nature Reviews Genetics* 9, no. 5 (2008): 406-411.
[79] Roski, Bo-Linn, and Andrews, 1119.
[80] Omer Tene and Jules Polonetsky, "Privacy in the Age of Big Data: A Time for Big Decisions," *Stanford Law Review Online* 64 (2011): 63.
[81] Mittelstadt and Floridi, 314.
[82] Ibid., 315.
[83] Ibid., 315.
[84] Lisa M. Lee, Charles M. Heilig, and Angela White, "Ethical Justification for Conducting Public Health Surveillance without Patient Consent," *American Journal of Public Health* 102, no. 1 (2012): 38.

[85] Ibid.
[86] Charles A. Shoniregun, Kudakwashe Dube, and Fredrick Mtenzi, *Electronic Healthcare Information Security* (New York, NY: Springer, 2010), 34.
[87] Ibid.
[88] Ibid.
[89] Ibid.
[90] Ibid.
[91] Ibid.
[92] Ibid.
[93] Ibid., 35.
[94] Ibid., 35.
[95] Ibid., 35.
[96] Ibid., 35.
[97] Ibid., 35.
[98] Ibid., 35.
[99] Ibid., 35.
[100] Ibid., 35.
[101] Lisa M. Lee and Stephen B. Thacker, "Public Health Surveillance and Knowing About Health in the Context of Growing Sources of Health Data," *American Journal of Preventive Medicine* 41, no. 6 (2011): 636.
[102] Centers for Disease Control and Prevention, "Distinguishing Public Health Research and Public Health Nonresearch," ed. Department of Health & Human Services (2010), 1.
[103] Lee, Heilig, and White, 38.
[104] Stephen B. Thacker, "Historical Development," in *Principles and Practice of Public Health Surveillance*, ed. Lisa M. Lee, Steven M. Teutsch, and Michael E. St. Louis (New York: Oxford University Press, 2010), 1-17.
[105] Lee, Heilig, and White, 41.
[106] Ibid.
[107] Lisa M. Lee and Lawrence O. Gostin, "Ethical Collection, Storage, and Use of Public Health Data: A Proposal for a National Privacy Protection," *JAMA: The Journal of the American Medical Association* 302, no. 1 (2009): 82-84.
[108] Lee, Heilig, and White, 41.
[109] Centers for Disease Control and Prevention, 1.
[110] Lee, Heilig, and White, 41.
[111] Ibid.
[112] Ibid.
[113] Ibid., 42.
[114] SL Groseclose, RR German, and P. Nsubuga, "Evaluating Public Health Surveillance," in *Principles and Practice of Public Health Surveillance*, ed. Lisa M. Lee, et al. (New York: Oxford University Press, 2010), 166.
[115] Lee, Heilig, and White, 42.
[116] Ibid.
[117] Ibid.
[118] Ibid.
[119] Ibid.

[120] Ibid.
[121] Ibid.
[122] Ibid.
[123] Ibid.
[124] Ibid.
[125] Lee and Gostin, 82-84.
[126] Lee, Heilig, and White, "42.
[127] Ibid.
[128] Amy L. Fairchild, et al., *Searching Eyes : Privacy, the State, and Disease Surveillance in America* (Berkeley: University of California Press, 2007), 72.
[129] William H. Foege, Robert C. Hogan, and Ladene H. Newton, "Surveillance Projects for Selected Diseases," *International Journal of Epidemiology* 5, no. 1 (1976): 29-37.
[130] Martha F. Rogers, Mary Glenn Fowler, and Mary Lou Lindegren, "Revised Recommendations for Hiv Screening of Pregnant Women," *Morbidity and Mortality Weekly Report: Recommendations and Reports* 50, no. RR-19 (2001): 59-86.
[131] Ibid.
[132] Lee, Heilig, and White, 43.
[133] Ibid.

Literature

Aristotle. 1988. *The Politics.* Cambridge Texts in the History of Political Thought. Cambridge: Cambridge University Press.

Callahan, Daniel. 2003. "Individual Good and Common Good: A Communitarian Approach to Bioethics." *Perspectives in Biology and Medicine* 46, no. 4: 496.

Dupré, Louis. 1993. "The Common Good and the Open Society." *The Review of Politics* 55, no. 4: 687-712.

Fairchild, Amy L., Ronald Bayer, James Keith Colgrove, and Daniel Wolfe. 2007. *Searching Eyes : Privacy, the State, and Disease Surveillance in America.* Berkeley: University of California Press.

Foege, William H., Robert C. Hogan, and Ladene H. Newton. 1976. . "Surveillance Projects for Selected Diseases." *International Journal of Epidemiology* 5, no. 1 (01): 29-37.

Fox, Renée C. 1989. *The Sociology of Medicine : A Participant Observer's View.* Prentice Hall Foundations of Modern Sociology Series. Englewood Cliffs, N.J.: Prentice Hall.

Gallagher, Ryan. 2013. "Software That Tracks People on Social Media Created by Defense Firm." *The Guardian.*

Groseclose, SL, RR German, and P. Nsubuga. 2010. "Evaluating Public Health Surveillance." In *Principles and Practice of Public Health*

Surveillance, edited by Lisa M. Lee, SM Teutsch, Stephen B. Thacker and Michael E. St. Louis, New York: Oxford University Press, 166-197.

Have, Henk ten. 2016. *Global Bioethics: An Introduction.* London; New York, NY: Routledge, Taylor & Francis Group.

Knoppers, Bartha Maria, and Ruth Chadwick. 2005. "Human Genetic Research: Emerging Trends in Ethics." *Nature Reviews Genetics* 6, no. 1: 75-79.

Lee, Lisa M., and Lawrence O. Gostin.2009. "Ethical Collection, Storage, and Use of Public Health Data: A Proposal for a National Privacy Protection." *JAMA: The Journal of the American Medical Association* 302, no. 1 (07/01): 82-84.

Lee, Lisa M., Charles M. Heilig, and Angela White. 2012. "Ethical Justification for Conducting Public Health Surveillance without Patient Consent." *American Journal of Public Health* 102, no. 1 (Jan): 38-44.

Lee, Lisa M., and Stephen B. Thacker. 2011. "Public Health Surveillance and Knowing About Health in the Context of Growing Sources of Health Data." *American Journal of Preventive Medicine* 41, no. 6: 636-640.

Lunshof, Jeantine E., Ruth Chadwick, Daniel B. Vorhaus, and George M. Church. 2008. "From Genetic Privacy to Open Consent." *Nature Reviews Genetics* 9, no. 5 (05/01): 406.

McAfee, Andrew, and Erik Brynjolfsson. 2012. "Big Data: The Management Revolution; Exploiting Vast New Flows of Information Can Radically Improve Your Company's Performance. But First You'll Have to Change Your Decision-Making Culture." *Harvard Business Review* 90, no. 10: 60.

Mittelstadt, Brent D., and Luciano Floridi. 2016. "The Ethics of Big Data: Current and Foreseeable Issues in Biomedical Contexts." *Science and Engineering Ethics* 22, no. 2 (Apr): 303-341.

Prevention, Centers for Disease Control and. 2010. "Distinguishing Public Health Research and Public Health Nonresearch." edited by Department of Health & Human Services.

Raghupathi, Wullianallur, and Viju Raghupathi. 2014. "Big Data Analytics in Healthcare: Promise and Potential." *Health Information Science and Systems* 2, no. 3: 1-10.

Reich, Warren T., ed. 1995. *Encyclopedia of Bioethics*. Washington, DC: Simon & Schuster.

Rendtorff, Jacob D. 2008. "The Limitations and Accomplishments of Autonomy as a Basic Principle in Bioethics and Biolaw." Chap. 6 In *Autonomy and Human Rights in Health Care: An International*

Perspective, edited by David N. Weisstub and Guillermo Diaz Pintos, Dordrecht: Springer.

Rogers, Martha F., Mary Glenn Fowler, and Mary Lou Lindegren. 2001. "Revised Recommendations for Hiv Screening of Pregnant Women." *Morbidity and Mortality Weekly Report: Recommendations and Reports* 50, no. RR-19: 59-85.

Roski, J., G. W. Bo-Linn, and T. A. Andrews. 2014. "Creating Value in Health Care through Big Data: Opportunities and Policy Implications." *Health Affairs* 33, no. 7 (Jul): 1115-1122.

Shoniregun, Charles A., Kudakwashe Dube, and Fredrick Mtenzi. 2010. *Electronic Healthcare Information Security.* New York, NY: Springer.

Smith, Thomas W. 1999. "Aristotle on the Conditions for and Limits of the Common Good." *The American Political Science Review* 93, no. 3: 625-636.

Sutrop, Margit. 2011. "Changing Ethical Frameworks: From Individual Rights to the Common Good?". *Cambridge Quarterly of Healthcare Ethics* 20, no. 4: 533-545.

Takala, Tuija. 2007. "Setting a Dangerous Precedent: Ethical Issues in Human Genetic Database Research." *Medical Law International* 8, no. 2: 105-138.

Tauber, Alfred I. 2005. *Patient Autonomy and the Ethics of Responsibility.* Basic Bioethics. Cambridge, MA: MIT Press.

Tene, Omer, and Jules Polonetsky. 2011. "Privacy in the Age of Big Data: A Time for Big Decisions." *Stanford Law Review Online* 64: 63-69.

Thacker, Stephen B. 2010. "Historical Development." In *Principles and Practice of Public Health Surveillance*, edited by Lisa M. Lee, Steven M. Teutsch and Michael E. St. Louis, New York: Oxford University Press, 1-17.

Thomasma, David C. 2008. "Evolving Bioethics and International Human Rights." Chap. 1 In *Autonomy and Human Rights in Health Care: An International Perspective*, edited by David N. Weisstub and Guillermo Diaz Pintos, Dordrecht: Springer.

Zwitter, Andrej. 2014. "Big Data Ethics." *Big Data & Society* 1, no. 2: 1-6.

CHAPTER SEVEN

THE RESILIENCE OF JUSTICE IN PUBLIC HEALTH IN DEALING WITH A PANDEMIC

SAMUEL KWAME DANSO

Introduction

Public health issues should be of grave concern to everyone worldwide. Number three of the seventeen Sustainable Development Goals (SDGs) of the United Nations (UN) emphasizes the health and well-being promotion of people of all ages.[1] Most countries in the third world lag on this goal, which must be done in no time. This section of the chapter will focus on ethical issues in promoting public health. Areas in focus comprise aspects of public health and its promotion, the lifestyle of the populace with specific mention of choices, and poverty. The final part is about environmental issues dealing with climate change and pollution.

Public Health and Ethics

Undoubtedly, there is a considerable cost involved in health care and its promotion. Public health should not be sacrificed at any expense. Ensuring public health must be a priority and a persistent obligation of those in charge of the population and all stakeholders. It is considered a moral issue rather than legal for the populace to be able to access health care.[2] An attempt to promote public health and well-being calls for various ethical concerns. A conscious effort should be made to uphold ethical practices in all spheres to facilitate public health and well-being. As a society, there is always the need to ensure maximum good, as championed by the utilitarians. Health, freedom, and happiness are intrinsically good and valuable, so every reasonable person should cherish them.[3] This section focuses on aspects of public health and how it can be promoted for the betterment of all.

Aspects of Public Health

Public health encompasses three core aspects: the entire population, preventive nature, and collective effort. Firstly, public health is an issue of the whole population rather than individual members. Secondly, public health concentrates more on prevention than curative. Finally, collective force is required to bring about public health.[4] These aspects carry ethical implications. Specific ailments are contagious, and it would be unethical for anyone not to come on board to help curtail the spread. In addition, it becomes everyone's moral duty to cooperate when it comes to the prevention of diseases in any society. Socrates, a great ancient philosopher, saw ethics as how people should live and the reasons for that. In contemporary times, ethics is seen from two dimensions: normativity and speculation. Normativity involves examining theories systematically, whereas speculative focuses on the reflective nature of ethics.[5] The aspects of public health should call for concern and severe reflection to put all hands on deck.

To further illustrate these aspects of public health, one would admit that there is a clear connection between the health of individuals and the population. Such a relationship is so complex, indicating that the overall health of the individuals in society also means that the population is healthy. In this vein, it becomes imperative that the community ensures the prevention of anything that could harm the people. Should there be any outbreak of an epidemic, society again comes in to help alleviate it. In all these, the state assumes a paternal role in public health issues.[6]

Promoting Public Health

Promoting public health and ensuring equitable healthcare access is justice. Justice is one of the four principles of health care promulgated by Beauchamp and Childress. As a health care principle, respect is about people's decision-making autonomy. The emphasis on justice concerns the fair and equitable distribution of health care. Beneficence, as a principle, seeks to control risks that could adversely affect people's lives. Finally, the nonmaleficence principle is to do with the prevention of harm.[7] Beauchamp and Childress's four healthcare principles equally apply to promoting public health. By its very nature, public health concerns the entire society (group), still not undermining the individuals' health. It is the individuals that constitute the population. As one of the principles, justice aims to achieve fairness and equity for the people to have healthcare. For Aristotle, justice implies apportioning to people what is due to them.[8] Anything less than this will be injustice and unethical. Public health should not be denied to the people.

The most prominent and influential strategy to promote public health is encouraging health literacy. Research has proved that public libraries have been effective in promoting health literacy. Through public libraries, people have access to relevant health information. Despite electronic access to information via cell phones and computers, it has been noted that patronage in the public library has increased in recent years, as confirmed by the American Library Association.[9] Libraries in the US have been identified as public places to partner with health authorities to promote health education. Libraries can be easily accessed in the neighborhoods to provide credible, confidential, and accessible health information. Health information through social media might be miss leading and not factual. It is noted that in 2014, 58% of the US public libraries helped people obtain online health information. About 60% assisted in finding resources on health insurance. In addition, 48% of the libraries helped clients to get an understanding of specific health-related topics.[10]

People's Lifestyle

One of the core functions of public health is to do with intelligence, known as PHI (Public Health Intelligence). Intelligence is all about credible information. PHI seeks to identify, connect, collect, analyze, synthesize, assess, interpret, and generate insightful information necessary for dissemination as a vital resource in making informed decisions that help improve public health.[11] The lifestyle of people has many implications for the general well-being of society. Lifestyle choices should also welcome one's responsibilities. This is described in the consequentialist approach to morality. However, it is also a public concern.[12] Two critical factors for consideration in the lifestyle of people are poverty level and life choices. A high level of poverty in a society usually dictates the lifestyle of the people. Similarly, people's life choices equally affect them and humanity, hence the need for PHI.

Poverty

Eradicating all forms of poverty is goal number one of the SDGs under the UN. The subsequent goals are equally connected to the first. Ending hunger and ensuring good health and well-being of all people as goals number two and three, respectively. The UN cap for measuring poverty is US$1.25 per person per day.[13] The effect of poverty on people's health is enormous. Poverty is closely associated with malnutrition, lack of potable water, and unsafe poor shelter. These issues also result in diarrheal and malaria, leading to maternal and infant mortality. Simply put, poverty causes poor health and a high mortality rate, especially among children. These are common in poor developing countries. Among developed countries, poverty can still be found with its associated problems. As a result of high industrial operations, some areas are so polluted that people develop respiratory diseases. In addition, increased dropout rates among students, delinquency, and aggressive behavior are typical of children living in poor communities in industrialized nations.[14]

Lifestyle Choices

People generally make several decisions and make choices each day. Some day-to-day thoughts and actions are wrong, and others are good. The consequence-based or utilitarianism theory explains better. This theory accepts the utility principle, which opines that people's actions should result in maximum value rather than disvalue.[15] In other words, the actions and choices of individuals should promote the good of the society in which health is prominent. Generally, lifestyle is influenced by certain factors, including economic, cultural, political, geographical, and religious. These factors can dictate a person's choices, subsequently impacting society. Variables in lifestyle comprise exercise, diet, sleep, sexual behavior, use of technology, medication and substance abuse, and recreation.[16]

Environmental Issues and Public Health

Environmental issues are so huge to discuss when it comes to public health. The connection between nature and humanity is the focus of this aspect. Public health should extend beyond a geographical area and be seen as a global issue.[17] An area's activities can have a resultant effect on the other side of the globe concerning global warming. This then calls for the responsibility of the entire public rather than the individual affected community.[18] Two critical aspects comprise climate change and pollution. The way to deal with such issues calls for ethical considerations.

Climate Change

One can readily admit that climate change has become the most discussed topic across continents. It seriously has been a significant threat to public health in recent years. Extreme weather conditions with associated natural disasters: flooding, drought, the rising level of the sea, desertification, hurricanes, and other storms often draw people's attention to the need for urgent action on climate change.[19] Such events have serious consequences, which result in malnutrition, respiratory sickness, diarrhea, malaria, and neonatal mortality. These are the well-known five issues leading to infant mortality across the world. It is highly predicted that these will never improve due to worsening climatic conditions. Children are often the most vulnerable to all the severe effects of climate change. Climate change-related issues broadly affect children's mental health, growth, and development.[20] It is not only children who are adversely impacted by climate change and its associated problems; adults, too, are at risk. This calls for a consented effort to combat climate change. Government support in promoting inter-agency harmonization and designing programs and strategies are much needed. As a society, participation from the entire population is equally necessary to combat climate change through awareness creation and ensuring stakeholders' commitments. Other equally essential ways are the need for a paradigm shift in how climate change has been perceived and the need for capacity building through innovative ways.[21]

Pollution

Polluting the environment goes a long way to affect the elements of nature, namely, air, water, and food, which should benefit human beings. Fetuses, babies, and children are at the most significant risk of environmental pollution. The vulnerability of children to pollution is significantly more than adults. The simple fact is that they consume relatively more water and food and breathe more air considering the pound-by-pound weight of the body compared to adults. As a result, they consume more proportion of toxic in the environment.[22] Environmental pollution is said to be a significant cause of disease outbreaks. A recent study has proved that high population density associated with pollution contributes to the rapid spread of diseases. A case in point is the COVID-19 pandemic in Europe. Densely populated areas with high carbon emissions recorded a high pandemic rate. Similar findings were also recorded in China.[23]

One cannot talk about the pollution issue in society without mentioning smoking. The subject of smoking has been a controversial one to discuss. Smoking advocates think of it as a way of securing their pleasure.[24] As such, the discussion continues that it would be rude to express an opinion significantly if it does not harm the other. Again, where there are moral uncertainties concerning the practice, one should not pass judgment. However, from the ethical perspective, an individual's autonomous choices should be evaluated to ensure moral correctness.[25] This is a concern for public health because one's personal choices could be a menace to society. From an ethical point of view, public health cannot be quiet on such issues even if the repercussion goes to the individual. Such individuals are still part of the larger community. A community effort is needed to ensure public health.

Conclusion

In conclusion to this section, public health is an important area that calls for crucial attention worldwide. As noted, public health centers on preventing harm, focusing on the entire society, and needing a consented effort to bring success in health promotion. In these aspects, ethical issues are much involved. People's health could be at stake if practical measures, like promoting health literacy, are not seriously considered. Public health is closely connected to environmental issues, hence the need to take climate change and pollution seriously.

Section 2: Applying the Principle of Justice in the Promotion of Public Health

Introduction

The concept of justice permeates all spheres of life: social, political, economic, religious, moral, health, and the like. According to Beauchamp and Childress, four essential principles are needed for an analytical and ethical framework. Among these principles is Justice. In healthcare, the principle of justice is mainly concerned with the fair and equitable distribution of benefits, costs, and risks. Another principle is respect, which aims at autonomy when making decisions. There is also beneficence in the way of controlling risks and costs. Nonmaleficence, as a principle, is about preventing harm.[26]

Ethics based on principles is said to have been developed by W. D. Ross. Ross advised that psychologists, and for that matter, all moral agents, should follow behavioral principles like truthfulness, doing no harm, and showing kindness to others. The principle-based approach has been applied to healthcare by Beauchamp and Childress, American ethicists. The five principles are beneficence, respect/autonomy, nonmaleficence, justice, and newly added veracity for professionals.[27] These principles are integral aspects of bioethics. This section of the chapter aims to apply this concept of justice in public health in its preventive nature and collective efforts. How justice relates to environmental issues will also be discussed. The final aspect of the section is to do with egalitarianism with a specific focus on poverty among the people and government interventions.

Justice in Public Health

By its very nature, public health is to do with justice. Similarly, the addition to the healthcare principles termed public beneficence was propagated by Knapp and VandeCreek. Public beneficence emphasizes the obligation an individual has to the public. It also has to do with seeking the well-being of other people.[28] Therefore, justice and public beneficence principles are closely linked and applicable to public health ethics. Public beneficence thrives in three areas; firstly, where an individual's behavior threatens the public. Secondly, public beneficence must supersede beneficence, autonomy, and non-maleficence. Finally, it is recommended that psychologists promote public beneficence for it to succeed.[29]

Public health encompasses three critical areas of prevention concerning the whole population and collective effort. Issues of public health concern all people rather than individuals, hence, the need for a consented effort to achieve public health goals. Prevention is at the heart of public health more than curative.[30] With people in society contributing their quota to attain the aims of public health is also justice. Justice in healthcare calls for ensuring that people know health issues, especially what is prevalent in society at any point in time. Justice provides health literacy whereby people can access vital information and facilities and use them to make informed health decisions.[31] A further discussion on the preventive goal and collective effort follows.

Prevention Goal

To ensure the prevention of health hazards in society under public health, there is a need for Public Health Intelligence (PHI). PHI needs information necessary for processing and making crucial informed decisions. Intelligence deals with plausible information required for decision-making. PHI collects, identifies, analyzes, connects, synthesizes, interprets, assesses, and disseminates information much needed for public health education and decisions.[32] PHI, therefore, plays a critical intermediary role in collecting, processing, and making available what has been acquired. One of the core duties of public health is prevention, more than combating an outbreak or curing diseases.[33]

There are three major domains that PHI emphasizes in all its dealings, namely, protection, improvement, and healthcare. Health Protection deals with imminent threats to health among the people. Secondly, Health Improvement is always looking at the best way to achieve the general well-being of the people. Finally, Healthcare in Public Health is concerned with improving opportunities for available healthcare.[34] Besides the intelligence needed for critical decision-making, there is also the need for well-trained professionals. The professionals are there to study the trends of events in the environment and be abreast with health-related issues, such as food insecurities, around their jurisdiction.[35] Consequently, to ensure justice prevails in the prevention aspect of public health, there is a need for credible information necessary for decision-making and trained personnel to act on them. Having vital health information and not disseminating it amounts to injustice, which does not benefit anyone.

Collective Effort

To effectively bring all people on board in public health, justice calls for ensuring everyone is not left out and that there are enough resources accessible to the people. Generally, people within the brackets of the aged, the less educated, and the minorities be they ethnic, tribal, or racial, are often health illiterate and are deficient in resources for healthcare access.[36] When a particular group of people in society lack the necessary permit, public health promotion will be a deficit and will be against the principle of justice. Collective effort cannot prevail. It is worth noting that factors such as languagc, education, culture, and resources can promote people's literacy in health.[37] Health information should be translated into the language people are familiar with to bring them on board the health promotion campaign. Whenever people have a clear direction about the entire society's focus on health, their maximum cooperation can be assured.

Justice and the Environmental Issues

The most life essentials are food, water, shelter, and fuel for cooking and heating. The sources of these necessities are often questionable in ethical terms. The process of improving the life of humanity also comes with various health implications. The need to apply fertilizers and other chemicals to the farms may result in causing pollution to the source of drinking water. Materials used in providing shelter and accommodation also affect the environment.[38] Epidemiology teaches that the environment plays a considerable role in disease outbreaks. The environment forms the basis for public health and epidemiology. Public health is closely connected to environmental health. Several causes of mortality and morbidity are linked to the environment, namely cardiovascular, respiratory, parasitic, and infectious diseases. Others are cancer, motor accidents, mental disorders, etc.[39] Such life dilemmas call for various ethical considerations. For the environment to properly serve man's needs, it is crucial to care for the environment. Additional issues to consider are climate change factors, pollution, and justice done to everyone.

Climate Change Factors

Climate change and its implications might seem natural, but there is so much human involvement. Since human beings are a huge factor in climate change and its consequences, there should be justice. A recent international study has revealed four critical areas in which climate change generates adverse environmental effects. They comprise excessive temperatures, carbon dioxide concentrations, rise in sea levels, and weather-related events.[40] These factors are undoubtedly health threats and should be deemed public health issues. More practically, the factors translate to cardiovascular illnesses, vector-borne diseases, mental health, greenhouse emissions, and allergens.[41] Any aspect caused by humans should be probed, and justice should be served because human life is at stake in such cases.

Pollution and Justice

Pollution of the environment adversely affects nature and its beauty. Water, air, and food vital for life-sustaining turn toxic due to uncontrolled pollution levels. When such a situation happens, infants are the most vulnerable.[42] Several environmental movements have recently emerged to use legal actions to seek justice in society. Such activities need the strong support of the health community. With such interventions, companies and even governments can face legal consequences. In addition, courts can impose injunctions to bring imminent harm to a halt. Also, authorities are called to review and withdraw permits when judged necessary to do so. Public health experts testify in courts on the implications of some toxin activities in society. All should come on board to realize that environmental rights are a human rights issue.[43] It is the role of the courts of competent jurisdiction to ensure justice is served to all.

The Concept of Egalitarianism

Egalitarianism was initially applied to the belief that there should be human equality. In its current form, egalitarianism applies to a philosophy advocating that there should be no inequalities in people's social life.[44] Egalitarianism perfectly applies to public health. Public health deals with the masses, the entire population, and even the global level. Inequality in the health system is likely to result in an epidemic, considering highly infectious diseases which are airborne. Typical examples are Tuberculosis and COVID-19. Upon various studies on the COVID-19 pandemic, it has become well-known that there are three modes of transmission. Airborne is one of the routes whereby a person with the virus emits aerosol within the range. Other transmission routes are direct and close person-to-person contact through fluids and hand contamination of the infected surfaces. Based on such intelligence that authorities like the WHO (World Health Organization), CDC (Center for Disease Control) in the US, and HAS (Haute Autorit'e de Sant'e) in France came out with the recommendations like social distancing, frequent handwashing,[45] and sanitizing. Justice demands that all people globally cooperate with the authorities in combating a pandemic by responding positively to health recommendations.

The infrastructural disparity between rich and poor communities has dire implications for public health. Arguably, such inequalities have consequences for the public health system. The inadequate infrastructure, lack of maintenance, and flawed rehabilitation system are recipes for problems in public health. In effect, urban planning should always have public health in mind. A moral lesson implies that building infrastructure should consider general health advantages before economic gain.[46]

Poverty in Society

The first among the United Nations (UN) Sustainable Development Goals (SDGs) is poverty eradication. Being the first on the list shows how acute poverty is among people worldwide. According to the UN description, anybody living on US$1.25 or less a day is poor.[47] Fighting poverty in the world is a just course. Poverty is connected to various causal factors, some political, social, cultural, geographical, economic, and the rest. Justice is not served in most cases and areas, so poverty still prevails. Research in the US has shown that deprivation can shorten a person's life expectancy between 10 and 15 years. This supports the assertion that poverty affects an individual's health at every stage of life. Poverty becomes a threat to public health, which calls for the intervention of higher authorities.[48]

Public health cannot achieve success without a drastic reduction in poverty. This implies that public health should frantically seek to address the issue of poverty, which is a health threat. Public health should work alongside social and economic forces. Well-trained healthcare professionals should be advocates for practices and policies involving intersectoral players.[49] The UN has identified that the SDGs cannot be achieved unless health is at the heart of all issues. Economic, social, and environmental dimensions have direct impacts on health. Health should be the central element of the SDGs. All forms of diseases (communicable as well as non-communicable) should be dealt with to ensure the SDGs' realization.[50]

Government Interventions

The crucial role governments could play in egalitarianism is to be sensitively responsible in focusing on the opportunity to ensure everyone has access to quality healthcare. It is incumbent on every government to create the opportunity for people to have an equal life expectancy, but not to deprive some people.[51] The people and the government have a moral duty to any action or inaction on public health issues. In all situations, justice must prevail for the betterment of all people. Public health ethics should be viewed on consequentialist, deontological, and virtual-based paradigms.[52] Consequentialism judges any action based on its outcome. Government should take just and moral steps. Deontology ensures that people have the right to make ethical choices. The virtue-based paradigm judges an individual's character to be morally right.[53] The government's role in ensuring ethical frameworks is paramount.

On the issue of achieving the good of the public, government intervention cannot be overemphasized. Making sure people can easily access resources necessary for healthy living and considering health as a necessity and not an item to purchase is a vital role of the government[54]. The government's responsibility is to ensure social justice for all people regarding health improvement and disease prevention. The government does this by engaging in environmental, behavioral, and social factors threatening life and health.

Conclusion

The theme of justice is so prevalent in ethics. Bioethics is no exception and is much needed to ensure the equitable distribution of health facilities. Public health deals with the whole population and calls for collective efforts to achieve its mandate of preventing harm. Individuals' role in bringing about public health is also justice. The issue of the environment demands collective action from individuals, groups, institutions, and national and international to bring about success. When the environment is not conducive, people's health will be adversely impacted. Egalitarianism calls for a fair and equitable distribution of resources to ensure access to healthcare is a priority to all people irrespective of tribe, race, income, or geographical connections.

Section 3: Pandemic Discourse

Introduction

The chapter's final section focuses on applying critical issues discussed earlier to a pandemic situation. The main issues involve public health ethics, the environment, the principle of justice, governments and politics, and religion. This application explores how collective effort, a tenet of public health, could actively support or hinder mitigation efforts during a pandemic. To reiterate, public health deals with three key aspects, namely, general population, preventive nature, and collective effort,[55] which call for the collaboration of people in society. By definition, a pandemic affects many people simultaneously in a community or even the entire global society.[56]

Collective effort is much needed to curtail a pandemic; unfortunately, it could impede fighting it. The entire block of society also has various religious, economic, political, and sociocultural subgroups. The complex nature of any given society cannot be underestimated. People constituting a society may have their values firmly in both formal and informal settings.[57] How ethical such groups should support public health and contribute to curtailing any form of a pandemic is critically essential. It is characteristic of a group to have leaders or some form of authority that members listen to and take orders. What has become the pandemic of our time is Covid-19. Covid-19 is the short form of a new coronavirus disease detected in 2019. Covid-19 was first seen in the Chinese city of Wuhan in the province of Hubei. Studies have found that the local seafood market is the epicenter of the virus. When one is infected, the virus mainly attacks the respiratory tract, but other vital organs, like the lungs, can also be affected. Covid-19 has a high rate of infection; hence in no time, it became a global pandemic.[58]

The subsequent discussion explores the religious, economic, political (governmental), and environmental influences on the fight against the pandemic. These aspects could positively and negatively impact the need to combat the plague. Ethical consideration for the principle of justice in all aspects is also crucial.

Religious Influence on a Pandemic

From time immemorial, religion has been a significant force that binds people together, creating a religious body. Through religion, people have a form of belief that translates into their moral and ethical conduct. In a broader sense, religion is described as a system based on spirituality, faith, and worship which connects the human person as having inner life and soul to engage in different types of practices.[59] Various religions have other explanations as to why there are calamities, misfortunes, and diseases. Some attribute them to witchcraft, demons, or evil spirits in general. In polytheistic religions, specific deities are believed to be the forces behind epidemics. Similarly, other gods are believed to avert calamities. In another development, natural disasters like floods, earthquakes, and drought can be attributed to the anger of a superpower.[60] That is, religion creates some form of value system in people.

Also, a sense of devotion fosters discipline among the people of a particular religion. Another feature that religion has is to make people prosocial.[61] To some people, religion and culture are intrinsically bound. With such a background, religion should ideally forge for the common good in society. However, the reality is different. According to some research, what accounts for such deviation is mainly due to the contemporary sociocultural setting.[62] Political and judicial institutions contribute to such variation. These institutions often derail the trust religious people have in them, thereby not being cooperative in promoting the excellent everyday society requires.[63] How religious organizations play a role in a pandemic is further discussed by looking at the support they offer and how they become a stumbling block.

Positive Influence of Religion

It is an undeniable fact that the religious bond is so strong. Leveraging such characteristics of religious bodies makes a vast difference in dealing with a pandemic. Various studies can prove a positive correlation between mental health and religiosity. In addition, inculcating religiosity into healthcare plans yields higher positive results.[64] At the pandemic's peak, much research was conducted from religious perspectives to seek views. A study in the US realized that out of the 1,800 practicing Muslims, about 84% believed the pandemic was Allah testing the people on earth.[65] Such perception has been consistent with what has previously been held by some Hindu, Muslim, and Christian apologists in an attempt to explain the causes of the tsunami, floods, and disasters that have caused the death of many people.[66] With this background, it is assumed that religious people would always think seriously about a pandemic, readily offering support to help curtail the situation.

Various studies can also confirm that religion's role in times of pandemic is enormous. Firstly, the mortality rate is reduced among faith-connected people even when one is infected. Secondly, people's immune system improves with a strong religious connection. Again, the stress level is minimized for religious people during a pandemic.[67] Most Christian denominations complied with the public health protocols to fight the Covid-19 pandemic's spread. Many religious and worship centers saw the need to close as required by the CDC to avoid the vast spread of the virus. At the same time, churches were reaching out to people stuck in their homes with messages of encouragement and prayers through online services to provide for their spiritual needs. Similarly, food and other essentials were also sent to people in their homes to continue the works of charity and supply their material needs.[68]

Negative Influence of Religion on a Pandemic

In contemporary times, it has been noted that the significant adverse influence of religion on a pandemic is based on the premise that people have their religious freedom.[69] This typically emerged in the US during the Covid-19 pandemic. Some sects of Christians, mostly the evangelicals, were defiant to restrictions because, to some extent, they violated freedom of religion. In Greece, it has been established that some priests insisted and gave the consecrated wine (known among believers as the Blood of Jesus) to worshippers with the same spoon not washed between uses.[70] Besides championing religious freedom during the pandemic, some religious people vehemently opposed the vaccination program for various reasons. One of the very reasons they do not trust science is that they believe in God. A movement called 'anti-vax' claims mandatory vaccination is also against a person's healthcare principles and freedom to choose.[71]

Some religious leaders rejected measures prescribed by public health authorities. A case in point is where a Christian religious leader in the US who was initially infected and recovered could simultaneously describe lockdowns (affecting the in-person church services) as putting an innocent person in prison.[72] People were strongly advised to observe protocols such as physical/social distancing to avoid the rapid spread of infection. However, churches were still gathering, and many members sang and shouted in enclosed and clustered areas. A typical example is a mega church in Korea at the pandemic's peak, where the congregants opined that the government was using public health to restrict religion.[73] It is equally the view of some believers, as a case in point with a 60-year-old man, that Christians are special people and cannot be infected or even fall seriously sick to the pandemic. They blame the politicians and doctors for not giving the necessary medication.[74]

Economic Influence on a Pandemic

Research has shown a strong connection between economics, mental health, and public health during a pandemic.[75] Whenever there is a pandemic, economic activities are generally affected. Measures like the lockdowns imply many businesses cannot operate. Businesses had no option but to lay off employees. People are virtually home, and productivity generally is severely affected. The global supply chain is heavily disrupted. These are some economic impacts the Covid-19 pandemic brought to the global economy. A prolonged economic downturn due to inactivity equally impacted the health of many people.[76]

Economic Support to a Pandemic

During the Covid-19 pandemic, governments in most countries offered economic relief to the citizenry through stimulus packages to ensure liquidity circulation. Reports indicate that 13.3% of the US GDP was devoted. Japan provided as much as 21% of its GDP. Germany and Australia offered more than 10% of their GDP as a stimulus package to low-income workers. In countries like India, Thailand, China, Singapore, and the Philippines, various forms of support were explicitly given to migrants.[77] Such economic interventions are egalitarian and an act of justice to offer adequate support to the needy. The positive impact is to help people cooperate and respond to the pandemic mitigation efforts.

Another economic benefit derived from the pandemic, which also helped in the mitigation effort, is the practice of remote working. People had the privilege of working from their homes, saving time and the cost of travel. This practice has offered flexibility to workers as it has been in demand for years.[78] Remote working is practically only possible in some forms of work. Remote working became the best alternative to retain workers and ensure productivity continues during the Covid-19 pandemic. Remote working helps prevent the virus's spread as workers stay home with families rather than mingle with others and families. This, of course, has economic implications for workers, employers, and the nation. Trust, respect, and fairness are required of both employees and employers in the practice of remote working.[79] In this situation, the principle of justice is being applied. When both parties are satisfied, public health also improves.

Economic Challenges in the Pandemic

Economic recession is being experienced across the globe as a residual of the Covid-19 pandemic. Inflation and rising prices of goods, including food, have become the norm. In the US, the CPI (Consumer Price Index) rose to a record high in 40 years at the rate of 9.1% as of June 2022, a 12-month rate of inflation.[80] This is mainly attributable to the recession from the Covid-19 pandemic. Four critical areas which emanated from the pandemic responsible for the inflation comprise energy and food prices, as in the case of Europe. Secondly, the prices of durable goods, as in the US, cause inflation. Thirdly, fiscal stimulus and the labor market also contribute to inflation. Finally, industrialization is another contributory factor.[81] The tourism sector, a significant source of income for many nations, has also suffered tremendous losses due to the Covid-19 pandemic. Other global challenges still face tourism, even in the post-Covid-19 recovery era.[82] The economic downturn resulting from the Covid-19 pandemic has been a severe one. In the US, the most hit areas are agriculture, food services, and retailing. Medicaid, the health insurance, costs can lead to eligibility reduction, minimized cover benefits, and low provider rates.[83]

Government's Impact

The principle of Justice ensures an equitable and fair share of the nation's resources. No other body could better do this besides the government of a country. The government needs to ensure that all people can have equal healthcare provisions.[84] On this note, politics should not have any place in public health. Should that happen, the non-political people and those in opposition lose complete trust and get confused about any regulations the government might put in place to combat a pandemic.[85]

Positive Impact

In times of a pandemic, the government (the political authority) is expected, first and foremost, to deliver precise and consistent information to the citizenry. Government should communicate clearly to the public.[86] Information about the pandemic and measures put in place to curtail it should all be well given to the people. The government should take PHI seriously to meet this requirement. Another key area the government must play a role in is the supply of resources, especially for those affected by the pandemic who need to quarantine. It is equally essential for a government to alleviate the people from anxiety, fear, loneliness, and boredom associated with the pandemic.[87] During the Covid-19 pandemic, governments around the globe offered many intervention programs to help the citizenry. The government of Ghana provided free water and electricity across the country and sometimes food when people were under strict lockdown in some key cities. It was through the interventions of the World Bank with the support of $100 million that the government of Ghana could do that for the citizenry.[88]

Negative Impact

Public health is argued to have a political face and can be used for political gains. When public health becomes political and used as government machinery, ethics will be compromised and undermined.[89] Politics, both national and international, can be blamed for health insecurity in the world. Health insecurity was exposed at the height of the pandemic when countries, including the developed ones, struggled to get an adequate supply of logistics: nurses, doctors, vaccines, testing kits, advanced ventilators, elemental oxygen, and hospitals.[90] Governments around the world need to do more to improve health security. In all these, the world needs to adopt a workable strategy that will take public health seriously. If, after 100 years of the global Spanish Flu pandemic, the world is still unable to timely combat Covid-19, then the WHO and political leaders have a long way to go.[91]

The Environment and Pandemic

The issues of the environment are so crucial to sustaining life. Climate change, associated with global warming, has contributed to increased viruses and parasites affecting humans.[92] Three reasons have been assigned for this assertion. Firstly, warm temperatures facilitate the rapid incubation of viruses. Secondly, agents like mosquitoes become hungrier in hot temperatures, infecting many people with malaria parasites. Finally, mosquitoes expand exponentially in warm climates.[93] Most infectious diseases which result in epidemics or pandemics are transmitted by some living organisms called vectors. Typically, these vectors are stinging or biting insects that transmit pathogens to people. Mosquitoes carry malaria parasites. Likewise, ticks bring Lyme disease by transmitting the bacteria.[94] There are three types of flu, A, B, and C, which according to scientists, infect human beings. These types of flu also have the potential to mutate and cause outbreaks as epidemics. Types A and C infect both humans and animals, including birds. Type B only infects humans.[95] It is essential to know that other living organisms which humans share in the environment can easily transmit viruses to humans.

Environment and Mitigating Efforts

One positive impact the Covid-19 pandemic has brought into the world is the reduction of greenhouse gas emissions. This reduction resulted from the protocols which sought to curb the pandemic, namely, the lockdown of businesses, ban on travel, and stay-home orders, among others. On the other hand, such measures had ripple effects on the global economy with consequences such as job and income losses.[96] The secondary and tertiary sectors were heavily impacted when the lockdowns occurred. The manufacturing industries, education, food services, finance, and others caused a reduction in movements, implying less fuel consumption for transportation. Hence, the environment is spared some pollution that adversely harms the ecosystem.[97]

As airplanes were grounded at the height of the pandemic, there was a drastic reduction in transportation, including international travel. Factories shut down their operations as well. It has been well noted that the decline in greenhouse gas emissions during the pandemic is so significant at a record high. The impact has both social and environmental benefits. Thousands of human lives are saved with such a tremendous reduction in emissions. Such reduction is equally essential for the environment.[98] It is, therefore, imperative to ensure human knowledge, awareness, and environmental protection, which are vital to human survival and existence. Humanity should take serious responsibility to act expeditiously on ecological degradation.[99]

Environment Facilitating Pandemic

The Covid-19 pandemic has adversely affected social, economic, and environmental activities. The systemic economic inequality is likely to deepen. The disparity between the poor and the rich will continue to widen. The flawed healthcare system in low-income countries is on the verge of declining further. With Covid-19, human vulnerability to zoonotic diseases is revealed.[100] Despite advancements in technology, science, and medicine, Covid-19 has exposed the world's vulnerability in health security.

The UN Human Development Report of 1994 discussed seven security areas of priority. If not all, a significant number of these security concerns have been exposed. They comprise "economic security, health security, political security, community security, personal security, food security, and environmental security."[101] All these security concerns eventually end up harming the environment. Only a quarter of the world's population is secure economically. Covid-19 has instead worsened the situation. Reports indicate that there is enough food to supply the world's population. However, poor distribution systems, wars, and inadequate income to purchase food are the reasons for food insecurity worldwide.[102] Justice demands equal distribution, so governments and non-governmental bodies should play a pivotal role in ensuring food security and, at the same time, wastage that goes to destroy the environment. In his encyclical, *Laudato Si*, Pope Francis emphatically calls on everyone to come on board to protect the earth (environment), for it is "our common home."[103]

Conclusion

To conclude this chapter, it is apt to reiterate some of the critical lessons the Covid-19 pandemic has shown the world. One important lesson the pandemic has emphasized is to do with global interconnectedness. The virus was first reported in China, and in no time, cases were being reported in different countries across the globe. It is also a great lesson to learn that public health calls for a concerted effort because it should concern everyone. Prevention is the focus of public health; however, should there be any outbreak, public health offers swift measures to deal with the situation. The principle of justice, alongside egalitarianism and distributive justice, is key to promoting and ensuring public health. Religious, sociocultural, political, and economic groups should facilitate the collective efforts needed to promote public health and curtail pandemics. The need to care for the environment should be everyone's responsibility, a call for public beneficence. The effort to achieve the SDGs must be an all-hands-on-deck approach. Human life unequivocally depends on the environment; all should act now and prudently.

Notes

[1] *United Nations*, Department of Economic and Social Affairs: Sustainable Development. Accessed, September 15, 2022. https://sdgs.un.org/goals/ goal3
[2] Linda Farber Post and Jeffrey Blustein. *Handbook for Health Care Ethics Committees,* Third edition (Baltimore: John Hopkins University Press, 2021), 216.
[3] Tom L. Beauchamp and James F. Childress, *Principles of Biomedical Ethics* (New York: Oxford University Press, Incorporated, 2001), 341.
[4] Angus Dawson, "Resetting the Parameters: Public health as the foundation for public health ethics," in *Public Health Ethics: Key Concepts and Issues in Policy and Practice,* (New York: Cambridge University Press, 2012), 3.
[5] Anthony Kessel and Carolyn Stephens, "Environment, Ethics and Public Health," in *Public Health Ethics: Key Concepts and Issues in Policy and Practice*, ed. Angus Dawson (New York: Cambridge University Press, 2012), 154.
[6] Dawson, 3-6.
[7] Beauchamp and Childress. 12.
[8] Thomas R. McCormick, *Principles of Bioethics*, School of Medicine, University of Washington, 2013,
https://depts.washington.edu/bhdept/ethics-medicine/bioethics-topics/articles/principles-bioethics
[9] Mary Grace Flaherty, *Promoting Individual and Community Health at the Library* (Chicago: American Library Association, 2018), 1-2.
http://ebookcentral.proquest.com/lib/duquesne-ebooks/detail.action?docID=5516322.
[10] Ibid., 6-8.
[11] WHO, *2021 Annual Global Public Health Intelligence Report*, under the Creative Commons AttributionNonCommercial-ShareAlike 3.0 IGO licence (CC BY-NC-SA 3.0 IGO; https://creativecommons.org/licenses /by-nc-sa/3.0/igo
[12] Kessel and Stephens, 157.
[13] United Nations, "Transforming our world: the 2030 Agenda for Sustainable Development," under *Department of Economic and Social Affairs*. Retrieved, September 18, 2022. https://sdgs.un.org/2030agenda
[14] Alexis Lieberman and Joav Merrick, *Poverty and Children: A Public Health Concern* (Nova Science Publishers, Incorporated, 2009), 1-2.

http://ebookcentral.proquest.com/lib/duquesne-ebooks/detail.action?docID=3020203
[15] Beauchamp, and Childress, 340.
[16] D. Darius Farhud, "Impact of Lifestyle on Health," *Iranian Journal of Public Health, 44*(2015; 11), 1442-1444. https://doi.org/https://www.ncbi .nlm.nih.gov/pmc/articles/PMC4703222/
[17] Kessel and Stephens, 155.
[18] Ibid., 156.
[19] Emmet Power, Niamh McCarthy, Ina Kelly, et al., "Climate Change and Mental Health: Time for Action and Advocacy." *Irish Journal of Psychological Medicine*, (2022) 1–3. doi:10.1017/ipm.2021.70.
[20] F. Vergunst, & H. L. Berry, "Climate Change and Children's Mental Health: A Developmental Perspective," *Clinical Psychological Science (*2021*)*. https://doi.org/10.1177/21677026211040787
[21] R. Zhang, X Tang, J. Liu, *et al.* "From concept to action: a united, holistic and One Health approach to responding to the climate change crisis," in *Infect Dis Poverty* 11, 17 (2022). https://doi.org/10.1186/s40249 -022-00941-9
[22] Leonardo Trasande, "Children's Environmental Health: Mercury and Lead," in *Environmental Policy and Public Health: Air Pollution, Global Climate Change, and Wilderness*, ed. William N. Rom, (San Francisco: Jossey-Bass, 2012), 102-103.
[23] Javier Cifuentes-Faura, "Is Environmental Pollution Associated with an Increased Number of COVID-19 Cases in Europe?" *International Journal of Environmental Research and Public Health* 19, 2 (2022): 703. https://doi.org/10.3390/ijerph19020703
[24] Richard Ashcroft, "Smoking, health and ethics," in *Public Health Ethics: Key Concepts and Issues in Policy and Practice*, ed. Angus Dawson, (New York: Cambridge University Press, 2012), 85.
[25] Ibid., 88-89.
[26] Beauchamp and Childress, 12.
[27] Samuel Knapp, Michael Gottlieb, and Mitchell M. Handelsman, "Talking to Patients Who Fail to Follow Public Health Measures during the COVID-19 Pandemic: A Principle-Based Approach," *Practice Innovations* 6, 3 (2021): 141.
[28] Ibid., 141.
[29] Ibid., 142-43.
[30] Dawson, 3.
[31] M. Gabriela Wilson, J. Marion Ball, Peter Szczesny, et al, "Health Intelligence Atlas: A Core Tool for Public Health Intelligence," *Applied Clinical Informatics,* 2021, 12 (4): 944-53. Doi:10.1055/s-0041-1735973
[32] WHO, *2021 Annual Global Public Health Intelligence Report*, under the Creative Commons AttributionNonCommercial-ShareAlike 3.0 IGO licence (CC BY-NC-SA 3.0 IGO; https://creativecommons.org/licenses /by-nc-sa/3.0/igo
[33] Dawson, 3.
[34] Julie Sin, 'Public Health: Three Key Domains,' *Commissioning and a Population Approach to Health Services Decision-Making* (Oxford, 2020; online edn, Oxford

Academic, 21 Jan. 2021), https://doi-org.authenticate.library.duq. edu/10.1093/oso/9780198840732.003.0005, accessed 10 Oct. 2022. Pg52

[35] Carly R. Levy, Lynelle M. Philips, Carolyn J.Murray, et al. "Addressing Gaps in Public Health Education to Advance Environmental Justice: Time for Action," *American Journal of Public Health*, 112 (1), (2022): 70.

[36] Wilson, Ball, Szczensy, et al, 944-53.

[37] Ibid., 944-53.

[38] Kathryn H. Jacobsen, *Introduction to Global Health*, Vol. Third edition (Burlington, MA: Jones & Bartlett Learning, 2019), 65.

[39] Levy, Philips, Murray, et al., 71-72.

[40] Walter Leal Filho, Linda Ternova, Muhammad Muddassir, et al., "An Analysis of Climate Change and Health Hazards: Results from an International Study," *International Journal of Climate Change Strategies and Management* 14, no.4 (July 15, 2022): 376.

[41] Ibid., 378-79.

[42] Trasande, 102.

[43] David W. Patterson, Richard Harvey, Vlatka Matkovic, et al., "Post COP26: Legal Action Now Part of Public Health's Environment and Climate Change Toolbox," *European Journal of Public Health* 32, no.4 (2022): 519

[44] *Merriam-Webster.com Dictionary*, s.v. "egalitarianism," accessed October 13, 2022, https://www.merriam-webster.com/dictionary/ egalitarianism

[45] Bertrand R. Rowe, Andre Canosa, Amina Meslem, and Frantz Rowe, "Increased Airborne Transmission of COVID-19 with New Variants, Implications for Health Policies." *Building and Environment* 219 (July 1, 2022), 109131-32.

[46] Marcus D. Hendricks, "Leveraging Critical Infrastructure Within an Environmental Justice Framework for Public Health Prevention," *American Journal of Public Health*, 112(7), (2022): 973-74.

[47] United Nations, "Transforming our world: the 2030 Agenda for Sustainable Development," under *Department of Economic and Social Affairs*. Retrieved, September 18, 2022. https://sdgs.un.org/2030agenda

[48] James H Prince, Jagdish Khubchandani, and Fern J Webb, "Poverty and Health Disparities: What Can Public Health Professionals Do?" *Health Promotion Practice* 19, no 2 (March 2018): 171-72.

[49] Ibid., 174.

[50] Amiya Kumar Sahoo, Hari Charan Behera, and Ajit Kumar Behura, "Philosophy of Sustainable Development: Understanding Public Health," *Environment, Development & Sustainability* 24, no. 10 (October 2022): 12250.

[51] Christoph Jedan, "Proportionate Universalism and Public Health: An analysis of WRR Policy Brief From Disparity to Potential," *NTT* 75(1), (2021): 83.

[52] Sahoo Behera, and Behura, 12254.

[53] Hugh LaFollette, *Ethics in Practice: An Anthology*. Vol. Fourth Edition. (Hoboken: Wiley-Blackwell, 2014), 25-26.

[54] Sandro Galea, *Well: What We Need to Talk About When We Talk About Health*, (Oxford: Oxford University Press, 2019), 168.

[55] Dawson, 3

[56] *Merriam-Webster.com Dictionary*, "pandemic," accessed October 7, 2022,

https://www.merriam-webster.com/dictionary/pandemic
[57] Giuliana Spadaro, Caroline Graf, Shuxian Jin, Sakura Arai, et al. "Cross-Cultural Variation in Cooperation: A Meta-Analysis." In *Journal of Personality and Social Psychology* 123, 5 (2022): 1025.
[58] Anushka Bhargava, and Praveen Dahiya, "COVID-19 Pandemic: Assessment of Current Strategies and Socio-Economic Impact," *Journal of Health Management* 24, 3 (2022): 466.
[59] Dacre Knight, Daniel V Dudenkov, and William P Cheshire Jr. "Religion in the US during the Time of a Pandemic: A Medical Perspective," *Journal of Religion and Health* 60,5 (2021): 3178.
[60] Graeme Lang, "Religious Responses to a Pandemic: Explanation, Compliance, and Defiance," *Journal for the Study of Religion, Nature & Culture* 16, 1 (2022): 24-25.
[61] Spadaro, Graf, Jin, Arai, et al., 1027.
[62] Ibid.,1028.
[63] Ibid., 1028-29.
[64] Tommy DeRossett, Donna J. LaVoie, and Destiny Brooks, "Religious Coping amidst a Pandemic: Impact on COVID-19-Related Anxiety, *Journal of Religion and Health* 60, 5 (2021): 3162.
[65] Lang, 26.
[66] Ibid., 26.
[67] Knight, Dudenkov, and Cheshire Jr. "Religion in the US during the Time of a Pandemic: A Medical Perspective," *Journal of Religion and Health* 60,5 (2021): 3183-84.
[68] Ibid., 3180.
[69] Lang, 31.
[70] Ibid.
[71] Ibid.
[72] Knight, Dudenkov, and Cheshire Jr., 3180.
[73] Lang, 30.
[74] Knight, Dudenkov, and Cheshire Jr., 3181.
[75] Ibid., 3189.
[76] Helia Mojtabavi, Nasirudin Javidi, AnneFrederique Naviaux, et al., "Exploration of the Epidemiological and Emotional Impact of Quarantine and Isolation During the COVID-19 Pandemic," *Covid-19 and Its Global Economic Impact*, Vol. 1318 (Cham: Springer International Publishing, 2021), 698.
[77] Bhargava and Dahiya, 473.
[78] Olatunji David Adekoya, Toyin Ajibade Adisa, and Opeoluwa Aiyenitaju, "Going Forward: Remote Working in the Post-COVID-19 Era," *Employee Relations* 44, 6 (2022): 1417.
[79] Ibid., 1423.
[80] Bart Hobijn, Russell A. Miles, James Royal, and Jing Zhang. "What is Driving U.S. Inflation amid a Global Inflation Surge?" *Chicago Fed Letter* 470 (2022): 1.
[81] Ibid., 1,5.

[82] Czeslaw Adamiak, and Barbara Szyda, "Combining Conventional Statistics and Bid Data to Map Global Tourism Destinations Before COVID-19," *Journal of Travel Research* 61, 8 (2022): 1848

[83] Nicole Huberfeld, Sarah H. Gordon, and David K. Jones. "Federalism Complicates the Response to the COVID-19 Health and Economic Crisis: What Can Be Done?" *Journal of Health Politics, Policy & Law* 45, 6 (2020): 956-57. Bhargava and Dahiya, 474.

[84] Jedan83.

[85] Knight, Dudenkov, and Cheshire Jr., 3181.

[86] Mojtabavi, Javidi, Naviaux, et al., 698.

[87] Ibid.

[88] *World Bank*. "World Bank Group Supports Ghana's Covid-19 Response," Press Release. Assessed, November 12, 2022.

[89] Frias, Lopez, Francisco Javier, and Donald B. Thompson. "Solidarity and Public Health." *Medicine, Health Care and Philosophy: A European Journal*, 6,9 (2022), 376.

[90] Michael Share, "COVID-19: The Pandemic That Swept the World: Its Political and Economic Effects on Human Security." *Intercultural Communication Studies* 30, 1 (2021): 64

[91] Ibid.

[92] Connie Goldsmith, *Pandemic: How Climate, the Environment, and Superbugs Increase the Risk*, (Minneapolis, MN: Twenty-First Century Books, 2018), 40.

[93] Ibid.

[94] Ibid., 39.

[95] Ibid., 100.

[96] Manfred Lenzen, Mengyu Li, Arunima Malik, Francesco Pomponi, et al., "Global Socio-Economic Losses and Environmental Gains from the Coronavirus Pandemic," *PloS One* 15, 7 (2020): 5.

[97] Bhargava and Dahiya, 474.

[98] Lenzen, Li, Malik Pomponi, et al., 8.

[99] Ivan Efreaim A. Gozum, Jr., Abelardo E. Garcial, and John Lu Allan M. Nucum, "Laudato Si and the Role of the Church in Promoting Environmental Awareness Toward a Better Health-Related Quality of Life," *Journal of Religion and Health* (2022): 6.

[100] Ibid., 8-9.

[101] Share, 58-59.

[102] Ibid., 60.

[103] Gozum, Garcial, and Nucum, 7.

Literature

Adamiak, Czeslaw, and Barbara Szyda. 2022. "Combining Conventional Statistics and Bid Data to Map Global Tourism Destinations Before COVID-19." *Journal of Travel Research* 61, no. 8: 1848-71. doi: 10.1177/00472875211051418

Adekoya, Olatunji David, Toyin Ajibade Adisa, and Opeoluwa Aiyenitaju. 2022. "Going Forward: Remote Working in the Post-COVID-19 Era." Employee Relations 44, no. 6: 1410-27. doi: 10.1108/ER-04-2021-0161.

Ashcroft, Richard, 2012. "Smoking, health and ethics," in Angus Dawson *Public Health Ethics: Key Concepts and Issues in Policy and Practice*. New York: Cambridge University Press.

Beauchamp, Tom L., and James F. Childress. 2001. *Principles of Biomedical Ethics*. New York: Oxford University Press, Incorporated.

Bhargava, Anushka, and Praveen Dahiya. 2022. "COVID-19 Pandemic: Assessment of Current Strategies and Socio-Economic Impact." *Journal of Health Management* 24, no. 3: 466-77. doi: 10.1177/09720634221109295.

Cifuentes-Faura, Javier, 2022. "Is Environmental Pollution Associated with an Increased Number of COVID-19 Cases in Europe?" *International Journal of Environmental Research and Public Health* 19, no. 2: 703. https://doi.org/10.3390/ijerph19020703

Dawson, Angus. 2012. "Resetting the Parameters: Public health as the foundation for public health ethics," in *Public Health Ethics: Key Concepts and Issues in Policy and Practice.* New York: Cambridge University Press.

DeRossett, Tommy, Donna J. LaVoie, and Destiny Brooks. 2021. "Religious Coping amidst a Pandemic: Impact on COVID-19-Related Anxiety. In *Journal of Religion and Health* 60, no. 5: 3161-76.

Farhud, D. Darius, 2015. "Impact of Lifestyle on Health." *Iranian Journal of Public Health*, *44*(2015; 11), 1442-1444. https://doi.org/https://www.ncbi.nlm.nih.gov/pmc/articles/PMC4703222/

Flaherty, Mary Grace. 2018. *Promoting Individual and Community Health at the Library*, Chicago: American Library Association.

Frias, Lopez, Francisco Javier, and Donald B. Thompson. 2022. "Solidarity and Public Health." *Medicine, Health Care, and Philosophy: A European Journal*, 6, no. 9: 1-12. doi: 10.1007/s11019-022-10084-1

Galea, Sandro, 2019. *Well: What We Need to Talk About When We Talk About Health*, (Oxford: Oxford University Press).

Goldsmith, Connie. 2018. *Pandemic: How Climate, the Environment, and Superbugs Increase the Risk*. Minneapolis, MN: Twenty-First Century Books.

Gozum, Ivan Efreaim A., Jr., Abelardo E. Garcial, and John Lu Allan M. Nucum. 2022. "Laudato Si and the Role of the Church in Promoting

Environmental Awareness Toward a Better Health-Related Quality of Life." *Journal of Religion and Health*: 1–15. doi. 10.1007/s10943-01631-4.

Hendricks, Marcus D. 2022. "Leveraging Critical Infrastructure Within an Environmental Justice Framework for Public Health Prevention," *American Journal of Public Health*, 112(7): 972-74.

Hobijn, Bart, Russell A. Miles, James Royal, and Jing Zhang. 2022. "What is Driving U.S. Inflation amid a Global Inflation Surge?" *Chicago Fed Letter* 470: 1-6. doi:10.21033/cfl-2022-470.

Huberfeld, Nicole, Sarah H. Gordon, and David K. Jones. 2020. "Federalism Complicates the Response to the COVID-19 Health and Economic Crisis: What Can Be Done?" *Journal of Health Politics, Policy & Law* 45, no. 6: 951-65. doi: 10.1215/036168-8641493.

Jacobsen, Kathryn H. 2019. *Introduction to Global Health*, Vol. Third edition. Burlington, MA: Jones & Bartlett Learning.

Jedan, Christoph, 2021. "Proportionate Universalism and Public Health: An analysis of WRR Policy Brief From Disparity to Potential," *NTT* 75, no. 1: 73-93.

Kessel, Anthony, and Stephens, Carolyn, 2012. "Environment, Ethics and Public Health." In Angus Dawson *Public Health Ethics: Key Concepts and Issues in Policy and Practice*. New York: Cambridge University Press.

Knapp, Samuel, Michael Gottlieb, and Mitchell M. Handelsman. 2021. "Talking to Patients Who Fail to Follow Public Health Measures during the COVID-19 Pandemic: A Principle-Based Approach." *Practice Innovations* 6, no. 3: 139-48. doi: 10.1037/pin0000144.

Knight, Dacre, Daniel V Dudenkov, and William P Cheshire Jr. 2021. "Religion in the US during the Time of a Pandemic: A Medical Perspective." *Journal of Religion and Health* 60, no. 5: 3177-92. doi:10.1007/s10943-021-01366-8.

LaFollette, Hugh. 2014. *Ethics in Practice: An Anthology*. Vol. Fourth Edition. Hoboken: Wiley-Blackwell.

Lang, Graeme. 2022. "Religious Responses to a Pandemic: Explanation, Compliance, and Defiance." Journal for the Study of Religion, Nature & Culture 16, no. 1: 23-49. doi:10.1558/jsrnc.19456.

Leal Filho, Walter, Linda Ternova, Muhammad Muddassir, et al. 2022. "An Analysis of Climate Change and Health Hazards: Results from an International Study," *International Journal of Climate Change Strategies and Management* 14, no. 4: 375-98. doi:10.1108/IJCCSM-08-2021-0090.

Lenzen, Manfred, Mengyu Li, Arunima Malik, Francesco Pomponi, et al. 2020. "Global Socio-Economic Losses and Environmental Gains from the Coronavirus Pandemic." *PloS One* 15, no. 7: e025654. doi: 10. 1371/journal.pone.0235654.

Levy, Carly R., Philips, Lynelle M., Murray, Carolyn J., et al. 2022. "Addressing Gaps in Public. Health Education to Advance Environmental Justice: Time for Action," *American Journal of Public Health*, 112, no. 1: 69-74.

Lieberman, Alexis, and Joav Merrick. 2009. *Poverty and Children: A Public Health Concern*, Nova Science Publishers, Incorporated.

McCormick, Thomas R. 2013. *Principles of Bioethics*, School of Medicine, University of Washington. https://depts.washington.edu/bhdept/ethics-medicine/bioethics-topics/articles/principles-bioethics

Merriam-Webster.com Dictionary, N.d. s.v. "egalitarianism," accessed October 13, 2022, https://www.merriam-webster.com/dictionary/egalitarianism

Merriam-Webster.com Dictionary. N. *d. s. v.* "Pandemic," accessed October 7, 2022, https://www.merriam-webster.com/dictionary/pandemic

Mojtabavi, Helia, Nasirudin Javidi, AnneFrederique Naviaux, et al.2021. "Exploration of the Epidemiological and Emotional Impact of Quarantine and Isolation During the COVID-19 Pandemic." *Covid-19 and Its Global Economic Impact*. Vol. 1318. Cham: Springer International Publishing. doi: 10.1007/978-3-030-63761-3_46

Patterson, David W, Richard Harvey, Vlatka Matkovic, et al. 2022. "Post COP26: Legal Action Now Part of Public Health's Environment and Climate Change Toolbox," *European Journal of Public Health* 32, no.4: 519-520.

Post, Linda Farber, and Jeffrey Blustein. 2021. *Handbook for Health Care Ethics Committees*. Third Edition, Baltimore: John Hopkins University Press.

Power, Emmet, Niamh McCarthy, Ina Kelly, et al. 2022. "Climate Change and Mental Health: Time for Action and Advocacy." *Irish Journal of Psychological Medicine*, 1–3. doi:10.1017/ipm.2021.70

Prince, James H, Jagdish Khubchandani, and Fern J Webb. 2018. "Poverty and Health Disparities: What Can Public Health Professionals Do?" *Health Promotion Practice* 19, no. 2: 170-74. doi:10.1177/ 1524839918755143.

Rowe, Bertrand R., Andre Canosa, Amina Meslem, and Frantz Rowe, 2022. "Increased Airborne Transmission of COVID-19 with New

Variants, Implications for Health Policies." *Building and Environment* 219. DOI: 10.1016/j.buildenv.2022.109132.

Sahoo, Amiya Kumar, Hari Charan Behera, and Ajit Kumar Behura. 2022. "Philosophy of Sustainable Development: Understanding Public Health," *Environment, Development & Sustainability* 24, no. 10: 12248-62. doi:10.1007/s10668-021-01945-5.

Share, Michael. 2021. "COVID-19: The Pandemic That Swept the World: Its Political and Economic Effects on Human Security." *Intercultural Communication Studies* 30, no. 1: 49-66.

Sin, Julie. 2021. "Public Health: Three Key Domains." In *Commissioning and a Population Approach to Health Services Decision-Making.* Oxford, 2020; online edn, Oxford Academic. https://doi-org.authenticate.library.duq.edu/10.1093/oso/9780198840732.003.0005,

Spadaro, Giuliana, Caroline Graf, Shuxian Jin, Sakura Arai, et al. 2022. "Cross-Cultural Variation in Cooperation: A Meta-Analysis." In *Journal of Personality and Social Psychology* 123, no. 5: 1024-88. doi:10.1037/pspi0000389.supp

Trasande, Leonardo. 2012. "Children's Environmental Health: Mercury and Lead," in William N. Rom, *Environmental Policy and Public Health: Air Pollution, Global Climate Change, and Wilderness*. San Francisco: Jossey-Bass.

United Nations. N. d. Department of Economic and Social Affairs: Sustainable Development. Retrieved, September 15, 2022. https://sdgs.un.org/goals /goal3

United Nations. N. d. "Transforming our world: the 2030 Agenda for Sustainable Development," under the *Department of Economic and Social Affairs*. Retrieved, September 18, 2022. https://sdgs.un.org/2030 agenda

Vergunst, F., & Berry, H. L. 2021. Climate Change and Children's Mental Health: A Developmental Perspective. *Clinical Psychological Science*. https://doi.org/10.1177/21677026211040787

Wilson, M. Gabriela, Ball, J. Marion, Szczesny, Peter, et al. 2021. "Health Intelligence Atlas: A Core Tool for Public Health Intelligence," *Applied Clinical Informatics,* 12, no. 4: 944-53. Doi:10.1055/s-0041-1735973

WHO. N. d. *2021 Annual Global Public Health Intelligence Report*, under the Creative Commons AttributionNonCommercial-ShareAlike 3.0 IGO licence (CC BY-NC-SA 3.0 IGO; https://creativecommons.org/ licenses/by-nc-sa/3.0/igo

World Bank. 2020. "World Bank Group Supports Ghana's Covid-19 Response," Press Release. Assessed, November 12, 2022. https://www.worldbank .org/en/news/press-release/2020/04/02/world-bank-group-supports-ghanas-covid-19-response

Zhang, R., Tang, X., Liu, J. et al.2022. "*From* concept to action: a united, holistic and One Health approach to responding to the climate change crisis," in *Infect Dis Poverty* 11, no. 17. https://doi.org/ 10.1186/s40249-022-00941-9

CHAPTER EIGHT

THE RESILIENCE OF DOUBLE EFFECT REASONING IN HEALTHCARE

SCOTT N. HELSTAD

Introduction

One of the guiding principles of the Christian life is to do good and to avoid evil, which in clinical ethics is known as the principle of beneficence. The Christian understanding of beneficence is rooted in Christ's proclamation that "in everything, do to others what you would have them do to you."[1] On the surface, this guiding principle seems easy to fulfill; however, the day-to-day implementation of beneficence is often debated, especially when an action has two outcomes, one good and one bad. Fortunately, the Catholic intellectual tradition has a rich theological and philosophical history that can offer light on this difficult topic. In his book, *Summa Theologica*, St. Thomas Aquinas introduces double-effect reasoning (DER), more commonly known as the principle of double effect. Both terms will be used interchangeably. The principle's main goal is to maintain the pursuit of good, even in difficult circumstances that may also bring about evil consequences.

In this chapter, the historical development of the principle of double effect will be examined, first by beginning with the origins of the principle of double effect in Aquinas's Summa Theologica and ending the examination with a modern understanding of the principle compiled by Jean-Pierre Gury. The second section will properly describe the principle of double-effect in contemporary understanding. Then examples of the usage of double-effect reasoning will be provided from recent Catholic social and moral teaching. This will be accompanied by an analysis of two critiques of the principle of double effect, which will assist in better understanding the application of the reasoning. Then the final section will assess the application of the principle of double effect in the infamous "Phoenix case" in 2009. The facts of the case will first be provided. Then various views within Catholic bioethics will be explored. In conclusion, this chapter explains the principle of double effect, its complexities, and its modern application.

Origin of the Principle of Double Effect

To begin the analysis of the principle of double effect, it is valuable to consider the following cases:

A twenty-five-year-old woman is admitted to a Catholic hospital because she is experiencing pain around her pelvis. The proper tests are ordered and after some investigative work by the physicians, her diagnostics come back positive for ovarian cancer. At this point in the scenario, there is not a clear moral dilemma. Routinely, the oncologist would treat her with chemotherapy, and hopefully, months later she would be in remission, cancer-free. However, think about how much the case would change if the woman not only found out that she had ovarian cancer at the hospital but also discovered that she was pregnant. In this scenario, the oncologist would have to inform her that the chemotherapy will almost certainly terminate the pregnancy and that the viable options for her own health can only be held off for a short time. Questions that the patient and clinicians would have to consider in this decision-making process would be: is it morally acceptable to treat the mother for cancer, while knowingly killing the unborn child? Would this termination of pregnancy be considered an abortion?

Another case to consider is of an eighty-seven-year-old man dying of heart failure who has recently been transferred from the hospital to hospice care. Due to several complications, the man is in severe pain that is causing him unbearable discomfort in his final days of life. The attending physician wants to relieve his pain so that he can have a comfortable and dignified death. The physician begins to contemplate increasing the man's pain medication, yet becomes concerned that the increase in sedatives could take away his consciousness and hasten his death. The physician wonders, would this treatment be a form of euthanasia? Would the physician be held morally or legally culpable?

In both of the cases, an ethical dilemma developed due to decisions arising from a difficult diagnosis and prognosis. Difficult, because the necessary treatment for each of the patients seems to be bound with morally evil consequences – the death of the unborn child and the quickening of the death of the elderly man. In both cases, double-effect reasoning can be applied to better assess the ethical solutions for the dilemma. Double effect is the principle that states the acceptability of a person doing a good (or neutral) action, even if there are foreseeable negative consequences.[2] For the principle of double effect to be applied, four conditions must be met to make the application valid.[3] The first condition is that the act itself is indifferent or good. The second condition is that the agent must intend only the good effect, not the bad effect. The third condition is that the good effect does not result due to the bad effect. The last condition is that the good effect must be proportionate to the bad effect.[4] It is worth emphasizing again that all four conditions must be met for an action to have the double effect principle applied to it for moral permissibility. The conditions do not have to be and are not met in the presented order; rather, all four conditions are to be met simultaneously. To better understand double-effect reasoning it is beneficial to understand the origin and continual development of the principle.[5]

Double Effect Reasoning's Origin in Thomas Aquinas

In *Summa Theologica*, Thomas Aquinas is credited with the creation of double-effect reasoning. However, nowhere in the plethora of Aquinas's writings is the principle of double effect ever explicitly written. Instead, succeeding Thomistic thinkers articulated the principle from numerous articles on various topics like homicidal self-defense, the consideration of scandal, and the permissibility of alluring dress.[6] In II-II q.64 a.7, Aquinas considers if it is lawful for a private individual to kill a man in self-defense. Aquinas refers to St. Augustine, who argues that one must forfeit their own life rather than undergo the action of taking another's life. Aquinas dissents from Augustine's claim by stating that one act could have two effects. One effect is the intended and the other is the *praeter intentionem*, the "in addition" to intention. He continues that moral acts receive their character from their intention, not the unintentional outcomes. Thus, for self-defense, one could have the intention to preserve one's own life, but consequently take the life of another. This is justifiable, due to one's own obligation to preserve one's own life. However, one must not use more force than necessary to preserve one's life.[7] In Article 7, Aquinas states that a private individual may not intentionally kill an assailant, but through an honest desire of self-defense, one may unintentionally cause the death of the aggressor, resulting in the *praeter intentionem.* To Aquinas, this also still follows Augustine's reasoning, because Aquinas holds the belief that Augustine only addressed intentional killing, not unintentional.[8]

To further understand what Aquinas implicitly stated on double-effect reasoning in Article 7, it is helpful to refer to Article 8, where Aquinas asserts that an agent is not responsible for the accidental death of another.

This conclusion seems redundant to Article 7; however, Aquinas is making a distinction between unintentional consequences, *praeter intentionem*, and accidental consequences, *per accidens.* To understand this distinction, an assessment of "intention" and "act" must be made first. Intention means to purposely do something. Intention can be understood in the example of one's health. A person's end goal is to be healthy when they are sick, so one chooses the means of medication to attain the end of health. Revealing that intent incorporates the means and the end together. Giving intent the moral relevance to the action. That being said, in Aquinas's assessment of an action, intention is only the third of three criteria determining if it is a good or bad act. The first two criteria are what the agent does and the circumstances in which the agent does it. So, if there is an accidental outcome, *per accidens*, the agent is not held accountable. [9] An example of this would be that if a child were to fall and die during a game of tag, the tagger would not be responsible for the accidental death while playing tag.

The provided explanation of intention and act regarding accidental consequences still does not reveal the permissibility of an unintentional consequence. Returning to II-II q.64 a.7, Aquinas states that a private individual is permitted to kill an aggressor when it is a proportionate use of force in relation to the aggressor and is not the intention of the agent. This is unlike a public official, who through using minimal force, is permitted to kill an assailant. (Aquinas makes a case for the difference between private individuals' and public officials' intentionality to kill in an earlier article; overly simplified, the difference is due to the public official's duty to serve the common good).[10] So, it is understood that the intentionality of self-defense is the crux of the moral action in this case.

Aquinas affirms that a private individual cannot intentionally kill, unlike a public official; along with the fact that private individuals are not responsible for accidental deaths. This provides another option to categorize a consequence that explains, *praeter intentionem.* In the previous example of an accidental death, the circumstance of the death is crucial. For example, when playing tag, it is reasonable to imply that a park or playground is not a usual place for a death to occur, along with the fact that there is not an additional object of violence that can be used to harm another. For example, a sword at the park would increase the risk of death if it were present. Nor in either player's mind is there an intention for the game to end in the death of the opponent. So, if a person trips and falls it is labeled an accidental death. Now, think about if someone is being attacked, and both the aggressor and the agent have swords. The agent in self-defense has the intention of self-preservation, and with reason, knows that sword fighting could end in death. Even if they did not choose to be in a dual, by participating the agent makes an acceptance of risking the aggressor's life through self-defense. So, this is what is meant by *praeter intentionem*; in this example, the killing with the foreknowledge and a willingness to risk an aggressor's life, without the intention to kill them. Another term for *praeter intentionem* is referred to as a risked consequence.[11]

To better explain this, the role of a firefighter can better articulate an example of a risked consequence. A firefighter knowingly puts himself in harm's way when entering a burning building. A firefighter does not intentionally try to get killed in a fire, assumably that they are not suicidal. A firefighter does not accidentally get killed in a burning building, because he is accepting the risk of burning or a structural collapse when he enters a dangerous situation. Rather, the firefighter's action accepts reasonable risked consequences that could cause his death. This is different than the playground example because a playground has far fewer risks than a burning building does. However, suppose a plane crashes into a burning building and kills a firefighter. That would be an accidental death because it was unpredictable and not foreseen that a plane could crash and kill the firefighter.

With intention in mind, one who endangers their own life for a good purpose does not intend their own death. Similarly, another distinction that is made pertaining to risked homicide and accidental death, which has already been mentioned, is the concept of foresight. Foresight strengthens the accidental tag example because one does not foresee their friend dying during a game of tag, whereas, in an attack, it is reasonable for one to conclude that death is possible. Regarding self-defense, one may try to preserve their own life by risking an assailant's life when it is not a direct intention to kill them; however, death can be foreseen as a consequence of defending oneself. Death is thc foreseeable indirect consequence of an action, which serves as a key distinction in double-effect reasoning.[12]

It is beneficial to draw the analysis of act and intention from Thomas Aquinas to a close with an exert from the *Summa Theologica*:

> "Nothing prevents one act from having two effect, of which only one is intended, the other being *praeter intentionem*. Now moral acts receive their character according to that which is intended, not, however from that which is *praeter intentionem*, since this is accidental (can be translated to foreseen)."[13]

Aquinas solidifies double effect reasoning when he states that actions like adultery and fornication are not ordered toward the conservation of one own's life, so they are never permissible. Unlike homicidal self-defense, where the intention is not to kill the assailant, it is a foreseeable outcome for the preservation of one's own life. That being said, Aquinas does not apply double-effect reasoning to inevitable evil outcomes, he only applies it to potential foreseen consequences.[14] The development of certain foreseen consequences will come from succeeding Thomist scholars.

To conclude the analysis of the origin of the principle of double effect, it is necessary to state what Thomas A. Cavanaugh's syntheses as Thomas Aquinas's implicit double-effect reasoning, which is revealed in Aquinas's consideration of homicidal self-defense, scandal, and alluring dress:

1) The act is the least harmful option capable of achieving the end;
2) considered independently of the evil effect, the act is permissible;
3) the agent is more obliged to pursue the good than to avoid the evil; and
4) the agent intends the good and does not intend the evil as a means or as an end.[15]

The Thomistic Tradition's Development of Double Effect Reasoning

Since the origin of double effect reasoning in Thomas Aquinas's *Summa Theologica*, many scholars in the Thomistic tradition have affirmed and developed the principle. The first thinker who developed double-effect reasoning is St. Antoninus (1389-1459), who repeats Aquinas's criteria of the principle of double effect in regard to self-defense and scandal. Antoninus then applies the principle to alluring dress. He claims that one cannot wear clothing that intentionally invokes lust; however, in accordance with local tradition one may dress to attract the opposite sex, even if, it could be foreseen to stir lust of the passions of an ungoverned person.[16]

The work of Antoninus is continued by Cardinal Cajetan (1469-1534). He reiterates Aquinas's premise that intention has moral importance, whereas, unintentional consequences have less moral significance. Cajetan then provides an essential example case for the development of the principle of double effect. It is that a physician intends the health of a patient by prescribing medication, but the physician does not intend the side effects of the medication. The most often used modern example is the oncologist who intends to heal a patient from cancer, though also causes nausea and a weakening of the patient's immune system. In Cajetan's account of Aquinas' I-II q.20 a.5, he affirms that foreseen bad effects do not make good actions evil. He provides the effective, unique, and comical example of Christ revealing Himself as the Messiah. Through the revelation of His own divinity, Christ provoked the disbelief of many. Cajetan guarantees that in accordance with Aquinas's reasoning, Christ did not cause scandal.[17]

The line of thinking continues through prominent Dominican scholars and was adopted into the scholarly work of the Society of Jesus. The next landmark development of double-effect reasoning was in 1647, when Domingo de Sta Teresa developed the first set of conditions that must be met for double-effect reasoning to be applied. His list is the model for the modern version and the provided example of Cavanaugh's synthesis of Aquinas's original conditions. His conditions are that there must be a proportionately good effect; that becomes before or conjoined with the evil effect; the permitted evil is only foreseen, not intended; and one cannot remove the good effect without significant challenges.[18]

Domingo de Sta Teresa's account of the principle of double effect is effective, though there are numerous critiques (that will be addressed later) on the choice of wordage and the potential overlap of conditions. Modern scholars believe that the criteria provided could be three more clearly articulated conditions, rather than four ambiguous conditions.[19]

Before addressing the Catholic Church's usage of double effect reasoning, a quick review of John de Lugo's contribution is beneficial. De Lugo is credited with one of the major transitions to allow application for the modern era. He provided the case of a man fleeing on horseback from an unjust attacker. Upon fleeing the attacker on a narrow trail there is an infant laying on the trail, that will inevitably be trampled by the horse if the man proceeds. This example is not logically perfect; however, he proposed this case to demonstrate the permissibility of a woman having a tumor removed which will result in the death of an unborn child. This intellectual proposition was later affirmed by St. Alphonsus Liguori and then adopted into the modern understanding of double effect reasoning by the moral theologian Jean-Pierre Gury in 1850.[20]

Gury's criteria may seem familiar:

1) The act in itself is good and indifferent;
2) the agent intends the good effect and not the evil effect;
3) the good effect is not produced by the evil effect; and
4) there is a proportionately grave reason for causing the evil effect.[21]

Modern Understanding and Usage of Double Effect Reasoning

The previous section focused on introducing the principle of double effect from the standpoint of its origin. This was done through an analysis of the roots of double-effect reasoning in Thomas Aquinas's *Summa Theologica*. Aquinas does not explicitly write out the principle of double effect as we know it today; rather, he employs double effect reasoning to better evaluate the moral permissibility of self-defense, how people are allowed to dress, and when to consider scandal when choosing an action. After Aquinas's death, scholars within the Catholic and Thomistic tradition identified and developed his reasoning into the principle of double effect. As was presented, the principle of double effect has been heavily relied upon in the Catholic tradition. This is because real-life scenarios (particularly in healthcare) are incredibly complex. The line between what is deemed morally good and morally evil is often blurred by different facts and viewpoints that are presented while making a decision. The purpose of double-effect reasoning is to assist a person through this difficult decision-making process, especially when the person's aim is to do good and avoid evil, even though the action they have to carry out may have evil consequences.[22] The specific question that the principle of double effect addresses is: is it morally permissible to perform an action that has two or more effects as a result; some of which the effects are good and intended, and some effects are bad and not intended?[23]

Situations with moral tension like these are often difficult to make judgments on. However, due to double-effect reasoning's Thomistic roots, the Catholic Church has embraced double-effect reasoning as a philosophical tool to be applied when assessing one's moral life. The Church often implements and refers to the principle of double effect when difficult cases arise, and the Church promotes using it when it is absolutely necessary. In this section, the contemporary understanding and usage of the principle of double effect will be stated and analyzed to better understand how to apply the principle in Catholic healthcare. This section will also include an evaluation of two counterarguments against the principle of double effect, to better understand where people misunderstand its application.

Contemporary Catholic Usage and Understanding of Double Effect Reasoning

In 1850, Jesuit moral theologian, Jean-Pierre Gury adapted previous versions of the criteria for the principle of double effect into the understanding that is used today.[24] (For the four criteria, refer to the previous section). However, double-effect reasoning is not just an abstract philosophical position, it is used in Catholic hospitals in tough situations and has been mentioned in numerous magisterial pronouncements. A brief contemporary-historical look at double-effect reasoning is valuable to understanding and seeing the essentialness of the principle.

In 1956, Pope Pius XII was asked by the Italian Society of Anesthesiology, "Can narcotics be used even if the lessening of pain will probably be accompanied by a shortening of life?" In 1957, Pope Pius XII responded that the usage of narcotics is lawful, even if it hastens death. In that case, the usage of narcotics intends to alleviate pain, not hasten a patient's death. Also, Pope Pius XII notes that the pain of the patient must be reasonably proportionate to the dosage of narcotics that is being prescribed. Pope Pius XII does not explicitly list the criteria from the principle of double effect, but his answer to the Italian Society of Anesthesiology clearly draws on double effect reasoning. In 1980, Pope John Paul II affirmed Pope Pius XII's implementation of double-effect reasoning in his condemnation of euthanasia. Pope John Paul II wrote that, terminal sedation "retains its full force" and that it is distinctly different from euthanasia in the explanation provided by Pope Pius XII.[25]

Double effect reasoning is also found in various works of other popes. A unique usage of double effect reasoning by Joseph Cardinal Ratzinger in 2004, before becoming Pope Benedict XVI, is worth considering. Ratzinger wrote a curial document titled, "Worthiness to Receive Holy Communion," where he is guiding the faithful to consciously consider the seriousness of being in union with the Church for the reception of the Eucharist. He mentions the dilemma of cooperating with evil in politics. Specifically for the guidance of voters, Ratzinger writes that if a voter intentionally votes for a candidate that is for the legalization or expansion of abortion or euthanasia this is clearly wrong. However, if a person votes for the same candidate, but for different reasons other than their stance on abortion and euthanasia, this could be permissible as long as their intention is clear and there are proportionate reasons not to vote for the opposing candidates. Once again, this is an effective example of double-effect reasoning being applied for Catholics to utilize while navigating the complexity of life.[26]

The previous examples provided show that multiple well-respected pontiffs utilized double-effect reasoning. However, it is even more beneficial to review what the magisterium conveys on the principle and its usage. In the *Catechism of the Catholic Church,* the principle of double effect is once again not explicitly written on; however, like Aquinas and the recent Popes, the type of reasoning is present. While using the four conditions that must be met to apply double-effect reasoning, the Catechism has a considerable amount of wisdom to verify the correct usage.

Of the first criterion the catechism states, "The act in itself must be good or indifferent." The Church writes that a morally good act does not just involve the goodness of the object, but also of the end and of the circumstances.[27] This is directly developed from Thomas Aquinas's evaluation of moral acts. The object, intention, and circumstances are the three sources that must be considered when determining the morality of human actions. The sources must be considered together, not separate from each other. Meaning that an evil end can corrupt a typically good act. For example, praying and fasting in order to be seen, is a morally bad act.[28] The Church also states that some actions are judged by the "object of choice," because the action can never be morally permissible. For example, fornication can never be a good moral act; so, there is not a case where premarital sex could have double-effect reasoning applied for justification.[29] This notion can once again find its roots in the Summa Theologica.

In regard to the second criterion, "the agent must intend the good effect and not intend the evil effect," the Church asserts that an effect can be foreseen without being willed by the moral agent.[30] To better understand this, the Catechism provides the example of the death of a person who was aiding someone else in danger. (Recall the firefighter example in section 1). The death of the person was foreseeable and tolerated, but in no way was it intended by the aider. An example contrary to the previous example would be a death caused by a drunk driver. Double effect reasoning could not be implied to this, because the driver took on moral culpability by choosing to drive while impaired; this is an immoral act.[31] This would not be the case for the person aiding another, which is a good moral act. Complementary to what the Church has already said on this criterion, it is also stated that a person may not do evil so that good may result from it.[32] This is the third condition that Gury presents in his criteria for the principle of double effect.

Looking at "means to an end" encapsulates a few different aspects that need to be assessed when trying to apply double-effect reasoning. The Church is very clear that evil actions cannot be done for the potential good effects. This, consequently, leads to the third criterion, "the good effect cannot be produced by the evil effect."[33] In the Catechism the example of a mother becoming fatigued caring for her child is presented. It seems like an odd example, but it is easy to distinguish the causality. The mother caring for her (potentially sick) child is an example of a good moral act; however, as she is caring for her child, she is unintentionally fatiguing herself. (Fatigue in this case is a side-effect, which is also considered as a consequence). This consequence is neutral or not good, not necessarily an evil consequence; either way, the consequence is not desirable but was foreseen when the action of caring for a sick child was undergone. Instantaneously, the mother is caring for the child, while also fatiguing herself.[34] To better explain causality, the example of chemotherapy could be used because it shows that causality is not actually about time, but the cause. Chemotherapy is a treatment to kill cancerous cells in the body, therefore it is used in healthcare today to effectively treat patients. However, chemotherapy makes a person incredibly sick due to its poisonous nature, which is undesirable. The implantation of chemotherapy simultaneously kills the cancerous cells (treating the patient), while also causing sickness. Often, the sickness shows effects before the healing of the patient. If causality were contingent on time this would not be permissible; however, it is not because the cause of the healing began at the same time as the undesired, but foreseen, consequence of the sickness.[35]

Analyzing the last of the four criteria is important because it will lead to a better understanding of the principle through potential philosophical misunderstandings. The condition states that there must be a proportionately grave reason for allowing the causation of the evil effect.[36] This is the principle of proportionality, which inhibits the toleration of bad effects that are not proportionate to the good effects of a given action.[37] The principle of proportionality is often used to discern the permissibility of tactics during wartime. For example, it would be forbidden for an army to bomb innocent civilians for the aim of scaring the military combatants (as was the case in WWII with the dropping of two atomic bombs). Another example can be depicted in the abortion argument. The principle of proportionality would not allow or deem proportionate the killing of unborn babies for the sake of a better quality of life argument.

Two Critiques of the Principle of Double Effect

The abortion example, of course, is very controversial today; however, it exemplifies the crux of the principle of proportionality and has the potential to resolve two misunderstood critiques of double-effect reasoning. Especially rooting in the fourth criterion, critics misuse and argue that the principle of double effect is rooted in consequentialism and proportionalism. Consequentialism is the view that the morality of an action can be judged solely by its positive or negative consequences.[38] Consequentialism in many ways opposes a deontological approach to ethics. Consequentialism has results of the action as the highest principle. This often leads to many exceptions being justified, in regard to traditional Catholic moral teaching.[39] An extreme example of consequentialism would be justifying a direct abortion for the sake of fiscal reasons of a family. The principle of double effect, when misunderstood could seem to be a tool for consequentialists, because the fourth condition of double effect reason does in fact weigh the outcomes of effect; however, what is not taken into account for this claim is that the previous three conditions must already be met, along with the fact that the principle of double effect when considering the proportionate effects, goes well beyond the numbers and values of what cannot always be counted.[40] For example, looking at the atomic bomb scenario, taking a civilian's life is not justifiable over (say) ten combatants' life. The principle of double effect does not subscribe to a "1 to 1" utilitarian ratio of proportionality. Rather, it is rooted in Aquinas and the Catholic traditions' precedence to primacy to certain goods. An example of this can be found in the case of abortion arguments. The Catholic understanding of proportionality would not allow consequentialist thinking that would allow the killing of innocent life, even if it benefitted the quality of life of (say) five other people. Simplified, the dignity of a human person is more important than improving quality of life, regarding a subjective standard.

The second misunderstanding that arises from the fourth condition is proportionalism which has seeped into Catholic moral theology since Peter Knauer argued for its value in the 1960s.[41] Proportionalism runs a fine line between incorrect and correct moral theology. Proportionalism agrees that there is an objective standard of values and disvalues; however, the main difference is that proportionalism denies that an act can be outright immoral in all cases, meaning a deemed morally bad act, can be used to promote a good outcome. This is similar to consequentialism but also differs due to the acknowledgment that the action is wrong.[42] An example of this would be a vasectomy. Proportionalism and orthodox Catholic moral theology would agree that a man being intentionally sterile is a moral evil. Yet a proportionalist would make the argument that a vasectomy is morally permissible due to the objective good of sexual intercourse and allocating more resources to a family's current children.[43]

In contrast to the proportionalist argument, the principle of double effect could still justify sterilization. However, the sterilization would have to be (for example) a foreseen consequence of an orchiectomy to remove cancerous testicles. The act of an orchiectomy is morally neutral, meeting the first condition. The intention of the procedure would be removing the cancer, not the sterilization; this would meet the second condition. The removal of the cancerous testicle(s) and the sterilization are simultaneous, the removal of the cancer is not a result of the sterilization; this would meet the third condition. The removal of cancer, to sustain one's life, is proportionate to the bad effect of sterilization, which would meet the fourth condition of the principle of proportionality.

This usage of the principle of double effect is affirmed in the ERDs when it is said, "Procedures that induce sterility are permitted when their direct effect is the cure or alleviation of a present and serious pathology and a simpler treatment is not available."[44]

Proportionalism is especially difficult to parse out from Catholic moral theology because it affirms the objective value of many of the same goods that orthodox Catholic theologians argue for. The issue is the prioritization of the goods, along with the denial of the utilization of intrinsically evil actions. That is why Pope John Paul II had to authoritatively correct both consequentialism and proportionalism during his papacy. John Paul II declared that both philosophical stances are not in accordance with the fullness of truth. Consequentialism is condemned for the evaluation of moral rightness being a result of the effect. Proportionalism is condemned for the comparison of goods and values with the aim of finding the greater good and lesser evils, while also disregarding the status of moral evil on particular acts. [45]

In conclusion, double-effect reasoning is not a tool of consequentialism or proportionalism. It is evident that critiques against the principle of double effect can be made especially from the fourth condition. However, those arguments are made from the assumption that the first three criteria do not have to be met. This is, of course, incorrect, because all four of the conditions must be met for the implementation of double-effect reasoning to be morally permissible. Evidently, the principle of double effect is difficult to understand and even harder to apply which will be revealed in the upcoming section.

The "Phoenix Case"

As shown in the two prior sections, the principle of double effect has had a long history of development and is quite complex to grasp. Through the explanation of the critiques of consequentialism and proportionalism, even at a philosophical level double effect reasoning raises many debates on its legitimacy in the Catholic intellectual tradition. That is evident even without mentioning many of the divergent Catholic interpretations of the principle that arose throughout the previous centuries opposing Gury's conclusion. However, the application of the principle of double effect is even more difficult to apply in practical scenarios, particularly in Catholic healthcare today. In this section, a summary of the Phoenix case will be provided. Then a brief overview of Catholic moral theologians' and bioethicists' evaluations of the case will be analyzed.

Double Effect Reasoning in "*The Ethical and Religious Directives*"

The *Ethical and Religious Directives for Catholic Health Care Services*, which is often abbreviated as the ERDs is a document written by the United States Conference of Catholic Bishops (USCCB). The ERDs offer moral guidance, rooted in the Church's moral and theological teachings, pertaining to a plethora of healthcare aspects and topics.[46] The ERDs aim to continually affirm the ethical standards of health care rooted in the Church's tradition of safeguarding and promoting the dignity of the human person. Along with providing the faithful with authoritative guidance on the standard of moral issues in healthcare today. The directives in the ERDs are rooted in natural law in light of Christian revelation. The directives in the ERDs are focused on the Catholic understanding of the human person, human acts, and the goal of human activity.[47]

Much of the conflict that will be seen in the Phoenix case is a result of various interpretations of the directives in the ERDs. Before beginning an analysis of the case, it is beneficial to provide a summary of a few principles and directives that are contested in the Phoenix case from the ERDs.

The first concept that needs to be understood is the difference between "direct" and "indirect" actions in Catholic healthcare. These terms are often applied to actions upon reflection of the purpose and morality of the act in healthcare.[48] Direct/indirect is a judgment made on the intention of an agent's act.[49] This terminology is often present when analyzing the application of double-effect reasoning. "Indirect" is often used as an act that is permitted using the principle of double effect. "Direct" is often used when an action does not meet a condition in double effect reasoning, making it a forbidden action.[50] Indirect acts are actions that meet all four criteria of double effect reasoning, meaning that the act has foreseen evil consequences, but is not intentional.[51] For example, a hysterectomy of a pregnant woman who has cervical cancer is considered in Catholic bioethics to be an indirect abortion.[52] Another example is when a patient's pain is treated with a barbiturate drip that leads to the unintentional hastening of death, this is considered an indirect form of euthanasia but is more commonly known as terminal sedation.[53] (Which was affirmed by Pope Pius XII and be seen in the previous section). Direct acts are actions that are intentionally bringing about the deemed evil consequence.[54] For example, a standard abortion, where the fetus is removed in various ways from the mother's body, is a direct abortion. It is considered a direct abortion because the act itself is deemed evil, which goes against the first condition of double effect reasoning.[55] Another example, in contrast to terminal sedation, would be (direct) euthanasia. Which would be the treatment of a patient's pain being the eradication of suffering through the administering of a lethal drug that causes death.[56] In conclusion, the difference between indirect and direct acts can be distinguished in the moral judgment of the act (condition one) and the intentionality of the act (condition two).

Specifically, in the ERDs double effect reasoning and direct/indirect distinctions are found in two prominent directives that guide the Phoenix case. Both are found in "Part Four: Issues in Care for the Beginning of Life." In this section, the USCCB guides the faithful through standards of care and restrictions that need to be in place to promote the sanctity of human life through Christian marriage and family.[57] The first directive to consider is Directive 45, which is on abortion. Directive 45 states that a (direct) abortion is the intentional termination of a pregnancy before viability or the destruction of a fetus. The directive continues that any procedure that has the sole immediate effect of a termination of pregnancy is an abortion. It specifies that in a moral context, this includes termination as early as a point of time after conception. The directive then says that Catholic health institutions cannot provide an abortion and Catholic institutions need to be aware of the principle of material cooperation. Which is the principle that actions can be judged as morally evil if a person's (or institution's) actions cooperate or in some way assist in another's morally evil action.[58] The Directive clarifies that especially with abortions, Catholic institutions need to be careful considering the danger of scandal when associating with abortion providers.[59] For example, a Catholic hospital's leaders should reflect upon the possibility of cooperation and scandal when deciding to host an STI testing event with a pregnancy center that provides abortions.

The next Directive that involves double-effect reasoning is Directive 47. Directive 47 states that: "operations, treatments, and medications that have as their direct purpose the cure of a proportionately serious pathological condition of a pregnant woman are permitted when they cannot be safely postponed until the unborn child is viable, even if they will result in the death of the unborn child."[60]

The ERDs reveal the difficulty of the dichotomy of direct/indirect abortions within Catholic healthcare. It is clearly stated, along with tradition traced back to first-century Christians, that (direct) abortions are never permissible.[61] However, if a procedure or medication has the aim to cure a proportionately bad condition, then in the case of treating a pregnant woman, the treatment is permissible, even with the result of the death of the unborn child. This directive is effectively the principle of double effect applied within the context of an extremely difficult clinical case during pregnancy. As is evident, the four conditions are applied in this directive.

The act must be an operation, treatment, or medication that is administered for the promotion of the health of the woman. The sole intention cannot be to abort the unborn child; rather instead, the sole intention must be to heal the woman from the disease or condition. Causality is assumed by fulfilling the first condition, but to clarify, the good of the treatment cannot result from the abortion.[62] Lastly, the procedure must be aligned with the principle of proportionality.[63]

In conclusion, the ERDs, have an important task to communicate the Catholic moral teachings about patients, clinicians, and institutions. The directives are well formulated and offer significant guidance for any reader. However, the practical application of the directives often can be difficult in complex clinical settings. The seemingly black-and-white directives can become grey quickly. As was hopefully conveyed through the dichotomy of direct/indirect acts. On the surface or philosophically it is easy to make those distinctions; however, with the various time constraints that can arise in clinical settings and people's varying intentions, it becomes near impossible to judge. This is why the infamous Phoenix case is so fascinating and divisive among Catholic moral theologians and ethicists.

The Phoenix Case: Medical Overview

In October 2009, a twenty-seven-year-old woman, with a medical history of moderate pulmonary hypertension went in to see her pulmonologist after experiencing worsening symptoms.[64] The woman was a Catholic mother of four children.[65] After routine tests, the woman found out that she was seven and a half weeks pregnant. The pulmonologist recommended that the pregnancy be terminated because the morbidity rate of pregnant women with pulmonary hypertension is ten to fifteen percent. Other sources have a mortality rating as high as forty percent for the average case.[66] However, in her case, the pulmonologist believed that she had a fifty-fifty chance of survival. The mother decided that she was not going to have an abortion.[67] It is unknown what other treatments were recommended to her at this time.

Three weeks later, in November 2009, the woman was admitted into the emergency room at St. Joseph's Hospital and Medical Center, due to her worsening symptoms. She was eleven weeks pregnant at this time. The medical records state that she now had severe pulmonary hypertension that was limiting cardiac output. She had right heart failure and cardiogenic shock. It was noted that it was life-threatening now, and a physician noted that the chance of mortality for her was nearing one hundred percent. Surgery was absolutely necessary for her survival.[68]

The cause of the patient's heart failure and cardiogenic shock is believed to have been the physiological changes that occur at ten weeks of pregnancy worsening the patient's previous pulmonary hypertension.[69] Specifically, the development of the placenta had exacerbated the worsening of the condition.[70] The unborn child was not at a viable age to live outside the womb and the unborn child, due to heart failure and cardiogenic shock, could not remain in the womb because the maternal environment was threatening it. This conundrum resulted in both the mother and the unborn child being in the dying process. However, as described, the unborn child's life was not going to able to be saved.[71]

An ethics committee was called and evaluated the options available for the woman. The task for the committee was to address problems caused by the placental issue. Would a dilation and curettage be morally permissible in Catholic teaching? Specifically, is it aligned with the Catholic tradition and the directives in the *Ethical and Religious Directives for Catholic Health Care Services*. The ethics committee decided that it was not a direct abortion, but an indirect abortion. This decision was made and announced to the physicians by a Catholic religious sister.[72] With the consent of the patient, a dilatation and curettage (D&C) was performed. [73] A D&C is a procedure that is often used to remove tissue from the uterus.[74] However, a percentage of abortions also use this procedure, which makes the Phoenix case more controversial.[75] After the procedure, the woman was able to be discharged to return home to her family.[76]

The Phoenix Case: Aftermath Overview

The aftermath of the Phoenix case has been quite complex and controversial. This has led to the Phoenix case being one of the most important cases for Catholic bioethics in recent years. The aftermath began in November 2010, when a reviewal process was undergoing the permissibility of the actions taken a year earlier. The Bishop of the diocese requested that a respected Catholic bioethicist complete a moral analysis in defense of the actions in the case. Upon completion, the bishop disagreed with the defense and rejected it. Then it was announced that the religious sister was automatically excommunicated for the approval of the direct abortion.[77] A few months after the ex-communication, in December 2010, the bishop underwent conversations and negotiations with the hospital and its parent company, Catholic Healthcare West. The content of the meeting has not been released, but the negotiations ended without an agreement. This led to the bishop revoking the official Catholic status of the hospital that the Phoenix case took place in.[78]

Besides the actual actions that happened in the Phoenix Case, there have been numerous controversies that have originated in the aftermath of the case. A source of controversy is the claim that the religious sister was automatically excommunicated by approving the D&C procedure. The question at hand is whether Canon Law, particularly 1323 and 1324, which focuses on the formal cooperation in a moral evil, was actually met. An example of these conditions being met would be direct assistance in an abortion. So, two things are being debated; the first being the approval of an abortion with direct assistance, and the second is the ethical question of was the D&C in this case a direct or indirect abortion.[79]

Another controversy is the removal of the Catholic status of the hospital. The claim was made by the bishop for a few reasons. The first reason is, as he interprets it, the direct abortion that occurred a year prior. However, the bishop also cites that the hospital had been formally cooperating with a program that offers abortions, contraceptives, and sterilizations at other hospitals.[80] It has also been claimed that the hospital had previously taken part in numerous other abortions, but, it was noted by the source, that the circumstances of the abortions are not clear. However, the health system governing the hospital said it would not change there stance on these circumstances.[81] These accusations have been heavily contested by the hospital's parent company, along with a few prominent Catholic associations and bioethicists. The President of the health system made numerous public statements affirming the system's commitment to the common good and Catholic teaching. Along with the continual endorsement from the Catholic Health Association.[82]

The Phoenix Case: Moral Analyses Overview

A consistent theme in reading various moral analyses is the disclaimer that the majority of authors make regarding the lack of complete knowledge of the medical facts of the procedure that was approved in the November of 2009. It is important to recognize that the majority of the facts that are being used in the analyses are from Therese Lysaught's moral analysis of the case commissioned by the bishop. Along with the information made public by Catholic Healthcare West. At the time that many of the analyses were written the bishop of the Phoenix Diocese, nor the religious sister, who was at the time of the case the Vice President of Mission Integration, were making many public comments.[83]

The crux of the main debate on the Phoenix Case is whether the D&C was a direct or indirect abortion. In more detail, the question is whether a D&C is a direct or indirect abortion when the development of the placenta exacerbates an underlying condition (pulmonary hypertension) that becomes so severe as to threaten both the life of the mother and the life of the unborn child.[84] The two main sides of this debate are generally represented by the bishop and by the religious sister. The position of the bishop is that the procedure causing the death of the unborn child constitutes a direct abortion, which is always immoral. The position of the sister is that the procedure causing the death of the unborn child constitutes an indirect abortion, because it was an unintended consequence, and the action was made to save the life of the mother.[85]

From these two standpoints, many similar viewpoints have come to the surface. In the defense of St. Joseph's Hospital and the religious sister's decision are many influential voices in Catholic healthcare like the Catholic Health Association and Dr. Therese Lysaught,[86] whose viewpoint will be briefly discussed. The opposing standpoint, which stems from the bishop's perspective, has many prominent voices in Catholic healthcare like the National Catholic Bioethics Center.[87] Fr. Nicanor Austriaco's viewpoint will be presented as a representation of the various opposing moral analyses, which focus on the differing interpretations of organ ownership.[88] Lastly, there are a few Catholic ethicists who propose, in a sense, a middle ground to the Phoenix Case that is a beneficial understanding of the different moral analyses. Two incredibly convincing arguments are from Dr. Gerard Magill and Dr. Kevin Flannery.[89]

As mentioned earlier, Dr. Therese Lysaught was commissioned by the Diocese of Phoenix to conduct a moral analysis in defense of the Phoenix Case. Her analysis concludes that the procedure was an indirect abortion, not a direct abortion. She arrives at this conclusion by following the ethical framework of Martin Rhonheimer and Germain Grisez.[90] She concludes that the principle of double effect does not apply to the Phoenix case's D&C procedure.[91] Rather, she makes the argument that the D&C, being a direct assault on the fetus, does not constitute the moral object of the action. In her opinion, the moral object of the D&C was to save the mother's life, while also recognizing that the death of the fetus is a foreseen consequence. It is also noted in her analysis that the fetus was already in the process of dying, which benefits the justification of moral judgment.[92]

Lysaught's moral analysis has been heavily critiqued by many moral theologians and bioethicists within the Catholic tradition. A fascinating critique and counterargument is presented by Fr. Nicanor Austriaco, which was published in the National Catholic Bioethics Quarterly. In his piece, Austriaco makes the argument that the procedure was a clear example of a direct abortion. (It is to be noted, that in his analysis he claims that placentectomy was used, not a D&C).[93] He argues that the child was within a threatening environment (the womb) due to the heart failure and shock; however, the fetus was not the threat to the mother. So when the physicians decided to remove the placenta, it was a direct act on the unborn child. The basis of his argument seems to be similar to the principle of double effect regarding ovarian cancer. However, he differentiates from the removal of the ovary, because he has deemed the placenta to be an organ of the unborn child. He states that the procedure in the Phoenix case was similar, in moral judgment, to that of a Mayan priest taking the heart out of a sacrificial victim.[94]

Dr. Magill is critical of the arguments made by Lysaught and Austriaco. Of Lysaught's analysis, Magill points out that labeling the procedure a direct abortion would be at odds with the Committee on Doctrine's requirement that all procedures must not directly attack the life of a fetus. Another critique is that not applying double-effect reasoning causes multiple problems that weaken her analysis. The first problem is that confusion is created by Lysaught using double-effect reasoning terminology, but she does not actually use the principle. The second problem is that Lysaught's classification of the death of the fetus as "nondirect" and an "accessory consequence" is wrong because such classifications are not consistent with each other; an accessory consequence would not be considered nondirect.[95] Of Austriaco, Magill disagrees with the assertion that the placenta is an organ of the fetus alone. Magill states that the placenta is a shared organ between the mother and the unborn child. This directly refutes Austriaco's claim that it is a direct abortion, because the placenta, in a sense, is owned by the mother as well.[96] So, the removal of the placenta is similar to that of procedures done to treat ovarian cancer or an ectopic pregnancy.

Magill's moral analysis involves applying double-effect reasoning to the Phoenix case. His argument is justified by the principle of double effect because the fetus is not directly targeted in the D&C.[97] He compares the procedure to that of a cancerous uterus. The Phoenix case intention was to remove the placenta, not to kill the fetus.[98] He writes that the D&C is not morally evil, because it removes the pathological circumstance, which satisfies the first condition of double effect reasoning. To meet the second condition, he shows that the removal of the placenta simultaneously heals the patient and kills the fetus. It is not the killing of the fetus that causes the healing of the mother, it is the removal of the placenta. Thirdly, his argument satisfies the third criterion by claiming that the intention of the procedure was not to kill the fetus but to heal the mother. This claim is supported by the patient's faith and refusal to terminate the pregnancy when it was suggested by the pulmonologist.[99] Lastly, Magill applies the principle of proportionality to the act, which meets the fourth condition of the principle of double effect. The saving of a life in imminent danger is proportionate to the unintended, but foreseen, death of the unborn child.[100] This is a difficult claim to make; however, it is consistent with the historical development of double-effect reasoning.

Conclusion

The principle of double effect has a long story of development in the Catholic intellectual tradition. Beginning with St. Thomas Aquinas the principle originated and then was developed up to the modern understanding by Jean-Pierre Gury. In the recent century, double-effect reasoning has been endorsed by numerous popes and has been used in numerous magisterial documents. As exemplified in the Phoenix case, it is a complex principle to apply because numerous philosophical standpoints critique its usage, along with the fact that the practical agreements (like the "ownership" of the placenta) can cause minuet differences that drastically change the judgment of moral permissibility of an action. However, in conclusion, it is an incredibly important and effective principle within Catholic healthcare, that has the continuous aim to do good and avoid evil.

Notes

[1] Mathew 7:12, NIV

[2] The National Catholic Bioethics Center, *NCBC Commentary on the 'Phoenix Case,* 2 (Philadelphia, PA: 2010).

[3] Ibid.

[4] Thomas A. Cavanaugh, "The History of DER," *Double Effect Reasoning, 26* (New York, NY: Oxford University Press, 2006).

[5] The National Catholic Bioethics Center 2.

[6] Cavanaugh "The History of DER,"1-2.

[7] Aquinas, Thomas. II-II q.64 a.7, *Summa Theologica.*

[8] Cavanaugh "The History of DER," 3-5.

[9] Ibid., 7-9.

[10] Aquinas, Thomas. II-II q.64 a.2 and a. 3, *Summa Theologica.*

[11] Cavanaugh, *"The History of DER,"* 3, 9-10.

[12] Ibid., 3, 9-11.

[13] Aquinas, Thomas *Summa Theologica,* II-II q.43 a.3

[14] Cavanaugh, *"The History of DER,"* 11-12.

[15] Ibid., 12.

[16] Ibid., 15-16.

[17] Ibid., 17-19.

[18] Ibid., 21-22.

[19] Ibid., 36.

[20] Ibid., 23-25.
[21] F.J. Conell, *Principle of Double Effect,* (Encyclopedia.com) Accessed December 7, 2022.
https://www.encyclopedia.com/religion/encyclopedias-almanacs-transcripts-and-maps/double-effect-principle
[22] Cavanaugh, "The History of DER," 201.
[23] David F. Kelly, Gerard Magill, and Henk ten Have, *Contemporary Catholic Health Care Ethics*, (Washington, DC: Georgetown University Press, 2013), 104.
[24] Cavanaugh, *"The History of DER,"*, 23-25.
[25] Ibid., 197.
[26] Eternal Word Television Network, *The Principle of Double Effect.* Accessed November 1, 2022.
https://www.ewtn.com/catholicism/answers/principle-of-double-effect-23224
[27] Kelly, Magill, Ten Have, 105.
[28] Mathew 6:16-18, NIV
[29] Catholic Church (promulgated by Pope John Paul II), *Catechism of the Catholic Church*, Second Edition. (Washington, DC: United States Conference of Catholic Bishops, 2016), 1755.
[30] Kelly, Magill, Ten Have, 106.
[31] Catholic Church, *Catechism of the Catholic Church*, 1737.
[32] Catholic Church, *Catechism of the Catholic Church,* 1756
[33] Kelly, Magill, Ten Have, 107.
[34] Catholic Church, *Catechism of the Catholic Church,* 1737
[35] EWTN, *The Principle of Double Effect.*
[36] Kelly, Magill, Ten Have, 107-108.
[37] Cavanaugh, *"The History of DER,"* 181.
[38] University of Texas, *Ethics Unwrapped.* Accessed November 1, 2022. https://ethicsunwrapped.utexas.edu/glossary/consequentialism
[39] Kelly, Magill, Ten Have, 62.
[40] Cavanaugh, *"The History of DER,"* 31-32.
[41] Kelly, Magill, Ten Have, 86.
[42] Ibid., 65.
[43] Cavanaugh, *"The History of DER,"* 38-40.
[44] ERDs, 24.
[45] Pope John Paul II, *Veritatis Splendor*, 75. Accessed November 1, 2022.
https://www.vatican.va/content/john-paul-ii/en/encyclicals/documents/hf_jp-ii_enc_06081993_veritatis-splendor.html
[46] Catholic Health Association, *Ethical and Religious Directives.* Accessed December 2, 2022. https://www.chausa.org/ethics/ethical-and-religious-directives#:~:text=The%20Ethical%20and%20Religious%20Directives,aspects%20of%20health%20care%20delivery.
[47] United States Conference of Catholic Bishops, 2.
[48] Kelly, Magill, Ten Have, 107.
[49] Cavanaugh, *"The History of DER,"* 63.
[50] Kelly, Magill, Ten Have, 108.
[51] Cavanaugh, *"The History of DER,"* 63.

[52] Kelly, Magill, Ten Have, 108.
[53] Cavanaugh, *"The History of DER,"* 63.
[54] Ibid.
[55] Kelly, Magill, Ten Have, 108.
[56] Cavanaugh, *"The History of DER,"* 63.
[57]United States Conference of Catholic Bishops, 20.
[58] Kelly, Magill, Ten Have, 119.
[59]United States Conference of Catholic Bishops, 23.
[60]Ibid.
[61] Catholic Church, *Catechism of the Catholic Church,* 2271.
[62] Cavanaugh, *"The History of DER,"* 36.
[63] United States Conference of Catholic Bishops, 23.
[64] Therese Lysaught, *Moral Analysis of Procedure at Phoenix Hospital*, (Chicago, IL: Loyola University Chicago, 2011), 537.
[65] Gerard Magill, *Quaestio Disputata*, (Theological Studies: 2011), 848-849.
[66] Austriaco, Nicanor Pier Giorgio (O.P.), *Abortion in a Case of Pulmonary Hypertension,* (Philadelphia, PA: The National Catholic Bioethics Quarterly), 2011: 505.
[67] Lysaught, 538.
[68] Ibid.
[69] Ibid.
[70] Magill, 849.
[71] Lysaught, 539.
[72]Ibid.
[73] Cavanaugh, Thomas A. *Double-effect Reasoning, Craniotomy, and Vital* Conflicts, (Philadelphia, PA: The National Catholic Bioethics Quarterly), 455.
[74] Mayo Clinic, *Dilation and curettage,* Accessed December 7, 2022. https://www.mayoclinic.org/tests-procedures/dilation-andcurettage/about/pac20384910#:~:text=Dilation%20and%20curettage%20(D%26C)%20is,after%20a%20miscarriage%20or%20abortion.
[75] Guttmacher Institute, *Medication Abortion Now Accounts for More Than Half of All US Abortions.* Accessed December 7, 2022. https://www.guttmacher.org/article/2022/02/medication-abortion-now-accounts-more-half-all-us-abortions
[76] Thomas A. Cavanaugh, *Double-effect Reasoning, Craniotomy, and Vital* Conflicts, (Philadelphia, PA: The National Catholic Bioethics Quarterly), 455.
[77] Magill, 849.
[78] Marianne Medlin, *Catholic Health Association Defies Phoenix Bishop Over Abortion Case*, (Catholic News Agency: 2010). Accessed December 1, 2022. https://www.catholicnewsagency.com/news/21663/catholic-health-association-defies-phoenix-bishop-over-abortion-case
[79] Magill, 850.
[80] Medlin.
[81] NCBC, 3.
[82] Medlin.
[83] The National Catholic Bioethics Center, 3.

[84] Magill, 851.
[85] Ibid.
[86] Medlin.
[87] The National Catholic Bioethics Center, 1-4.
[88] Austriaco, 497.
[89] Kelly, Magill, Ten Have, 112-113.
[90] Lysaught,543-546.
[91] Lysaught, 547-548.
[92] Magill, 861.
[93] Austriaco, 505.
[94] Ibid., 507-508.
[95] Magill, 861-862.
[96] Kelly, Magill, Ten Have, 113.
[97] Ibid.
[98] Magill, 868.
[99] Cavanaugh, "Double-effect Reasoning, Craniotomy, and Vital Conflicts," 455.
[100] Magill, 871-872.

Literature

Austriaco, Nicanor Pier Giorgio (O.P.). 2011. *Abortion in a Case of Pulmonary Hypertension,* (Philadelphia, PA: The National Catholic Bioethics Quarterly): 493-508.

Catholic Church (promulgated by Pope John Paul II). 2016. *Catechism of the Catholic Church*, Second Edition. (Washington, DC: United States Conference of Catholic Bishops).

Catholic Health Association. 2018. *Ethical and Religious Directives.* Accessed December 2, 2022. https://www.chausa.org/ethics/ethical-and-religiousdirectives#:~:text=The%20Ethical%20and%20Religious%20Directives,aspects%20of%20health%20care%20delivery.

Cavanaugh, Thomas A. 2011. "Double-effect Reasoning, Craniotomy, and Vital Conflicts." *The National Catholic Bioethics Quarterly.* 11, no. 3: 443-453.

Cavanaugh, Thomas A. 2006. "The History of DER," *Double Effect Reasoning.* (New York, NY: Oxford University Press).

Conell, F.J. N. d. *Principle of Double Effect,* (Encyclopedia.com) Accessed December 7, 2022. https://www.encyclopedia.com/religion/encyclopedias-almanacs-transcripts-and-maps/double-effect-principle

Eternal Word Television Network, N. d. *The Principle of Double Effect.* Accessed November 1, 2022. https://www.ewtn.com/catholicism/answers/principle-of-double-effect-23224

Guttmacher Institute, N. d. *Medication Abortion Now Accounts for More Than Half of All US Abortions.* Accessed December 7, 2022. https://www.guttmacher.org/article/2022/02/medication-abortion-now-accounts-more-half-all-us-abortions

Kelly, David F., Gerard Magill, and Henk Ten Have. 2013. *Contemporary Catholic Health Care Ethics*. (Washington, DC: Georgetown University Press).

Lysaught, Therese. 2011. Moral Analysis of Procedure at Phoenix Hospital, (Chicago, IL: Loyola University Chicago), 537-552.

Mayo Clinic, N. d. *Dilation and curettage,* Accessed December 7, 2022. https://www.mayoclinic.org/tests-procedures/dilation-andcurettage/about/pac20384910#:~:text=Dilation%20and%20curettage%20(D%26C)%20is,after%20a%20miscarriage%20or%20abortion.

Medlin, Marianne. 2010. Catholic Health Association Defies Phoenix Bishop Over Abortion Case, (Catholic News Agency). Accessed December 1, 2022. https://www.catholicnewsagency.com/news/21663/catholic-health-association-defies-phoenix-bishop-over-abortion-case

National Catholic Bioethics Center. 2010. "NCBC Commentary on the 'Phoenix Case.' (Philadelphia, PA).

John Paul II. 1993. *Veritatis Splendor*. Accessed November 1, 2022. https://www.vatican.va/content/john-paulii/en/encyclicals/documents/hf_jp-ii_enc_06081993_veritatis-splendor.html

University of Texas. N. d. *Ethics Unwrapped.* Accessed November 1, 2022. https://ethicsunwrapped.utexas.edu/glossary/consequentialism

CHAPTER NINE

RESILIENCE IN THE DIAGNOSIS AND TREATMENT OF PSYCHIATRIC CONDITIONS

GABRIELLA AGOSTARO

Introduction

Psychiatry became recognized as a distinct specialty of medicine through the late 18th century and early 19th century.[1] The development of this specialized field came from the intersection of three primary developments in patients presenting with mental illnesses. The first development is the mind as an active faculty or intermediary between the body and the Christian soul.[2] The second development is the rejection of humoral theories of insanity, which focused on yellow bile, black bile, phlegm, and blood.[3] Instead, the focus was turned to the brain as the site of madness.[4] The final development resulted from the increase of asylum for patients presenting with psychiatric symptoms.[5]

10.5% of the global burden of disease in the Americas can be attributed to mental health and substance abuse disorders.[6] In North America, the treatment gap for patients with moderate to severe mental health disorders was 53.2%.[7] The early onset of psychiatric conditions and the primary lack of knowledge in psychiatry have led to a substantial gap in care throughout medical history. As a result of the lack of knowledge and enhanced research methods in the field of psychiatry, individuals suffering from psychiatric and psychological conditions have often been stigmatized and ridiculed for their disorders.

This chapter will analyze current advancements in the diagnostic and treatment process of psychiatric conditions throughout the United States. Diagnostic qualifications for patients presenting with psychiatric symptoms will be explored to establish the proper patients suitable for psychiatric treatment. Additionally, therapeutic interventions, medication protocols, and institutionalization protocols will be explored to provide a foundation for the treatment of psychiatric conditions in the United States.

Diagnosis

A proper diagnostic protocol for psychiatric conditions is critical for individuals seeking assistance for their psychiatric symptoms. An appropriate psychiatric diagnosis may require persistence and may result from the patient seeking assistance from numerous practitioners because a specific psychiatric diagnosis can be difficult to pinpoint. While the diagnostic process may require time and patience, this is an essential first step for patients seeking mediation for their psychiatric suffering. Diagnostic techniques vary across providers, but the traditional methods of practice result from physical examinations, laboratory testing, and psychiatric examinations.

Physical Examinations

Physical examinations are a crucial aspect of the diagnostic process for psychiatric conditions. As patients seek answers for their tendencies and symptoms, practitioners must explore physiological processes that may be displayed by the patient. In some cases, the patient may have a detrimental physical condition that mimics the symptoms of a psychiatric disorder. Physical examinations enable both physicians and psychiatrists to analyze the physical symptoms that may otherwise go unnoticed.

While symptoms may be psychiatric or physical, physiatrists must ensure that they can recognize the common symptoms resulting from a physical condition to guarantee proper treatment protocols. In a case report exploring the role of physical examinations in psychiatry, a patient was suspected of having Neuroleptic Malignant Syndrome by the attending care team. Neuroleptic Malignant Syndrome is a reaction that occurs as a result of taking antipsychotic medication.[8] In the case report examination, the consulting psychiatrist recommended a broader differential diagnosis, that included the Herpes Simplex Virus.[9] Although the attending physicians were adamant that the patient had a condition resulting from antipsychotic medication, the psychiatrist's recommendation for the treatment of Herpes Simplex Virus was ultimately correct.[10] This case demonstrates the continued need for physical examinations from psychiatrists and traditional physicians. Traditional care teams often attribute psychiatric symptoms to psychiatric disorders. Without the frequent physical examinations performed by psychiatrists, biological etiology may be overlooked, and psychiatric symptoms may be misdiagnosed.

Patients with psychiatric conditions often have other physical medical conditions associated with their diagnosis. Patients diagnosed with schizophrenia, bipolar, or major depressive disorder have an increased likelihood of cardiovascular morbidity.[11] As a result of diagnostic errors and the medical conditions that can simultaneously occur with a psychiatric illness, psychiatrists must continue to perform thorough physical examinations throughout their practice. In patients who have psychiatric illnesses that are associated with additional comorbidities, the psychiatrist must continue to monitor patients to ensure that other physical symptoms do not become problematic.

While Physical examinations are essential to obtain a complete diagnostic picture, there are still no standardized guidelines for psychiatrists when performing physical examinations. Mortality rates are 2-4 times higher in patients with psychiatric conditions compared to the general population.[12] Given the comorbidities that occur with psychiatric disorders and the continued misdiagnosis resulting from uneducated care teams, standardized guidelines for physical examinations conducted by psychiatrists must happen in the future to combat these difficulties.

Laboratory Tests

The use of laboratory tests in the field of psychiatry is essential for the proper diagnosis of patients with psychiatric symptoms. Laboratory tests clarify the physiological processes that regularly occur throughout the human body. Similar to physical examinations, laboratory tests enable clinicians to assess physiological characteristics to understand the processes that may contribute to psychiatric symptoms. This method of testing assists in developing an appropriate treatment plan for the patient and ensures that physical conditions that may be present in the patient are not being overlooked.

As society has become more advanced, scientists have begun to understand reactions that occur throughout the body that influence daily function. Numerous metals and chemicals have been associated with decreased function in patients suffering from psychiatric symptoms. Patients with hyponatremia or low levels of sodium in the blood can experience temporary psychosis.[13] Hyponatremia is a condition that occurs commonly, especially in patients admitted to the hospital. Although hyponatremia can, in most cases, be treated, the treatment depends on early detection through blood tests and clinical indicators.[14] If clinicians are unaware of the warning signs for hyponatremia and do not perform the necessary laboratory testing, patients may go untreated.

Hypocalcemia is a similar disorder that can potentially create psychiatric symptoms in patients if left untreated. Hypocalcemia occurs when the patient has significantly reduced calcium levels in their blood. Hypocalcemia patients will display temporary psychiatric symptoms that include psychosis, hallucinations, and paranoid delusions.[15] Similar to hyponatremia, hypocalcemia can be detectable through clinical indicators and laboratory testing procedures.

Various physiological conditions can present in patients similar to those presenting with psychiatric disorders. While psychiatric conditions have the potential to be troublesome to distinguish from physical conditions, medical personnel must continue to evaluate patients critically. This ensures that the clinician is assessing the patient for a true psychiatric disorder or medical mimic. Laboratory tests that assist clinicians in developing these diagnoses include thyroid function testing, chemistry panel, complete blood count, C-reactive protein, etc.[16] While physicians and psychiatrists continue to diagnose and treat patients, laboratory tests are necessary for the diagnostic process of patients with psychiatric symptoms and the continued safety of the patient.

Psychiatric Evaluations and Intake Assessments

Psychiatric evaluations are a fundamental aspect of the diagnostic process in psychiatry. When clients present with psychiatric symptoms, the clinician conducts a psychiatric evaluation to determine the nature of the patient's symptoms. Psychiatric evaluations enable the clinician to better understand the client's needs and enable them to develop a plan to assist the patient with treatment.

The primary goal of a psychiatric evaluation is to obtain information from the patient. Psychiatrists will interview the patient to understand the habits and symptoms that they traditionally display. The interview consists of numerous techniques, which may include mental status evaluations, behavior pattern examinations, and thorough patient histories. The mental status evaluation will traditionally consist of numerous screening questions. These screening questions include tests for cognitive impairment, practical reasoning, social judgment, and decision-making.[17] Mental status examinations allow the clinician to understand if the patient's mental capacity contributes to psychiatric symptoms or if the patient's lack of mental capacity results from psychiatric symptoms.

Behavior patterns exhibited by the patient are an aspect of the interview process that allows the clinician to understand the associated behaviors that influence the patient's symptoms. A review of behavior patterns may include questions regarding depression, anxiety, substance abuse, suicidal thoughts, etc.[18] The behavior patterns interview assists the clinician in recognizing which behaviors may put the patient or other individuals close to the patient at risk of harm.

The in-depth review of patient history is crucial to the patient intake interview process. This conversation includes details regarding current and previous diagnoses and psychiatric treatment methods previously utilized.[19] A thorough understanding of a patient's psychiatric history is vital to the diagnostic process of the patient. While psychiatric diagnoses can be complex, a history enables the clinician to interpret past traumas and determine if they contribute to the patient's current symptoms.[20] Additionally, many psychiatric conditions have hereditary associations. These hereditary associations have been linked to disorders including bipolar disorder and schizophrenia. Through patient history discussions, clinicians can assess family history to determine if a patient is at a heightened risk for developing a specific psychiatric condition.

During the initial psychiatric evaluation and intake assessment, patients may be given psychiatric questionnaires to evaluate the prevalence of certain psychiatric disorders. The Hamilton Anxiety Scale (HAM-A) is a common assessment utilized to measure generalized anxiety disorder.[21] The HAM-A scale explores various physical and emotional symptoms that can be associated with generalized anxiety disorder.

Similarly, the Beck Depression Inventory (BDI) is utilized to assess the prevalence and severity of depression in patients. Patients tested using the BDI may have a history of psychiatric conditions or may not have a history of psychiatric conditions.[22] The BDI test is a self-reporting questionnaire that requires test takers to report symptoms on a four-point scale.[23] Exams such as the HAM-A and BDI allow clinicians to ascertain additional information to develop a more in-depth understanding of the patient's symptoms and level of discomfort with their current condition.

While psychiatric evaluations have been standardized in some areas, they lack continuity in other regards. Standardized psychiatric interviews and assessments such as the HAM-A and the BDI test are utilized by numerous clinicians throughout the field of psychiatry. While these techniques are used by many professionals, standardized guidelines are not available regarding their use. Instead, psychiatrists rely upon loose recommendations to direct their use in clinical practice, which can result in varying data regarding their efficacy. As a result, clinicians have worked to more adequately standardize psychiatric assessments to improve the standard of care for patients presenting with psychiatric symptoms.[24]

Psychosocial Treatment Methods

Psychiatric conditions have been stigmatized and undertreated for numerous years. While many individuals have shied away from psychosocial treatment methods, various therapy methods have recently become more widely utilized. Psychotherapy, art/music therapy, and psychoeducation are methods that have become increasingly utilized and are worthy of further exploration.

Psychotherapy

Psychotherapy is the primary approach to mental health management and treatment that mental health professionals provide.[25] Psychotherapy, also referred to as "talk therapy," enables patients to discuss experiences and issues they have experienced throughout their lives. This therapy involves an interpersonal relationship between the patient and the therapist. It requires the therapist to recall psychological methodologies to assist the patient in mediating their difficulties.[26]

Cognitive behavioral therapy is a widely practiced sub-specialty of psychotherapy that explores the intersection of thoughts, feelings, and behaviors. The primary goal of cognitive behavioral therapy is to focus on current symptoms and analyze the negative thought patterns associated with them.[27] Therapists then work with the patient to adjust the thought processes related to the negative thought patterns the patient experiences. This process enables the individual to be aware of their thinking and reframe and control their thought processes.[28] Cognitive behavioral therapy does not traditionally focus on past events. It is primarily practiced for a select number of sessions allowing the therapist to focus on primary symptoms displayed by the patient.

Cognitive-behavioral therapy utilizes techniques such as establishing collaboration and Socratic questioning to enhance the interpersonal relationship between the patient and the therapist. Establishing collaboration in cognitive behavioral therapy enables the therapist to gather information from the patient, encourage patient involvement in decision-making, and obtain client feedback from sessions.[29] Socratic questioning allows the therapist to perform guided therapy and ask discovery-oriented questions.[30] Discovery-oriented questions can be thought-provoking, allowing the patient to take charge when addressing their troubles in therapy sessions.

The efficacy of psychotherapy techniques is contingent upon numerous factors. A patient's willingness to participate in therapeutic interventions can have a significant impact on the effectiveness of treatment. A recent study explored client preferences in overall psychotherapy outcomes. As a result of this study, it was found that therapists who incorporated patient preferences in psychotherapy sessions had a reduced rate of patient dropouts compared to patients who had non-preferred treatments.[31] Studies such as this demonstrate that patients who are actively involved in therapeutic decision-making are more likely to continue with the treatment protocols.

Psychotherapy techniques are effective in treating patients with various mental disorders. When examining the efficacy of psychotherapy, treatment methods can be broken down into common and special factors. Common factors include the bond between the patient and the therapist and agreement about the goals and tasks of therapy.[32] Specific factors include the specific type of psychotherapy and the specific type of psychological disorder. While psychotherapy has provided promising results, it is not yet clear how psychotherapy works to enhance behavior in patients presenting with psychological conditions. The efficacy of cognitive-behavioral therapy could come from specific common factors, specific special factors, or a combination of both.[33]

Art and Music Therapy

Art and music therapy are specialized disciplines of therapy that focus on the creative processes of the patient. The early specialization of art and music therapy primarily focused on the aspects of healing.[34] Art as healing and music as healing were conducted in group sessions focused primarily on diversion from hardships that these creative processes could employ.[35] Currently, art and music therapy have developed into a process in which the therapist takes a more active role in the discoveries that are formed due to the creative process developed by the patient.

Music therapy can be practiced and utilized through numerous techniques. Some of the primary methods practiced include active interventions, receptive interventions, and combined interventions. Active interventions allow patients to take part in playing music actively; this could consist of playing instruments or singing.[36] Receptive interventions will enable the participant to listen to music; this could result from recorded music or live music entertainment.[37] Finally, combined interventions will allow the participant to actively play and listen to music.[38]

Music therapy, in conjunction with other traditional psychiatric therapeutic interventions, such as talk therapy, has been beneficial in alleviating symptoms associated with numerous debilitating conditions. In a randomized controlled experimental trial exploring the emotional well-being of students in grades eight and nine, students in the experimental group with music therapy sessions displayed decreased anxiety symptoms and increased emotional competence.[39] Similarly, in a study exploring the effectiveness of music therapy for women with severe depression, it was found that the women exposed to music therapy had a significantly decreased depression score.[40]

Art therapy can be utilized and practiced through traditional techniques and more advanced technological techniques. Traditionally, art therapy has been practiced using raw materials such as paper, glue, paint, colored pencils, crayons, etc. As a result of new technological innovations, art therapy has now extended to techniques that utilize enhanced technology or electronic arts.[41] These advancements may include digital photography and photo editing software, videography and video editing software, computer painting, or online art design studios. With improvements in digital technology, art therapy techniques have reached and assisted a more diverse patient audience.

The utilization of art therapy in conjunction with additional therapeutic methods has improved symptoms in patients suffering from psychiatric conditions. A randomized controlled trial explored the effect of art therapy on patients with personality disorders.[42] In patients with personality disorders, art therapy has indicated better mental health and self-regulation.[43] Similarly, art therapy has been explored in studies that analyze anxiety levels in women. In a study that measured the effectiveness of art therapy, it was shown that the therapy was effective in reducing anxiety levels in self-reported emotion control, working memory, and task monitoring.[44]

Art therapy and music therapy have both been beneficial in improving patients' symptoms with both psychological and psychiatric conditions. While art and music therapy have been valuable for numerous patients, an approach focused on a more standardized methodology may improve the symptoms of a broader range of patients and provide more generalizable results for this discipline of therapy.[45]

Psychoeducation

Psychoeducation was developed as a result of the Deinstitutionalize movement of the 1950s and 1960s, which involved the closing of mental health institutions to integrate individuals with mental health conditions into society.[46] Psychoeducation is a classification of psychosocial therapy that focuses on teaching patients and families about their psychiatric conditions.[47] This technique could include sessions that deeply inform the patient of the symptoms associated with the condition. Additionally, patients may be given information regarding medication and management strategies.[48]

Psychoeducation techniques may be practiced with an individual patient, family, or in a group setting. While psychoeducation can be practiced in numerous environments, the models of psychoeducation are as follows:

1) Information model
2) Skills training model
3) Supportive model
4) Comprehensive model[49]

The information model of psychoeducation requires the clinician to focus on providing information associated with a psychiatric condition.[50] This information could be divulged to both the patient and their family and aims to give individuals a better understanding of the condition that they're facing.

The skill training model of psychoeducation allows the clinician to develop skills with the client. This teaching method enables the patient and their family members to better accept the condition and deal with the symptoms more effectively.[51] For example, skills may include mechanisms to deal with heightened anxiety, such as diversion techniques or deep breathing methods.

The supportive model of psychoeducation involves a supportive model of thinking. For example, the supportive psychoeducation model may include involvement in support groups with family members and other patients.[52] The supportive model of psychoeducation aims to guide patients and their families in exploring their thoughts and feelings in a supportive environment with like-minded individuals.

The comprehensive psychoeducation model uses a combination of the information model, the skills training model, and the supportive model to assist patients.[53] Through this psychoeducation method, patients can learn about their conditions, explore skills and mechanisms to help with symptoms and obtain support from family members and like-minded peers.

Psychoeducation techniques have helped mediate symptoms associated with numerous conditions. In a study exploring the implementation of psychoeducation into metacognitive therapy for patients with schizophrenia, it was found that the implementation of psychoeducation could contribute to an individual perception of their illness.[54] This heightened perception can promote recovery in patients diagnosed with schizophrenia disorder. Similarly, patients with borderline personality disorder reported that psychoeducation techniques were helpful with their treatment process.[55]

Medications

Prescription medication can be utilized to assist with the symptoms that come as a result of psychiatric conditions. Medication is frequently only prescribed if patient symptoms are severe and other therapeutic attempts have been ineffective in reducing the discomfort experienced by the patient. Medication for psychiatric conditions may only be prescribed by a psychiatrist or another attending physician with a medical degree. Therapists and counselors without a medical degree can still conduct therapeutic sessions with patients with psychiatric conditions but must refer a patient who meets the criteria for prescription medication to a psychiatrist for all medication inquiries.

Antipsychotics

Antipsychotics are a medication used to treat disorders that bring about psychotic symptoms. These conditions may include schizophrenia, bipolar disorder, borderline personality disorder, etc. Antipsychotics are categorized as first-generation antipsychotics or second-generation antipsychotics.

First-generation antipsychotics were the original class of antipsychotics created for patients presenting with psychiatric symptoms. This class of medication works as a dopamine receptor antagonist. Dopamine receptor antagonists reduce the effect of dopamine within the brain. This process is done through a block of dopamine receptors in the brain, which can be beneficial in altering the behaviors exhibited by those experiencing extreme psychosis.[56] First-generation antipsychotics include medications such as haloperidol, chlorpromazine, and fluphenazine.[57] While first-generation antipsychotic medications are not used as frequently as second-generation antipsychotics, they are still a beneficial treatment option for a variety of patients.

Second-generation antipsychotics are a newer class of antipsychotics that were developed as an improved option for patients suffering from psychosis. Second-generation antipsychotics work to block dopamine receptors and 5HT2A receptors in the brain.[58] Second-generation antipsychotics include amisulpride, aripiprazole, clozapine, olanzapine, quetiapine, and risperidone.[59]

Antipsychotics are frequently utilized in clinical practice, but these medications have been shown to produce numerous side effects. Side effects of antipsychotic medication may include weight gain, sleep disorders, severe sedation, difficulties with social activities, movement disorders, etc.[60]

Since side effects can influence the quality of life of a patient prescribed antipsychotics, clinicians have worked to decrease the likelihood of debilitating symptoms. A reduction of side effects has been combated by assessing the actual benefits of the medication to the patient, lowering medication dosage, and exploring the usage of concomitant medications to prevent general symptoms that may occur, such as constipation.[61] While side effects are prevalent in both first and second-generation antipsychotics, second-generation antipsychotics traditionally have side effects that are not as severe as first-generation antipsychotics.[62] The lower severity of side effects can contribute to higher treatment adherence rates in those prescribed second-generation antipsychotics.[63]

Antipsychotics are the primary course of treatment and have the potential to improve the quality of life in patients experiencing psychosis substantially. When exploring the type of antipsychotic recommendation for a patient, clinicians must analyze the best fit for medication on an individualized patient basis. While first-generation antipsychotics tend to be less tolerable than second-generation antipsychotics, first-generation antipsychotics are more affordable, which can be a beneficial option for patients suffering financially.[64] Additionally, while some patients may not tolerate side effects associated with first-generation antipsychotics, others may have no problems with these medications.

Benzodiazepines

Benzodiazepines are traditionally utilized to treat the symptoms associated with anxiety-related disorders and are currently prescribed to over 5% of the population.[65] This category of medication works to enhance the effect of Gamma-aminobutyric acid (GABA) production in the brain. GABA is a neurotransmitter in the brain that works to produce a calming effect throughout the human body.[66]

The common types of benzodiazepines are alprazolam (Xanax), Clonazepam (Klonopin), Diazepam (Valium), Lorazepam (Ativan), and Midazolam (Versed).[67] In patients with decreased levels of GABA, benzodiazepines enhance the sedating effect that is missing due to a lack of GABA production, enabling the patient to have decreased anxiety symptoms. The process enhanced by benzodiazepines can be especially useful in patients suffering from Generalized Anxiety Disorder and other anxiety-related disorders such as Post-Traumatic Stress Disorder or Obsessive-Compulsive Disorder.

Benzodiazepines have extensive benefits for patients suffering from psychiatric conditions, but they also have significant risk factors that the user and clinician should consider when weighing the benefits of Benzodiazepine use. In pregnant women, this medication can potentially cause premature birth and low birth weight.[68] Traditional users can experience fatigue, memory loss, and drug dependency or addiction.[69]

Benzodiazepine addiction has become an issue that has influenced the lives of numerous patients prescribed benzodiazepines. A total of 12.6% of the population in the United States reported benzodiazepine use in a 2015 and 2016 survey.[70] Misuse of benzodiazepines accounted for 17.2% of the 12.6% of the population that utilizes benzodiazepines in the United States.[71] Similarly, in a recent study that explored benzodiazepine usage from prescriptions written for patients perioperatively, 19.5% of patients continued to use benzodiazepines after surgery persistently.[72]

To combat the issue of addiction among benzodiazepine users, clinicians have worked to improve prescribing practices for patients. While benzodiazepines have traditionally been utilized as a long-term medication, research suggests that this medication should only be prescribed for a short period.[73] Short-term benzodiazepine use is beneficial for the treatment of psychiatric symptoms and also reduces the likelihood of dependency.

Patients on benzodiazepines for an extended period have reported that they have not been informed of the risks that could occur through the long-term use of the medication.[74] They also reported difficulty in discontinuing the medication and also reported symptoms that remained even after the drug was stopped.[75]

To assist patients with the discontinuation of benzodiazepines, clinicians have learned that this category of medication should be discontinued through a tapering process. Medication tapers must be tailored to the specific patient in an attempt to prevent withdrawal symptoms and reduce potential addictive properties.[76]

Antidepressants

Depression is a common psychiatric condition that affects over 4.4% of the global population.[77] Antidepressants are traditionally utilized for the mediation of depressive disorders and other anxiety disorders that are not resolved with conventional therapy sessions. These disorders may include Major Depressive Disorder, Bipolar Depression, and Generalized Anxiety Disorder.

The types of antidepressants that are frequently used in practice are selective serotonin reuptake inhibitors (SSRIs), serotonin-norepinephrine reuptake inhibitors (SNRIs), and norepinephrine and specific serotonergic antidepressants (NASSAs). SSRIs work to increase serotonin levels that are deficient in patients experiencing depression. This occurs by inhibiting the presynaptic reuptake of serotonin in the brain, therefore increasing serotonin at the postsynaptic membrane.[78] SSRIs are utilized the most frequently for depressive disorders due to their efficacy and a lower likelihood of side effects. The primary SSRIs utilized in the United States are vilazodone, citalopram, paroxetine, sertraline, fluoxetine, escitalopram, and fluvoxamine.[79]

Serotonin-norepinephrine reuptake inhibitors (SNRIs) are a type of antidepressant that works similarly to SSRIs. SNRIs work by inhibiting serotonin and norepinephrine reuptake in the brain.[80] This class of antidepressants has been beneficial in assisting with symptoms associated with depression, anxiety disorders, and even chronic pain. The common SNRIs approved for the treatment of depression are desvenlafaxine, duloxetine, levomilnacipran, and milnacipran.[81]

The primary side effects associated with SNRIs are insomnia, tiredness, nausea, change of appetite, and dry mouth.[82] Clinicians prescribing SNRIs must be aware of the potential side effects that can result from the withdrawal of these types of medications. In patients prescribed SNRIs, it was reported that withdrawal symptoms could last up to a few weeks.[83] Given the potential of withdrawal symptoms, a taper-down method must be utilized in patients prescribed SNRIs.

Norepinephrine and specific serotonergic antidepressants (NASSAs) are another class of antidepressants that work similarly to SSRIs. NASSAs work to increase norepinephrine and serotonin between nerve synapses and block serotonin receptors. NASSAs are typically prescribed to patients who have not experienced relief from their symptoms after using typical SSRIs or SNRIs. This type of antidepressant can also benefit patients experiencing adverse side effects associated with SSRIs and SNRIs. The primary NASSAs utilized in practice include mirtazapine, mianserin, and setiptiline.[84]

Similar to patients prescribed SSRIs or SNRIs, patients prescribed NASSAs have a high likelihood of experiencing withdrawal symptoms. After discontinuation of NASSAs, patients reported symptoms of dizziness, vomiting, anxiety, and panic attacks.[85] These reported symptoms were prevalent regardless of whether the medication was stopped abruptly or used a taper-down method.[86]

As antidepressant use has continued to increase, practitioners have developed techniques to assist with deprescribing methods. Clinicians have continued to create enhanced drug tapering methods to help patients attempting to discontinue a psychiatric medication.[87] Patients experiencing withdrawal symptoms can utilize a taper program over a few months which slowly reduces the potency of the drug.[88]

Institutionalization

Psychiatric conditions can often be debilitating for those affected and are frequently difficult for clinicians to treat. The adverse effects of these disorders can not only be difficult for the patient but also for family members, especially when affected individuals become difficult to reason with or care for. Given the difficulties associated with psychiatric disorders, institutionalization has assisted in giving families and patients the extra support they may require.

Hospital Psychiatric Hold

Hospital psychiatric holds have become more prevalent in the United States as the mental health field has developed and psychiatric care has become prioritized. A psychiatric hold aims to assist an individual experiencing extreme psychiatric symptoms that may be a danger to themselves or others. Psychiatric symptoms that warrant a psychiatric hold may include severe depression, hallucinations, and delusions. The two types of psychiatric holds are voluntary psychiatric hold and involuntary psychiatric hold. A voluntary psychiatry hold or voluntary commitment is when a patient willingly seeks assistance from emergency medical personnel for their psychiatric symptoms.

When a patient presents at the emergency department for a voluntary psychiatric hold, they must agree to give 72-hour written notice before leaving the facility.[89] This 72-hour time frame gives mental health professionals time to assess the patient and develop a plan of action for the patient after they are released from the hospital.[90] This may include transfer to a psychiatric hospital or outpatient therapy. Additionally, this

72-hour requirement can give patients time to settle down if they are experiencing a psychiatric crisis, or are experiencing difficulties through the therapeutic process.

Patients experiencing extreme psychiatric symptoms often have trouble with judgment and insight and lose the ability to act autonomously. Since patients who need psychiatric assistance have difficulty recognizing when they need help, involuntary psychiatric holds have become an essential aspect of psychiatric care.[91]

When a patient is recommended for an involuntary psychiatric hold, they are required to be admitted to the emergency room and must be evaluated by a physician to determine if an involuntary psychiatric hold is necessary. If a psychiatric hold is deemed necessary, the patient will be admitted to the hospital for treatment for a maximum of 120 hours.[92] For the patient to be considered for a longer length of stay, the patient must be reevaluated, and in some states, a court order must be provided to extend the length of stay. While policy and legislation regarding involuntary psychiatric holds differ by state, the primary criteria is that an individual must be a danger to themselves or others.[93]

Patients on an involuntary psychiatric hold may have no history of psychiatric illness. Some patients admitted to a psychiatric department may come due to suicidal thoughts resulting from health conditions such as chronic pain or other life-altering events.[94] The primary goal of involuntary psychiatric holds is to assist in ensuring the well-being and safety of individuals suffering from psychiatric conditions and those that surround them.

In a study that explored patients' beliefs about their need for psychiatric hospitalization, 52% of patients who initially said they did not need hospitalization revised their belief upon discharge and said they did need hospitalization.[95] This study assists in clarifying the concept that during extreme psychiatric crises, patients may not truly be able to act autonomously regarding decision-making. Similarly, involuntary hospitalization does not influence patient length of stay or readmission risks.[96]

State-run Institutions

In 1955, state-run institutions provided beds for over 560,000 patients.[97] When mental illness was extremely stigmatized, individuals with psychiatric problems and developmental disabilities were housed in state-run mental institutions. At the time, the goal of this type of institutionalization was to house patients with psychiatric conditions far away from the general population.

Many patients who were institutionalized away from society had the potential to be functional in society through assistance with therapy and medication protocols. Other patients were institutionalized but had no real hope of living a functional independent life without the direct supervision of medical personnel.

Throughout history, many state-run institutions have lacked quality patient care. In early history, patients in mental institutions were often experimented on without their consent and were left to fend for themselves. As a result of a lack of oversight and federal funding, state-run institutions quickly underwent harsh criticism.[98] In the 1960s state-run mental health institutions became known as a place that did not provide adequate or effective treatment for patients.[99] These institutions required extensive policy adjustment, but instead, the United States went through a rapid deinstitutionalization movement.

The deinstitutionalization movement in the United States worked to integrate individuals who were institutionalized with mental conditions into the general population. Society began to become more accepting of individuals with mental illnesses and recognized that many could be reintegrated into society. The Community Mental Health Construction Act passed in 1963, made federal grants available to states who developed assistance programs for patients with psychiatric conditions.[100] As a result of the 1963 act and the introduction of the federal Medicaid system, individuals with psychiatric conditions became a partial responsibility of the federal government, rather than the primary responsibility of the state, further motivating states to close down state-run mental health institutions.[101]

As a result of the deinstitutionalization movement, individuals who can function in society with therapy and psychiatric medication have lived productive lives. On the other hand, individuals who have been unable to live everyday life with therapy and medication have substantially struggled. Individuals with severe psychiatric conditions cannot function without the assistance of others, traditionally cannot stay employed, have increased incarceration rates, and also have an increased rate of homelessness.[102]

While state-run mental institutions received a bad reputation for the lack of quality care in early history, they did serve as a crucial home and resource for individuals who could not be integrated into society or function without the assistance of medical personnel. While state-run mental health institutions still exist in some states, the number of institutions and the number of patients admitted have been reduced dramatically. The closure of these institutions has left a substantial gap in long-term care methods for patients suffering from extreme psychiatric conditions.

Forensic Psychiatric Institutions

Forensic psychiatry involves the assessment and treatment of individuals who have psychiatric conditions that have led to criminal offenses or have the potential to lead to criminal offenses.[103] Forensic psychiatric institutions could be within a specialized secure hospital, a prison system, or within the community

Patients housed in a forensic psychiatric institution for a short-term period have a 36% chance of recidivism within two years of release.[104] Patients with substance abuse disorders and antisocial personality disorder are associated with higher rates of recidivism.[105] The primary goal of forensic psychiatric institutions is to treat admitted patients and reduce the likelihood of recidivism after patients are released from the institution. To decrease rates of recidivism, forensic institutions have seen that continued contact with patients after treatment can be beneficial.[106] As new research continues to be developed, clinicians can utilize this data to direct future care.

In an attempt to predict recidivism rates before they occur, researchers have explored various practices that enable a clinician to assess a patient's likelihood of recidivism after care in a forensic psychiatric institution.[107] Neuroimaging has been recently explored for predicting recidivism rates. Through measuring cerebral blood flow throughout various regions of the brain through neuroimaging clinicians may have the capability of predicting recidivism rates in the future.[108]

Most patients admitted to forensic psychiatric institutions are legally required to stay within the institution. Given the lack of autonomy that many of these patients face as a result of their institutionalization, nurses have worked to enhance decision-making processes for the patients wherever they see suitable.[109] While the patients have limited opportunity for decision-making within the institution and many patients are not willing to respond to decision-making opportunities from the nursing staff, the staff still aims to allow patients to act autonomously whenever feasible.

Forensic psychiatric institutions provide a level of care that is significantly different than traditional psychiatric hospitalizations or institutionalization. Patients in forensic psychiatric institutions are often legally required to stay in the institution for an extended duration of time. Given these specific requirements, the standard of care in forensic psychiatric institutions must be different. For example, in forensic psychiatric institutions, a heavy focus on the environment the patient resides in must be considered in an attempt to establish an environment that remains therapeutic.[110]

The safety of individuals institutionalized in forensic psychiatric units is essential for effective psychiatric care for those who are institutionalized. In a study exploring patient risk for victimization, it was found that 31.3% of patients experienced victimization at some point during their time of institutionalization.[111] As a result of this study, forensic psychiatric institutions must continue to develop and enforce strict policies and procedures for patients and staff to abide by.

Researchers have recently begun to explore the usage of transitional housing in the discipline of forensic psychiatry. Patients who have exceeded the use of a forensic psychiatry institution and do not require close constant care will be beneficial candidates for transitional housing. Transitional housing can be used as a step-down facility for patients who no longer require intense psychiatric care, but also are not ready to begin an independent home treatment plan.[112] Transitional housing can create a beneficial stepping stone for some psychiatric patients, but since this method of living can be challenging for patients with a history of forensic psychiatric conditions a thorough screening process must be conducted.

Conclusion

The field of psychiatry has progressed rapidly over the past 100 years. Diagnostic qualifications and procedures for psychiatric conditions have been advanced. These methods of practice include the utilization of physical examinations, laboratory tests, and psychiatric evaluations. A combination of these three methods ensures that patients are diagnosed properly and physical symptoms do not go overlooked.

In the realm of treatment procedures, therapeutic techniques have been developed and enhanced. Psychotherapy methods have continued to be a dominating force in the field of psychiatry. Additionally, art and music therapy and psychoeducation have begun to show great positive improvements in patient care and should continue to be studied for efficacy in patient care.

Prescribing procedures and medication protocols in the field of psychiatry have also been enhanced. Medications that were once prescribed frequently by psychiatrists are now being recognized for their addictive properties. While some medications may be beneficial for patients experiencing psychiatric conditions, others may come with symptoms that do not outweigh the benefits.

Finally, developed institutionalization procedures for psychiatric care provide substantial promise for patients in need of care. Psychiatric hospital holds provide the most feasible option for patients experiencing psychiatric symptoms now, but do not provide substantial long-term care.

While diagnostic and treatment protocols have been developed substantially, patients with psychiatric conditions are still not receiving adequate care. For patients suffering from treatment-resistant psychiatric disorders, or who are unable to function in society independently, a gap in substantial long-term care is prevalent and must be addressed.

Notes

[1] Kenneth S. Kendler, Kathryn Tabb, and John Wright, "The Emergence of Psychiatry: 1650–1850," *American Journal of Psychiatry* 179, no. 5 (May 2022): 329–35.
[2] Kendler, Tabb, and Wright.
[3] Christos Tsagkaris and Konstantinos Kalachanis, "The Hippocratic Account of Mental Health: Humors and Human Temperament," *Mental Health: Global Challenges Journal* 3, no. 1 (September 20, 2020): 33–37.
[4] Kendler, Tabb, and Wright.
[5] Kendler, Tabb, and Wright.
[6] Robert Kohn, et al., "Mental Health in the Americas: An Overview of the Treatment Gap," *Revista Panamericana de Salud Pública* 42 (October 18, 2018).
[7] Kohn, et al.
[8] Brian D. Berman, "Neuroleptic Malignant Syndrome," *The Neurohospitalist* 1, no. 1 (January 2011): 41–47.
[9] Michael Esang, Sabina Goldstein, and Ravina Dhami, "The Role of Physical Examinations in Psychiatry as Illustrated in a Case of Neuroleptic Malignant Syndrome Versus Viral Encephalitis: A Case Report and Literature Review," *Cureus* 11, no. 6 (n.d.): e4840.
[10] Esang, Goldstein, and Dhami.
[11] Andrew Peter Vanezis and Dudley Manns, "Physical Examinations of Mental Health Service Users," *Progress in Neurology and Psychiatry* 14, no. 4 (2010): 19–23.
[12] Gill Garden, "Physical Examination in Psychiatric Practice," *Advances in Psychiatric Treatment* 11, no. 2 (March 2005): 142–49.
[13] Rakesh Kumar Singh and Suprakash Chaudhury, "Hyponatremia-Induced Psychosis in an Industrial Setting," *Industrial Psychiatry Journal* 18, no. 2 (2009): 137–38.
[14] Xu Zhang and Xiao-Ying Li, "Prevalence of Hyponatremia among Older Inpatients in a General Hospital," *European Geriatric Medicine* 11, no. 4 (August 2020): 685–92.

[15] Andrea Hall, Talha Bin Farooq, and Renato Alcaraz, Jr, "Hypocalcemia-Induced Reversible Psychosis," *Cureus* 14, no. 1 (January 2, 2022).
[16] Jerry McKee and Nancy Brahm, "Medical Mimics: Differential Diagnostic Considerations for Psychiatric Symptoms," *The Mental Health Clinician* 6, no. 6 (November 3, 2016): 289–96.
[17] Edward L. Zuckerman, *Clinician's Thesaurus, 8th Edition: The Guide to Conducting Interviews and Writing Psychological Reports* (Guilford Publications, 2019).
[18] Zuckerman.
[19] "Practice Guidelines for the Psychiatric Evaluation of Adults, Third Edition," in *The American Psychiatric Association Practice Guidelines for the Psychiatric Evaluation of Adults*, Third, Practice Guidelines (American Psychiatric Association, 2015).
[20] "Practice Guidelines for the Psychiatric Evaluation of Adults, Third Edition."
[21] Amanda Baker, et al., "Anxiety Symptoms Questionnaire (ASQ): Development and Validation," *General Psychiatry* 32, no. 6 (December 18, 2019): e100144.
[22] Gordon Jackson-Koku, "Beck Depression Inventory," *Occupational Medicine* 66, no. 2 (March 2016): 174–75.
[23] Jackson-Koku.
[24] Mark Zimmerman, "What Should the Standard of Care for Psychiatric Evaluations Be?," *The Journal of Nervous and Mental Disease* 191, no. 5 (May 2003): 281–86.
[25] Cosima Locher, Sibylle Meier, and Jens Gaab, "Psychotherapy: A World of Meanings," *Frontiers in Psychology* 10 (March 22, 2019): 460.
[26] T. Munder, et al., "Is Psychotherapy Effective? A Re-Analysis of Treatments for Depression," *Epidemiology and Psychiatric Sciences* 28, no. 03 (June 2019): 268–74.
[27] *Cognitive Behavioral Therapy*, *InformedHealth.Org [Internet]* (Institute for Quality and Efficiency in Health Care (IQWiG), 2016), http://www. ncbi.nlm.nih.gov/books/NBK279297/.
[28] *Cognitive Behavioral Therapy*.
[29] Annika Okamoto, et al., "The Therapeutic Relationship in Cognitive–Behavioral Therapy: Essential Features and Common Challenges," *Practice Innovations* 4, no. 2 (June 2019): 112–23.
[30] Okamoto, et al.
[31] Joshua K. Swift, et al., "The Impact of Accommodating Client Preference in Psychotherapy: A Meta-Analysis," *Journal of Clinical Psychology* 74, no. 11 (2018): 1924–37.
[32] Pim Cuijpers, Mirjam Reijnders, and Marcus J.H. Huibers, "The Role of Common Factors in Psychotherapy Outcomes," *Annual Review of Clinical Psychology* 15, no. 1 (May 7, 2019): 207–31.
[33] Cuijpers, Reijnders, and Huibers.
[34] Andrea Gilroy and Colin Lee, *Art and Music: Therapy and Research* (Routledge, 2019).
[35] Gilroy and Lee.

[36] Maria Facchini and Chiara Ruini, "The Role of Music Therapy in the Treatment of Children with Cancer: A Systematic Review of Literature," *Complementary Therapies in Clinical Practice* 42 (February 2021): 101289.
[37] Facchini and Ruini.
[38] Facchini and Ruini.
[39] Sylvia Y. C. L. Kwok, "Integrating Positive Psychology and Elements of Music Therapy to Alleviate Adolescent Anxiety," *Research on Social Work Practice* 29, no. 6 (September 1, 2019): 663–76.
[40] Najmeh Hamid and Ali Biat, "The Effectiveness of Music Therapy on Depression and Happiness of Depressed Women," *NeuroQuantology* 17, no. 12 (2019): 19–26.
[41] Ms Cathy A. Malchiodi, *The Handbook of Art Therapy and Digital Technology* (Jessica Kingsley Publishers, 2018).
[42] Suzanne Haeyen, et al., "Efficacy of Art Therapy in Individuals With Personality Disorders Cluster B/C: A Randomized Controlled Trial," *Journal of Personality Disorders* 32, no. 4 (August 2018): 527–42.
[43] Haeyen, et al.
[44] Annemarie Abbing, et al., "Anxiety Reduction through Art Therapy in Women. Exploring Stress Regulation and Executive Functioning as Underlying Neurocognitive Mechanisms," *PLOS ONE* 14, no. 12 (December 3, 2019): e0225200.
[45] Mathew Chiang, William Bernard Reid-Varley, and Xiaoduo Fan, "Creative Art Therapy for Mental Illness," *Psychiatry Research* 275 (May 2019): 129–36.
[46] Sujit Sarkhel, O. P. Singh, and Manu Arora, "Clinical Practice Guidelines for Psychoeducation in Psychiatric Disorders General Principles of Psychoeducation," *Indian Journal of Psychiatry* 62, no. Suppl 2 (January 2020): S319–23.
[47] Sarkhel, Singh, and Arora.
[48] Yasaman Mottaghipour and Maryam Tabatabaee, "Family and Patient Psychoeducation for Severe Mental Disorder in Iran: A Review," *Iranian Journal of Psychiatry* 14, no. 1 (January 2019): 84–108.
[49] Sarkhel, Singh, and Arora, "Clinical Practice Guidelines for Psychoeducation in Psychiatric Disorders General Principles of Psychoeducation."
[50] Sarkhel, Singh, and Arora.
[51] Sarkhel, Singh, and Arora.
[52] Sarkhel, Singh, and Arora.
[53] Sarkhel, Singh, and Arora.
[54] Libby Igra, et al., "'Making Sense of My Diagnosis': Assimilating Psychoeducation into Metacognitive Psychotherapy for Individuals with Schizophrenia," *Journal of Psychotherapy Integration* 31, no. 3 (September 2021): 277–90.
[55] Ingrid Tande Ditlefsen, et al., "'Yes, There Is Actually Hope!'—A Qualitative Investigation of How Patients Experience Mentalization-Based Psychoeducation Tailored for Borderline Personality Disorder," *Journal of Psychotherapy Integration* 31, no. 3 (September 2021): 257–76.
[56] "APA Dictionary of Psychology," accessed July 10, 2022, https://dictionary.apa.org/.

[57] Yirgalem Shewakena Wubeshet, Oumer Sada Mohammed, and Tigestu Alemu Desse, "Prevalence and Management Practice of First Generation Antipsychotics Induced Side Effects among Schizophrenic Patients at Amanuel Mental Specialized Hospital, Central Ethiopia: Cross-Sectional Study," *BMC Psychiatry* 19, no. 1 (December 2019): 32.
[58] Hans-Jürgen Möller, "Antipsychotic and Antidepressive Effects of Second Generation Antipsychotics: Two Different Pharmacological Mechanisms?," *European Archives of Psychiatry and Clinical Neuroscience* 255, no. 3 (June 2005): 190–201.
[59] Stefan Leucht, et al., "Second-Generation versus First-Generation Antipsychotic Drugs for Schizophrenia: A Meta-Analysis," *The Lancet* 373, no. 9657 (January 2009): 31–41.
[60] Wubeshet, Mohammed, and Desse.
[61] "Management of Common Adverse Effects of Antipsychotic Medications," accessed July 26, 2022.
[62] Merhawi Bahta, et al., "Impact of Adverse Reactions to First-Generation Antipsychotics on Treatment Adherence in Outpatients with Schizophrenia: A Cross-Sectional Study," *Annals of General Psychiatry* 20, no. 1 (December 2021): 27.
[63] Bahta, et al.
[64] Stefan Leucht, Maximilian Huhn, and John M. Davis, "Should 'Typical', First-Generation Antipsychotics No Longer Be Generally Used in the Treatment of Schizophrenia?," *European Archives of Psychiatry and Clinical Neuroscience* 271, no. 8 (December 1, 2021): 1411–13.
[65] Donovan T. Maust, Lewei A. Lin, and Frederic C. Blow, "Benzodiazepine Use and Misuse Among Adults in the United States," *Psychiatric Services* 70, no. 2 (February 2019): 97–106, https://doi.org /10.1176/appi.ps.201800321.
[66] Mary J. Allen, Sarah Sabir, and Sandeep Sharma, "GABA Receptor," in *StatPearls* (Treasure Island (FL): StatPearls Publishing, 2022).
[67] Edilma Sanabria, et al., "Benzodiazepines: Their Use Either as Essential Medicines or as Toxics Substances," *Toxics* 9, no. 2 (February 2021): 25.
[68] Amber N. Edinoff, et al., "Benzodiazepines: Uses, Dangers, and Clinical Considerations," *Neurology International* 13, no. 4 (November 10, 2021): 594–607.
[69] Edinoff, et al.
[70] Maust, Lin, and Blow, "Benzodiazepine Use and Misuse Among Adults in the United States."
[71] Maust, Lin, and Blow.
[72] Jason D. Wright, et al., "Association of New Perioperative Benzodiazepine Use With Persistent Benzodiazepine Use," *JAMA Network Open* 4, no. 6 (June 3, 2021): e2112478.
[73] Carrie M. Silvernail and Steven L. Wright, "Surviving Benzodiazepines: A Patient's and Clinician's Perspectives," *Advances in Therapy* 39, no. 5 (May 2022): 1871–80.

[74] Alistair J. Reid Finlayson, et al., "Experiences with Benzodiazepine Use, Tapering, and Discontinuation: An Internet Survey," *Therapeutic Advances in Psychopharmacology* 12 (January 1, 2022): 20451253221082384.
[75] Reid Finlayson, et al.
[76] Joseph V. Pergolizzi Jr, Jo Ann LeQuang, and Robert B. Raffa, "Benzodiazepines: Thinking Outside the Black Box," *Journal of Clinical Pharmacy and Therapeutics* 46, no. 3 (2021): 554–59.
[77] Amber N. Edinoff, et al., "Selective Serotonin Reuptake Inhibitors and Adverse Effects: A Narrative Review," *Neurology International* 13, no. 3 (September 2021): 387–401.
[78] Edinoff, et al.
[79] Andrew Chu and Roopma Wadhwa, "Selective Serotonin Reuptake Inhibitors," in *StatPearls* (Treasure Island (FL): StatPearls Publishing, 2022).
[80] Randy A. Sansone and Lori A. Sansone, "Serotonin Norepinephrine Reuptake Inhibitors: A Pharmacological Comparison," *Innovations in Clinical Neuroscience* 11, no. 3–4 (2014): 37–42.
[81] Sansone and Sansone.
[82] Daniel Santarsieri and Thomas L Schwartz, "Antidepressant Efficacy and Side-Effect Burden: A Quick Guide for Clinicians," *Drugs in Context* 4 (October 8, 2015): 212290, https://doi.org/10.7573/dic.212290.
[83] Giovanni A. Fava, et al., "Withdrawal Symptoms after Serotonin-Noradrenaline Reuptake Inhibitor Discontinuation: Systematic Review," *Psychotherapy and Psychosomatics* 87, no. 4 (2018): 195–203.
[84] Michele Protti, et al., "New-Generation, Non-SSRI Antidepressants: Drug-Drug Interactions and Therapeutic Drug Monitoring. Part 2: NaSSAs, NRIs, SNDRIs, MASSAs, NDRIs, and Others," *Medicinal Research Reviews* 40, no. 5 (2020): 1794–1832.
[85] Fava, et al.
[86] Fiammetta Cosci, "Withdrawal Symptoms after Discontinuation of a Noradrenergic and Specific Serotonergic Antidepressant: A Case Report and Review of the Literature," *Personalized Medicine in Psychiatry* 1–2 (March 2017): 81–84.
[87] Tony Kendrick, "Strategies to Reduce Use of Antidepressants," *British Journal of Clinical Pharmacology* 87, no. 1 (2021): 23–33.
[88] Kendrick.
[89] "Voluntary and Involuntary Commitment," Erie County, PA, accessed July 29, 2022.
[90] "Reasonable or Random: 72-Hour Limits to Psychiatric Holds," accessed July 29, 2022.
[91] Megan Testa and Sara G. West, "Civil Commitment in the United States," *Psychiatry (Edgmont)* 7, no. 10 (October 2010): 30–40.
[92] "Voluntary and Involuntary Commitment."
[93] Leslie C. Hedman, et al., "State Laws on Emergency Holds for Mental Health Stabilization," *Psychiatric Services* 67, no. 5 (May 2016): 529–35.

[94] Christian Lachner, et al., "Are Pre-Existing Psychiatric Disorders the Only Reason for Involuntary Holds in the Emergency Department?," *International Journal of Emergency Medicine* 13, no. 1 (February 3, 2020).
[95] William Gardner, et al., "Patients' Revisions of Their Beliefs About the Need for Hospitalization," *American Journal of Psychiatry* 156, no. 9 (September 1999): 1385–91.
[96] Tyler A. Durns, et al., "Effects of Temporary Psychiatric Holds on Length of Stay and Readmission Risk among Persons Admitted for Psychotic Disorders," *International Journal of Law and Psychiatry* 76 (May 2021): 101695.
[97] Fuller Torrey, et al., "No Room at the Inn- Trends and Consequences of Closing Public Psychiatric Hospitals-2005-2010" (The Treatment Advocacy Center, July 19, 2012).
[98] Meighan B. Haupt, "The Vital Role of State Psychiatric Hospitals" (National Association of State Mental Health Program Directors, July 2014).
[99] Testa and West, "Civil Commitment in the United States."
[100] Daniel Yohanna, "Deinstitutionalization of People with Mental Illness: Causes and Consequences," *AMA Journal of Ethics* 15, no. 10 (October 1, 2013): 886–91.
[101] Yohanna.
[102] Antonio Ventriglio, et al., "Homelessness and Mental Health: A Challenge," *International Journal of Social Psychiatry* 61, no. 7 (November 1, 2015): 621–22.
[103] Paul E. Mullen, "Forensic Mental Health," *The British Journal of Psychiatry* 176, no. 4 (April 2000): 307–11.
[104] Petra Schaftenaar, et al., "Relational Caring and Contact after Treatment. An Evaluation Study on Criminal Recidivism," *International Journal of Law and Psychiatry* 60 (September 1, 2018): 45–50.
[105] Nirvana Morgan and Giada Del Fabbro, "Factors Associated with Recidivism at a South African Forensic Psychiatric Hospital," *South African Journal of Psychiatry* 24 (May 31, 2018).
[106] Schaftenaar, et al.
[107] Carl Delfin, et al., "Prediction of Recidivism in a Long-Term Follow-up of Forensic Psychiatric Patients: Incremental Effects of Neuroimaging Data," ed. Stephan Doering, *PLOS ONE* 14, no. 5 (May 16, 2019): e0217127.
[108] Delfin, et al.
[109] Emilie Magnusson, Anna Karin Axelsson, and Malin Lindroth, "'We Try' – How Nurses Work with Patient Participation in Forensic Psychiatric Care," *Scandinavian Journal of Caring Sciences* 34, no. 3 (2020): 690–97.
[110] Allan Seppänen, et al., "Modern Forensic Psychiatric Hospital Design: Clinical, Legal and Structural Aspects," *International Journal of Mental Health Systems* 12 (October 20, 2018): 58.
[111] Hali Griswold, et al., "Assessing Risk for Victimization in a Forensic Psychiatric Setting Using the Short-Term Assessment of Risk and Treatability," *Violence and Victims* 33, no. 6 (2018): 1012–35.
[112] Clark Patrick Heard, et al., "Transitional Housing in Forensic Mental Health: Considering Consumer Lived Experience," *Health & Justice* 7, no. 1 (May 20, 2019): 8.

Literature

Abbing, Annemarie, Leo de Sonneville, Erik Baars, Daniëlle Bourne, and Hanna Swaab. 2019. "Anxiety Reduction through Art Therapy in Women. Exploring Stress Regulation and Executive Functioning as Underlying Neurocognitive Mechanisms." *PLOS ONE* 14, no. 12 (December 3,): e0225200. https://doi.org/10.1371/journal.pone.0225200.

Allen, Mary J., Sarah Sabir, and Sandeep Sharma. 2022. "GABA Receptor." In *StatPearls*. Treasure Island (FL): StatPearls Publishing. http://www.ncbi.nlm.nih.gov/books/NBK526124/.

"APA Dictionary of Psychology." N.d. Accessed July 10, 2022. https://dictionary.apa.org/.

Bahta, Merhawi, Azieb Ogbaghebriel, Mulugeta Russom, Eyasu H. Tesfamariam, and Tzeggai Berhe. 2021. "Impact of Adverse Reactions to First-Generation Antipsychotics on Treatment Adherence in Outpatients with Schizophrenia: A Cross-Sectional Study." *Annals of General Psychiatry* 20, no. 1 (December): 27. https://doi.org/10.1186/s12991-021-00348-0.

Baker, Amanda, Naomi Simon, Aparna Keshaviah, Amy Farabaugh, Thilo Deckersbach, John J Worthington, Elizabeth Hoge, Maurizio Fava, and Mark P Pollack. 2019. "Anxiety Symptoms Questionnaire (ASQ): Development and Validation." *General Psychiatry* 32, no. 6 (December 18): e100144. https://doi.org/10.1136/gpsych-2019-100144.

Berman, Brian D. 2011. "Neuroleptic Malignant Syndrome." *The Neurohospitalist* 1, no. 1 (January): 41–47. https://doi.org/10.1177/1941875210386491.

Chiang, Mathew, William Bernard Reid-Varley, and Xiaoduo Fan. 2019. "Creative Art Therapy for Mental Illness." *Psychiatry Research* 275 (May): 129–36. https://doi.org/10.1016/j.psychres.2019.03.025.

Chu, Andrew, and Roopma Wadhwa. 2022. "Selective Serotonin Reuptake Inhibitors." In *StatPearls*. Treasure Island (FL): StatPearls Publishing. http://www.ncbi.nlm.nih.gov/books/NBK554406/.

Cognitive Behavioral Therapy. InformedHealth.Org [Internet]. 2016. Institute for Quality and Efficiency in Health Care (IQWiG). http://www.ncbi.nlm.nih.gov/books/NBK279297/.

Cosci, Fiammetta. 2017. "Withdrawal Symptoms after Discontinuation of a Noradrenergic and Specific Serotonergic Antidepressant: A Case Report and Review of the Literature." *Personalized Medicine in Psychiatry* 1–2 (March): 81–84.

https://doi.org/10.1016/j.pmip.2016.11.001.
Cuijpers, Pim, Mirjam Reijnders, and Marcus J.H. Huibers. 2019. "The Role of Common Factors in Psychotherapy Outcomes." *Annual Review of Clinical Psychology* 15, no. 1 (May 7): 207–31. https://doi.org/10.1146/annurev-clinpsy-050718-095424.

Delfin, Carl, Hedvig Krona, Peter Andiné, Erik Ryding, Märta Wallinius, and Björn Hofvander. 2019. "Prediction of Recidivism in a Long-Term Follow-up of Forensic Psychiatric Patients: Incremental Effects of Neuroimaging Data." Edited by Stephan Doering. *PLOS ONE* 14, no. 5 (May 16): e0217127. https://doi.org/10.1371/journal.pone. 0217127.

Ditlefsen, Ingrid Tande, Helene Amundsen Nissen-Lie, Agnes Andenæs, Eivind Normann-Eide, Merete Selsbak Johansen, and Elfrida Hartveit Kvarstein. 2021. "'Yes, There Is Actually Hope!'—A Qualitative Investigation of How Patients Experience Mentalization-Based Psychoeducation Tailored for Borderline Personality Disorder." *Journal of Psychotherapy Integration* 31, no. 3 (September): 257–76. https://doi.org/10.1037/int0000243.

Durns, Tyler A., Patrick H. O'Connell, Anna Shvartsur, Jessica S. Grey, and Brent M. Kious. 2021. "Effects of Temporary Psychiatric Holds on Length of Stay and Readmission Risk among Persons Admitted for Psychotic Disorders." *International Journal of Law and Psychiatry* 76 (May): 101695. https://doi.org/10.1016/j.ijlp.2021.101695.

Edinoff, Amber N., Haseeb A. Akuly, Tony A. Hanna, Carolina O. Ochoa, Shelby J. Patti, Yahya A. Ghaffar, Alan D. Kaye, et al. 2021. "Selective Serotonin Reuptake Inhibitors and Adverse Effects: A Narrative Review." *Neurology International* 13, no. 3 (September): 387–401. https://doi.org/10.3390/neurolint13030038.

Edinoff, Amber N., Catherine A. Nix, Janice Hollier, Caroline E. Sagrera, Blake M. Delacroix, Tunde Abubakar, Elyse M. Cornett, Adam M. Kaye, and Alan D. Kaye. 2021. "Benzodiazepines: Uses, Dangers, and Clinical Considerations." *Neurology International* 13, no. 4 (November 10): 594–607. https://doi.org/10.3390/neurolint 13040059.

Esang, Michael, Sabina Goldstein, and Ravina Dhami. N.d. "The Role of Physical Examinations in Psychiatry as Illustrated in a Case of Neuroleptic Malignant Syndrome Versus Viral Encephalitis: A Case Report and Literature Review." *Cureus* 11, no. 6: e4840. https://doi.org/10.7759/cureus.4840.

Facchini, Maria, and Chiara Ruini. 2021. "The Role of Music Therapy in the Treatment of Children with Cancer: A Systematic Review of Literature." *Complementary Therapies in Clinical Practice* 42 (February): 101289. https://doi.org/10.1016/j.ctcp.2020.101289.

Fava, Giovanni A., Giada Benasi, Marcella Lucente, Emanuela Offidani, Fiammetta Cosci, and Jenny Guidi. 2018. "Withdrawal Symptoms after Serotonin-Noradrenaline Reuptake Inhibitor Discontinuation: Systematic Review." *Psychotherapy and Psychosomatics* 87, no. 4: 195–203. https://doi.org/10.1159/000491524.

Garden, Gill. 2005. "Physical Examination in Psychiatric Practice." *Advances in Psychiatric Treatment* 11, no. 2 (March): 142–49. https://doi.org/ 10.1192/apt.11.2.142.

Gardner, William, Charles W. Lidz, Steven K. Hoge, John Monahan, Marlene M. Eisenberg, Nancy S. Bennett, Edward P. Mulvey, and Loren H. Roth. 1999. "Patients' Revisions of Their Beliefs About the Need for Hospitalization." *American Journal of Psychiatry* 156, no. 9 (September): 1385–91. https://doi.org/10.1176/ajp.156.9.1385.

Gilroy, Andrea, and Colin Lee. 2019. *Art and Music: Therapy and Research*. Routledge.

Griswold, Hali, Debbie Green, Keith Cruise, Brian Belfi, Jonathan Lam, Laura Grossi, Elena Cucco, and Elisabeth Iskander. 2018. "Assessing Risk for Victimization in a Forensic Psychiatric Setting Using the Short-Term Assessment of Risk and Treatability." *Violence and Victims* 33, no. 6: 1012–35. https://doi.org/10.1891/0886-6708.33.6.1012.

Haeyen, Suzanne, Susan van Hooren, William van der Veld, and Giel Hutschemaekers. 2018. "Efficacy of Art Therapy in Individuals With Personality Disorders Cluster B/C: A Randomized Controlled Trial." *Journal of Personality Disorders* 32, no. 4 (August): 527–42. https://doi.org/10.1521/pedi_2017_31_312.

Hall, Andrea, Talha Bin Farooq, and Renato Alcaraz Jr. 2022. "Hypocalcemia-Induced Reversible Psychosis." *Cureus* 14, no. 1 (January 2). https://doi.org/10.7759/cureus.20874.

Hamid, Najmeh, and Ali Biat. 2019. "The Effectiveness of Music Therapy on Depression and Happiness of Depressed Women." *NeuroQuantology* 17, no. 12: 19–26. https://doi.org/10.14704/nq.2019.17. 12.NQ19110.

Haupt, Meighan B. 2014. "The Vital Role of State Psychiatric Hospitals." National Association of State Mental Health Program Directors, July.

Heard, Clark Patrick, Jared Scott, Allan Tetzlaff, and Heather Lumley. 2019. "Transitional Housing in Forensic Mental Health: Considering Consumer Lived Experience." *Health & Justice* 7, no. 1 (May 20): 8. https://doi.org/10.1186/s40352-019-0091-z.

Hedman, Leslie C., John Petrila, William H. Fisher, Jeffrey W. Swanson, Deirdre A. Dingman, and Scott Burris. 2016. "State Laws on Emergency

Holds for Mental Health Stabilization." *Psychiatric Services* 67, no. 5 (May): 529–35. https://doi.org/10.1176/appi.ps.201500205.

Igra, Libby, David Roe, Adi Lavi-Rotenberg, Paul H. Lysaker, and Ilanit Hasson-Ohayon. 2021. "'Making Sense of My Diagnosis': Assimilating Psychoeducation into Metacognitive Psychotherapy for Individuals with Schizophrenia." *Journal of Psychotherapy Integration* 31, no. 3 (September): 277–90. https://doi.org/10.1037/int0000207.

Jackson-Koku, Gordon. 2016. "Beck Depression Inventory." *Occupational Medicine* 66, no. 2 (March): 174–75. https://doi.org/10.1093/occmed/kqv087.

Kendler, Kenneth S., Kathryn Tabb, and John Wright. 2022. "The Emergence of Psychiatry: 1650–1850." *American Journal of Psychiatry* 179, no. 5 (May): 329–35. https://doi.org/10.1176/appi.ajp.21060614.

Kendrick, Tony. 2021. "Strategies to Reduce Use of Antidepressants." *British Journal of Clinical Pharmacology* 87, no. 1: 23–33. https://doi.org/10.1111/bcp.14475.

Kohn, Robert, Ali Ahsan Ali, Victor Puac-Polanco, Chantal Figueroa, Victor López-Soto, Kristen Morgan, Sandra Saldivia, and Benjamín Vicente. 2018. "Mental Health in the Americas: An Overview of the Treatment Gap." *Revista Panamericana de Salud Pública* 42 (October 18): e165. https://doi.org/10.26633/rpsp.2018.165.

Kwok, Sylvia Y. C. L. 2019. "Integrating Positive Psychology and Elements of Music Therapy to Alleviate Adolescent Anxiety." *Research on Social Work Practice* 29, no. 6 (September 1): 663–76. https://doi.org/10.1177/1049731518773423.

Lachner, Christian, Michael J. Maniaci, Tyler F. Vadeboncoeur, Nancy L. Dawson, Teresa A. Rummans, Archana Roy, Lorrina L. Hall, and M. Caroline Burton. 2020. "Are Pre-Existing Psychiatric Disorders the Only Reason for Involuntary Holds in the Emergency Department?" *International Journal of Emergency Medicine* 13, no. 1 (February 3). https://doi.org/10.1186/s12245-020-0265-4.

Leucht, Stefan, Caroline Corves, Dieter Arbter, Rolf R Engel, Chunbo Li, and John M Davis. 2009. "Second-Generation versus First-Generation Antipsychotic Drugs for Schizophrenia: A Meta-Analysis." *The Lancet* 373, no. 9657 (January): 31–41. https://doi.org/10.1016/S0140-6736(08)61764-X.

Leucht, Stefan, Maximilian Huhn, and John M. Davis. 2021. "Should 'Typical', First-Generation Antipsychotics No Longer Be Generally Used in the Treatment of Schizophrenia?" *European Archives of*

Psychiatry and Clinical Neuroscience 271, no. 8 (December 1): 1411–13. https://doi.org/10.1007/s00406-021-01335-y.

Locher, Cosima, Sibylle Meier, and Jens Gaab. 2019. "Psychotherapy: A World of Meanings." *Frontiers in Psychology* 10 (March 22): 460. https://doi.org/10.3389/fpsyg.2019.00460.

Magnusson, Emilie, Anna Karin Axelsson, and Malin Lindroth. 2020. "'We Try' – How Nurses Work with Patient Participation in Forensic Psychiatric Care." *Scandinavian Journal of Caring Sciences* 34, no. 3: 690–97. https://doi.org/10.1111/scs.12773.

Malchiodi, Ms Cathy A. 2018. *The Handbook of Art Therapy and Digital Technology*. Jessica Kingsley Publishers.

"Management of Common Adverse Effects of Antipsychotic Medications." N. d. Accessed July 26, 2022. https://onlinelibrary.wiley.com/ doi/epdf/10.1002/wps.20567.

Maust, Donovan T., Lewei A. Lin, and Frederic C. Blow. 2019. "Benzodiazepine Use and Misuse Among Adults in the United States." *Psychiatric Services* 70, no. 2 (February 2019): 97–106. https://doi.org/10.1176/appi.ps.201800321.

McKee, Jerry, and Nancy Brahm. 2016. "Medical Mimics: Differential Diagnostic Considerations for Psychiatric Symptoms." *The Mental Health Clinician* 6, no. 6 (November 3): 289–96. https://doi.org/10.9740/mhc.2016.11.289.

Möller, Hans-Jürgen. 2005. "Antipsychotic and Antidepressive Effects of Second Generation Antipsychotics: Two Different Pharmacological Mechanisms?" *European Archives of Psychiatry and Clinical Neuroscience* 255, no. 3 (June): 190–201. https://doi.org/10.1007/s00406-005-0587-5.

Morgan, Nirvana, and Giada Del Fabbro. 2018. "Factors Associated with Recidivism at a South African Forensic Psychiatric Hospital." *South African Journal of Psychiatry* 24 (May 31). https://www.ajol.info/index.php/sajpsyc/article/view/172414.

Mottaghipour, Yasaman, and Maryam Tabatabaee. 2019. "Family and Patient Psychoeducation for Severe Mental Disorder in Iran: A Review." *Iranian Journal of Psychiatry* 14, no. 1 (January): 84–108.

Mullen, Paul E. 2000. "Forensic Mental Health." *The British Journal of Psychiatry* 176, no. 4 (April): 307–11. https://doi.org/10.1192/bjp.176.4.307.

Munder, T., C. Flückiger, F. Leichsenring, A. A. Abbass, M. J. Hilsenroth, P. Luyten, S. Rabung, C. Steinert, and B. E. Wampold. 2019. "Is Psychotherapy Effective? A Re-Analysis of Treatments for Depression." *Epidemiology and Psychiatric Sciences* 28, no. 03 (June): 268–74.

https://doi.org/10.1017/S2045796018000355.
Okamoto, Annika, Frank M. Dattilio, Keith S. Dobson, and Nikolaos Kazantzis. 2019. “The Therapeutic Relationship in Cognitive–Behavioral Therapy: Essential Features and Common Challenges.” *Practice Innovations* 4, no. 2 (June): 112–23. https://doi.org/10.1037/ pri0000088.
Pergolizzi Jr, Joseph V., Jo Ann LeQuang, and Robert B. Raffa. 2021. “Benzodiazepines: Thinking Outside the Black Box.” *Journal of Clinical Pharmacy and Therapeutics* 46, no. 3: 554–59. https://doi.org/10.1111/jcpt.13333.
“Practice Guidelines for the Psychiatric Evaluation of Adults, Third Edition.”2015. In *The American Psychiatric Association Practice Guidelines for the Psychiatric Evaluation of Adults*, Third. Practice Guidelines. American Psychiatric Association. https://doi.org/10.1176/appi. books.9780890426760.pe02.
Protti, Michele, Roberto Mandrioli, Camilla Marasca, Andrea Cavalli, Alessandro Serretti, and Laura Mercolini. 2020. “New-Generation, Non-SSRI Antidepressants: Drug-Drug Interactions and Therapeutic Drug Monitoring. Part 2: NaSSAs, NRIs, SNDRIs, MASSAs, NDRIs, and Others.” *Medicinal Research Reviews* 40, no. 5: 1794–1832. https://doi.org/10.1002/med.21671.
“Reasonable or Random: 72-Hour Limits to Psychiatric Holds.” N.d. Accessed July 29, 2022. https://doi.org/10.1176/appi.ps.202000284.
Reid Finlayson, Alistair J., Jane Macoubrie, Christy Huff, D.E. Foster, and Peter R. Martin. 2022. “Experiences with Benzodiazepine Use, Tapering, and Discontinuation: An Internet Survey.” *Therapeutic Advances in Psychopharmacology* 12 (January 1): 204512532210 82384. https://doi.org/10.1177/20451253221082386.
Sanabria, Edilma, Ronald Edgardo Cuenca, Miguel Ángel Esteso, and Mauricio Maldonado. 2021. “Benzodiazepines: Their Use Either as Essential Medicines or as Toxics Substances.” *Toxics* 9, no. 2 (February): 25. https://doi.org/10.3390/toxics9020025.
Sansone, Randy A., and Lori A. Sansone. 2014. “Serotonin Norepinephrine Reuptake Inhibitors: A Pharmacological Comparison.” *Innovations in Clinical Neuroscience* 11, no. 3–4: 37–42.
Santarsieri, Daniel, and Thomas L Schwartz. 2015. “Antidepressant Efficacy and Side-Effect Burden: A Quick Guide for Clinicians.” *Drugs in Context* 4 (October 8): 212290. https://doi.org/10.7573/dic.212290.
Sarkhel, Sujit, O. P. Singh, and Manu Arora. 2020. “Clinical Practice Guidelines for Psychoeducation in Psychiatric Disorders General

Principles of Psychoeducation." *Indian Journal of Psychiatry* 62, no. Suppl 2 (January): S319–23. https://doi.org/10.4103/psychiatry.IndianJ Psychiatry_780_19.

Schaftenaar, Petra, Ivo van Outheusden, Geert-Jan Stams, and Andries Baart. 2018. "Relational Caring and Contact after Treatment. An Evaluation Study on Criminal Recidivism." *International Journal of Law and Psychiatry* 60 (September 1): 45–50. https://doi.org/10. 1016/j.ijlp.2018.07.011.

Seppänen, Allan, Iida Törmänen, Christopher Shaw, and Harry Kennedy. 2018. "Modern Forensic Psychiatric Hospital Design: Clinical, Legal and Structural Aspects." *International Journal of Mental Health Systems* 12 (October 20): 58. https://doi.org/10.1186/s13033-018-0238-7.

Silvernail, Carrie M., and Steven L. Wright. 2022. "Surviving Benzodiazepines: A Patient's and Clinician's Perspectives." *Advances in Therapy* 39, no. 5 (May): 1871–80. https://doi.org/10.1007/s12325-022-02055-y.

Singh, Rakesh Kumar, and Suprakash Chaudhury. 2009. "Hyponatremia-Induced Psychosis in an Industrial Setting." *Industrial Psychiatry Journal* 18, no. 2: 137–38. https://doi.org/10.4103/0972-6748. 62277.

Swift, Joshua K., Jennifer L. Callahan, Mick Cooper, and Susannah R. Parkin. 2018. "The Impact of Accommodating Client Preference in Psychotherapy: A Meta-Analysis." *Journal of Clinical Psychology* 74, no. 11: 1924–37. https://doi.org/10.1002/jclp.22680.

Testa, Megan, and Sara G. West. 2010. "Civil Commitment in the United States." *Psychiatry (Edgmont)* 7, no. 10 (October): 30–40.

Torrey, Fuller, Doris Fuller, Jeffrey Geller, Carla Jacobs, and Kristina Ragosta. 2012. "No Room at the Inn- Trends and Consequences of Closing Public Psychiatric Hospitals-2005-2010." The Treatment Advocacy Center, July 19.

Tsagkaris, Christos, and Konstantinos Kalachanis. 2020. "The Hippocratic Account of Mental Health: Humors and Human Temperament." *Mental Health: Global Challenges Journal* 3, no. 1 (September 20): 33–37. https://doi.org/10.32437/mhgcj.v3i1.83.

Vanezis, Andrew Peter, and Dudley Manns. 2010. "Physical Examinations of Mental Health Service Users." *Progress in Neurology and Psychiatry* 14, no. 4: 19–23. https://doi.org/10.1002/pnp.168.

Ventriglio, Antonio, Massimo Mari, Antonello Bellomo, and Dinesh Bhugra. 2015. "Homelessness and Mental Health: A Challenge." *International Journal of Social Psychiatry* 61, no. 7 (November 1): 621–22. https://doi.org/10.1177/0020764015585680.

Erie County, PA. N.d. “Voluntary and Involuntary Commitment.” Accessed July 29, 2022. https://eriecountypa.gov/departments/human-services/ mental-health/voluntary-and-involuntary-commitment/.

Wright, Jason D., Jacob C. Cogan, Yongmei Huang, Ana I. Tergas, Caryn M. St. Clair, June Y. Hou, Fady Khoury-Collado, et al.2021. “Association of New Perioperative Benzodiazepine Use With Persistent Benzodiazepine Use.” *JAMA Network Open* 4, no. 6 (June 3): e2112478. https://doi.org/10.1001/jamanetworkopen.2021.12478.

Wubeshet, Yirgalem Shewakena, Oumer Sada Mohammed, and Tigestu Alemu Desse. 2019. “Prevalence and Management Practice of First Generation Antipsychotics Induced Side Effects among Schizophrenic Patients at Amanuel Mental Specialized Hospital, Central Ethiopia: Cross-Sectional Study.” *BMC Psychiatry* 19, no. 1 (December): 32. https://doi.org/10.1186/s12888-018-1999-x.

Yohanna, Daniel. 2013. “Deinstitutionalization of People with Mental Illness: Causes and Consequences.” *AMA Journal of Ethics* 15, no. 10 (October 1): 886–91. https://doi.org/10.1001/virtualmentor.2013.15.10.mhst1-1310.

Zhang, Xu, and Xiao-Ying Li. 2020. “Prevalence of Hyponatremia among Older Inpatients in a General Hospital.” *European Geriatric Medicine* 11, no. 4 (August): 685–92. https://doi.org/10.1007/s41999-020-00320-3.

Zimmerman, Mark. 2003. “What Should the Standard of Care for Psychiatric Diagnostic Evalutions Be?" *The Journal of Nervous and Mental Disease* 191, no. 5 (May): 281–86. https://doi.org/10.1097/01.NMD.0000066149.40946.FA.

Zuckerman, Edward L. 2019. *Clinician’s Thesaurus, 8th Edition: The Guide to Conducting Interviews and Writing Psychological Reports*. Guilford Publications.

CHAPTER TEN

SUBSTANCE ABUSE TREATMENT AND RESILIENCE DECREASE RECIDIVISM IN INCARCERATED PATIENTS

ASHLEY MAZUR-WILLIAMS

Introduction

Approximately 7.1 million adults in the United States require criminal justice supervision. This substantial increase is due to more stringent laws and penalties for drug offenses. More than half of all prisoners meet the diagnostic criteria for drug abuse or dependence.[1] Drugs are a significant concern in many prisons around the world.[2] A national survey of New Mexico's jails, in 2002 by the Bureau of Justice Statistics, showed that 70% of inmates met the criteria for substance abuse, as opposed to the 9% of the United States general population.[3] This influx of prisoners with substance abuse concerns has led to prisons and jails becoming increasingly deadly places in recent years, with the one cause of death that has climbed dramatically being overdoses.[4] From 2001 to 2018, the number of people who have died of drug or alcohol intoxication in state prisons increased by more than 600% and jails by over 200%, according to the Bureau of Justice Statistics.[5] The year 2018 brought the highest number of prison deaths on record since the collection of this data began, over 20 years ago.[6] Since the Bureau of Justice Statistics had begun collecting data in 2001, no manner of death has increased more than drug overdoses and alcohol intoxication.[7] This chapter will describe the link between drug abuse and crimes, genetics, mental health, substance abuse diagnosis, and mental health and resilience. The second section of this chapter will describe the prevalence of overdoses in prison systems, how drugs enter the prisons, and why the inmates continue to use drugs while incarcerated. The third section of this chapter explains why the prison system has significant delays in the treatment of substance abuse and drug overdoses, including both facility and inmate limitations. The final section of this chapter will explain the treatment and various programs offered in the prisons and how to decrease recidivism by building resilience through proper substance abuse treatment. We often consider someone as resilient if they can overcome a tragic or traumatic event.[8] Treatment of substance abuse disorders and fostering resilience leads to a decrease in crime rates and recidivism.

Drug Abuse and Criminal Behaviors Cause & Effect

The link between criminal behaviors and drug use is very well documented and has several dimensions.[9] When inmates are asked, they often report being under the influence of drugs or alcohol during the time they committed their crime(s) resulting in their sentence.[10] In 2003, 75% of male arrestees and 74% of female arrestees tested positive for drugs.[11] There are three groups that those with drug-crime associations fall into. The first group is called 'forward causation', which means that drug use causes criminal behaviors due to economic necessity and the need to fund their drug use or because of psychopharmacological changes precipitated by drug ingestion. The second category is 'reverse causation', which is that criminal involvement leads to drug use because opportunities to use drugs increases with involvement in criminal behavior. The third category is 'confounding', which is that crime and drug use share a common set of causes but that there is no direct causal relationship.[12] Most studies that discuss these categories of cause and effect find that offending rates are substantially higher after drug use initiation, though the severity of the crimes are often low.[13] The level of addictiveness in the substance being used affects criminal behaviors also.

The seriousness of the crimes also decreased substantially for those who were positive for drug use.[14] Those dependent on heroin or other opioids are disproportionately more likely to be involved in criminal activity, especially acquisitive offending, or crimes committed for financial gain.[15] Opioid use disorder (OUD) is more severe and advanced in those in the criminal justice system than those in the general population.[16] Those currently using drugs are more likely to commit crime(s) and this theory can be explained by three concepts. The first is that offenses may be simple violations of laws prohibiting contact or use of illegal drugs. This concept includes charges like drug possession or paraphernalia possession. The second concept says that the crime must be linked to drugs because of the motivation to get money to support their drug use. This includes criminal charges like theft, burglary, or drug possession with the intent to sell. The third is that crimes may be linked to a drug-using lifestyle, like exposure to situations that encourage deviant lifestyles. This concept includes criminal charges such as trespassing, other crimes due to homelessness, and driving under the influence.[17] There is a link between substance abuse and criminality, which begins with the link between an individual's genetics, mental health, and substance abuse.

Genetics, Mental Health, and Substance Abuse

Genes are the building blocks of human cells and contribute to almost every human disease. Once a disease occurs, genes influence the severity and progression of the disease(s) including mental health disorders. Genes are powerful tools when researching mental illness. There is no question that the susceptibility for mental illness is determined through genetic composition. Genes are critical to the development of the brain's function and synapses. There are variations in sequences of DNA that affects behavior potentially increasing the risk of mental illness, through the change in architecture, function, or adaptability of neural circuits in the brain.[18] Genes can be affected by the environment and manipulated through the changes that occur in the neural pathways of the brain.

The brain is the most complex object of investigation in the history of biological sciences; developing through non-linear gene-gene and gene-environment interactions, which create the complex interconnections of millions of neurons. Every time something new is learned, the active neurons in the brain alter the synaptic architecture on the circuit in which the learning has occurred, through the creation of new synapses, old ones pruned, and existing synapses can be strengthened or weakened, through a process called plasticity. Plasticity is how the brain learns to function and process things differently based on experiences.[19] There is increasing evidence that addiction is a treatable disease of the brain, though most individuals do not receive treatment.[20] Diseases of the brain, whatever their pathophysiological cause, affect behavior through alteration of the brain function. Addiction and post-traumatic stress disorders have a similar process in the brain. Both conditions represent the infringement of normal learning processes regarding reward and fear.[21] The reward and fear cortex of our brain, or fight or flight responses, are altered when a person has certain mental health issues, including substance abuse.

Mental Health and Resilience

Suffering traumatic events, disasters, or psychological distress are part of life, although having these experiences cause negative emotions. In the 1620s, the word resilience was first utilized as a word to signify rebound. Though in the 19th century, the word began to be used as a figurative phrase, regarding people or groups of people. Studies on resilience began with children who had mental disorders. More recently, the focus on adults who lived through adverse childhood experiences, experienced terror, survived an accident, lived in a risky environment, or worked in a dangerous position such as firefighters, paramedics, or policemen has increased worldwide. There is still not a single final definition for the word resilience, though it most commonly is defined as 'the capacity of dynamic systems to adapt successfully to disturbances that threaten its ability to function and continue developing, or a class of phenomena characterized by good outcomes in spite of serious threats to adaptation or development'. [22]

Researchers describe mental resilience as the ability to have a quick and effective recovery after stress or trauma.[23] Resilience is what we use to get over the effects of adversity and/or risk and not only help us bounce back but sometimes take advantage of difficult situations.[24] Overcoming negative impacts is important for our development.[25] The two criteria for resilience include the threat side of the interference, meaning there must be a significant risk or adversity in the person's life; the other is about the person's success in development or the quality of adaptation.[26] Resilient people tend to share similar characteristics like optimism, altruism, moral compass, faith, spirituality, humor, having a role model, social support, facing fears, meaning or purpose in life, and training.[27] These behavioral characteristics aid in the development of protective factors.

Protective factors play a major role in the risk situations in a person's life, and these protective strengths show the person's resilience.[28] Resilience factors provide what is known as mental 'immunity', which helps to promote health and wellness.[29] Even though negative experiences lead to negative emotions, these experiences also help develop and foster resilience.[30] Though it should be noted that not all environmental challenges or stressors can be buffered by natural, individual resilience and can lead to the development of flaws in resilience factors.[31] These flaws can lead to the development of mental health concerns and substance abuse. Resilience is a concept that has recently attracted attention in regard to substance abuse.[32]

Resilience has been regarded as a critical component of mental health recovery, especially in substance abuse. There are two categories of resilience that can be enhanced in everyone. The first category is 'innate resilience' which is a product of inherent factors. The second is 'acquired resilience' which can be modified or strengthened by environmental factors. Substance abuse can affect a person's self-control and self-esteem; therefore, it is important to provide those individuals with the support that fosters their social support and self-esteem.[33] Resilience and the perception of social support play a significant role in the consumption of alcohol or drug use. Those who perceived less social support overall were more likely to have consumed drugs or alcohol in the three months before incarceration. Inmates having less emotional support, or fewer positive social interactions caused lower resilience and therefore limited the inmates' protective factors against substance abuse in prison. These findings have new implications regarding the programs that include social support by building and/or strengthening already existing friend and family networks, at a global level and in terms of social and affective interactions.[34] Resilience is a crucial psychological trait for relapse prevention.[35] Limitations on resilience are one of the factors that lead to increased substance abuse and overdoses.

Overdose Prevalence

In 2018, state prisons reported 4,135 deaths, not including executions, which is the highest number of deaths since the Bureau began collecting data in 2001.[36] There are two categories of death, 'natural' and 'unnatural' death. Natural deaths are those attributed to illness or age. Unnatural deaths are attributed to those caused by suicide, homicide, accident, executions, and drug or alcohol intoxication.[37] In 2018, 77% of all prison deaths were considered 'natural', though these numbers are moving in the opposite direction, and 'unnatural' deaths are becoming more prevalent, especially drug overdoses.[38] Before the fentanyl overdose numbers had begun to rise, overdoses never ranked higher than fifth among the causes of death for prisoners. In 2019, only cancer killed more people in California prisons than drug overdoses.[39] Deaths are not the only concern; emergency department visits and overnight hospitalizations for drug overdoses among prisoners have spiked by more than 50% between 2014 and 2017.[40] These overdoses frequently occur because of the tolerance changes that occur during incarceration resulting from the inconsistency in the amount of drugs the person has access to. The individuals' tolerance to their drug of choice has even more serious implications for the release of those with substance abuse histories.

The access to drugs on and off leads to rapid changes in tolerance which places the users at a high risk for overdose, both inside prison and upon release.[41] The limitation of opioid use, while incarcerated, leads to a diminished physiological tolerance, therefore contributing to the risk of drug overdose following incarceration release. The drug user is used to using a certain amount of their drug of choice, and when they leave prison, they tend to try to use the same amount they used to before becoming incarcerated. These individuals have nearly 13 times the risk of death as the general population with overdose as the leading cause.[42] The increased numbers highlight the persistent issues seen due to the way the United States Criminal Justice System handles cases involving people with substance abuse disorders. The criminal justice system incarcerates people with addiction problems and then does not provide them with adequate treatment options.[43] Though, in some cases, due drug accessibility, people develop a stronger tolerance to their substance of choice while in prison-.

How Drugs Enter the Prisons

Studies have shown that a third of all prisoners claim that it is nearly effortless to get ahold of illegal substances while in prison than outside of prison, and even simpler to obtain than soap.[44] The identification of drugs that are illegally used and legally prescribed, how drug use and sales are sanctioned, how drug profits are generated, how drugs enter the prison systems, drug use, and overdose are important public health treatment priorities in prisons. The concern with the data that is available on these topics is that it is limited in reliability due to the sensitivity of the nature of the topic, the availability of surveillance within the prisons, and the reluctance of the prisoners to self-report due to the fear of repercussions. There are also ethical issues in terms of compromising prisoner confidentiality and anonymity for those involved.[45] Drugs get into prisons in a variety of ways including through visitors, packages, letters, correctional staff, incoming mail, and visitations.[46]

Paper products were one of the primary ways that the prisoners were able to smuggle drugs into the prison. Their friends and family tuck strips of paper soaked with drugs into the mail, books, or other paper items brought or sent into the correctional facility.[47] Certain drugs have a potent smell that must be disguised and that can be done with perfumes and other scented sprays, leading to the disposal of any paper items that have a smell to them. Certain substances like synthetic cannabis can be sprayed onto paper and then the paper is smoked or eaten, which is one of the most complicated methods to be detected.[48] Another drug that was easily snuck into the facilities was suboxone strips, they were put into the seal of the envelopes which were then licked by the inmates to ingest the drug. These strips could also be tucked between two pieces of paper and therefore hidden from detection. Many facilities have made the rule that the inmate cannot be provided with any external envelopes. Family members have even been known to put drugs in paint and have their children paint a picture with the contaminated paint.[49] Besides paper products, drugs can be brought into the prison through visitors in many other ways.

One of the most common routes for drugs to enter the prison system is through visitation.[50] Visitors can bring drugs in balloons that can be swallowed, or carried in a person, such as a lawyer's shoe, and concealing drugs in body orifices, baby's clothing, diapers, or various food or drink containers. The drugs can be passed through a handshake, kiss, or even simply over the table.[51] Many of the state prisons have cracked down on the allowance of in-person visitations to hopefully decrease the influx of contraband.[52] There are minimal consequences for the visitors who bring drugs into the prisons. The prison very rarely notifies the police, and they typically send the visitor away and do not let them have physical contact visits or they must undergo a strip search if they are allowed to have physical contact at their following visits.[53] These limited consequences have led to an increase in drug smuggling because of the lack of fear of repercussions; though changes are finally being made and laws are being more strictly enforced.

Small packages containing drugs can be thrown over the walls and fences of the prisons. Typically, prisons have a large amount of land outside of their walls, providing a safe place for people to throw packages over without being seen. Prisoners often have free movement throughout the outside recreational areas of the prison, allowing them the ability to walk to the walls and pick the packages up that have been thrown over. These package drops are coordinated using contraband mobile phones that were also smuggled into the prisons. This coordination often includes other inmates causing a form of distraction such as a loud argument or fight to divert the guards' attention. Once the drugs have been thrown over, the inmates crowd around the package, giving the inmate in charge of collection the ability to hide the drugs, typically in their rectums or vaginas. The packaging that is used for the drugs to be delivered can include tennis balls, arrows, or even dead birds. When larger packages need to be delivered, packages can be dropped in by drones, especially when there are shortages of guards or large areas that are not monitored.[54]

Newly convicted, incoming, or transferring prisoners can also bring drugs in by hiding them in their body cavities.[55] Drugs can be brought into the prisons by prisoners moving between courts, jails, and other prisons. In some cases, prisoners will attempt to get themselves recalled to the courts so they can smuggle drugs back into the prison with them. Prisoners sometimes hide their drugs in their body cavities, specifically their rectum or vagina, but in many cases, searches are not comprehensive leading to the prisoners being able to bring drugs in through much simpler, less invasive ways, once trust is built. Some inmates have brought in drugs that have been sewn into the lining of their underwear, in the waistband of their pants, or even in their shoes.[56] Some prison guards also look the other way when prisoners are smuggling drugs into the prisons, especially if being coerced in any way.

One of the most concerning methods of drugs entering prisons is through corrupt prison staff including guards, nurses, trainers, suppliers, prison tutors, etc. One prison official said that bringing drugs into the prison system would be simple because day staff would be searched typically around once a year and night staff were never searched. Many prisons do not have any form of drug-sniffing dog to search the staff either upon their arrival into the prison for their shifts.[57] The prisons with the lower-paid correctional officers tended to have increased rates of narcotics smuggling.[58] The correctional staff can be bribed due to the need for increased income. The manipulation or blackmailing of officers can occur, leading to the smuggling of contraband and drugs. One inmate describes the process in which this is done, beginning with the inmate learning small personal details about the correctional staff. The inmate begins by asking the staff to bring them something small and building up to larger and more personal items while continuing to build their 'relationship' with the staff member. This continues to build until the staff is in a position where they cannot refuse because of the fear of disciplinary action for the infractions they already committed and the threats of the inmates to expose the criminal behavior.[59] Getting the drugs into the prison is the first step and then moving them throughout the prison is typically a straightforward process.

Once the drugs enter the prison they are moved in various ways throughout the facility. Once inside the inmates use their distribution enterprise to sell their drugs without the risk of drug charges. When inmates are locked in their cells, they create makeshift ropes also known as 'kites' or 'lines', which is how they pass the drugs between the cells. Prisoners often divert street drugs but also their legally prescribed medications as well. This is done by the prisoner placing the medication in their mouth, sipping water in front of the nurse, and then spitting the medication back into the cup to then take back to their cell to sell the medication to another inmate.[60] The drug trade in prison is one of the most common and profitable enterprises inside correctional facilities, especially prisons. These enterprises are extremely lucrative and require hierarchies, just like any other enterprise outside of prison, which is one of the main reasons that drug use continues in the prison system.

Why They Continue to Use

Prisons are high-risk environments for drug use initiation, re-initiation after abstinence, and the use of riskier methods of drug use.[61] Several factors lead to inmates' drug abuse continuation while they are incarcerated. These factors include boredom, isolation, lack of treatment options, being miserable and depressed leading to self-medication for both mental and physical health needs going unmet. [62] Incarceration is not only difficult for someone who has mental health concerns, but it also creates and exacerbates disconnection, despair, and overall psychological distress.[63] These feelings of disconnection, despair, and distress lead to a significant decrease in resilience, therefore leading to the (re-)initiation of drug use. Drugs like heroin and cannabis are used as a way of coping or self-medication and are preferred due to their sedative and analgesic effects to relieve stress, facilitate sleep, deal with the boredom of 'doing time', and ultimately help to relieve the pain of imprisonment.[64] Drugs are used in place of resilience when resilience is limited by trauma or lack of social support. When prisoners become involved with the drug sales enterprise, they feel that sense of social support and that they are part of a bigger social network, leading to the illusion of resilience.

The sales of drugs in prison are bound by masculinity, self-identity, and internal hierarchies, therefore playing a role in creating social and personal interactions in the prison system. There are two systems for selling drugs in prison which include established enterprises and separate suppliers. Established enterprises are organized and related to drug sellers and gang structures outside of the prison systems. Separate suppliers are more individualistic and opportunistic and therefore, many prisoners have taken this strategy for their sales of drugs. The sale of drugs enhances the status, feeling of being part of a larger social network, and capital of the prisoner through the ability to bring drugs into the prisons, making connections to external drug and criminal networks, and displaying courage. The sale of drugs increases the level of respect the person has in prison.[65] Prisons are perfect places for the spreading of drug use because of the simplicity of drug users to establish social relationships and pass on their drug habit.[66] The drugs in the prison system have a powerful influence on the prison environment and are a profitable commodity.[67]

The lack of substance abuse disorder and mental health treatments in correctional facilities leads to the continuation of drug use. Integrating treatment into the criminal justice system would provide treatment to individuals who otherwise would not receive treatment, improving their medical outcomes and decreasing recidivism or reincarceration rates.[68] Most inmates do not have access to rehabilitative programs and therefore, there is an increase in recidivism rates.[69] 'Even when there are treatments available, it is not of the highest caliber', says Rebecca Fealk, program coordinator for the Arizona Chapter of the American Friends Service Committee, which is an organization that promotes criminal justice reform. She says that the inmates who were in treatment programs explain the program as a classroom in which they are given a packet of papers to fill out, referencing making good choices and which coping mechanisms they would need to not continue or begin using drugs again.[70] Many state laws obligate the state to rehabilitate their offenders, and auditors have found that many inmates need help to treat their drug addictions and are not able to enroll in their prison programs.[71] Certain states have begun rolling out comprehensive treatment programs through the use of medications for the treatment of opioid addiction including Suboxone and Methadone, though this rollout has been very slow.[72] The lack of treatment and delays in care within the correctional facilities have led to an increase in drug use and overdoses.

Delays in Care

Prisons are inherently dangerous places that expose incarcerated individuals to unbearable mental and physical conditions.[73] The security risks and preventive measures of the inmates and the guards lead to delayed responses in overdose situations. The highest security areas of the facility require officers to find backup before they can open a cell to check on a prisoner, even if the prisoner appears to be overdosing. Finding a second officer or an available supervisor to enter the cell, because they were busy or because they do not want to fill out the paperwork.[74] The guards are not proactive when an overdose occurs, especially due to their fear of repercussions and lack of appropriate training. In 1990 and 1995, the correctional facilities that operated primarily as substance abuse treatment institutions nearly doubled.[75] In 2000, 96% of all American correctional facilities offered counseling, with 90% offering drug and alcohol counseling.[76] The downside to the limitations on psychiatric and drug treatment facilities and the use of prisons as treatment facilities is problematic because of the limitations on required resources for proper treatment due to the system being ill-equipped. There are many resource limitations that prisons have when attempting to provide care to those with a substance abuse history, mental health diagnoses, and especially overdoses. These limitations include security risks, overcrowding, and lack of substance abuse or behavioral health trained staff, as well as the fear of the inmates to ask for assistance due to the repercussions from both the guards and/or the other inmates.

Facility Limitations

One of the most significant causes for limitations in enrollment in drug dependency programs for inmates is the significant overcrowding in the correctional facilities, per a 2012 report by the Bureau of Prisons.[77] Due to overcrowding and lack of resources, the Department of Corrections uses a ranking system based on the inmate's need, risk to recidivate, and time remaining on their sentence, to determine eligibility for substance abuse program enrollment.[78] Those who can qualify for an early release by completing the substance abuse program are sent to the front of the line for treatment.[79] Those with substance abuse concerns and longer sentences tend to wait many years for treatment.[80] The level of addiction of potential risk for overdose due to drug of choice, years of use, etc. are not considered when determining who is qualified for treatment. Outside of overcrowding, there is a limitation on access to drug treatment programs, because of inadequate resources and staff shortages.[81]

As of May 2021, the state of Arizona alone had 12 vacant positions for substance abuse counselors and was struggling to retain counselors, especially in rural areas.[82] One of the main issues in providing treatment programs to inmates is the inability to find the staff to perform the work. This can be rectified through the increase in funding for staff to work in these facilities. One of the benefits that was considered by the state of Arizona was the creation of a loan repayment program for substance abuse counselors, therapists, psychologists, or other mental health professionals who are willing to work for the Department of Corrections for at least five years.[83] These benefits can increase the number of individuals who are willing to work in the correctional setting. The increase in trained staff will allow for an increase in treatment programs allowing for more access to the programs. There are significant delays in care due to the staff and prison system itself, as well as the inmates having the fear to request help when necessary due to system or fellow inmate backlash.

Inmate Limitations

When there is an overdose behind bars, getting the individual who overdosed the help they need is difficult for various reasons from the prisoner code prohibiting asking for help, or because understaffed or apathetic correctional officers do not respond quickly or at all, or the fear of repercussions from the officers or other inmates. Substance abuse leads to increased risk to those in prison because the inmates are often alone and may also be reluctant to call for help if there is a problem. One inmate said, “You’d be watching someone turn blue and be on the fence about whether you should call for help or not because of the consequences of, what’s this going to mean for me and everyone else.” Inmates view seeking assistance for their overdosing cell mate as snitching because if the staff found drugs on anyone, the entire unit would be placed on lockdown. Lockdown means the loss of visitation, communication with loved ones, cell searches, and closer monitoring. They also lose their ability to purchase items from the commissary, use the phones, and are only allowed out for a shower and one hour of exercise. The inmates are afraid of other inmates and the consequences if they bring attention to their pod or yard.

One ex-inmate was interviewed relating to the substance abuse and overdose prevalence in the prison where he was incarcerated. He said that the inmates who overdosed were considered a burden by the other inmates. The inmates were afraid of “bringing heat to the yard.” He said that when someone overdosed, the other inmates would take the overdosing person back to their room and tuck them into bed. They would try to get them into their room without being noticed and then wait for checks to occur for the correctional officers to find the person, typically dead at this point. He said that unless the person overdosed or “fell out” in the yard, they typically died. If someone overdosed in the yard, the guards typically noticed because the person would be out in the open, so there was nowhere to hide the person who was overdosing. The prisoners were afraid of the repercussions of drugs being found in their yard, or suspected to be “on their yard”, and they rarely would ask for help when someone was overdosing. Even if they do seek medical help, the care is scarce and subpar.[84] The following section will describe the treatment program options and the limitations those programs have in reducing drug use and overdoses.

Treatment Programs

In 2006, the National Institute on Drug Abuse published 13 principles of drug abuse treatment for criminal justice populations. The principles state that drug addiction is a brain disease that affecting behavior, recovery from drug addiction requires effective treatment, and management of the problem over time is required. The next principle states that the treatment must last long enough to produce stable behavioral changes, the assessment is the first step in treatment, and tailoring services to fit the needs of the individual is an important part of effective substance abuse treatment for the criminal justice populations. The next principle says that drug use during treatment should be carefully monitored, and that criminal justice supervision should incorporate treatment planning for those with a substance abuse history, and that treatment providers should be aware of correctional supervision requirements. The following principle is that continuity of care is essential for those with substance abuse re-entering their community. The next principle states that a balance of rewards and sanctions encourages pro-social behavior and participation in treatment followed by ensuring that offenders with co-occurring drug abuse and mental health problems often require integrated treatment plans. Medications are an important part of treatment for most offenders with a substance abuse history and the final principle is that treatment planning for substance abuse offenders who are re-entering or already living in the community should include strategies to prevent and treat serious, chronic medical conditions including HIV/AIDS, hepatitis B and C, and tuberculosis.[85] These key principles, if followed, can help to significantly reduce relapse and recidivism.

Many individuals requiring substance abuse treatment do not have an opportunity for treatment until they encounter the criminal justice system.[86] After coming into contact with the criminal justice system, many inmates are finally diagnosed with a mental health and substance abuse diagnosis.[87] Once substance abuse treatment has begun in the prison system, follow through to after-release is imperative to success and limitation on recidivism. Almost all probationers (99%) had one or more conditions to their sentencing required by the court or probation agency, among such conditions were drug testing and substance abuse treatment.[88] Studies have shown that offenders who have ceased their drug use have reduced rates of reoffending and re-arrest.[89] Increases in admission to substance abuse treatment are associated with a reduction in crime rates, therefore reducing (re-)incarceration rates.[90] Substance abuse treatment is more cost-effective than prison or other punitive measures.[91] Keeping inmates, especially those with a substance abuse history, leads to a limitation on unnecessary (re-)incarcerations, therefore limiting costs to the prison systems.

Research from the Federal Center for Substance Abuse Treatment and Treatment Research Institute at the University of Pennsylvania, in 2005, found that substance abuse treatment results in significant crime reductions and substance use, as well as improves the person's ability to function in health and social areas.[92],[93] These social area improvements lead to improvements in resilience, decreasing relapse. There are many different methods to measure resilience and choosing the appropriate scale could provide enough information to identify the strength of the person's resilience. The scale should consider the person's perception of self, future, social competence & support, and family cohesion.[94] Using the information about resilience can help to plan more precisely and help the inmate meet their goal of reintegration.[95] Criminal justice professionals, substance abuse treatment providers, and other social service agencies can optimize their resources to benefit the health, safety, and well-being of the individuals and communities they serve, by working together.[96] Treatments need to begin in prison and continue upon release.

Treatments in Prison

The use of neuroscientific findings and evidence-based principles of addiction treatment, if implemented in the criminal justice system, could improve public health and decrease criminal behaviors and recidivism.[97] In correctional facilities, substance abuse treatment includes detoxification, professional counseling, a residential stay, or maintenance drug programs such as Alcoholics Anonymous (AA), Narcotics Anonymous (NA), and other self-help groups, and drug or alcohol education or awareness.[98] Due to limitations on options, many times the inmates attempt to lead classes for the other inmates to try to kick their addiction issues, though the effectiveness of these classes is debated.[99] Certain states are attempting to change sentencing laws when substance abuse is involved in the commission of a crime. For example, Arizona is trying to pass a sentencing reform law, though if this reform is going to tie early releases to substance abuse counseling or other programs completed in prison, the Department of Corrections will need to provide more options to their inmates.[100] Only 28% of jails surveyed in 2011 reported having any form of substance abuse treatment programs, with only 18% of jails reporting having consistent funding provided to them.[101]

A substance abuse program coordinator, Rebecca Fealk, says that money is not the issue in correctional facilities, it is the way they use their budgets. She says that the Department of Corrections for the state of Arizona has a budget of $1.1B but the budget needs to be prioritized for substance abuse programs.[102] Prisons had been recently allocated $5M for the fiscal year of 2022 to expand their substance abuse programs. This budget can be used to help increase the staff necessary for these programs to be successful, including training for those who work in prisons but are not behavioral health professionals. The prison systems need to offer treatment programs that are equivalent to those offered outside the walls. There are limited program options and space for treatment programs and these programs are not prioritized.

The program options that can be offered to the inmates for substance abuse treatment are limited. The first treatment program is the Drug Abuse Education program which entails a series of classes that educate inmates on substance abuse and the effects it has on their mental and physical health. This program is often utilized while awaiting admittance into a long-term treatment program. The second program is the Nonresidential Drug Abuse Treatment which is a 12-week cognitive behavioral therapy (CBT) program involving group sessions. These programs address criminal lifestyles while giving the inmate an opportunity to increase skills in the areas of rational thinking, communication, and reentry into society. [103] Treatment programs that have participants attending group meetings and self-help group treatment focusing on self-esteem and community relationship building lead to an increase in resilience, therefore decreasing substance abuse relapse.[104] Those who are enrolled in this program have a short sentence and do not meet the criteria for Residential Drug Abuse Programs (RDAP), are waiting for admission into RDAP, are being transitioned back to their community, or currently have a positive urinalysis test.[105] The RDAP is the most intensive program provided by the Federal Bureau of Prisons. In this program, the inmates live in their own community, separate from the rest of the general prison population. These inmates are scheduled to work, school, or vocational activities half of the day and substance abuse treatment the other half of the day. These activities lead to an increase in resilience and confidence for reentry back into their community. These programs typically last nine months. Research has shown that inmates who are involved in RDAP are significantly less likely to recidivate and have a drug relapse than those who do not take part in RDAP.[106] Drug treatment within the criminal justice system, whether it is mandated or optional, through drug courts, transitional or aftercare programs has been shown to reduce re-arrest and new arrest rates, as well as drug use in general.[107]

Treatment to Decrease Recidivism Through Resilience

Most people who spend time in prison or jail will be released one day, though unfortunately, many of those people return to their life of crime and usually end up going back to prison.[108] Untreated substance-abusing offenders are more likely than treated offenders to relapse and return to criminal behaviors, bringing about re-arrest, then reincarceration, jeopardizing public health and public safety, and therefore taxing the resources of the criminal justice system even further.[109] Substance abuse treatment helps in the transition from the criminal justice system to the community.[110] Treating drug-involved offenders provides a unique opportunity to decrease substance abuse and recidivism due to drug use-associated criminal behaviors.[111] Substance abuse treatment can be incorporated into the criminal justice system in various ways including requiring treatment as a condition of probation, drug courts that blend judicial monitoring and sanctions with treatment, treatment in prison followed by community-based treatment after release, and treatment while under probation or parole supervision.[112]

A report by the National Center on Addiction and Substance Abuse, in 2010, showed that offenders who receive a full course of evidence-based treatment and recovery services have the best outcomes, including reduction in relapse and recidivism rates. Essential elements to the success of the program include screening the individual to determine the extent and nature of risky substance use or addiction and a comprehensive assessment of the patient's substance-related problem and treatment requirements. They must also provide an individualized treatment plan that is tailored to the unique needs of the offender, aftercare plan with community supervision, case management, and integrated services like medical, psychiatric, housing, childcare, social support, vocational and employment assistance. Finally, there must be monitoring of substance use and relapse episodes followed by prompt rewards and/or sanctions.[113] This report notes that there must be trained correctional staff for the above practices to be implemented appropriately.[114]

Combining correctional and community-based treatment opportunities reduces an offender's risk of recidivism, decreases substance abuse, improves prospects for employment, and increases pro-social behaviors.[115] These pro-social behaviors are the backbone of resilience. Without feeling the support necessary, the inmate has a significant limitation on their resilience, therefore increasing their relapse and recidivism risk. After completion of the drug rehabilitation program in prison, the inmate who is being released can partake in a program for outpatient/community treatment. Community Treatment Services (CTS) gives continued care to offenders who have been released, though placed in Residential Reentry Centers or on home confinement.[116] These services are necessary because the most vulnerable time for offenders to relapse back into drug use or criminal behaviors is immediate upon release. This is also the time for the highest risk of overdose and death.[117] A study of Delaware prison inmates who completed a drug treatment program while incarcerated showed that 5 years after completion, 59.6% of those who graduated from the aftercare program did not have any new arrests.[118] Everyone progresses through substance abuse treatment at different rates, though one of the most reliable findings in treatment research is that lasting reductions in criminal activity and drug abuse are related to the length of treatment. Better outcomes are associated with treatments that last longer than 90 days, with the greatest reduction occurring in those who complete treatment.[119] Substance abuse treatment programs that included in-prison drug treatment, a transitional program composed of a combination of work release, drug treatment, and aftercare services provide an effective environment for successful community reentry, therefore increasing resilience.[120]

Conclusion

Positive adaptation is judged by the positive behavior desired by society for people of a particular age, and the absence of undesirable traits and behaviors including emotional distress, mental illness, or criminal behavior.[121] These positive adaptations are caused by an individual's resilience. A person must have a certain level of resilience to be successful in their substance abuse and mental health treatment, as well as overall life. Being incarcerated is one of the difficult situations because an individual has to be strong physically and mentally, especially to reintegrate into society upon release, and not re-offend.[122] Resilience is imperative throughout the incarceration and release process. Even if the individual does not know how to access these resilience factors, they are present in everyone and therefore can have those factors strengthened, allowing development or adaptation during substance abuse and mental health treatment programs.[123] Treatment of substance abuse disorders and fostering of resilience leads to a decrease in crime rates and recidivism. This chapter explained the link between drug abuse and criminal behaviors, the genetics, mental health, and substance abuse diagnosis, and how resilience affects a persons' mental health. The second section of this chapter described the prevalence of overdoses in prison, how drugs enter the prisons, and why inmates continue to use drugs while incarcerated. The following section includes the delays in care seen by prison systems and both facility and inmate limitations on proper care. The final section describes the treatment programs offered in prison systems and on transition back to the community to decrease recidivism by building resilience.

Notes

[1] Redonna K. Chandler PhD, Bennett W.Fletcher PhD, , & Nora D.Volkow MD, "Treating Drug Abuse and Addiction in the Criminal Justice System: Improving Public Health and Safety," *JAMA: Journal of the American Medical Association* 301, no. 2 (2009). 183.
[2] Torsten Kolind and Karen L. Duke, "Drugs in Prison: Exploring Use, Control, Treatment, and Policy," *Drugs: Educattion, Prevention, and Policy* 23, no. 2 (2016), 89.
[3] University of New Mexico, "Jail Based Substance Abuse Treatment Literature Review " *Institute for Social Research* (2011), 1.
[4] Beth Schwartzapfel and Jimmy Jenkins, "Overdose Death in State Prisons Have Jumped Dramatically Since 2001," *NPR* 2021, 1.
[5] Ibid.
[6] BethSchwartzapfel and Jimmy Jenkins Inside the Nation's Overdose Crisis in Prisons and Jails, *The Marshal Project* 2021,1.
[7] Leah Wang & Wendy Sawyer,"New Data: State Prisons are Increasingly Deadly Places," in *Prison Policy Initiative* (2021), 1.
[8] Peter Gergely Paszti, "Measuring Resilience in Prisoners," no. 2 (2020), 73.
[9] University of Mexico, 1.
[10] Ibid.
[11] Ibid.
[12] Matthias Pierce, Karen Hayhurst, Sheila M. Bird, Matthew Hickman, Toby Seddon, ,Graham Dunn, , & Tim Millar, "Insights into the Link Between Drug Use and Criminality: Lifetime Offending of Criminally-Active Opiate Users," *Drug and Alcohol Dependence* 179 (2017), 309.
[13] Pierce, et al., 310.
[14] Ibid., 313.
[15] Ibid., 309.
[16] Eliana Kaplowitz, Ashley Q.Truong, Alexndria Macmadu, Meghan Peterson, Lauren Brinkley-Rubinstein , Nathan Potter, Traci C. Green , Jennifer G. Clarke, and Josiah D. Rich, "Fentanyl-Related Overdose During Incarceration: A Comprehensive Review, " *Health and Justice* 9, no. 13 (2021), 1.
[17] University of Mexico, "Jail Based Substance Abuse Treatment Literature Review," 1.

[18] Steven E. Hyman, "The Genetics of Mental Illness: Impliccatioons for Practice " *Bulletin of the World Health Organization* 78, no. 4 (2000), 456.
[19] Ibid., 455.
[20] Chandler, Fletcher and Volkow,183.
[21] Hyman, 456.
[22] Paszti, 73.
[23] Dmitry M. Davydov, Robert Stewart, Karen Ritchie, & Isabelle Chaudieu, "Resilience and Mental Health, " *Clinical Psychoology Review* 30, no. 5 (2010), 481.
[24] Paszti, "Measuring Resilience in Prisoners,"74.
[25] Ibid., 84.
Ibid., 78.
[27] Ibid., 81.
[28] Ibid., 84.
[29] Davydov, et al., 1.
[30] Paszti, 74.
[31] Davydov, et al., 490.
[32] Ayako Yamashita, Shin-ichi Yoshioka, & Yuki Yajima, "Resilience and Related Factors as Predictors of Relapse Risk in Patients with Substance Use Disorder: A Cross-Sectional Study " *Substance Abuse Treatment, Prevention, and Policy* 16, no. 40 (2021), 2.
[33] Yamashita, et al., 7.
[34] F. Caravaca-Sanchez, M.Garcia-Jarillo, "Perceived Social Support, Resilience, and Consumption of Psychoactive Substances Amongst Inmates in Prisons," *Esp. Sanid Penit.* 22, 2 (2020), 77.
[35] Yamashita, et al., 7.
[36] Wang and Sawyer, 1.
[37] Ibid.
[38] Ibid.
[39] Schwartzapfel and Jenkins, "Overdose Death," 1.
[40] Ibid.
[41] Ibid.
[42] Kaplowitz, et al., 1.
[43] Schwartzapfel and Jenkins, "Overdose Death," 1.
[44] AndrewO'Hagan and Rachel Hardwick, "Behind Bars: The Truth About Drugs in Prison," *Forensic Research & Criminology International Journal* 5, no. 3 (2017),310.
[45] Kolind and Duke, 89.
[46] Wang and Sawyer, 1.
[47] Ibid.
[48] O'Hagan, 310.
[49] Ibid., 312.
[50] Ibid., 310.
[51] Ibid., 310.
[52] Wang and Sawyer, 1.
[53] O'Hagan, 311.

[54] Ibid.
[55] Schwartzapfel and Jenkins, "Overdose Death,"1.
[56] O'Hagan, 312.
[57] Ibid.
[58] Schwartzapfel and Jenkins, "Overdose Death,"1.
[59] O'Hagan, 312.
[60]Ibid.
[61] Kolind and Duke, 89.
[62] Schwartzapfel and Jenkins, "Overdose Death," 1.
[63] Wang and Sawyer, 1.
[64] Kolind and Duke, 90.
Ibid.
[66] O'Hagan, 309.
[67] Kolind and Duke, 90.
[68] Chandler, et al., 184.
[69] 12 News, "Arizona Prisons Not Providing Treatment Programs to Most Inmatess, Report Finds " *12 News* (Arizona) 2021. Page 1.
[70] Jeremy Duda, "Sentencing Reform Debate Shines Light on Lack of Substance Abuse Treatment in Prisons " *AZ Mirror*, 2020 1.
[71] 12 News, 1.
[72] Schwartzapfel and Jenkins, "Overdose Death,"1.
[73] Wang and Sawyer, 1.
[74] Schwartzapfel and Jenkins, "Overdose Death,"1.
[75] Bureau of Justice Statistics, "Drug Treatment Under Correctional Supervision " *Bureau of Justice Statistics* (2021),1.
[76] Ibid.
[77] Sean Smith, "Drug Treatment Programs of the Federal Bureau of Prisons Exist, but Need More Availability," *American Legislative Exchange Council* (2017), 1.
[78] Duda, 1.
[79] Ibid.
[80] Ibid.
[81] Smith, 1.
[82] 12 News, 1.
[83] Duda, 1.
[84] Schwartzapfel and Jenkins, "Overdose Death,"1.
[85] University of New Mexico, 3.
[86] Ibid., 2.
[87]Ibid.
[88] Bureau of Justice Statistics, 1.
[89] University of New Mexico, 2.
[90] Ibid.
[91] Ibid.
[92] Ibid.
[93] Ibid.
[94] Paszti, 81.
[95] Ibid.

[96] University of New Mexico, 4.
[97] Chandler, et al., 183.
[98] Bureau of Justice Statistics, 1.
[99] Duda, 1.
[100] Ibid.
[101] University of New Mexico, 1.
[102] Duda, 1.
[103] Smith, 1.
[104] Yamashita, et al., 2.
[105] Smith, 1.
[106] Ibid.
[107] University of New Mexico, 2.
[108] Paszti, 74.
[109] University of New Mexico, 4.
[110] Ibid., 2.
[111] Chandler, et al., 183.
[112] University of New Mexico, 4.
[113] Ibid., 3
[114] Ibid., 4.
[115] Ibid., 5.
[116] Smith, 1.
[117] Ibid.
[118] University of New Mexico, 3.
[119] Ibid., 4.
[120] Ibid., 2.
[121] Paszti, 78.
[122] Ibid., 84.
[123] Ibid.

Literature

Bureau of Justice Statistics. 2021. "Drug Treatment under Correctional Supervision." 1.

Caravaca-Sanchez, F. and M. Garcia-Jarillo. 2020. "Perceived Social Support, Resilience, and Consumption of Psychoactive Substances Amongst Inmates in Prisons." *Esp. Sanid Penit.* 22, 2: 75-79.

Chandler PhD, Redonna K., Bennett W. Fletcher PhD, & Nora D. Volkow MD. 2009. "Treating Drug Abuse and Addiction in the Criminal Justice System: Improving Public Health and Safety ". *JAMA: Journal of the American Medical Association* 301, no. 2: 183-90.

Davydov, Dmitry M., Robert Stewart, Karen Ritchie, & Isabelle Chaudieu. 2010. "Resilience and Mental Health ". *Clinical Psychoology Review* 30, no. 5: 479-95.

Duda, Jeremy. 2020. "Sentencing Reform Debate Shines Light on Lack of Substance Abuse Treatment in Prisons " *AZ Mirror*, 1.

Hyman, Steven E. 2000. "The Genetics of Mental Illness: Impliccatioons for Practice ". *Bulletin of the World Health Organization* 78, no. 4: 455-63.

Kaplowitz, Eliana, Ashley Q. Truong, Alexandria Macmadu, Meghan Peterson, Lauren Brinkley-Rubinstein, Nathan Potter, Traci C. Green, Jennifer G. Clarke, and Josiah D. Rich.2021. "Fentanyl-Related Overdose During Incarceration: A Comprehensive Review." *Health and Justice* 9, no. 13: 1-11.

Kolind, Torsten &. Karen Duke.2016. "Drugs in Prison: Exploring Use, Control, Treatment, and Policy ". *Drugs: Educattion, Prevention, and Policy* 23, no. 2: 89-92.

12 News. 2021. "Arizona Prisons Not Providing Treatment Programs to Most Inmatess, Report Finds " *12 News* (Arizona), 1.

O'Hagan, Andrew & Rachel Hardwick. 2017. "Behind Bars: The Truth About Drugs in Prison." *Forensic Research & Criminology International Journal* 5, no. 3: 309-20.

Paszti, Peter Gergely. 2020. "Measuring Resilience in Prisoners ". no. 2: 73-86.

Pierce, Matthias, Karen Hayhurst,Sheila M. Bird, Matthew Hickman, Toby Seddon,Graham Dunn, & Tim Millar. 2017. "Insights into the Link between Drug Use and Criminality: Lifetime Offending of Criminally-Active Opiate Users ". *Drug and Alcohol Dependence* 179: 309-16.

Schwartzapfel, Beth & Jimmy Jenkins. 2021. "Inside the Nation's Overdose Crisis in Prisons and Jails " *The Marshall Project*.

—. 2021. "Overdose Death in State Prisons Have Jumped Dramatically since 2001." *NPR*.

Smith, Sean. 2017. "Drug Treatment Programs of the Federal Bureau of Prisons Exist, but Need More Availability." *American Legislative Exchange Council*: 1.

University of New Mexico. 2011. "Jail Based Substance Abuse Treatment Literature Review." *Institute for Social Research: 1-9.*

Wang, Leah & Wendy Sawyer. 2021. "New Data: State Prisons Are Increasingly Deadly Places." In *Prison Policy Initiative*, 1.

Yamashita, Ayako, Shin-ichi Yoshioka, & Yuki Yajima. 2021."Resilience and Related Factors as Predictors of Relapse Risk in Patients with Substance Use Disorder: A Cross-Sectional Study ". *Substance Abuse Treatment, Prevention, and Policy* 16, no. 40: 1-9.

CHAPTER ELEVEN

THE RESILIENCE OF CLINICAL SPIRITUAL CARE IN A PLURALISTIC HEALTH CARE ENVIRONMENT

ROBERT ALAN PRICE

Introduction

The concept of the critical importance of ethics in health care is likely as ancient as health care itself. There has been and continues to be a cornucopia overflowing with a rich diversity of approaches to express and reflect what is the best ethical conduct and deportment in the provision of health care. As varied as the approaches are, so are their sources of inspiration.

There are discussions of a common morality that is veritably universal amongst humankind, regardless of time and place.[1] Over the centuries, this concept has led to the enshrinement of these moral precepts into various documents from the sacred texts of various world religions to very recent modern documents such as the United Nations Educational, Scientific and Cultural Organization (UNESCO)'s "Universal Declaration on Bioethics and Human Rights" (adopted 19 October 2005).[2]

There have been attempts to enact legislation to preserve the integrity and efficacy of health care, from the Code of Hammurabi's establishment of rates of pay for providers and harsh penalties for malpractice (the removal of the surgeon's hands!)[3] to modern court decisions that assert control over the appropriate medical courses of action for specific health care scenarios, such as the famous case of Claire Conroy, where the Supreme Court of New Jersey ruled that life-sustaining treatments may be withheld or withdrawn from a patient lacking decision-making capacity when it is known that the patient would have declined the life-sustaining treatments if the patient possessed capacity and could articulate his or her preferences at the time.[4]

No comprehensive discussion of bioethics can ignore the import of the ancient yet relevant Hippocratic Oath; in earlier, less modernized versions, the Oath emphasized that the provider must first and foremost be a good human being, and a discussion of personal ethics preceded the teaching of the medical arts.[5]

There have been more modern treatises written by professionals, such as the early *Medical Ethics* by physician Thomas Percival in 1803[6] to the influential textbook by philosopher Tom L. Beauchamp and theologian James F. Childress, *Principles of Biomedical Ethics*, whose "pivotal moral principles" have dominated (though not without their detractors[7]) much of the discussion in modern American health care ethics education in the last four decades.[8]

There are numerous extant codes of professional ethics for each health care-related field, oftentimes with each state licensing board or professional board creating its own and requiring acknowledgment, affirmation, and/or agreement by its members. Clinical chaplains as a progressively professionalizing field have likewise created several professional ethics expectations, which will be explored below. Unfortunately, these codes focus on professional or business ethics and fall short of approaching health care ethics theory, which, as discussed below, has the potential for adverse outcomes for both patients and clinical chaplains. This study will begin by reviewing the ethics educational and clinical preparation of chaplains as of this writing, including a brief look at the ethics content of some of the most prominent professional organizations' certification materials. This will aid in understanding the ethics orientation of many clinical professionals in the field.

Regarding health care ethics theory, it is the popular Principlism approach espoused by Beauchamp and Childress that will be briefly focused upon below, concentrating our study on the four primary ethical principles they identify: autonomy, nonmaleficence, beneficence, and justice. This study will then focus on how clinical chaplain care engages these four primary ethical principles, with the following discussion topics: the patient's autonomy and the chaplain's faith, nonmaleficence and proselytization, beneficence and clinical chaplain care, and finally, justice and patients of other faiths.

Professional Ethics vis-à-vis Health Care Ethics

Professional Confusion and Unconsciousness

There is a great deal of confusion and a general lack of awareness in many health care professions regarding the differences between professional ethics and health care ethics, with many professional fields confounding or at the least combining the two. Clinical chaplaincy, as the specific subject field of this chapter, is particularly affected by this oversight. In most cases, it is an unconsciousness of the matter, a case of "you don't know what you don't know," as the major professional codes of ethics and educational training pathways commonly fail to provide or require training specifically in health care ethics.

Ethics in the Educational Preparation for Clinical Chaplaincy

There is an extensive degree of variation in the required educational preparation for clinical chaplains in the United States. For this study, we will choose as our exemplar the largest and most established organization employing clinical chaplains, the U.S. Department of Veterans Affairs (VA). The VA operates the nation's largest integrated health care network, the Veterans Health Administration (VHA), and employs the largest number of professional health care chaplains, averaging between 850-1000 clinical chaplains as of this writing.[9]

As employees of a U.S. government agency, numerous federal regulations prescribe the policies and procedures for all matters pertaining to VHA clinical chaplains, and the qualifications for a full-performance level clinical chaplain in the VA are publicly available. According to "VA Handbook 5005," the human resources policy establishing staffing regulations and guidance for the VA, these are the selected minimum requirements relevant to the present discussion for a candidate to be appointed as a VHA Clinical Chaplain: The candidate must (a) have earned the Master of Divinity or equivalent degree from an accredited school (must be "Council for Higher Education [CHEA]" or "Association of Theological Schools [ATS]" and recognized by the U.S. Department of Education"), (b) the candidate must have earned a minimum of "four units of [Clinical Pastoral Education {CPE}] from a CPE center that is accredited by an organization recognized by the U.S. Department of Education" (see below), (c) the candidate must "possess and maintain a full and active ecclesiastical endorsement" from a U.S. government-approved endorsing organization, and (d) the candidate must be a fully

Board Certified Chaplain (BCC), obtaining said credential from the Board of Chaplaincy Certification, Inc. (BCCI) or an equivalent organization with a standing reciprocity agreement with BCCI.[10] For candidates lacking the M.Div. degree, historically, the VA National Chaplain Center (NCC) would convene a workgroup to assess their submissions for educational equivalency, whereupon NCC would issue the successful candidate a *Letter of Equivalency for M.Div.* The current policy of the new National VA Chaplain Service is to require the candidate's board-credentialing agency (i.e. BCCI or reciprocal organization) to certify educational equivalency.[11]

Examining each of these four areas of qualification, there is a dearth of health care ethics education in the requirements of each.

First, the Master of Divinity (M.Div.) degree is the normative graduate degree for entry-level Christian ministry positions, requiring "a minimum of 72 semester credits."[12] While the ATS standards of accreditation delineate numerous required subjects, including "professional ethics," there is understandably no requirement for any other ethics topics or types, as this degree is intended to prepare candidates for a broad range of ministries.[13] Within the whole of the accreditation manual, this is the sole reference to ethics. Individual schools may indeed offer ethics education, but evaluating the ten largest ATS-accredited M.Div. programs as of the 2021-2022 academic year (by total headcount, in order from largest enrollment, Liberty University John W. Rawlings School of Divinity [2,429], Southern Baptist Theological Seminary [1,923], Southeastern Baptist Theological Seminary [857], Midwestern Baptist Theological Seminary [749], Asbury Theological Seminary [658], Regent University School of Divinity [617], New Orleans Baptist Theological Seminary [577], Fuller Theological Seminary [569], Southwestern Baptist Theological Seminary [557], and Dallas Theological Seminary [548]),[14] none of them required more than 3 hours of ethics as part of their normative programs, with most making ethics optional or not offered at all. These ten (10) programs, out of two hundred seventy-two (272) total, represent a third (32.6%) of all current M.Div. students.[15]

Most of these schools offered specialized tracks that allow for additional ethics preparation, but only two (Southern Baptist Theological Seminary and Dallas Theological Seminary) have any specific bioethics instruction. Liberty University John W. Rawlings School of Divinity offered one class on "Chaplain Ethics for Advisement and Pastoral Care," and one on "Healthcare Chaplaincy."[16] Southern Baptist Theological Seminary required nine (9) hours of ethics classes in their Ethics and Philosophy track and offered an optional Biomedical Ethics class to fulfill three (3) of these hours.[17] Southeastern Baptist Theological Seminary had an ethics track requiring twelve (12) hours, but none of these were specifically health care ethics.[18] Midwestern Baptist Theological Seminary did not offer any specific chaplaincy track or additional ethics education options.[19] Asbury Theological Seminary offered a chaplaincy specialty track, but no health care ethics courses.[20] Regent University School of Divinity offered a "Chaplain Ministry" concentration, but only one general course in "Christian Ethics" was available.[21] New Orleans Baptist Theological Seminary offered a concentration in "Chaplaincy and Pastoral Care," with only one "Christian Ethics" or "Biblical Ethics" course offered.[22] Fuller Theological Seminary offered a concentration in "Chaplaincy," but no classes on health care ethics were available.[23] Southwestern Baptist Theological Seminary offered concentrations in both "Chaplaincy" and "Ethics, Philosophy, and Apologetics," but again, no classes on health care ethics were available (and the chaplaincy concentration requires no ethics courses!).[24] Dallas Theological Seminary is unique amongst its peers as it did not offer a Master of Divinity degree; however, its available 79-hour Master of Arts in Chaplaincy and Ministry Care could be considered equivalent to a traditional M.Div. for these purposes. This program offered one "Bioethics and Pastoral Care" course.[25] So from the above, it can be readily seen that the ten largest accredited programs in the United States do not provide even a minimal foundational health care ethics education (with most programs offering no programming whatsoever in bioethics), which the future clinical chaplains will require if they are to best serve patients and families in health care contexts.

Ethics in the Clinical Preparation for Health Care Chaplaincy

Following the completion of the Master of Divinity degree or its equivalent, chaplaincy candidates must complete clinical training. While numerous organizations provide clinical education for religious ministry professionals aspiring to best serve in the pluralistic health care environment, out of brevity, this current study will limit our consideration to one of the larger, more prolific organizations in the United States, the Association for Clinical Pastoral Education (ACPE). "ACPE is the only agency recognized by the United States Department of Education (USDE) to accredit programs of Clinical Pastoral Education (CPE) in the United States. ACPE currently accredits over 300 CPE programs offered in hospitals, prisons, churches, seminaries, community-based organizations, and other settings" nationwide.[26]

In each of the CPE programs in which this author has been involved, the topic of "ethics" was covered in a single didactic session or not at all. Discussions with several other professionals involved in CPE have consistently resulted in similar experiences being described. In all but one case, the "ethics" was professional ethics, i.e. professional boundaries, and not specifically health care ethics. This correlates with the ACPE objectives and outcomes, divided into two levels for clinical pastoral education students, and organized under the categories of pastoral formation, pastoral competence, and pastoral reflection.[27]

No Level I outcomes explicitly discuss health care ethics, but a few could be argued to imply clinical ethics. One Level 2 outcome directly references health care ethics. One objective overtly mentions clinical ethics.

Under the category of "Pastoral Formation," there is no ethics content stated or implied. Under the category of "Pastoral Competence," Objective O6 seeks "to develop students' ability to make effective use of…applied clinical ethics in their pastoral care of persons and groups." Level II outcome L2.2 explicitly articulates an ethics element: "provide pastoral ministry with diverse people, taking into consideration multiple elements of cultural and ethnic differences, social conditions, systems, justice and applied clinical ethics issues without imposing one's own perspectives." The "Pastoral Reflection" category likewise does not include any ethics content, professional or clinical. The Level II outcome L2.9 requires that "by the end of Level II, students will be able to demonstrate awareness of the Common Qualifications and Competencies for Professional Chaplains."[28] This document is the same in content as that aforementioned in the APC competencies section above, referred to here in the ACPE manual.[29]

Ethics in the Board Certification for Health Care Chaplaincy

Following the completion of the Master of Divinity degree or its equivalent and the attainment of at least four (4) units of CPE, the next qualification for clinical chaplains is appropriate professional board certification from the Board of Chaplaincy Certification, Inc. (BCCI) or an equivalent organization with a standing reciprocity agreement with BCCI.[30] The process to apply for Board Certified Chaplain with BCCI is extensive, with numerous education, training, and competencies requirements.[31] The requisite clinical training was just discussed in the previous section. Per BCCI standards, the educational requirement for fully board-certified chaplains is a minimum of 72 graduate hours, of which ethics are not a specific requirement.[32] In the discussion of educational equivalency, the educational requirement is again 72 graduate hours, and again, ethics is not mentioned.[33] In the document delineating required competencies, under "Section I: Integration of Theory and Practice Competencies" standard "ITP4," candidates must in their written materials for submission, "incorporate a working knowledge of different ethical theories appropriate to one's professional context." Under "Section II: Professional Identity and Conduct Competencies" standard "PIC7," they must demonstrate how they "function within the [Association of Professional Chaplains (APC)] APC Code of Ethics." The APC professional code of ethics will be further discussed below. Finally, under "Section IV: Organizational Leadership Competencies" standard "OL4," candidates must demonstrate how they "promote, facilitate, and support ethical decision-making in one's workplace."[34] How candidates were to have obtained this knowledge content or experiences is not suggested.

Professional Consciousness and Conscientiousness

Professional Ethics

As do many other professional fields, clinical or health care chaplaincy has several codes of ethics that have been established by a number of their various professional organizations; the APC Code of Ethics is one of the exemplars. Especially relevant to the subject matter of this present study is section "130 Ethical Standards for Members," and particularly subsections 130.11, 130.12, and 130.13, wherein it is stipulated that the expected standard is associated with clinical chaplains

> "shall serve all persons without discrimination regardless of religion, faith group, race, ethnicity, sexual orientation, gender, age, or disability…shall demonstrate respect for the opinions, beliefs and professional endeavors of…those with whom they have contact in their professional role as chaplain.…[and] shall affirm the religious and spiritual freedom of all persons and refrain from imposing doctrinal positions or spiritual practices on persons whom they encounter in their professional role as chaplain.[35]

Particularly addressing (albeit concisely) pluralism and proselytization here (subjects which will be discussed in more detail below), APC demonstrates high standards and expectations for both quality and health care ethics of their professional clinical chaplain members with these explicit subjects. The code addresses additional topics of professionalism or business ethics, such as a high standard of respect for colleagues, privacy matters, and strong prohibitions against any kind of exploitation of the pastoral relationship or sexual relationships of any kind with persons served (patients/clients/staff ministered to). This exemplar thus represents sound professional ethics but falls short in further addressing the clinical ethics of providing spiritual or pastoral care, the subject of the present work. As this is one of the most common "ethics" standards clinical chaplains in the United States are oriented to, the sparsity of bioethics content and reasoning embedded within may contribute to this author's perception of an overall lack of bioethics expertise amongst many clinical chaplains; in this way, this author finds this code to be inadequate.

Health Care Ethics

As the discussion above intimates, in addition to observing professional codes of ethics, clinical chaplains, as health care providers, should observe the same normative ethical standards for the delivery of care as any other health care provider. As clinical chaplain care is unique in many respects from the scopes of practice of other clinical providers, a specific set of guidelines is required to inform clinical chaplains of their roles and responsibilities from a bioethical point of view.

One of the only such examples in existence at present is the 2000 APC document "Guidelines for the Chaplain's Role in Health Care Ethics." The document includes eight principles, with several guidelines under each, for the chaplain's inclusion in health care ethics committees. While the majority of the document is favorably describing the values of chaplain inclusion in the health care ethics consultation team, it is overly myopic in limiting the chaplain's potential contributions to the consultation process itself.

Framing the intent to "emphasize chaplaincy care's unique perspective as integral to the ethical reflection process of a health care organization," the authors suggest, "chaplains play a crucial role in health care ethics reflection. Chaplains may be of assistance to health care ethics committees as they discuss the questions of philosophy, theology, spirituality, human values and morals, which are integral to ethical questions."[36] At this point in this document, the framers seem to be supporting the interdisciplinary value of the clinical chaplain on the health care ethics team; however, their following assertion undermines this clinical value, as it diminishes the chaplain's contribution:

> While some chaplains have education and/or training in ethics, their roles as chaplains differ from those of ethicists. Chaplains identify and clarify the patient's spiritual and moral perspectives as essential ingredients in the

process of health care ethics reflection. Integration of these perspectives with those of other healthcare disciplines fosters a holistic approach to healthcare ethics.[37]

Clinical chaplains trained in health care ethics need not be limited to identifying and clarifying the morality or spirituality of the patient and/or their loved ones or the health care team's attitudes or treatment decisions. Indeed, taking religious, spiritual, and cultural factors into account is a consideration all ethicists (irrespective of clinical background) should already be doing to the best of their abilities and understanding. It can be argued that a clinical chaplain is not necessarily required for this exercise, and the absence of chaplains on many health care ethics teams demonstrates this. Rather, the ethically trained clinical chaplain's role is to provide their unique expertise as a highly-educated religious ministry professional and a clinically trained spiritual care provider to the team, in addition to doing the work of a duly trained health care ethics consultant. This expertise uniquely qualifies ethically trained clinical chaplains to be clinical ethicists in their own right and to serve as productive members and valuable contributors to health care ethics consultation teams overall. This APC document does not make this assertion strongly enough. Furthermore, the expectations for chaplains gaining health care expertise are substandard. While the document states chaplains should seek out "continuing education in health care ethics and ethics consultation in order to achieve a working knowledge of basic principles, ethical decision-making, current issues, and developing trends" this statement in itself implies or expects a lack of expertise. A "working knowledge of basic principles" does not make one a trained health care ethicist, and many contexts in which clinical chaplains are to serve, such as VHA, expect far more of their chaplains' expertise in this area. This guiding document provides numerous opportunities for improvement, and this study will seek to begin to discuss some of the knowledge and practical applications needed for health care ethics expertise in clinical chaplaincy.

Principlism in Health Care Ethics

This study now turns to a discussion of the four normative principles for health care ethics, as presented by Beauchamp and Childress in their *Principles of Biomedical Ethics*, in its eighth edition as of this writing.[38] To best respect the dignity of all persons served within the health care environment, an individual's autonomy should be acknowledged and respected, the complementary principles of nonmaleficence and beneficence should be endeavored, and the crucial principle of justice should be treated as sacrosanct. Beyond this, the underlying values from which the principles are derived should drive the delivery of all high-quality health care processes and encounters. This idealized reality must be the aim of any truly ethical health care environment. These four principles will be concisely explored below, to be followed by an examination of how clinical chaplaincy engages these principles in spiritual care delivery.

Principle I: Patient Autonomy

Patient Autonomy Defined

The term "autonomy" linguistically comes to the English language from the ancient Greek *autos* meaning "him/her/itself"[39] and *nomos* meaning "anything established, anything received by usage, a custom, a law, a command."[40] The intended meaning of this derived word in its bioethical context would ultimately be "self-ruling" or "self-directing." Generally speaking, for a person to be autonomous in an idealized scenario, he or she must possess both liberty and agency. Liberty requires freedom from the control of others in the decision-making process, while agency requires the ability or "capacity" to make the decision.[41] According to Beauchamp and Childress' theory of autonomy in nonideal health care settings, there are three requisite conditions: "intentionality," "understanding," and "non-control."[42] To further unpack these concepts, "intentionality" would mean purposeful actions or decisions, regardless of whether or not the actor or decision maker possesses conflicting thoughts about having to make said decision or take such action. The concept of "understanding" simply means that the actor or decision maker possesses an adequate awareness of the situation and ramifications of the decision; this is especially critical in a health care environment, where a risk-to-benefit ratio must be considered for most treatment decisions as standard practice. "Noncontrol" is meant to convey that there are no external forces, like doctors or concerned family members, forcing the actor or decision maker to exercise an external willpower-derived decision or action, rather than that which reflects the freely derived choice of the individual's own will.[43]

Ethical Considerations Limiting Patient Autonomy

There are numerous potential limitations or degrees of autonomous decision-making based on the above conditions.[44] For example, is a person who chooses to adhere to a certain religion truly acting autonomously when the choice may be based upon the experience of severe consequences, such as social disdain or even eternal punishment, if they do not choose that religion? In a more health care-oriented example, is a person that declines certain medical treatments, such as a Jehovah's Witness who refuses blood products, able to fully exercise autonomy when this decision is based upon the precepts of their faith tradition?[45] A reasonable theory of autonomy must provide allowances for persons to choose to invest confidence and compliance with an external authority whom they respect and trust, such as an attending physician or a religious tradition, otherwise, autonomy would be rare indeed.[46]

Furthermore, cultural considerations must be made to maximally integrate this concept into patient care. For example, in cultures where individual autonomy is not valued over all others, respecting a patient's autonomy may include respecting a fully capable and capacitated patient's wishes to limit their own autonomy and empower others, such as a spouse or family member(s), to be more fully informed of the medical situation and to make health care decisions on their behalf.[47]

Finally, despite its historical position of favor amongst much of the field (particularly in American bioethics), autonomy is but one principle among many other worthy principles, and not foremost among all, trumping all other principles and values.[48] Each situation should be weighed and balanced, and multiple principles and values should be considered before the most appropriate ethical decision can be identified.

Principle II: Nonmaleficence

Do No Harm

Nonmaleficence comes to the English language via a combination of the Latin roots *non* (meaning no or do not) and *maleficencia* (meaning evildoing, itself combined from the Latin words *male*, meaning badly, wrongly, and *facio*, meaning to do, act).[49] This principle is attributed to the Latin phrase, "*primum non nocere*," itself commonly and mistakenly believed to be derived from the (Attic Greek) Hippocratic oath.[50] It translates as, "Above all, do no harm."

Regardless of origin, this bioethics principle reflects a longstanding prohibition against providers causing harm or bad outcomes for patients. Some philosophers combine this principle with the next we will examine, *beneficence*, or doing good. There is a significant distinction in these principles, however, and thus this writing will maintain the separation. Perhaps the most meaningful difference is the behaviors involved in observing these principles; nonmaleficence involves the intentional and willful avoidance of any harm causation, while beneficence involves further decisive action(s) taken to commit good and cumber evil, as discussed below.[51]

Critical Distinctions between "Harm" and "Wrong"

A likewise meaningful distinction must be made between the concepts of patient "harm" and a patient being "wronged." Beauchamp and Childress discuss the notion of "wronging" meaning "violating someone's rights," while the notion of harm is far broader and may occur without a person being wronged. Persons may be harmed by disease, normative bodily progressions such as the aging process, natural disasters (unless we are attributing these to "Acts of God," which would be an altogether different ethics discussion), and events the person had already assented to voluntarily.[52] For this discussion of ethical conduct and principles, this writing will limit the discussion to how persons may be harmed by the (either willful or oblivious) actions or inactions of health care providers.

In bioethics, and medicine in general, the concept of harm is often vague and can be difficult to be clearly defined. Beauchamp and Childress present harm as "a thwarting, defeating, or setting back of some party's interests," crediting Feinburg in their definition.[53] The *Encyclopedia of Bioethics* clarifies harm must be "person-regarding," resultant of or "linked to human action…from intention or negligence," and further clarifies that the object of the harm is a "victim," not merely a person not deriving a benefit; in this way further, it is not analogous to beneficence.[54]

Principle III: Beneficence

Do Good

The ethical principle of beneficence, or doing good, comes to English from the Latin term, *beneficentia*, a combination derived from the roots *bene*, meaning "well," and *facio*, meaning, "to do."[55] As stated above, beneficence requires further intentional action to cause positive outcomes for a person, rather than merely the avoidance of causing harm, with commissions to prevent or remove harm coupled with the causing or promoting of good outcomes.[56] In principlism, this normatively myopically focuses on the individual patient, in contrast to the utilitarian theory of ethics, in which doing the most good for the most people is the central focus.[57]

Though controversial, a predominant perspective holds that this principle necessarily invokes a moral duty in health care, in that if a provider *can* do good, that provider *must* work to do that good, within reasonable limitations set by the consequences to other moral priorities given the benevolent endeavor (especially a sacrifice of time and resources) is made.[58]

Paternalism vice Beneficence

Of import to the present discussion is the matter of paternalism in health care. The *Oxford English Dictionary* defines the term, as "The policy or practice of restricting the freedoms and responsibilities of subordinates or dependants in what is considered or claimed to be their best interests."[59] It must be noted that ensuring the good of the patient does not mean trespassing upon said patient's rights and liberties when he or she possesses decision-making capacity to choose to receive or not to receive treatment.[60] This exercise of respect for persons observes the principles of autonomy, nonmaleficence, and beneficence simultaneously.

Principle IV: Justice

Treat Like Cases Alike

The final bioethical principle to be discussed in this writing comes to English via both the French *justice* and the Latin *iūstitia*. The *Oxford English Dictionary* defines the term partly, as "The quality of being just or right, as a human or divine attribute; moral uprightness; just behaviour or dealing as a concept or principle (one of the four cardinal virtues…)" and "To treat or represent with due fairness or appreciation; to deal with in a manner that is right or appropriate."[61] Much historical and contemporary thought regarding justice harkens back to Aristotelian theory, which asserts "Equals must be treated equally, and unequals must be treated unequally."[62] The Greek terms in *Nicomachean Ethics* commonly translated as "equity" and "equitable" are most often *ἐπιείκεια* and *ἐπιεικής*,[63] and convey the sense of reasonableness and fairness above all.[64] Beauchamp and Childress discuss six traditional theoretical approaches to justice, including perspectives from utilitarians, libertarians, egalitarians, communitarians, and health care-related capability and well-being theories.[65] Examined collectively, they present an imperative to treat like cases alike.

Equality and Equity

Added to the concept of justice must be the distinct but commonly synonymized terms *equality* (defined as the "condition of having equal dignity, rank, or privileges with others; the fact of being on an equal footing") and *equity* (defined as the "quality of being equal or fair; fairness, impartiality")[66] which have come to have even more significant meaningfulness in contemporary society of late. To observe the preceding three principles, justice must be likewise honored in order to best dignify and respect persons, irrespective of diversity factors such as gender, ethnicity, sexual identity, religious preferences, socioeconomic position, etc. of the patient. All human beings must be treated with the same dignity and respect as all other human beings in our care; an influential politician's daughter, or a wealthy and powerful hospital executive presenting as a patient, must be treated no differently than a person of color who is experiencing homelessness, or a mental health patient with a long history of experiencing addiction and episodes of acute mental health needs.

Clinical Chaplain Care Engaging Principlism in the Pluralistic Health Care Environment

The Patient's Autonomy and the Chaplain's Faith

As an exercise of their autonomy, patients have the right to choose to receive or not to receive spiritual care. Freedoms of religion and spirituality are recognized as universal human rights. In the 2005 United Nations Educational, Scientific and Cultural Organization (UNESCO) document "Universal Declaration on Bioethics and Human Rights" preamble, the authors dogmatically assert that "a person's identity includes biological, psychological, *social*, *cultural* and *spiritual* dimensions."[67] This is founded upon the 1948 United Nations (UN) document "Universal Declaration of Human Rights" which asserted respect and protections for "freedom of thought, conscience, and religion," elucidated in Articles 18 and 19.[68] As Ayton-Shenker writes, these are "natural-born rights for every human being, universally. They are not privileges."[69] In article 18, it is specified that all persons have the right to religious expression, which includes thought, changing one's faith or beliefs, and "freedom…to manifest his religion or belief in teaching, practice, worship and observance." Article 19 adds "the right to freedom of opinion and expression; this right includes freedom to hold opinions without interference."[70] In summation, a multinational group acknowledged that universally, patients have the right to believe whatever they elect to believe and to observe or practice whatever religion they choose for themselves. This would include the right to maintain their religion without any unwanted intrusion of any other person's beliefs and practices. This would further include the freedom *from* religion, as freedom of thought and opinion necessarily includes the freedom for rejecting religious ideology as much as embracing it. Specifically in the United States, the religion clauses of The First Amendment to the U.S. Constitution protect patients' freedoms for or from religion in much the same way.[71]

The Chaplain's Personal Faith

The religious ministry professional that is academically educated, clinically trained, and ecclesiastically endorsed to provide spiritual care to patients, their families, and staff as a clinical chaplain has invested years of preparation to serve in this complicated role. The clinical chaplain, as a human being, has a personal faith or system of belief that likely informed their choice of vocation. Must this be kept separate from their professional functioning?

Sulmasy writes, "The point of all…care is the person…spiritual care must focus on the spiritual needs of the sick person."[72] While the U.S. Constitution and the UN and UNESCO documents cited above do indeed afford the clinical chaplain as a person the same religious rights that the patient enjoys, it does not necessarily extend to the clinical chaplain as a professional, engaged in the provision of spiritual care. These are two separate issues. With many clinical chaplains and ecclesiastical endorsing organizations seeing the work of the chaplain as missional or evangelistic in nature, this can become ethically problematic for both the patient and the clinical chaplain. For example, perceiving an egregious violation of their own religious rights, many government-employed "evangelical chaplains [claim] that they are being made to subscribe to a theology in which they do not believe, the equal value of all religions."[73] This concept, pluralism, is indeed antithetical to many religious traditions. The *Oxford English Dictionary* defines *pluralism* as "[t]he presence or tolerance of a diversity of ethnic or cultural groups within a society or state; (the advocacy of) toleration or acceptance of the coexistence of differing views, values, cultures, etc."[74] While many currently serving clinical chaplains, along with the ecclesiastical organizations whose endorsement they must maintain, may not necessarily agree with the concept of pluralism and argue that their personal rights are being violated by its practice, the courts find otherwise:

> Increasingly, US courts are saying that the government can compel universalist religious commitments by chaplains because such commitments are founded not in a sectarian religious or theological position raising constitutional red flags, but, rather, in a view of the human person that understands him to be naturally spiritual, a view that should not be understood to be suspect under the establishment clause.[75]

Pluralism as a concept actually serves to maintain the religious clauses, and in the governmental context, preserves the establishment clause prohibiting the government from showing a preference for any one religion over another, or religion over non-religion. For example, a living example of the dangers of a governmental policy that does not value religious and cultural pluralism, the Dalai Lama leads Tibetan Buddhism and presently lives in India as a political refugee.[76] His attitude exemplifying compassion and universal tolerance is instructive: "It is up to us to recognize the necessity of different philosophical views and to acknowledge that every spiritual tradition is good since each one helps millions of people to progress and to suffer less…So we should have our own convictions, on the one hand, but on the other hand, we should keep our minds open and tolerant toward those who don't share them."[77] To practice ethically sound clinical chaplaincy, the religious ministry professional must be sensitive to the humanity and faith needs and preferences of the patient, irrespective of the faith needs and preferences of the clinical chaplain themselves. To impose upon the faith of the patient would be to trespass against the patient's autonomy and to commit proselytization, which will be discussed below. The Dalai Lama suggests,

> Many problems in the world arise because we have lost sight of the basic humanity that unites all members of the human family. We tend to forget that despite the diversity of race, religion, culture, language, and ideology, we all have an equal and fundamental right to peace and happiness. Each one of us wants to be happy and not to suffer. However, although we theoretically praise pluralism, unfortunately we often fail to put it into practice. In fact, our inability to embrace diversity has become a major source of conflict among peoples.[78]

Pluralism goes beyond mere tolerance of others, however, and requires the embrace of mutual humanity and intrinsic value.[79]

Nonmaleficence and Proselytization

Sharing Faith in Bad Faith

Nonmaleficence, mentioned previously, essentially means "do no harm." It was earlier developed that there is a conceptual difference between "harm" and "wrong" born to a patient. Beauchamp and Childress present that "wronging" a patient would mean "violating someone's rights," while harming the person is a far broader concept.[80] Continuing the above discussion, a clinical chaplain wrongs a patient whenever that chaplain acts out presuppositions or notions derived from their personal faith. For example, a patient who is of no faith or is of a different faith group than the chaplain is wronged when the clinical chaplain makes assumptions about the patient's faith and comments or proclaims accordingly. Furthermore, the patient may be harmed by this activity. In a more specific example, a Christian patient that is homosexual and in a same-sex relationship will be both wronged and harmed by a Christian clinical chaplain who preaches in chapel about marriage being only between a man and a woman, even though this statement may be within the orthodoxy of the chaplain's faith tradition.[81] For some patients, discussion of religion and spirituality in and of themselves may be harmful.[82]

Proselytization is De Facto Abusive

The *Oxford English Dictionary* defines the term, proselytize, as "to convert or attempt to convert from one opinion, religion, or party, etc., to another."[83] Particular to a pluralistic health care context in the United States, the regulations of the U.S. Department of Veterans Affairs (VA), Veterans Health Administration (VHA), define proselytizing as "attempting to impose religious beliefs or practices on an individual without the individual's consent."[84] "Consent" in this context is both a bioethical and legal term. "Informed consent" is a necessary aspect of most patient care procedures, reflecting respect for persons and patient autonomy.[85] When a clinical chaplain acts to impose their own values, theological opinions, or religious practices upon a patient, this clinical chaplain violates the patient's autonomy, wrongs the patient in violation of the principle of nonmaleficence, and may additionally cause harm to the patient, in several emotional, psychological, social, and/or spiritually abusive ways.[86]

Beneficence and Clinical Chaplain Care

Doing Good for Patients, their Families, and Their Conception of Deity if Held

Beneficence, or doing good, is most likely a chief motivating factor for most persons entering the helping professions, to which clinical chaplaincy belongs. Indeed, research into the efficacy of the incorporation of religious and spiritual care into the overall care of the patient has consistently evidenced positive associations.[87] As stated above, beneficence requires intentional action to cause positive outcomes for the patient. One prominent perspective sees this principle as invoking a moral duty in health care, in that if a provider *can* do good, that provider *must* work to do that good.[88] This is one of the areas in which many clinical chaplains may perceive themselves to be in a double bind, in that they have a duty to do good for the patient, and according to their theology, that necessarily means ensuring the patient acts, thinks, and believes in accordance with the chaplain's theology in order to have the best outcome (for example, going to heaven vice going to hell).[89] Of import again to the present discussion is the matter of paternalism in health care, and here specifically, clinical chaplain care.

Paternalistic Clinical Chaplain Care

The *Oxford English Dictionary* defines the term, *paternalism*, as "The policy or practice of restricting the freedoms and responsibilities of subordinates or dependents in what is considered or claimed to be their best interests."[90] Pellegrino and Thomasma clarify paternalism in health care ethics as any "medical decision to benefit a patient without the full consent" of said patient or authorized surrogate decision maker.[91] It must be noted that ensuring the good of the patient does not mean trespassing upon said patient's individual rights and liberties when he or she possesses the decision-making capacity to choose to receive or not to receive the care that the provider deems would be best.[92] To trespass upon the patient's autonomy in this manner is "morally offensive."[93] Rather, the patient determines what good is. The patient has the right to be "wrong" according to the clinical chaplain's theological opinion. The patient in a hospital bed, imminently dying, has the same rights as a healthy person to not have unwanted religious proclamations imposed upon them, especially in those moments of suffering.

Justice and Patients of Other Faiths

Treat Like Cases Alike

As previously discussed, Beauchamp and Childress examine six traditional theoretical approaches to justice, including perspectives from utilitarians, libertarians, egalitarians, communitarians, and health care-related capability and well-being theories.[94] Examined collectively, they present an imperative to treat like cases alike.

Many clinical chaplains are most comfortable providing care to persons of their own faith tradition, and are likely to confuse their roles, acting more like a religious ministry professional in the context of their faith group rather than a clinical chaplain in a pluralistic health care environment. There is often a misunderstanding, that if a patient is of a different faith group, a chaplain cannot provide care for them, but must refer to another religious ministry professional; this is categorically false. Sulmasy writes, "Human being is spiritual being…if spirituality is correctly defined as the characteristics of a person's life in relationship with questions of transcendence, then spiritual needs will be universal—encompassing even those who profess no religious belief."[95]

Patients of Other Faiths or No Faith

All human beings must be treated with the same dignity and respect as all other human beings in our care. Justice, along with the related concepts of equality and equity, demands respect for diversity and pluralism. Persons with faiths dissimilar to the clinical chaplain, or of no faith at all, may indeed have spiritual needs that the clinical chaplain is duty-bound to provide care for, in a respectful and dignified manner. Human beings in a pluralistic health care environment, who are likely experiencing great suffering, will almost universally benefit from non-judgmental compassion, a ministry of presence, sympathy, and empathy, which any clinical chaplain of any faith or no faith should be equipped to provide, exemplifying the virtue of justice.[96]

The Dalai Lama teaches, "…true compassion is universal. True compassion does not stem from the pleasure of feeling close to one person or another, but from the conviction that *other people are just like me* and want not to suffer but to be happy, and from a commitment to help them overcome what causes them to suffer...That is true, well thought-out compassion."[97] Treating like cases alike, fulfilling the ethical principle of justice, would necessarily require any clinical chaplain to demonstrate this kind of deliberative compassion towards any person in their care, irrespective of the similarities or differences in belief, culture, or any other example of diversity. This kind of compassion justly sees the patient as a person and fellow human being worthy of dignity and respect, not as a person of a different religion, or different skin tone, or different values system.

Conclusion

In this brief yet consequential discussion, this author has reviewed the current state of clinical chaplain preparation for clinical ethics, which was generally found to be lacking. The chapter then discussed the four normative principles for health care ethics, as presented by Beauchamp and Childress in their *Principles of Biomedical Ethics*, and how clinical chaplain care in a pluralistic health care environment may engage each. To best serve all persons cared for within the health care environment, the individual patient's autonomy regarding their religious and spiritual preferences must be acknowledged and respected, proselytization and spiritual abuse must be averted (nonmaleficence), the patient's ideal of their own good must be discovered and encouraged (beneficence), and equity, equality, justice, and goodness demand that all patients must be treated with respect and dignity, irrespective of their religious or spiritual beliefs and those of the clinical chaplain. Beyond this, the underlying values from which the principles are derived should drive the delivery of all high-quality health care processes and encounters. This idealized reality must be the aim of any truly ethical health care environment.

Revel writes,

> the principles or 'ends' of bioethics should be like a building where harmonious assembly of complementary components is essential[,]...are diverse[,] and encompass...respect of human rights and dignity[,]...respect for fundamental freedoms[,]... receiving treatment with intent to heal (beneficence), with justice and without maleficence...respect also for cultural pluralism, even for conflicting values that may impact either on the place of the individual in the collective, or on the definition of the beginning and end of life, or on what constitutes the quality of life.[98]

This author contends that like Revel's ideal of bioethics, so should clinical chaplaincy respect and dignify all of these areas of the patient's holistic self. The field has a considerable distance to cover if it shall approach meeting this ideal.

Notes

[1] David Degrazia and Tom L. Beauchamp, "Philosophy: Ethical Principles and Common Morality," in *Methods in Medical Ethics*, 2nd ed., ed. Jeremy Sugarman and Daniel P. Sulmasy, (Washington, DC: Georgetown University Press, 2010), 39-42.

[2] United Nations Educational, Scientific and Cultural Organization, "Universal Declaration on Bioethics and Human Rights," (Paris, France: United Nations Educational, Scientific and Cultural Organization, 2005), http://portal.unesco.org/en/ev.php-URL_ID=31058&URL_DO=DO_TOPIC&URL_SECTION=201.html.

[3] Code of Hammurabi; see edicts 215-223 of the 282 laws comprising the ancient code. Available at https://avalon.law.yale.edu/ancient/ hamframe.asp

[4] In re Conroy, N.J. 321, 486 A.2d 1209 (1985).

[5]Stavros A. Antoniou, et al., "Reflections of the Hippocratic Oath in Modern Medicine," *World Journal of Surgery* 34 (2010): 3078, doi: 10.1007/s00268-010-0604-3; see also Edmund D. Pellegrino, "Codes, Virtue, and Professionalism," in *Methods in Medical Ethics*, 2nd ed., ed. Jeremy Sugarman and Daniel P. Sulmasy, (Washington, DC: Georgetown University Press, 2010), 92-98.

[6] Tom L. Beauchamp and James F. Childress, *Principles of Biomedical Ethics*, 8th ed. (New York: Oxford University Press, 2019), 13, 35; see also Albert R. Jonsen, *The Birth of Bioethics* (New York: Oxford University Press, 1998), 7, 233, 255, 354.

[7] David F. Kelly, Gerard Magill, and Henk ten Have, *Contemporary Catholic Health Care Ethics*, 2nd ed. (Washington, D.C.: Georgetown University Press, 2013), 63-66.

[8] Beauchamp and Childress, 13.

[9] Matthew A. Cassady (Chaplain, Program Manager for Transitional Care and Education, National VA Chaplain Service), interview with author, July 9, 2022.

[10] U.S. Department of Veterans Affairs (VA), "VA Handbook 5005/135, Part II, Appendix G67," updated September 24, 2020, II-G67-1 - II-G67-3, https://www.va.gov/vapubs/viewPublication.asp?Pub_ID= 1358&FType=2.

[11] VA, "VA Chaplain Services – Ecclesiastical Endorsement – Patient Care Services," last updated October 11, 2022, https://www.patientcare.va.gov/chaplain/Ecclesiastical_Endorsement.asp.

[12] The Association of Theological Schools (ATS) Commission on Accrediting, "ATS Commission Standards of Accreditation: June 2020," approved June 2020, The Association of Theological Schools (ATS) (Pittsburgh: ATS, 2020), https://www.ats.edu/files/galleries/standards-of-accreditation.pdf, 5.

[13] ATS, "Standards of Accreditation," 5.

[14] ATS, "Annual Data Tables," ATS Data Tables 2021-2022: 2 – Enrollment, Table 2.12 Head Count and Full-Time Equivalent Enrollment by Degree Category, Fall 2021, https://www.ats.edu/files/galleries/2021-2022_Annual_Data_Tables.pdf.

[15] ATS, "Annual Data Tables," ATS Data Tables 2021-2022: 2 – Enrollment, Table 2.2 Head Count Enrollment; Table 2.12 Head Count and Full-Time Equivalent Enrollment by Degree Category, Fall 2021.

[16] Liberty University John W. Rawlings School of Divinity, "Master of Divinity in Professional Chaplaincy: Healthcare | Rawlings School of Divinity," Liberty University, accessed October 19, 2022, https://www.liberty.edu/divinity/masters/mdiv/healthcare-chaplaincy/.

[17] Southern Baptist Theological Seminary, "Master of Divinity (MDiv) in Ethics and Philosophy," Southern Baptist Theological Seminary, accessed October 19, 2022, https://www.sbts.edu/theology/degree-programs/mdiv/ethics-public-theology/.

[18] Southeastern Baptist Theological Seminary, "Program: Christian Ethics Track MDiv – Southeastern Baptist Theological Seminary – acalog ACMS," Southeastern Baptist Theological Seminary, accessed October 20, 2022, https://catalog.sebts.edu/preview_program.php?catoid=22&poid =4315.

[19] Midwestern Baptist Theological Seminary, "MDiv (Master of Divinity) – Standard Program | MBTS," Midwestern Baptist Theological Seminary, accessed October 19, 2022, https://www.mbts.edu/degrees /masters-studies/mdiv/.

[20] Asbury Theological Seminary, "Master of Divinity (M.Div.) - Asbury Theological Seminary - Asbury Theological Seminary," Asbury Theological Seminary, accessed October 19, 2022, https://asburyseminary.edu/academics/degrees/master-of-divinity/.

[21] Regent University, "Master of Divinity (M.Div.) – Regent University," Regent University, accessed October 20, 2022, https://www.regent.edu/program/mdiv-chaplain-ministry/#other-concentrations.

[22] New Orleans Baptist Theological Seminary, "Chaplaincy and Pastoral Care MDIV | Catalog," New Orleans Baptist Theological Seminary, accessed October 20, 2022, https://catalog.nobts.edu/graduate/masters/degrees/divinity/chaplaincy-pastoral-care.

[23] Fuller Theological Seminary, "MDiv Degree Outline | Fuller Seminary," Fuller Theological Seminary, accessed October 20, 2022, https://www.fuller.edu/master-of-divinity-degree-outline/.

[24] Southwestern Baptist Theological Seminary, "Master of Divinity Courses and Concentrations | MDiv Courses," Southwestern Baptist Theological Seminary, accessed October 20, 2022, https://swbts.edu/degree/master-of-divinity-concentrations/.

[25] Dallas Theological Seminary, "Master of Arts in Chaplaincy and Ministry Care - Dallas Theological Seminary," Dallas Theological Seminary, accessed October 20, 2022, https://www.dts.edu/academics/degrees-programs/master-of-arts-in-chaplaincy-and-ministry-care/.

[26] Association for Clinical Pastoral Education (ACPE), "CPE Center Accreditation," Association for Clinical Pastoral Education (ACPE), accessed October 19, 2022, https://acpe.edu/programs/accreditation.

[27] ACPE, "Objectives and Outcomes for Level I/Level II CPE," in "2020 Accreditation Manual," ACPE, accessed October 24, 2022,

https://www.manula.com/manuals/acpe/acpe-manuals/2016/en/topic/objectives-and-outcomes-for-level-i-level-ii-cpe.

[28] ACPE, "Objectives and Outcomes for Level I/Level II CPE," in "2020 Accreditation Manual," ACPE, accessed October 26, 2022, https://www.manula.com/manuals/acpe/acpe-manuals/2016/en/topic/objectives-and-outcomes-for-level-i-level-ii-cpe.

[29] ACPE, "Common Qualifications and Competencies for Professional Chaplains," in "2020 Accreditation Manual," ACPE.

[30] VA, "VA Handbook 5005/135, Part II, Appendix G67."

[31] Board of Chaplaincy Certification, Inc. (BCCI), "2021January28 Standards for BCCACC," Association of Professional Chaplains, accessed October 20, 2022, https://view.officeapps.live.com/op/view.aspx?src=https%3A%2F%2F033012b.membershipsoftware.org%2Ffiles%2F2021January28%2520Standards%2520for%2520BCCACC.docx&wdOrigin=BROWSELINK.

[32] BCCI, "BCCI Application Materials," Association of Professional Chaplains, accessed October 19, 2022, https://bcci.professionalchaplains.org/content.asp?admin=Y&pl=16&sl=16&contentid=30.

[33] BCCI, "Equivalency," Association of Professional Chaplains, accessed October 19, 2022, https://bcci.professionalchaplains.org/content.asp?pl=19&contentid=19.

[34] BCCI, "competencies_writing_guide," Association of Professional Chaplains, last updated May 2020, https://033012b.membershipsoftware.org/files/application_materials/competencies_writing_guide.pdf.

[35] Association of Professional Chaplains (APC), "Code of Ethics," Association of Professional Chaplains, revised September 24, 2000, https://www.professionalchaplains.org/Files/professional_standards/professional_ethics/apc_code_of_ethics.pdf.

[36] APC, "Guidelines for the Chaplain's Role in Health Care Ethics," approved by the APC Board of Directors 10/2000, https://033012b.membershipsoftware.org/files/application_materials/guidelines_for_chaplains_role_in_healthcare_ethics.pdf, 1.

[37] APC, "Guidelines," 1.

[38] Beauchamp and Childress, *Principles of Biomedical Ethics*, n.p.

[39] Thayer and Smith, "Greek Lexicon entry for Autos," *The NAS New Testament Greek Lexicon,* 1999. https://www.biblestudytools.com/lexicons/greek/nas/autos.html; see also Beauchamp and Childress, *Principles of Biomedical Ethics*, 99.

[40] Thayer and Smith, "Greek Lexicon entry for Nomos," *The NAS New Testament Greek Lexicon,* 1999, https://www.biblestudytools.com/lexicons/greek/nas/nomos.html; see also Beauchamp and Childress, *Principles of Biomedical Ethics*, 99.

[41] Beauchamp and Childress, 100.

[42] Ibid., 102.

[43] Ibid., 102-03.

[44] Ibid., 103-04.

[45] Hospital Information Services for Jehovah's Witnesses International Office, "Religious and Ethical Position on Medical Therapy and Related Matters," (Wallkill, NY: Watch Tower Bible and Tract Society of Pennsylvania, 2012), accessed November 2, 2022, https://www.jw.org/en/medical-library/strategies-downloads/religious-and-ethical-position-medical-therapy/.
[46] Beauchamp and Childress, 104; see also Rebecca Kulka, "Conscientious Autonomy: Displacing Decisions in Health Care," *Hastings Center Report* (March-April 2005): 37-41, doi: 10.1353/hcr.2005.0025.
[47] Sara Kolmes, Christine Ha, and Jordan Potter, "Responding to Cultural Limitations on Patient Autonomy: A Clinical Ethics Case Study," *HEC Forum* (August 9, 2022), doi: 10.1007/s10730-022-09490-y; see also Beauchamp and Childress, 104-05.
[48] Jonsen, 334-38; see also Beauchamp and Childress, 104-12.
[49] *Farlex Partner Medical Dictionary*. S.v. "nonmaleficence." Retrieved September 15, 2022 from https://medical-dictionary.thefreedictionary.com/nonmaleficence; see also *Oxford English Dictionary*, 3rd ed. (June 2005; last modified version published online March 2022), s.v. "maleficence," https://www-oed-com.authenticate.library.duq.edu/view/Entry/112856?redirectedFrom=maleficence#eid.
[50] Cedric M. Smith, "Origin and Uses of Primum Non Nocere--Above All, Do No Harm!" in *Journal of Clinical Pharmacology* 45 (2005): 371-77, doi: 10.1177/0091270004273680; see also Beauchamp and Childress, *Principles of Biomedical Ethics*, 155; Albert R. Jonsen, *A Short History of Medical Ethics* (New York: Oxford University Press, 2008), 1-3.
[51] Bettina Schoene-Seifert, "Harm," in *Encyclopedia of Bioethics*, 3rd ed., ed. Stephen G. Post (New York: Macmillan Reference USA, 2004), 2: 1034; see also Beauchamp and Childress, *Principles of Biomedical Ethics*, 156-58, 219.
[52] Beauchamp and Childress, 158-59.
[53] Joel Feinberg, *Harm to Others*, vol. 1 of *The Moral Limits of the Criminal Law* (New York: Oxford University Press, 1984), 32-36, 51-55, 77-78, as cited in Beauchamp and Childress, *Principles of Biomedical Ethics*, 158.
[54] Schoene-Seifert, "Harm," 1033-34.
[55] *Farlex Partner Medical Dictionary*. S.v. "beneficence." Retrieved September 15, 2022 from https://medical-dictionary.thefreedictionary.com/beneficence; see also *Oxford English Dictionary*, 3rd ed. (June 2005; last modified version published online December 2020), s.v. "beneficence," https://www-oed-com.authenticate.library.duq.edu/view/Entry/17679?redirectedFrom=beneficence#eid.
[56] Beauchamp and Childress, 157.
[57] Degrazia and Beauchamp, 43-44; see also Beauchamp and Childress, 388-94.
[58] Beauchamp and Childress, 219-24; see also Sir David Ross, *The Right and the Good* (Oxford: Clarendon, 1930), 21.
[59] *Oxford English Dictionary*, 3rd ed. (June 2005; last modified version published online March 2022), s.v. "pluralism,"

https://www-oed-com.authenticate.library.duq.edu/view/Entry/138757?redirectedFrom=paternalism#eid

60 Beauchamp and Childress, 230-43.

61 *Oxford English Dictionary*, 3rd ed. (June 2005; last modified version published online September 2022), s.v. "justice," https://www-oed-com.authenticate.library.duq.edu/view/Entry/102198?rskey=teRC8t&result=1#contentWrapper.

62 Beauchamp and Childress, 268.

63 Roger A. Shiner, "Aristotle's Theory of Equity," *Loyola of Los Angeles Law Review* 27, no. 4, (1994): 1247, https://digitalcommons.lmu.edu/llr/vol27/iss4/1; see also Beauchamp and Childress, 268.

64 Thayer and Smith, "Greek Lexicon entry for Epieikeia," *The NAS New Testament Greek Lexicon,* 1999, https://www.biblestudytools.com/lexicons/greek/nas/epieikeia.html.

65 Beauchamp and Childress, 271-81.

66 *Oxford English Dictionary*, 3rd ed. (June 2005; last modified version published online September 2022), s.v. "equity," https://www-oed-com.authenticate.library.duq.edu/view/Entry/63838?redirectedFrom=equity#eid.

67 UNESCO, "Universal Declaration on Bioethics and Human Rights," Preamble section, emphasis by this author.

68 United Nations (UN) General Assembly, "Universal Declaration of Human Rights," 217 (III) A (Paris, France: UN, 1948), n.p., https://www.un.org/en/about-us/universal-declaration-of-human-rights.

69 Diana Ayton-Shenker, "The Challenge of Human Rights and Cultural Diversity," UN Background Note DPI/1627/HR, March 1995 (New York: UN Department of Public Information, 1995), https://digitallibrary.un.org/record/205090.

70 UN, "Universal Declaration of Human Rights," articles 18-19.

71 US Constitution, First Amendment, https://constitution.congress.gov/constitution/amendment-1/.

72 Daniel P. Sulmasy, "Ethical Principles for Spiritual Care," in *Oxford Textbook of Spirituality in Healthcare*, ed. Mark Cobb, Christina M. Puchalski, and Bruce Rumbold (New York: Oxford University Press, 2014), 466.

73 Winnifred Fallers Sullivan, *A Ministry of Presence: Chaplaincy, Spiritual Care, and the Law* (Chicago: University of Chicago Press, 2014), 161, Proquest Ebook Central.

74 *Oxford English Dictionary*, 3rd ed. (September 2006; last modified version published online September 2022), s.v. "pluralism," https://www-oed-com.authenticate.library.duq.edu/view/Entry/146193?redirectedFrom=pluralism#eid.

75 Sullivan, 161-62.

76 His Holiness the 14th Dalai Lama of Tibet, "The Dalai Lama," The Office of His Holiness The Dalai Lama, accessed October 25, 2022, https://www.dalailama.com/the-dalai-lama.

[77] Dalai Lama, *My Spiritual Journey: Personal Reflections, Teachings, and Talks*, collected by Sofia Stril-Rever and trans. Charlotte Mandell (HarperCollins ebooks, 2011), 96, Kindle edition.
[78] Ibid., 135.
[79] See, for example, Aaron L. Mackler, *Introduction to Jewish and Catholic Bioethics: A Comparative Analysis* (Washington, D.C: Georgetown University Press, 2003), 229-31.
[80] Beauchamp and Childress, 158-59.
[81] See, for example, Southern Baptist Convention, "The Baptist Faith and Message 2000," (BFM) adopted June 14th, 2000, accessed October 12, 2022, https://bfm.sbc.net/bfm2000/, XVIII. The Family.
[82] Harold G. Koenig, *Spirituality in Patient Care: Why, How, When, and What* (Philadelphia: Templeton Foundation Press, 2002), 77-87.
[83] *Oxford English Dictionary*, 3rd ed. (June 2007; last modified version published online March 2022), s.v. "proselytize," https://www-oed-com.authenticate.library.duq.edu/view/Entry/152949?redirectedFrom=proselytize#eid.
[84] VA, VHA, "VHA Directive 1111," July 1, 2021, 3.
[85] Beauchamp and Childress, 118-23.
[86] Karen Lebacqz and Joseph D. Driskill, *Ethics and Spiritual Care: A Guide for Pastors, Chaplains, and Spiritual Directors* (Nashville: Abingdon Press, 2000), 127-52.
[87] Koenig, 8.
[88] Beauchamp and Childress, 219-24; see also Ross, 21; Sulmasy, 466.
[89] Michel Revel, "Article 12: Respect for Cultural Diversity and Pluralism," in *The UNESCO Universal Declaration on Bioethics and Human Rights: Background, Principles and Application*, Henk ten Have and M.S. Jean, eds., 199-209 (Paris: UNESCO Publishing, 2009), 203-07.
[90] *Oxford English Dictionary*, 3rd ed. (June 2005; last modified version published online March 2022), s.v. "pluralism," https://www-oed-com.authenticate.library.duq.edu/view/Entry/138757?redirectedFrom=paternalism#eid
[91] Edmund D. Pellegrino and David C. Thomasma, *For the Patient's Good: The Restoration of Beneficence in Healthcare* (New York: Oxford University Press, 1988), 10.
[92] Beauchamp and Childress, 230-43.
[93] Pellegrino and Thomasma, 87.
[94] Beauchamp and Childress, 271-81.
[95] Sulmasy, 466-67.
[96] Robin W. Lovin, *Christian Ethics: An Essential Guide* (Nashville: Abingdon Press, 2000), 73-74.
[97] Dalai Lama, 25, emphasis this author's.
[98] Revel, 207.

Literature

Antoniou, Stavros A., George A. Antoniou, Frank A. Granderath, Anna Mavroforou, Athanasios D. Giannoukas and Athanasios I. Antoniou. 2010. "Reflections of the Hippocratic Oath in Modern Medicine." *World Journal of Surgery* 34: 3075–3079, doi: 10.1007/s00268-010-0604-3.

Asbury Theological Seminary. N. d. "Master of Divinity (M.Div.) - Asbury Theological Seminary - Asbury Theological Seminary." Asbury Theological Seminary. Accessed October 19, 2022. https://asburyseminary.edu/academics/degrees/master-of-divinity/.

Association for Clinical Pastoral Education (ACPE). N. d. "CPE Center Accreditation." ACPE: The Standard Spiritual Care and Education. Accessed October 19, 2022. https://acpe.edu/programs/accreditation.

Association for Clinical Pastoral Education (ACPE). 2020. "Objectives and Outcomes for Level I/Level II CPE," in "2020 Accreditation Manual," ACPE, accessed October 24, 2022, https://www.manula.com/manuals/acpe/acpe-manuals/2016/en/topic/objectives-and-outcomes-for-level-i-level-ii-cpe.

Association of Professional Chaplains (APC). 2000. "Code of Ethics," Association of Professional Chaplains, revised September 24, 2000, https://www.professionalchaplains.org/Files/professional_standards/professional_ethics/apc_code_of_ethics.pdf.

Association of Professional Chaplains (APC). 2000. "Guidelines for the Chaplain's Role in Health Care Ethics." Approved by the APC Board of Directors 10/2000. https://033012b.membershipsoftware.org/files/application_materials/guidelines_for_chaplains_role_in_healthcare_ethics.pdf.

The Association of Theological Schools (ATS) Commission on Accrediting. 2022. "Annual Data Tables." ATS Data Tables 2021-2022. https://www.ats.edu/files/galleries/2021-2022_Annual_Data_Tables.pdf.

The Association of Theological Schools (ATS) Commission on Accrediting. 2020 "ATS Commission Standards of Accreditation: June 2020." Approved June. The Association of Theological Schools (ATS). Pittsburgh: The Association of Theological Schools, 2020. https://www.ats.edu/files/galleries/standards-of-accreditation.pdf.

Ayton-Shenker, Diana. 1995. "The Challenge of Human Rights and Cultural Diversity." United Nations Background Note DPI/1627/HR, March 1995. New York: United Nations Department of Public Information. https://digitallibrary.un.org/record/205090.

Beauchamp, Tom L. and James F. Childress. 2019. *Principles of Biomedical Ethics*. 8th ed. New York: Oxford University Press.

Board of Chaplaincy Certification, Inc. (BCCI). 2021. "2021January28 Standards for BCCACC," Association of Professional Chaplains, accessed October 20, 2022, https://view.officeapps.live.com/op/view.aspx?src=https%3A%2F%2F033012b.membershipsoftware.org%2Ffiles%2F2021January28%2520Standards%2520for%2520BCCACC.docx&wdOrigin=BROWSELINK.

Board of Chaplaincy Certification, Inc. (BCCI). N. d. "BCCI Application Materials," Association of Professional Chaplains, accessed October 19, 2022, https://bcci.professionalchaplains.org/content.asp?admin=Y&pl=16&sl =16&contentid=30.

Board of Chaplaincy Certification, Inc. (BCCI). 2020. "competencies_writing_guide," Association of Professional Chaplains, last updated May 2020, https://033012b.membershipsoftware.org/files/application_materials/competencies_writing_guide.pdf.

Board of Chaplaincy Certification, Inc. (BCCI). N. d. "Equivalency," Association of Professional Chaplains, accessed October 19, 2022, https://bcci.professionalchaplains.org/content.asp?pl=19&contentid=19

Dalai Lama, The. 2011. *My Spiritual Journey: Personal Reflections, Teachings, and Talks*. Collected by Sofia Stril-Rever and translated by Charlotte Mandell. HarperCollins ebooks, 2011. Kindle edition.

Dalai Lama, The.N. d. His Holiness the 14th Dalai Lama of Tibet, "The Dalai Lama," The Office of His Holiness The Dalai Lama, accessed October 25, 2022, https://www.dalailama.com/the-dalai-lama.

Dallas Theological Seminary. N. d. "Master of Arts in Chaplaincy and Ministry Care - Dallas Theological Seminary." Dallas Theological Seminary. Accessed October 20, 2022. https://www.dts.edu/academics/degrees-programs/master-of-arts-in-chaplaincy-and-ministry-care/.

Degrazia, David and Tom L. Beauchamp.2010. "Philosophy: Ethical Principles and Common Morality." In *Methods in Medical Ethics*, 2nd ed. Edited by Jeremy Sugarman and Daniel P. Sulmasy, 37-54. Washington, DC: Georgetown University Press.

Hospital Information Services for Jehovah's Witnesses International Office. 2012. "Religious and Ethical Position on Medical Therapy and Related Matters." Wallkill, NY: Watch Tower Bible and Tract Society of Pennsylvania. Accessed November 2, 2022.

https://www.jw.org/en/medical-library/strategies-downloads/religious-and-ethical-position-medical-therapy/.

Fuller Theological Seminary. N. d. "MDiv Degree Outline | Fuller Seminary." Fuller Theological Seminary. Accessed October 20, 2022. https://www.fuller. edu /master-of-divinity-degree-outline/.

Jonsen, Albert R. 2008. *A Short History of Medical Ethics.* New York: Oxford University Press.

Jonsen, Albert R. 1998. *The Birth of Bioethics*. New York: Oxford University Press.

Kelly, David F., Gerard Magill, and Henk ten Have. 2013. *Contemporary Catholic Health Care Ethics*, 2nd ed. Washington, D.C: Georgetown University Press.

Koenig, Harold G. 2002. *Spirituality in Patient Care: Why, How, When, and What.* Philadelphia: Templeton Foundation Press.

Kolmes, Sara, Christine Ha, and Jordan Potter.2022. "Responding to Cultural Limitations on Patient Autonomy: A Clinical Ethics Case Study." *HEC Forum* (August 9). doi: 10.1007/s10730-022-09490-y.

Kulka, Rebecca. 2005. "Conscientious Autonomy: Displacing Decisions in Health Care." *Hastings Center Report* (March-April): 37-41. doi: 10.1353/hcr.2005.0025.

Lebacqz, Karen and Joseph D. Driskill. 2000. *Ethics and Spiritual Care: A Guide for Pastors, Chaplains, and Spiritual Directors.* Nashville: Abingdon Press.

Liberty University John W. Rawlings School of Divinity. N. d. "Master of Divinity in Professional Chaplaincy: Healthcare | Rawlings School of Divinity." Liberty University. Accessed October 19, 2022. https://www.liberty.edu/divinity/masters/mdiv/healthcare-chaplaincy/.

Lovin, Robin W. 2000. *Christian Ethics: An Essential Guide.* Nashville: Abingdon Press.

Mackler, Aaron L. 2003. *Introduction to Jewish and Catholic Bioethics: A Comparative Analysis*. Washington, D.C: Georgetown University Press.

Midwestern Baptist Theological Seminary. N. d. "MDiv (Master of Divinity) – Standard Program |MBTS." Midwestern Baptist Theological Seminary. Accessed October 19, 2022. https://www.mbts.edu/degrees/masters-studies/mdiv/.

New Orleans Baptist Theological Seminary. N. d. "Chaplaincy and Pastoral Care MDIV | Catalog." New Orleans Baptist Theological Seminary. Accessed October 20, 2022. https://catalog.nobts.edu/graduate/masters/degrees/divinity/chaplaincy-pastoral-care.

Pellegrino, Edmund D. 2010. "Codes, Virtue, and Professionalism." In *Methods in Medical Ethics*, 2nd ed. Edited by Jeremy Sugarman and Daniel P. Sulmasy, 91-108. Washington, DC: Georgetown University Press.

Pellegrino, Edmund D. and David C. Thomasma. 1988. *For the Patient's Good: The Restoration of Beneficence in Healthcare.* New York: Oxford University Press.

Regent University. N. d. "Master of Divinity (M.Div.) – Regent University." Regent University. Accessed October 20, 2022. https://www.regent.edu/program/mdiv-chaplain-ministry/#other-concentrations.

Revel, Michel. 2009. "Article 12: Respect for Cultural Diversity and Pluralism." In *The UNESCO Universal Declaration on Bioethics and Human Rights: Background, Principles and Application*, edited by Henk ten Have and M.S. Jean, 199-209. Paris: UNESCO Publishing.

Ross, W. David. 1930. *The Right and the Good*. Oxford: Clarendon.

Schoene-Seifert, Bettina. 2004. "Harm." In *Encyclopedia of Bioethics.* 3rd ed. Edited by Stephen G. Post. 1033-38. Volume 2 of 5. New York: Macmillan Reference USA.

Shiner, Roger A. 1994. "Aristotle's Theory of Equity." *Loyola of Los Angeles Law Review* 27, no. 4: 1245-1264. https://digitalcommons.lmu.edu/llr/vol27/iss4/1.

Southeastern Baptist Theological Seminary. N. d. "Program: Christian Ethics Track MDiv – Southeastern Baptist Theological Seminary – acalog ACMS." Southeastern Baptist Theological Seminary. Accessed October 20, 2022. https://catalog.sebts.edu/preview_program.php?catoid=22&poid=4315.

Southern Baptist Convention. 2000. "The Baptist Faith and Message 2000." June 14th. https://bfm.sbc.net/bfm2000/.

Southern Baptist Theological Seminary. N. d. "Master of Divinity (MDiv) in Ethics and Philosophy." Southern Baptist Theological Seminary. Accessed October 19, 2022. https://www.sbts.edu/theology/degree-programs/mdiv/ ethics-public-theology/.

Southwestern Baptist Theological Seminary. N. d. "Master of Divinity Courses and Concentrations | MDiv Courses." Southwestern Baptist Theological Seminary. Accessed October 20, 2022. https://swbts.edu/degree/master-of-divinity-concentrations/.

Smith, Cedric M. 2005. "Origin and Uses of Primum Non Nocere--Above All, Do No Harm!" *Journal of Clinical Pharmacology* 45: 371-77. doi: 10.1177/0091270004273680.

Sullivan, Winnifred Fallers. 2014. *A Ministry of Presence: Chaplaincy, Spiritual Care, and the Law.* Chicago: University of Chicago Press. Proquest Ebook Central.

Sulmasy, Daniel P. 2014. "Ethical Principles for Spiritual Care." In *Oxford Textbook of Spirituality in Healthcare*. Edited by Mark Cobb, Christina M. Puchalski, and Bruce Rumbold, 465-70. New York: Oxford University Press.

United Nations Educational, Scientific and Cultural Organization. 2005. "Universal Declaration on Bioethics and Human Rights." Paris, France: United Nations Educational, Scientific and Cultural Organization. http://portal.unesco.org/en/ev.php-URL_ID=31058&URL_DO=DO_TOPIC&URL_SECTION=201.html.

United Nations General Assembly. 1948. "Universal Declaration of Human Rights" 217 (III) A. Paris: United Nations. https://www.un.org/en/about-us/universal-declaration-of-human-rights.

U.S. Department of Veterans Affairs (VA). N. d. "VA Handbook 5005/135, Part II, Appendix G67." Updated September 24, 2020. https://www.va.gov/vapubs/viewPublication.asp?Pub_ID=1358&FType=2.

U.S. Department of Veterans Affairs (VA). 2022. "VA Chaplain Services – Ecclesiastical Endorsement – Patient Care Services," last updated October 11, 2022, https://www.patientcare.va.gov/chaplain/Ecclesiastical_Endorsement.asp.

Veterans Health Administration (VHA). 2021. "VHA Directive 1111 Spiritual Care." July 1. https://www.va.gov/vhapublications/ViewPublication.asp?pub_ID=9350.

III. Resilience and Biotechnology

CHAPTER TWELVE

ECOLOGICAL RESILIENCY IN A BIOTECHNOLOGICAL ERA

FRANCIS SARPONG

Introduction

This chapter provides an ethical assessment of animals employed in xenotransplant interventions. It argues for ecological resiliency in employing biotechnological interventions to treat failing or failed human organs since uncontrolled use of animals can affect human existence. Research on animals has been utilized to find appropriate medical interventions for human suffering. The usage of animal organs to replace failed or failing organs are undergoing clinical trials. If successful, this intervention will increase access to organs for transplantation. However, xenotransplantation has several roadblocks, including the spread of zoonotic diseases. Ecological resiliency, therefore, requires interventions that enhance nature's benefits and human benefits.[1]

This project uses common morality as a principal source of moral justification to show that resiliency in common morality focuses on biotechnology breakthroughs' risk/benefit analysis. A risk/benefit assessment is meritorious irrespective of culture, group, or creed as a moral norm. A common concern for animal and human health is necessary for preventing the spread of zoonotic diseases. The first aspect of the chapter examines the biological context of xenotransplantation, drawing out its science and technology. It proceeds to discuss the life-saving and cost-saving benefits of xenotransplantation. In the final part, the rationale of the benefit of xenotransplant to humans is challenged, arguing for a sustainable consciousness in a biotechnological era.[2]

Xenotransplant and Biotechnologies

Organ shortage has augured exploring non-human alternative organs. For instance, surgeons have attempted to transplant animal organs into humans.[3] Such interventions are subject to rejection and infections in human recipients. Xenotransplant triggers a more severe immune response than allotransplantation.[4] Consequently, genetic engineers are racing to delete immunogenic components from xenografts and engineer animals to autonomously nurture organs made of whole human cells.[5] Xenotransplantation also raises ethical concerns, including the autonomy of recipients, their families, and society, and respect for human dignity. The issue of human dignity concerns the impact of xenotransplants on human ontology and teleology. This section shows that biotechnological interventions do not disregard human dignity.[6]

The discovery of similarities between humans and animals has encouraged the attempt to transplant animal organs into humans. Biologically, all humans and animals belong to the kingdom "animalia" sharing different degrees of semblance. Similar anatomy gives humans and animals comparable functional capacity. For instance, both humans and animals use their cognitive ability to fight or flee threats in their environment.[7] Moreover, humans and animals that reproduce sexually have similar characteristics. Sexual reproduction in humans and animals entails the fusion of a maternal ovule and a paternal spermatozoon. The resultant is a genotype and phenotype of the offspring. The DNA of each human and animal offspring is exceptional and unrepeatable. The DNA of all vertebrates shares more similar characteristics than differences. These similarities make cross-species sharing of tissues and organs feasible.[8]

Mathieu Jaboulay attempted the first xenotransplant in 1906. Jaboulay used pigs and goats as donors. Both transplants failed, resulting in the death of the patients.[9] In continuation of the exploration, scientists continue to examine transplanting baboon, chimpanzee, and pig organs in humans.[10] In 1963–64, transplanted chimpanzee kidneys functioned for about nine months. In 1964, scientists successfully transplanted a chimpanzee heart into a human being. Also, in 1992, a baboon liver transplant in a human being functioned for seventy days. Xenotransplant experts have identified more excellent prospects in transplanting pig organs into humans. Pigs are the most suitable non-human primates for transplant due to pig and human organs' anatomical and physiological semblance. Pigs and humans also have similar immune systems. It takes four to six months to rear pigs to maturity and harvests their organs. Also, considerable progress has been made in editing pig genes. These advances in xenotransplantation have the tools of biotechnology in their armory.[11]

Biotechnology is the engineering of creatures to improve living conditions. It uses processes and products to alter the phenomenon of life. It alters the lives of microbes, plants, animals, and human beings. Red/medical biotechnology, green biotechnology, blue biotechnology, and white or grey biotechnology exist. There is also gene and non-gene biotechnology. Non-gene biotechnology attempts to alter cells, tissues, or organisms to improve living conditions. On the other hand, gene biotechnology alters or transfers genes from one organism to another using genetic engineering. Biotechnologists employ several engineering techniques to alter organisms. These approaches include bioreactors, cell fusion, liposome-based delivery, cell or tissue culture, DNA fingerprinting, cloning, artificial insemination, ET technology, and stem cell technology. Biotechnology has prospects to relieve human suffering augured by cancer, Alzheimer's, mental illness, depression, loss of memory, and organ failure.[12]

In 1954, Joseph Murray employed genetic engineering to perform the initial kidney transplant using identical twins.[13] The shortage of human organs has necessitated xenotransplantation. Genetic engineers alter pig genes to make their organs transplantable in humans. They isolate individual pig and human genes and reintroduce them into pigs to grow organs suitable for xenotransplant. The basis of genetic engineering is the alteration of genetic material (hereditary material) or the combination of genes in an organism. Genetic researchers give the organism and its descendants' different traits by modifying them. This technology was practiced in earlier times by breeding plants and animals to produce favorable combinations of genes. Using this technology, genetic engineers have made the most economically essential flowers, vegetables, grains, cows, horses, dogs, and cats. During the 1970s and 1980s, researchers established ways to isolate individual genes and reintroduce them into cells or plants, animals, or other organisms.[14]

Transplantation of pig organs is undergoing human clinical trials made feasible by genetic engineering.[15] University of Maryland's surgeons have successfully transplanted a genetically modified pig heart into humans. Transplant surgeons at the University of Alabama, Birmingham, and NYU Langone Health have also succeeded in transplanting genetically modified pig kidneys into a brain-dead person.[16] Kidney xenotransplantation entails three phases: pre-transplantation, transplantation stage, and post-transplantation phase. During the pre-transplantation stage, the consent of the next-of-kin of the brain dead is sought and consequently enrolled in a clinical trial.

With normal hemodynamics, the recipient is medically sustained with phenylephrine, vasopressin, levothyroxine, and methylprednisolone. A crossmatch is established between the recipient and a thirteen-month-old pig donor with a functioning kidney. The donor's kidneys are surgically removed in the Xenotransplantation Procurement Campus (XPC) and conveyed on ice to the Donor Recovery Center (DRC). At the center, surgeons transplant the kidneys into the recipient and they are subsequently cared for to ensure continuous functioning of the kidneys.[17]

Despite progress, xenotransplantation risks organ rejection and infections in recipients.[18] Human anti-pig antibodies mitigate against the successful transplant of pig organs in humans.[19] These obstacles have opened an opportunity for genetic engineering of non-human donors and their human recipient counterparts.[20] Genetic engineers have successfully modified the genes of pigs to produce renal xenografts suitable for xenotransplantation. They can introduce human inhibitor genes (e.g., hDAF, hCD46), human anticoagulant genes (e.g., hTBM, hEPCR), and immunomodulatory genes (e.g., hCD47, hHO1) in pigs. They can also delete pig carbohydrate antigens and growth hormone receptor gene that impedes renal xenografts in humans. During the development of the genetically modified pig, engineers generate multi-cistronic vectors consisting of human genes such as DAF and CD46 or TBM, EPCR, CD47, and HO1.[21]

Despite the progress, xenotransplant is at risk of zoonosis.[22] It could facilitate the spread of infectious diseases among the human population.[23] Pig organs transplanted to humans have caused inflammation in their recipients.[24] Porcine endogenous retroviruses (PERVs) in the genomes of pig cells are highly infectious.[25] Recipients and people exposed to them risk contracting PERVs from pig organs.[26] The attempt to circumvent this problem has not been entirely successful.[27] Genetic engineers continue to record progress in avoiding zoonosis. In the first clinical porcine kidney xenotransplant, scientists engineered pigs bereft of infectious organisms, such as the porcine CMV and porcine endogenous retrovirus C. If successful xenotransplants have life-saving prospects.[28]

Life-saving Prospects

The virus unleashed from Wuhan's "wet" market resulted in the covid 19 pandemic. The human consumption of animals, microbes, and plants results in several extinct species, destruction of habitats, pollution, degradation, epidemics, and pandemics. These losses disturb the ecology endangering the human species. Humans consider themselves the subject of creation. Consequently, the human species utilize other species for their ends.[29] This idea is behind transplanting animal organs to humans with failing or failed organs. The ethical argument favoring xenotransplantation rests on beneficence to humanity. Xenotransplantation has life-saving and cost-saving prospects. It can also offer patients with failing or failed organs a quality of life equivalent to that enjoyed by the healthy. However, an understanding of nature's teleology grounds the survival of the human species since nature has the inherent capacity to sustain its teleology.[30]

Xenotransplants can eradicate the bottlenecks associated with organ allocation. No patient in need of a transplant may be excluded due to the scarcity of organs. Organ shortage is becoming a perennial problem denying patients of vital medical resources. In the absence of needed organs, organizations and states use different strategies to shortlist candidates for transplant.[31] These criteria are not without opposition. For instance, UNOS' probability of success criterion is criticized for being vague. It does not give the parameters encompassing a successful transplant. Perceived unjust allocation criteria can hinder living organ donation. Xenotransplantation may oust inclusion or exclusion criteria for organ allocation and distribution. Many patients who otherwise will be denied scarce organs could gain access when xenograft organs are available. Xenotransplantation thus has the potential to remove the bureaucracies of organ allocation.[32] Once organs are in abundance, these bureaucracies will become obsolete. The abundance will facilitate quick and selective treatments.[33]

Xenotransplantation can reduce human rights violations associated with organ procurements. Human organ trade is in ascendency without global regulations. The organ market generates over 1.7 billion in profit annually despite WHO status that abrogates profiting from organ donation. These profiteers traffic women, children, the poor, and the destitute and harvest their organs for their clandestine trade. Worldwide prohibition of organ trade has not successfully halted it.[34] Organ trade is fueled by an increase in the aging population, diabetes, vascular diseases, and organ failures. To find organs for their black market, traders use coercion, abduction, fraud, or deception to attract victims. Other times, organ traders abuse their power or authority over victims or exploit their victims' vulnerability and constrain consent to accept payment for organ donation.[35]

Shortage of organs has fueled illegal organ trade and procurement of executed prisoners' organs. Over five thousand transplants take place annually using illegally procured human organs. Organ trading is becoming a lucrative and profitable clandestine industry.[36] The assumption behind procuring executed prisoners' organs is a voluntary donation. Illegal procurement occurs when prisoners are half executed and their organs excised before allowing them to die.[37] Using organs procured from prisoners contravenes acceptable international medical practices. This restriction is under the pretext that prisoners lack genuine autonomy to consent to organ donation.[38] Without informed consent, procuring organs from prisoners treats them as a means to an end. Such treatment disregards their dignity and right to an autonomous decision about their bodies. Such disrespect for human dignity is morally unjustifiable. A wounded executed prisoner does not meet the international criteria for cadaveric organ donation. A cadaveric organ donor is a brain-dead individual.[39]

Xenotransplant is also a substitute in the face of human organ shortage. It has prospects to save the lives of patients with failing or failed organs who may not survive without transplant treatment. It will reduce the death rate of patients on waiting lists.[40] As of mid-February 2022, UNOS has 106,409 on its waitlist of transplant candidates. More than half of these patients are active expectant candidates. The timing of available organs has a significant impact on their well-being and lifespan as a quarter of transplant candidates die each year.[41] Xenotransplant may limit the transplant waitlist period. Patients wait for about four years before a donor organ becomes available. Xenotransplantation has the potential to clear the backlog of waiting patients.[42] It can provide available organs for transplants. Unrestricted accessibility to organs may match up to the medical needs of transplant. Biotechnology can generate needed organs for the treatment of organ failures. Gene editing can facilitate the production of human organs in non-human primates. A growing understanding of xenotransplantation advances the prospects of treating organ failures and saving the lives of patients with failing or failed organs.[43]

Also, a successful clinical trial of xenotransplant can pave the way for advancement in the procedure and increase the chances of saving additional patients with failing or failed organs. A clinical trial will offer laboratory and clinical teams insights into the prospects of xenotransplant.[44] Xenotransplant has shown promising signs. Pre-clinical trials have addressed several security and viability concerns, including eluding transmission of porcine retroviruses in the hosts. Genetically modified xenografts have successfully been transplanted in a brain-dead patient. The pig organs functioned uninterruptedly for over three days. The kidneys excreted urine during the period. These successes amount to advancement in xenotransplant proceedings. This new progress is a pointer to breakthroughs.[45] The persistent holdup in rolling out a clinical trial of xenotransplant has opportunity costs. The delay will siphon possible future advances, thereby eroding life-saving interventions for organ failure. Genetic engineering procedures have increased the chances of xenograft compatibility in human beings.[46]

Medical science has not recorded significant groundbreaking findings in recent times in counteracting sudden deaths and alleviating protracted suffering.[47] Prospects of genetic engineering were discovered by the Austrian monk Gregor Mendel during the 19th century. He identified the possibility of exchanging genetic information in an initial observation of pea plants. Fast forward to 1953, Watson and Crick identified the structure of DNA, namely, the double helix. In 1958, Arthur Kornberg discovered the possibility of producing DNA in a test tube. Later in 1960, a French scientist found messenger RNA (mRNA). Finally, in 1972, researchers discovered that human DNA is 99% similar to other primates, such as gorillas and chimpanzees. A few years later, in 1977, scientists sequenced extended segments of DNA. They used a genetically engineered bacteria to produce human growth protein, somatostatin. This discovery began the golden age of biotechnology. The year 1995 witnessed the initial baboon-to-human bone marrow transplant. Xenotransplantation research has proven to have potential gains. The primary tools for this intervention are accessible, fast-forwarding a route for further groundbreaking findings to reduce premature deaths, excruciating human suffering, and medical costs.[48]

Cost-saving Prospects and Equity

Organ transplant is less costly than other organ treatment interventions. For instance, organ transplant is more cost-efficient than dialysis for treating kidney failure and ventricular assist tools to treat failing hearts.[49] Delaying clinical trials of organ transplants has both financial costs and the cost to human lives. Many patients whom transplant interventions could have saved will succumb to premature death. The financial benefits and life-saving prospects justify organ transplantation intensifying.[50] However, a significant problem with transplant interventions in healthcare concerns increasing the practice of organ transplantation since transplant resources are scarce and costly. Heart xenotransplant costs about $1.5 million, including medications, hospital admission, surgery, and treatment. Xenotransplantation of genetically modified organs has a potential cost saving of $17,000 which otherwise would have been used for anti-rejection medication in allotransplants. These savings constitute about 0.01% of the cost of transplantation, leaving the cost of xenotransplantation on the high side.[51] Consequently, critics consider transplantation burdensome and starve other critical medical procedures.[52]

The cost of transplants is increasingly becoming more expensive due to the age of donors and recipients. There has been about a 300% increase in organ recipients above 65 years of age in the United States, representing a quarter of entire organ recipients. The aging population of organ recipients increases the costs of the intervention. Similarly, the cause of the organ donor's death also affects the cost of transplants.[53] For instance, transplants using cadaveric organs from donors with anoxic injury and elevated terminal creatinine attracted extra costs. These complexities spark center costs for transplants. Between 2002-2003 and 2012-2013, the transplant procedure cost increased by 10%. The cost varied depending on the recipient's expected post-transplant survival (EPTS). The cost is higher for patients with lower EPTS. The transplant's cost depends on the donor's race, body mass index (BMI), calculated panel reactive antibody (cPRA), health status, and gender. These donor factors are difficult to control due to the general organ shortage.[54]

Govind Persad, for instance, questions the opportunity costs associated with organ transplants. He challenges the beneficence associated with organ transplants in prioritizing the procedure over more cost-effective life-saving medical interventions. Per this assessment, saving patients with failing or diseased organs is less cost-saving than saving patients with other needs. The excruciating cost of transplant procedures and their life-saving prospects cast a slur on utilizing finite resources on such interventions at the detriment of other life-saving interventions with prospects of greater magnitude. The goal of public health, Persad shows, is maximum benefit to a broader population. He, therefore, considers the search for a more cost-effective medical intervention preceding transplant intervention.[55] However, xenotransplant has cost-saving prospects. About 20% of procured human kidneys have proven unviable. This waste can be ameliorated in the procurement of pig kidneys for xenotransplantation.[56] Additionally, the fallouts of organ transplants raised by critics are justified by equity. Equity in healthcare requires alleviating the inequalities occasioned by genetic and environmental inequalities.

Different factors contribute to organ failure, including gene defects. Insights from human gene mutation attribute a single-gene defect to heart failure.[57] Genetic defects affect people's health. Genetic differences can result in unequal life outcomes, such as organ failure in those with genetic defects. To some extent, inequalities in organ health can be inscribed in the human DNA. Genetic inequalities have an environmental solution since there is a relationship between genotypes and environmental interferences. For instance, genotypic traits can cause myopia or hyperopia, yet eyeglasses can offer good vision. Just as eyeglasses ensure that those with vision defects can see as much as those without lousy eyesight, transplantation interventions ensure that organ-failing patients can have an equivalent quality of life as do people with healthy organs. Attempts to naturalize inequality constrain the possibility of environmental interventions to restore nature's inequalities. The natural lottery can be ameliorated by biotechnological and economic assistance.[58]

The psychologist Leon Kamin also discovered that genetic inequalities could have environmental solutions when examining phenylketonuria (PKU).[59] PKU is a genetic disorder resulting in intellectual disability. This disability is treated with a diet deficient in phenylalanine without a gene therapy intervention.[60] Equity requires that each person receives sufficient resources for good health. Allocating equal resources to different needs will perpetuate the inequalities caused by the genetic lottery.[61] Per equity, more funds can be allocated to transplants than other medical needs according to the requirements of the transplant procedure. According to equity, a good action is performed (e.g., organ transplant), resulting in unintended consequences (e.g., the opportunity cost to other healthcare needs).[62] In the allocation of scarce resources, policymakers can intentionally allocate more resources to the sickest such as end-stage organ patients.[63] Patients with urgent medical need should be allocated more resources whereas patients with fewer medical needs should be allocated fewer resources when resources are scarce. The medical urgency criterion aligns with the formal principle of justice, according to which it is fair to treat equals equally and the unequal unequally. Inequalities in capabilities can detract from justice.[64]

Amartya Sen rightly admits that it will be unjust for two persons with different degrees of illness to be treated equally. Call the sickest person with organ disease "A" and the others "B." A suffers from a graver disability than B. A's disability makes him unequal to B. A is unequal to B concerning what he can do and be. A's disability may confine him to certain places and situations. Also, A may be more disadvantaged than B if A expends more than B to cope with his debility. As A spends more resources to manage his incapacity than B, he has reduced resources to his ends than B, with more resources to pursue similar ends. Thus, justice in allocating scarce-vital resources requires unequal access to resources to ensure equality in doing and being among patients with failing organs. Justice requires prioritizing the sickest patients in resource allocation and distribution over more minor sick patients to ensure all the patients are at par in well-being and capabilities. Justice requires giving resources to patients with more significant needs. Allocating resources based on the diversity of patient needs is justifiable because of equal capacities (doings) and functions (beings) for all patients.[65]

The clarion call for justice in healthcare resource allocation has generated different distribution approaches.[66] This project offers an approach that integrates equality and equity in healthcare. It shows that equity in healthcare resource distribution does not infringe on equal interest and esteem for patients. Justice requires the allocation of healthcare resources to ensure equal outcomes. Ensuring equal life outcomes such as good health does not infringe on justice. Justice in healthcare does not require allocating equal resources to patients. However, it requires sharing resources on a need basis to ensure equal capacities for flourishing. Then, those with equal capacities can be treated equally in other healthcare resource allocations. Beneficence, therefore, justifies the life-saving prospects and cost of xenotransplant interventions. Xenotransplantation offers patients with failing organ capacities, well-being, and a lifespan similar to people without organ diseases. However, the interspecies transplant also raises concerns about respect for animal rights – the extent of animal instrumentalization in promoting human wellbeing. This latter issue is the primary concern of the next section of this project.[67]

The Consumptive Rationale of Human Beneficence

The consumerist attitude of the 21st century entails utilizing resources for human beneficence, including employing animal organs to provide human therapies. Xenotransplantation has several consequences that require the careful interaction of humans and animals. Such carefulness is necessary for research on animals and public health policies. Uncontrolled consumption of animals for human medical intervention can have dire consequences. Research on animals should be guided by the prevention of transmission of diseases between human and animal populations.[68] The emergence and re-emergence of zoonotic epidemics and pandemics are warning signs of the dangers humanity faces in its interaction with animals. This awareness requires utilizing other creatures sustainably because several natural resources are finite. Additionally, other non-human creatures have purposes other than human interests. Finally, the survival of the human species relies on the survival of nature.[69]

The utilization of creation for human beneficence relies on the position of humanity in the created order. Humanity is understood as the ruler of creation. The human species as lords of creation stems from understanding its ontology. The Genesis account of creation, for instance, inspires a notion of human beings as the image of God. The creator of Genesis preceded his creation and is distinct from it. He also transcends his creation.[70] Following this conception, the human species as the image of God also perceive themselves as over and above creation and not necessarily parallel members of creation. Consequently, humans often assume superiority over other creatures. They do not consider themselves objects of creation like other creatures.[71] Adherents of such religious convictions attribute lower moral regard for animals than those without such beliefs. In their research, Stephen Kellert and Joyce Berry found regular church attendants ascribing low regard to animal rights than non-church attendants. Thus, an unfavorable correlation exists between certain religious beliefs and practices, and regard for animal rights.[72]

The humans of Genesis are "dominion bearers," assuming domination over other universe creatures. Subsequently, humans wield authority over birds of the air, sea fish, and terrestrial animals. Added to humanity's role as masters of creation to have dominion over it, they are responsible for nurturing the creation and sustaining the created order.[73] However, humanity as the image of God is also understood structurally or functionally. A structural view considers the "image of God" as human qualities that differentiate the human species from other creatures. These qualities include reason, mind, free will, consciousness, and conscience. These structural capacities endure over time. A functional view of humanity as an image of God has a relational undertone, perceiving the image of God in terms of the ability to enter into relationships. Per this view, God is trinitarian; and human relationships are modeled after the trinitarian relationship. Joshua Farris contests the relational view of the image of God. In his view, the image of God is more of a substantial reality than a functional capacity. He considers the functional capacity as a consequence of the substantial reality. In his view, there is a conceptual distinction between substantiality and functionality. He considers humanity as a substance possessing rich immaterial qualities. A departure from a functional view of humanity as an image of God distorts a harmonious relationship between humanity, animals, and the environment. Such distortion partly drives unsustainable consumption.[74]

Arguably, 21st-century consumption of nature is unsustainable. Yet, the sustenance of human life relies on the earth and its materials. The annihilation of the earth and its resources can result in human extinction.[75] Human dominion of nature leads to the irreversible destruction of natural habitats. Uncontrolled consumption leads to the removal of barriers placed in the created order. The quantitative amassment of resources impacts quality relations with nature and biodiversity. It disrupts the state of the created order. The quest for allografts is spurred by the use of wealth to produce transplant resources enough for all who need them. The human drive does not usually end with adequacy; uncontrolled consumption strives towards surplus and will continue to procure excess transplant resources. This attitude limits life to profit and surplus, whereas quality existence only comes next to manufacturing, profit, and accumulation. The pursuit of economic gains hurts ecosystem depletion. Such depletion affects the existence of humanity and the diversity of the created order. Unsustainable consumptive emphasis on individual wealth opposes public and nature's wealth and health. In the face of this opposition, humanity lacks the tool to respond to the threat to biodiversity.[76]

Humanity's unrestricted dominion of nature can disturb the created order. Unrestrained consumption of natural resources can have a devastating impact, signaled by the zoonosis threat. The Covid-19 pandemic started in a "wet" market in Wuhan. Preventing zoonosis requires improvement in the reportage of infectious diseases. Additionally, if shutting down the wet market is an appropriate step to avoid zoonosis, so is imposing a reasonable moratorium on animal research. The growth of human organs in pigs can be suspended for some years.[77] Similar to the wet market, xenotransplantation has the potential risk of transmitting infectious diseases from animals to humans without natural immunity. Any risk that threatens the survival of entire humanity is unacceptable and avoidable. A sufficient insight into the science of transplants is an essential requirement to maximize the success of such interventions. A reasonable moratorium can be placed on clinical trials of xenotransplantation to enhance the science of the intervention. A clinical trial of xenotransplantation requires a minimal risk level acceptable to stakeholders of each procedure.[78]

Most human diseases are contracted from animals. Covid-19, Ebola, H5N1and H1N1 were no exception.[79] Zoonotic diseases threaten public health in countries worldwide. They bring an undue burden on limited health resources as they require effective national, regional, and international surveillance and collaboration to contain and mitigate their further spread. During outbreaks of zoonotic diseases, under-resourced communities bear the most significant burden of sickness and death.[80] The lessons of the 1918-1919 HINI influenza pandemic were telling; it claimed almost 100 million lives. The outbreak of the H5N1 avian epizootics was another warning sign of an impending pandemic. Pandemic preparedness requires outrightly preventing an outbreak. However, these previous signs have not warned the world enough to prevent the Covid 19 pandemic.

Avoiding future outbreaks may require reducing the possibility of the exchange of viruses and bacteria between animals and humans.[81]

A "one health" approach is necessary to prevent zoonotic diseases. Humans, animals, and the environment continue to interact. Consequently, an outbreak of zoonotic diseases is foreseeable. However, ministries of health and agriculture can collaborate to prevent, contain, and mitigate the spread of infectious zoonotic diseases.[82] Health is not a preserve of humans; it is equally necessary to animals and the environment. Well-being is beneficial to humans, animals, and the environment. For instance, animal well-being is relevant in preventing disease spread among the human population. This sensitivity includes consideration of the link between the health of animals used in research and the health of humans. Ecological resiliency demands common well-being and health. The consequence of such a drive is the prevention of zoonotic diseases. The similarity between animals and humans in terms of functionality of physiology, brain, and genome compels a similar concern in health and well-being for both species. Promoting the health and well-being of animals and human has the propensity to prevent zoonosis and enhance the state of ecosystems.[83]

Ecological Resiliency

Ecological resiliency entails a harmonious relationship among creatures without irreparable damage. This harmony requires the preservation and protection of nature from indelible harm. It also demands sustainable development. The Brundtland Report of the United Nations World Commission on Environment and Development considers sustainable development as providing current needs without compromising the foreseeable needs of future generations. It outlined the Millennium Development Goals (MDGs) for 2015, including promoting ecological sustainability. Ecological resiliency safeguards the capacity of the earth's progeny to provide for their current and future social, economic, and environmental needs. It requires actions and practices that benefit human well-being and the well-being of biodiversity.[84] The United Nations Conference on Sustainable Development of 2012 upgraded the MDG into seventeen Sustainable Development Goals (SDGs). The third goal, promoting healthy living and well-being, requires preventing and mitigating epidemics and pandemics. Relatedly, the 12th goal seeks to promote responsible consumption of the earth's resources. The SDGs also aim to enhance current human development and future ecological resiliency. This vision entails disease prevention and control.[85]

Enhancing equity in health requires preventing zoonosis. Preventing zoonosis is beneficial to public health. It will also reduce the moral distress of policymakers responsible for mitigating pandemics and spearheading vaccine development and equity. Instead of focusing on human health, promoting universal health encompassing human health, animal health, and environmental health is more beneficial. Human health is a consequence of animal and environmental health.[86] Promoting animal health stems from supporting animal rights. This approach can prevent the transfer of viruses from animals to human populations. However, some scholars are dismissive of animal rights. Peter Carruthers argues that animals and their sentiency command no moral claims, such that some actions against animal interests are justifiable. He shows that unjustified omissions and actions meted out to animals are not based on their moral standing and their status as rights-bearers. In contraposition, he finds it morally obligatory to ascribe moral claims to infants and elderly senile lacking the use of reason. Differently put, he rejects any moral ascription to animals while upholding ascribing moral claims and rights to all humans no matter their rational capacity.[87]

On the other spectrum, some scholars uphold the rights of animals. Saskia Stucki offers normative arguments in support of animal rights. She considers animal rights as having functional significance, such as improving animal protection. Upholding animal rights can also protect animal interests and heighten the burden of proof for foreseeable animal infringements. A prima facie animal interest will protect animals from abuse arising from less competing human interests. It will also minimize the exploitation of animals as a means to human ends. Shielding animal rights restrain the harm meted out on animals during human-animal interactions.[88] Additionally, Tom Regan offers moral justification for animal rights with unrelenting support for the abolition of animal usage in science. He views the consideration of animals as resources as the ideology behind animal exploitation in science and healthcare research. The treatment of animals stems from convictions of their rights. If they lack any rights, they lack moral standing. Even if animals lack rights and moral standing, their plight, suffering, solitude, blamelessness, and indiscriminate extermination evoke a compassionate concern and protection.[89]

Substituted judgment employed in donation decisions of incapacitated humans can similarly be applied in xenotransplant interventions. Incapacitated and incompetent humans are assumed to lack the exercise of rationality and autonomy. Consequently, their surrogates represent and defend their autonomy. Several animals have proven rationality similar to the rational capacity of children and mentally retarded adults. A superior level of rationality that forbids an individual from using another as a means to an end equally obligates humans not to instrumentalize animals for human wellbeing.[90] Based on this reasoning, animals deserve inalienable rights and respect for their interests that are intuitively ascribed to human beings.[91] Xenotransplant abuses can spill over into allotransplant abuses. Disregard for animal rights has a high affinity to disrespect for the rights of disadvantaged social groups, including ethnic and racial minorities, illegal immigrants, the destitute, and the physically and mentally challenged individuals.[92]

Xenotransplant abuses also affect human well-being. Every abuse of animals infringes on the rights of humans. Animal abuse upsets humans and is so offensive to humanity. Mistreatment of animals infringes upon human property rights if animals are considered resources for human flourishing. For example, humans care for pigs as a source of animal protein. Consequently, the mistreatment of pigs affects human interest. Even if such mistreatment does not affect human interests, it still lacks moral justification. Animal mistreatment is morally reprehensible without a ripple effect on any human interest. Every form of cruelty to animals is morally unjustified. Humans have intrinsic value since they are "subjects of life." However, there are other co-subjects of life; plants and animals are other subjects of life possessing intrinsic value. The survival of the human race rests on the survival of animals and all other creatures. The fate of humans remains connected to our universe's fate.[93] The value of animals is independent of human interests. Animals have an inherent value similar to the intrinsic value ascribed to humans. The difference in value between animals and humans is grade and not type. A graduation conception of intrinsic value is more defensible than a hierarchical view of values.[94]

Genetic engineering in xenotransplantation has prospects for clinical treatments despite ethical concerns. Consequently, governments have allocated funds to enhance research and development in organ transplants. For instance, the US devotes about $250 billion towards technological advancement. Its Endless Frontier Act gives pride of place to research and development of biotechnology and medical technology.[95] Government commitments come with controlling influence. Government coercion can constrain voluntary consent to genetic interventions. Progress in biotechnological therapeutic interventions must be guided by laws and international treaties that constrain government monopoly, control, and manipulation of genetic interventions. Several risks have been entertained to ensure growth in organ transplants. Sound risks and benefits analysis is necessary to curtail the negative impacts of genetic interventions.[96]

Conclusion

Humans consider themselves as dominion bearers and subjects of the created order. Thus, they utilize other species for their ends. Using xenotransplantation, humans employ biotechnology to improve medical treatments. Xenotransplantation risks zoonosis. Zoonotic diseases can cause severe human suffering and death, as illustrated by the covid 19 pandemic. This concern provokes a rethinking about the connection humans have with the animals used for research in medicine. The exploitation of animals evokes ethical concerns about these actions and their consequences. Resiliency in ecology requires resiliency in biotechnology development to diminish risks. It requires enduring development for current and future generations. Ecological resiliency, therefore, relies on interventions aiming at human beneficence and nature's beneficence and understanding of humans as sustainers of the created order. Achieving a resilient ecology requires several concerted efforts, including resource conservation, biodiversity protection, energy conservation, renewable fossil-fuel-free energy, recycling waste, advocating equal and equitable opportunities for all, and adopting environmentally-friendly lifestyles.

Notes

[1] Ariel M. Tarazona, Maria C. Ceballos, and Donald M. Broom, "Human Relationships with Domestic and Other Animals: One Health, One Welfare, One Biology," *Animals* 10, no. 1 (2019): 43.
[2] Tom L. Beauchamp and James F. Childress, *Principles of Biomedical Ethics*, 8th edition (New York: Oxford University Press, 2019), 3-5, 444-458.
[3] Cara K. Black et al., "Solid Organ Transplantation in the 21st Century," *Annals of Translational Medicine* 6, no. 20 (2018): 1-9.
[4] David KC Cooper, Bernd Gollackner, and David H. Sachs, "Will the Pig Solve the Transplantation Backlog?" *Annual Review of Medicine* 53, no. 1 (2002): 133–47.
[5] Emily Waltz, "When Pig Organs Will Fly," *Nature Biotechnology* 35, no. 12 (2017): 1133–38.
[6] María Jorqui-Azofra, "Regulation of Clinical Xenotransplantation: A Reappraisal of the Legal, Ethical, and Social Aspects Involved," in *Xenotransplantation* (Springer, 2020): 315–58.

[7] Sebastian D. McBride and A. Jennifer Morton, "Indices of Comparative Cognition: Assessing Animal Models of Human Brain Function," *Experimental Brain Research* 236, no. 12 (2018): 3379–90.
[8] Michael K. Skinner, *Encyclopedia of Reproduction* (Academic Press, 2018).
[9] Clyde F. Barker and James F. Markmann, "Historical Overview of Transplantation," *Cold Spring Harbor Perspectives in Medicine* 3, no. 4 (2013): 1-15.
[10] Magdalena Hryhorowicz et al., "Genetically Modified Pigs as Organ Donors for Xenotransplantation," *Molecular Biotechnology* 59, no. 9 (2017): 435–44.
[11] D K C Cooper et al., "Xenotransplantation—the Current Status and Prospects," *British Medical Bulletin* 125, no. 1 (March 2018): 5–14.
[12] Saurabh Bhatia, "History, Scope and Development of Biotechnology," *Introduction to Pharmaceutical Biotechnology* 1 (2018): 1-6, 29-34.
[13] Ibid., 21-27.
[14] Ibid., 1-5, 29-34.
[15] Burcin Ekser, David KC Cooper, and A. Joseph Tector, "The Need for Xenotransplantation as a Source of Organs and Cells for Clinical Transplantation," *International Journal of Surgery* 23 (2015): 199–204.
[16] Roni Caryn Rabin, "Kidneys From a Genetically Altered Pig Are Implanted in a Brain-Dead Patient," *The New York Times*, January 20, 2022, sec. Health, https://www.nytimes.com/2022/01/20/health/transplants -pig-human-kidney.html.
[17] Paige M. Porrett et al., "First Clinical-Grade Porcine Kidney Xenotransplant Using a Human Decedent Model," *American Journal of Transplantation* (2022), 1-16.
[18] Hryhorowicz et al.
[19] David KC Cooper et al., "The Pathobiology of Pig-to-Primate Xenotransplantation: A Historical Review," *Xenotransplantation* 23, no. 2 (2016): 83–105.
[20] Cooper et al., "Xenotransplantation—the Current Status and Prospects."
[21] Porrett et al., 2.
[22] Cooper et al., "Xenotransplantation—the Current Status and Prospects," 11-12.
[23] Johannes Kögel and Georg Marckmann, "'Xenotransplantation Challenges Us as a Society' What Well-Informed Citizens Think about Xenotransplantation," *EMBO Reports* 21, no. 9 (2020): 1-3.
[24] Mohamed B. Ezzelarab et al., "Systemic Inflammation in Xenograft Recipients Precedes Activation of Coagulation," *Xenotransplantation* 22, no. 1 (2015): 32–47.
[25] Joachim Denner, "How Active Are Porcine Endogenous Retroviruses (PERVs)?," *Viruses* 8, no. 8 (2016): 215.
[26] Joachim Denner and Nicolas J. Mueller, "Preventing Transfer of Infectious Agents," *International Journal of Surgery* 23 (2015): 306–11.
[27] Joachim Denner, "Recent Progress in Xenotransplantation, with Emphasis on Virological Safety," no. 21 (2016): 717-727.
[28] Porrett et al., 3.
[29] James Mensch, "Subjectivity in the Age of Pandemics," *Open Theology* 7, no. 1 (2021): 83–90.

[30] Nezerith Cengiz and Christopher Simon Wareham, "Ethical Considerations in Xenotransplantation: A Review," *Current Opinion in Organ Transplantation* 25, no. 5 (2020): 483–88.
[31] N. Cengiz and C. S. Wareham, "Pig-to-Human Xenotransplantation: Overcoming Ethical Obstacles," *South African Journal of Bioethics and Law* 12, no. 2 (2019): 67.
[32] James F. Childress and J. Steinbock, "Fairness in the Allocation and Delivery of Health Care: A Case Study in Organ Transplantation," *Practical Reasoning in Bioethics* (1997), 214–36.
[33] Burcin Ekser et al., 199–204.
[34] Monir Moniruzzaman, "'The Heavier Selves': Embodied and Subjective Suffering of Organ Sellers in Bangladesh," *Ethos* 47, no. 2 (2019): 233–53.
[35] Frederike Ambagtsheer, "Understanding the Challenges to Investigating and Prosecuting Organ Trafficking: A Comparative Analysis of Two Cases," *Trends in Organized Crime* (2021): 1–28.
[36] Ibid.
[37] Norbert W. Paul et al., "Human Rights Violations in Organ Procurement Practice in China," *BMC Medical Ethics* 18, no. 1 (2017): 1–9.
[38] Arthur Caplan, "The Use of Prisoners as Sources of Organs–an Ethically Dubious Practice," *The American Journal of Bioethics* 11, no. 10 (2011): 1–5.
[39] Norbert W. Paul et al., 2–4.
[40] Taylor M. Coe et al., "Survival Benefit of Pig-to-Baboon Liver Xenotransplantation Utilizing Genetically Modified Swine," *Transplantation* 104, no. S3 (2020): 639–40.
[41] John Loike D and Alan Kadish, "Ethical Rejections of Xenotransplantation?" 1-4.
[42] "Organ Transplant Trends | More Transplants than Ever," UNOS, accessed February 15, 2022, https://unos.org/data/transplant-trends/.
[43] Michael F. Knoll, David KC Cooper, and Rita Bottino, "How the COVID-19 Pandemic May Impact Public Support for Clinical Xenotransplantation in the United States?" *Xenotransplantation* 27, no. 5 (2020): 1-3.
[44] David KC Cooper and Hidetaka Hara, "'You Cannot Stay in the Laboratory Forever': Taking Pig Kidney Xenotransplantation from the Laboratory to the Clinic," *EBioMedicine* 71 (2021): 1-9.
[45] Porrett, et al, "1-17.
[46] Amitai Etzioni, "For Faster Science: Accelerated Genetic Engineering.," *The Journal of Clinical Ethics* 32, no. 4 (2021): 349–57.
[47] Amitai Etzioni, "For Faster Science," 352.
[48] Bhatia and Goli, 18-27.
[49] Govind Persad, "Will More Organs Save More Lives? Cost-Effectiveness and the Ethics of Expanding Organ Procurement," *Bioethics* 33, no. 6 (2019): 687.
[50] Etzioni, 351-52.
[51] Loike and Kadish, 2.
[52] David A. Axelrod et al., "An Economic Assessment of Contemporary Kidney Transplant Practice," *American Journal of Transplantation* 18, no. 5 (2018): 1168–76.

[53] A. J. Matas et al., "OPTN/SRTR 2013 Annual Data Report: Kidney," *American Journal of Transplantation* 15, no. S2 (2015): 1–34.
[54] Axelrod, et al., 377–89.
[55] Govind Persad, "Will More Organs Save More Lives? Cost-effectiveness and the Ethics of Expanding Organ Procurement," *Bioethics* 33, no. 6 (2019): 684–90.
[56] David KC Cooper, "What Will Be the Cost of a Genetically-Engineered Pig Organ for Clinical Xenotransplantation?," *Xenotransplantation* 27, no. 5 (2020): e12606.
[57] Hiroyuki Morita, Jonathan Seidman, and Christine E. Seidman, "Genetic Causes of Human Heart Failure," *The Journal of Clinical Investigation* 115, no. 3 (2005): 518–26.
[58] Kathryn Paige Harden, *The Genetic Lottery: Why DNA Matters for Social Equality*, 1st edition (Princeton Oxford: Princeton University Press, 2021): 153-159.
[59] S. Scarr, *Race, Social Class, and Individual Differences in I.q.*, 1st edition (Hillsdale, N.J: Psychology Press, 1981), 482.
[60] Hiu Man Grisch-Chan et al., "State-of-the-Art 2019 on Gene Therapy for Phenylketonuria," *Human Gene Therapy* 30, no. 10 (2019): 1274–83.
[61] Harden, 160-173.
[62] C Petrini, "Preemptive Kidney Transplantation: An Ethical Challenge for Organ Allocation Policies," *Clin Ter* 168, no. 3 (2017): 192–93.
[63] Dana Kay Nelkin and Samuel C. Rickless, "So Close, Yet So Far: Why Solutions to the Closeness Problem for the Doctrine of Double Effect Fall Short: So Close, Yet So Far," *Noûs* 49, no. 2 (June 2015): 379.
[64] Beauchamp and Childress, 268.
[65] Amartya Sen, "Equality of What?" in *The Tanner Lectures on Human Values*, by Sterling M. McMurrin (Cambridge University Press, 2011), 197–220.
[66] Haylee Lane et al., "Equity in Healthcare Resource Allocation Decision Making: A Systematic Review," *Social Science & Medicine* 175 (2017): 11–27.
[67] Harry Brighouse and Ingrid Robeyns, "The Place of Capability in a Theory of Justice," in *Measuring Justice: Primary Goods and Capabilities* (Cambridge University Press, 2010).
[68] Tarazona, Ceballos, and Broom.
[69] John S. Mackenzie, et al., *One Health: The Human-Animal-Environment Interfaces in Emerging Infectious Diseases*, vol. 366 (Springer, 2013).
[70] *The New American Bible*, Genesis 1.
[71] Mensch, 83–85.
[72] Stephen R. Kellert, *Knowledge, Affection, and Basic Attitudes toward Animals in American Society: Phase III* (US Department of the Interior, Fish and Wildlife Service, 1980).
[73] "The New American Bible," Genesis 1:26.
[74] Joshua R. Farris, "An Immaterial Substance View: Imago Dei in Creation and Redemption," *The Heythrop Journal* 58, no. 1 (2017): 108–11.
[75] Derrick Jensen and Aric McBay, *What We Leave Behind* (Seven Stories Press, 2011).

[76] John Bellamy Foster, Brett Clark, and Richard York, *The Ecological Rift: Capitalism's War on the Earth* (NYU Press, 2011): 13-49.
[77] Mensch, 84–85.
[78] Chace Mitchell, et al., "Meta-Analysis of Public Perception toward Xenotransplantation," *Xenotransplantation* 27, no. 4 (2020): 1-12.
[79] Kristian G. Andersen, et al., "The Proximal Origin of SARS-CoV-2," *Nature Medicine* 26, no. 4 (2020): 450–52.
[80] Melody J. Maxwell, et al., "Building the Road to a Regional Zoonoses Strategy: A Survey of Zoonoses Programmes in the Americas," *PLoS One* 12, no. 3 (2017): 1-19.
[81] David M. Morens and Anthony S. Fauci, "The 1918 Influenza Pandemic: Insights for the 21st Century," *The Journal of Infectious Diseases* 195, no. 7 (2007): 1018–28.
[82] E. Paul J. Gibbs, "The Evolution of One Health: A Decade of Progress and Challenges for the Future," *Veterinary Record* 174, no. 4 (2014): 85–91.
[83] Tarazona, Ceballos, and Broom.
[84] Iris Borowy, *Defining Sustainable Development for Our Common Future: A History of the World Commission on Environment and Development (Brundtland Commission)* (London: Routledge, 2013).
[85] Michelle ML Lim, Peter Søgaard Jørgensen, and Carina A. Wyborn, "Reframing the Sustainable Development Goals to Achieve Sustainable Development in the Anthropocene—a Systems Approach," *Ecology and Society* 23, no. 3 (2018).
[86] Sonja Hartnack, "Animal Health Is Often Ignored, but Indispensable to the Human Right to Health," *International Journal for Equity in Health* 21, no. 1 (2022): 1–3.
[87] Peter Carruthers, *The Animals Issue: Moral Theory in Practice* (Cambridge University Press, 1992).
[88] Saskia Stucki, "Towards a Theory of Legal Animal Rights: Simple and Fundamental Rights," *Oxford Journal of Legal Studies* 40, no. 3 (2020): 533–60.
[89] Tom Regan, "The Case for Animal Rights," in *Advances in Animal Welfare Science 1986/87* (Springer, 1987), 179–89.
[90] Anton Leist and Peter Singer, *JM Coetzee and Ethics: Philosophical Perspectives on Literature* (Columbia University Press, 2010).
[91] Regan.
[92] Yon Soo Park and Benjamin Valentino, "Animals Are People Too: Explaining Variation in Respect for Animal Rights," *Hum. Rts. Q.* 41 (2019): 39.
[93] Regan.
[94] Peter Singer, "Speciesism and Moral Status," *Metaphilosophy* 40, no. 3–4 (2009): 567–81.
[95] Charles E. Schumer, "S.1260 - 117th Congress (2021-2022): United States Innovation and Competition Act of 2021," legislation, June 8, 2021, 2021/2022, https://www.congress.gov/bill/117th-congress/senate-bill/ 1260.
[96] Etzioni, 349–57.

Literature

Ambagtsheer, Frederike. 2021. "Understanding the Challenges to Investigating and Prosecuting Organ Trafficking: A Comparative Analysis of Two Cases." *Trends in Organized Crime*, 1–28.

Andersen, Kristian G., Andrew Rambaut, W. Ian Lipkin, Edward C. Holmes, and Robert F. Garry. 2020. "The Proximal Origin of SARS-CoV-2." *Nature Medicine* 26, no. 4: 450–52.

Axelrod, David A., Mark A. Schnitzler, Huiling Xiao, William Irish, Elizabeth Tuttle-Newhall, Su-Hsin Chang, Bertram L. Kasiske, Tarek Alhamad, and Krista L. Lentine. 2018. "An Economic Assessment of Contemporary Kidney Transplant Practice." *American Journal of Transplantation* 18, no. 5: 1168–76.

Barker, Clyde F., and James F. Markmann. 2013. "Historical Overview of Transplantation." *Cold Spring Harbor Perspectives in Medicine* 3, no. 4.

Beauchamp, Tom L., and James F. Childress. 2019. *Principles of Biomedical Ethics*. 8th edition. New York: Oxford University Press.

Bhatia, Saurabh. 2018. "History, Scope and Development of Biotechnology." *Introduction to Pharmaceutical Biotechnology* 1: 1–61.

Black, Cara K., Kareem M. Termanini, Oswaldo Aguirre, Jason S. Hawksworth, and Michael Sosin. 2018. "Solid Organ Transplantation in the 21st Century." *Annals of Translational Medicine* 6, no. 20.

Borowy, Iris. 2013. *Defining Sustainable Development for Our Common Future: A History of the World Commission on Environment and Development (Brundtland Commission)*. London: Routledge. https://doi.org/10.4324/9780203383797.

Brighouse, Harry, and Ingrid Robeyns. 2010. *Measuring Justice: Primary Goods and Capabilities*. Cambridge University Press.

Caplan, Arthur. "2011. The Use of Prisoners as Sources of Organs–an Ethically Dubious Practice." *The American Journal of Bioethics* 11, no. 10: 1–5.

Carruthers, Peter. 1992. *The Animals Issue: Moral Theory in Practice*. Cambridge University Press.

Cengiz, Nezerith, and Christopher Simon Wareham. 2020. "Ethical Considerations in Xenotransplantation: A Review." *Current Opinion in Organ Transplantation* 25, no. 5: 483–88.

Cengiz, N., and C. S. Wareham. 2019. "Pig-to-Human Xenotransplantation: Overcoming Ethical Obstacles." *South African Journal of Bioethics and Law* 12, no. 2: 66–70.

Childress, James F., and J. Steinbock. 1997. "Fairness in the Allocation and Delivery of Health Care: A Case Study in Organ Transplantation." *Practical Reasoning in Bioethics*, 214–36.

Coe, Taylor M., Danielle Detelich, Charles G. Rickert, Nikolaos Serifis, Rudy Matheson, Ivy Rosales, Heidi Yeh, Wenning Qin, Yinan Kan, and Jacob Layer. 2020. "Survival Benefit of Pig-to-Baboon Liver Xenotransplantation Utilizing Genetically Modified Swine." *Transplantation* 104, no. S3: S639–40.

Cooper, D K C, R Gaston, D Eckhoff, J Ladowski, T Yamamoto, L Wang, H Iwase, H Hara, M Tector, and A J Tector. 2018. "Xenotransplantation—the Current Status and Prospects." *British Medical Bulletin* 125, no. 1 (March): 5–14.

Cooper, David KC, and Hidetaka Hara. 2021. "'You Cannot Stay in the Laboratory Forever': Taking Pig Kidney Xenotransplantation from the Laboratory to the Clinic." *EBioMedicine* 71: 103562.

Cooper, David KC, Bernd Gollackner, and David H. Sachs. 2002. "Will the Pig Solve the Transplantation Backlog?" *Annual Review of Medicine* 53, no. 1: 133–47.

Cooper, David KC, Mohamed B. Ezzelarab, Hidetaka Hara, Hayato Iwase, Whayoung Lee, Martin Wijkstrom, and Rita Bottino. 2016. "The Pathobiology of Pig-to-Primate Xenotransplantation: A Historical Review." *Xenotransplantation* 23, no. 2: 83–105.

Cooper, David KC. 2020. "What Will Be the Cost of a Genetically-Engineered Pig Organ for Clinical Xenotransplantation?" *Xenotransplantation* 27, no. 5: e12606.

Denner, Joachim, and Nicolas J. Mueller. 2015. "Preventing Transfer of Infectious Agents." *International Journal of Surgery* 23: 306–11.

Denner, Joachim. 2016. "How Active Are Porcine Endogenous Retroviruses (PERVs)?" *Viruses* 8, no. 8: 215.

Denner, Joachim. 2016. "Recent Progress in Xenotransplantation, with Emphasis on Virological Safety," no. 21: 717–27.

Ekser, Burcin, David KC Cooper, and A. Joseph Tector. 2015. "The Need for Xenotransplantation as a Source of Organs and Cells for Clinical Transplantation." *International Journal of Surgery* 23: 199–204.

Etzioni, Amitai. 2021. "For Faster Science: Accelerated Genetic Engineering." *The Journal of Clinical Ethics* 32, no. 4: 349–57.

Ezzelarab, Mohamed B., Burcin Ekser, Agnes Azimzadeh, Chih Che Lin, Yuming Zhao, Rachael Rodriguez, Gabriel J. Echeverri, Hayato Iwase, Cassandra Long, and Hidetaka Hara. 2015. "Systemic Inflammation in Xenograft Recipients Precedes Activation of Coagulation." *Xenotransplantation* 22, no. 1: 32–47.

Farris, Joshua R. 2017. "An Immaterial Substance View: Imago Dei in Creation and Redemption." *The Heythrop Journal* 58, no. 1: 108–23.

Foster, John Bellamy, Brett Clark, and Richard York. 2011. *The Ecological Rift: Capitalism's War on the Earth*. NYU Press.

Gibbs, E. Paul J. 2014. "The Evolution of One Health: A Decade of Progress and Challenges for the Future." *Veterinary Record* 174, no. 4: 85–91.

Grisch-Chan, Hiu Man, Gerald Schwank, Cary O. Harding, and Beat Thöny. 2019. "State-of-the-Art 2019 on Gene Therapy for Phenylketonuria." *Human Gene Therapy* 30, no. 10: 1274–83.

Harden, Kathryn Paige. 2021. *The Genetic Lottery: Why DNA Matters for Social Equality*. 1st edition. Princeton Oxford: Princeton University Press.

Hartnack, Sonja. 2022. "Animal Health Is Often Ignored, but Indispensable to the Human Right to Health." *International Journal for Equity in Health* 21, no. 1: 1–3.

Hryhorowicz, Magdalena, Joanna Zeyland, Ryszard S\lomski, and Daniel Lipiński. 2017. "Genetically Modified Pigs as Organ Donors for Xenotransplantation." *Molecular Biotechnology* 59, no. 9: 435–44.

Jensen, Derrick, and Aric McBay. 2011. *What We Leave Behind*. Seven Stories Press.

Jorqui-Azofra, María. 2020. "Regulation of Clinical Xenotransplantation: A Reappraisal of the Legal, Ethical, and Social Aspects Involved." In *Xenotransplantation*: 315–58.

Kellert, Stephen R. 1980. *Knowledge, Affection, and Basic Attitudes toward Animals in American Society: Phase III*. US Department of the Interior, Fish and Wildlife Service.

Knoll, Michael F., David KC Cooper, and Rita Bottino. 2020. "How the COVID-19 Pandemic May Impact Public Support for Clinical Xenotransplantation in the United States?" *Xenotransplantation* 27, no. 5: e12623.

Kögel, Johannes, and Georg Marckmann. 2020. "'Xenotransplantation Challenges Us as a Society' What Well-Informed Citizens Think about Xenotransplantation." *EMBO Reports* 21, no. 9: e50274.

Lane, Haylee, Mitchell Sarkies, Jennifer Martin, and Terry Haines. 2017. "Equity in Healthcare Resource Allocation Decision Making: A Systematic Review." *Social Science & Medicine* 175: 11–27.

Leist, Anton, and Peter Singer. 2010. *JM Coetzee and Ethics: Philosophical Perspectives on Literature*. Columbia University Press.

Lim, Michelle ML, Peter Søgaard Jørgensen, and Carina A. Wyborn.2018. "Reframing the Sustainable Development Goals to Achieve

Sustainable Development in the Anthropocene—a Systems Approach." *Ecology and Society* 23, no. 3.

Loike, John D., and Alan Kadish. 2018. "Ethical Rejections of Xenotransplantation? The Potential and Challenges of Using Human-Pig Chimeras to Create Organs for Transplantation." *EMBO Reports* 19, no. 8: e46337.

Mackenzie, John S., Martyn Jeggo, Peter Daszak, and Juergen A. Richt. 2013. *One Health: The Human-Animal-Environment Interfaces in Emerging Infectious Diseases*. Vol. 366. Springer.

Matas, A. J., J. M. Smith, M. A. Skeans, B. Thompson, S. K. Gustafson, D. E. Stewart, W. S. Cherikh, J. L. Wainright, G. Boyle, and J. J. Snyder. 2015. "OPTN/SRTR 2013 Annual Data Report: Kidney." *American Journal of Transplantation* 15, no. S2: 1–34.

Maxwell, Melody J., Mary H. Freire de Carvalho, Armando E. Hoet, Marco AN Vigilato, Julio C. Pompei, Ottorino Cosivi, and Victor J. del Rio Vilas. 2017. "Building the Road to a Regional Zoonoses Strategy: A Survey of Zoonoses Programmes in the Americas." *PLoS One* 12, no. 3: e0174175.

McBride, Sebastian D., and A. Jennifer Morton. 2018. "Indices of Comparative Cognition: Assessing Animal Models of Human Brain Function." *Experimental Brain Research* 236, no. 12: 3379–90.

McMurrin, Sterling M. 2011. *The Tanner Lectures on Human Values*. Vol. 8. Cambridge University Press.

Mensch, James. 2021. "Subjectivity in the Age of Pandemics." *Open Theology* 7, no. 1: 83–90.

Mitchell, Chace, Alan Lipps, Luz Padilla, Zoie Werkheiser, David KC Cooper, and Wayne Paris. 2020. "Meta-Analysis of Public Perception toward Xenotransplantation." *Xenotransplantation* 27, no. 4: e12583.

Moniruzzaman, Monir. 2019. "'The Heavier Selves': Embodied and Subjective Suffering of Organ Sellers in Bangladesh." *Ethos* 47, no. 2: 233–53.

Morens, David M., and Anthony S. Fauci. 2007. "The 1918 Influenza Pandemic: Insights for the 21st Century." *The Journal of Infectious Diseases* 195, no. 7: 1018–28.

Morita, Hiroyuki, Jonathan Seidman, and Christine E. Seidman. 2005. "Genetic Causes of Human Heart Failure." *The Journal of Clinical Investigation* 115, no. 3: 518–26.

Nelkin, Dana Kay, and Samuel C. Rickless. 2015. "So Close, yet so Far: Why Solutions to the Closeness Problem for the Doctrine of Double Effect Fall Short." *Noûs* 49, no. 2: 376–409.

Park, Yon Soo, and Benjamin Valentino. 2019. "Animals Are People Too: Explaining Variation in Respect for Animal Rights." *Hum. Rts. Q.* 41: 39.

Paul, Norbert W., Arthur Caplan, Michael E. Shapiro, Charl Els, Kirk C. Allison, and Huige Li. 2017. "Human Rights Violations in Organ Procurement Practice in China." *BMC Medical Ethics* 18, no. 1: 1–9.

Persad, Govind. 2019. "Will More Organs Save More Lives? Cost-Effectiveness and the Ethics of Expanding Organ Procurement." *Bioethics* 33, no. 6: 684–90.

Petrini, Carlo. 2017. "Preemptive Kidney Transplantation: An Ethical Challenge for Organ Allocation Policies." *Clin Ter* 168, no. 3: e192-3.

Porrett, Paige M., Babak J. Orandi, Vineeta Kumar, Julie Houp, Douglas Anderson, A. Cozette Killian, Vera Hauptfeld-Dolejsek, et al. 2022. "First Clinical-Grade Porcine Kidney Xenotransplant Using a Human Decedent Model." *American Journal of Transplantation.* https://doi.org/10.1111/ajt.16930.

Rabin, Roni Caryn. 2022. "Kidneys From a Genetically Altered Pig Are Implanted in a Brain-Dead Patient." *The New York Times*, January 20, 2022, sec. Health. https://www.nytimes.com/2022/01/20/health/transplants-pig-human-kidney.html.

Regan, Tom. 1987. "The Case for Animal Rights." In *Advances in Animal Welfare Science 1986/87*, 179–89. Springer.

Scarr, S. 1981. *Race, Social Class, and Individual Differences in I.q.* 1st edition. Hillsdale, N.J: Psychology Press.

Schumer, Charles E. 2021. "S.1260 - 117th Congress (2021-2022): United States Innovation and Competition Act of 2021." Legislation, June 8, 2021. 2021/2022. https://www.congress.gov/bill/117th-congress/senate-bill/ 1260.

Singer, Peter. 2009. "Speciesism and Moral Status." *Metaphilosophy* 40, no. 3–4: 567–81.

Skinner, Michael K. 2018. *Encyclopedia of Reproduction.* Academic Press.

St, World Bible Publishing. 1987. *The New American Bible.* New American Bible Revised ed. edition. Iowa Falls, Iowa: Catholic World Press/World Bible Publishers.

Streiffer, Robert. 2019. "Human/Non-Human Chimeras." In *The Stanford Encyclopedia of Philosophy*, edited by Edward N. Zalta, Summer. Metaphysics Research Lab, Stanford University, 2019. https://plato.stanford.edu/archives/sum2019/entries/chimeras/.

Stucki, Saskia. 2020. "Towards a Theory of Legal Animal Rights: Simple and Fundamental Rights." *Oxford Journal of Legal Studies* 40, no. 3: 533–60.

Tarazona, Ariel M., Maria C. Ceballos, and Donald M. Broom. 2019. "Human Relationships with Domestic and Other Animals: One Health, One Welfare, One Biology." *Animals* 10, no. 1 (2019): 43.

UNOS. N. d. "Organ Transplant Trends | More Transplants than Ever." Accessed February 15, 2022. https://unos.org/data/transplant-trends/

Waltz, Emily. 2017. "When Pig Organs Will Fly." *Nature Biotechnology* 35, no. 12: 1133-38

CHAPTER THIRTEEN

THE HERMENEUTICS OF NUDGING: RESILIENT RECIPROCITY BETWEEN TRANSHUMANISM AND NUDGING

IAN M. DOHERTY

Introduction

The human condition is finite and weak. We are fraught with imperfections that are out of our control. Transhumanists seek to fix these imperfections and conquer the human condition by merging technology and biology. Popularizing the potential transhumanism offers is no easy feat. Its ideas and theories remain guarded in academic journals, online, and can require a certain philosophical acumen. Even then, people have to know the search terms to use. In order to popularize and convince people about transhumanism potential we can look to nudging. Humans do not always make the best decisions. We consistently misinterpret data, overestimate our abilities, and sometimes act on incorrect beliefs. Combatting and overcoming these deficiencies is a monumental task. Misinterpreting data, overestimating our beliefs, and making decisions based on incorrect beliefs can be catastrophic in the domain of healthcare. One way to overcome deficiencies in healthcare is through transhumanism. Engaging people through a behavioral science concept known as nudging may provide a path forward. This chapter will give an overview of transhumanism, and its prospects. Then it will define nudging, and show how they have been used in a healthcare setting. Lastly, the chapter will ask whether it is acceptable to nudge someone toward transhumanism, and how transhumanism challenges our ability to evaluate nudges. This chapter argues nudging individuals toward transhumanism raises fundamental questions about the role of medicine, virtue ethics, and how to build an inclusive society while pursuing an ideal body.

Foundations of Transhumanism

Transhumanism is not a formally defined discipline. Various academics and laypeople can be transhumanists by subscribing to certain ideals and changing their perception of the human condition and its constituent parts. Each academic discipline contributes to the movement through its methodology and style. While there are various approaches to the movement, the nature of transhumanism is indisputable. Transcending physical and mental limits were once tales found in myths, but now they are being seriously considered because of various technological advancements.

History of Transhumanism

Transhumanism is the ripe fruit of the post-Enlightenment humanism tree. Its roots are firmly ground in the soil of empirical science and critical reason. These epistemological sources stand in direct opposition to the once-dominant sources of knowledge of the time of revelation and religious authority. Around the time of the Renaissance, the idea of becoming a well-rounded person who is highly mature scientifically, morally, and spiritually began to take shape.[1] This outright rejection of traditional epistemological sources opened up a new way of thinking, one grounded in rational humanism. This movement sought to persuade people to use reason to make themselves in their image. Benjamin Franklin, Francis Bacon, Giovanni Pico della Mirandola, and other famous figures throughout history, conceived of different ways to become well-rounded people by using reason. The status quo broke once again when Charles Darwin published *Origin of the Species* in 1859. Darwin's work showed academics and laymen alike that humanity is on the continuum of evolution.

Around this same time, materialism began to take hold in the intellectual sphere. This approach sought to understand the world in a physical sense, reducing everything in the human body to physiological phenomena. The French physician and materialist philosopher, Julien Offray de La Mettrie saw human beings as an animal, or as a collection of springs that wind one another up.[2] Analogizing the human body this way posed interesting questions. If the body is a collection of springs, like a machine, then it should be possible to manipulate the body. Understanding the human body in this way allows one to exchange parts of themselves to be more efficient, stronger, powerful, etc. in a similar way to cars. While he was a trained physician and a philosopher, de La Mettrie applied his imagination to reconceive the human condition. Transhumanism found another home in literature. Here, transhumanism took off with the rise of science fiction. Authors and scientists allowed their imaginations to sit in the driver's seat as they took us through worlds where technology and biology were fully integrated. The reception of these works was mixed, oftentimes serving as a warning of what the world might turn into.[3] The first person to use the word "transhuman" was Julian Huxley in 1927 when he wrote that humanity ought to transcend itself. [4]

Once philosophers supplanted religious authority with reason the possibilities of what human beings could accomplish grew exponentially. Immanuel Kant and Friedrich Nietzsche exhaustively interrogated what humans could accomplish and know through reason alone. Nietzsche's Übermensch pushed people to strive for personal growth and cultural refinement. [5] Whereas Kant's call encouraged people to be bold and use their intelligence to reach new heights. A through line can be drawn from these calls to be bold and personally grow to the vision transhumanists have today. Transhumanism is a transitional phase in the process of becoming posthuman. Where posthuman is the point of no longer possessing any recognizably human features.[6] Bostrom and other prominent transhumanists view posthumanism in this light. But, this definition of posthuman is controversial and needs to be unpacked. Some posthuman thinkers, like Francesa Ferrando, believe a person can be posthuman today, someone just needs to change their mindset. Ferrando does not see time as a lynchpin in becoming posthuman. The journey to becoming posthuman can begin today by deconstructing humanism, anthropocentrism, and dualism.[7]

More recently, transhumanism made its way into academia. Here, philosophy, political science, computer science, biology, and many other departments theorized how transhumanism impacts their disciplines. There has been a noticeable growth in transhumanist books, academic journals, institutions, and conferences.[8] Transhumanists do not see the movement as a zero-sum game. They see it as part of a solution to existential risks like climate change, disease, and conflicts. The popularity of the movement shows people's belief in its ideas and what it can become. However, transhumanism is not without its critics, among them most notably are Leon Kass, Francis Fukuyama, and Paul Ramsey. This group of individuals and academics have been dubbed "bioconservatives" by transhumanists because they stand against transhumanism and its possibilities for a variety of reasons.[9] The criticisms range from violating dignity to robbing future generations of authentic experiences.

Prospects of Transhumanism

Transhumanism presents itself as a movement that will help humans transcend their physical and mental limitations. The potential of these technologies quickly conjures up feelings of joy and excitement, but as one sits with the possibilities, one encounters the potential negative implications of the technology. Some downsides are the loss of individuality, breaking down the barrier between being someone and something, and issues related to access to technologies, creating a second class of citizens.

While the prospects are largely related to physical and mental changes, some proposals offer a moral change. Ingmar Persson and Julian Savulescu argue biotechnology should be wielded in a way that makes humans morally better, even if we are no longer biologically human.[10] Their argument differs in marked ways from Bostrom and other prominent transhumanists. While Bostrom argues for physical technology to be implanted in the body, Persson and Savulescu propose to make humans more altruistic in the moral sense. They think this is an integral step to overcoming future existential challenges like climate change and resource depletion. [11] Of course, the proposal by Persson and Savulescu is idealistic. Questions about which morality is best are rarely settled. Throughout history, various leaders have tried to force their morality on populations.

Savulescu's ideas about genetic manipulation can be considered controversial. He took them a step further in persuading prospective parents to select the best child by arguing if we can find a gene linked to criminality, we ought to select against it. This justification is taken one step further than his original idea, the Principle of Procreative Beneficence.[12] There, he argues we ought to select children whom we expect will have the best life. In regards to criminality, we ought to select against genes that make for a worse life. [13] This proposal presupposes the good life and ignores the various influences that define a good life. Savulescu's proposal is a scientific solution to a sociopolitical problem.[14] Further, genetic reductionism is a theme that features prominently throughout transhumanism. Here, a person's genome factors into their prospects in life. Whether they are predisposed to erratic behavior or have a disability. Disability is put in a different light and is cast off as abnormal in an extreme sense. Questions about the normal body and normal genetic code are raised because of the new space disability occupies in opposition to transhumanism. Savulescu is silent about what selecting against genes linked to disability means for those who live with that disability.[15]

Numerous motivating factors drive proponents of transhumanism. Bostrom sees the threat of death as a crucial threat to our freedom. He sees transhumanism as the avenue to conquer death and rob it of its power.[16] The goal of conquering death is pie in the sky today, but it might be possible one day. Researchers are working today to slow aging. Some experiments conducted today involve various compounds injected into mice to better understand how they impact aging.[17] Humans have not participated in the chemical trials yet. There is an extremely high level of uncertainty about how the human body would digest the chemicals, or handle the enhancement. There are various diets that some researchers believe would slow aging. One of which is caloric restriction, which deals with having a nutritious, vitamin-rich diet that is low in calories.[18] The general efficacy of a diet like this is controversial. The diet might work for certain individuals, but not others. Further, there are too many variables to control in an experiment like this. In addition to life extension and anti-aging technologies, transhumanists seek to improve not only a person's capabilities but also someone's general happiness. The overarching tone of improving someone's capability and general happiness creates some issues. Rather than ensuring the world is problem-free and just for everybody, the issue is actually inward, in the human condition.

Overcoming genetic ailments used to be a pipedream. This does not seem to be the case anymore with how people view genetic engineering. Researchers and philosophers see it as a significant part of humanity's future. The human genome was mapped in 2000, but today researchers are working to unpack and better understand specific parts of the genome. Understanding specific parts give scientists a way to target treatment for a genetic condition or alter the gene entirely through genetic therapy. Genetic therapies exist in two forms: somatic and germ-line. Somatic gene therapy allows changes to be made to a person's DNA, but those changes do not carry over to the next generation. Germ-line therapy, on the other hand, is performed on sperm and egg cells and those changes can be inherited. [19] This technology is awesome and can give people the opportunity to have children without inheritable diseases, or selection for certain traits. This technology raises serious questions about genetic determinism which gives too much leeway to how we view people. The unresolved ethical questions about genetic engineering give people pause because it pushes humanity to the edge of creation.

Moreover, genetic engineering presents ethical issues related to self-understanding, where children may not relate to themselves in the same manner because they are not their own makers in the same way we are today.[20] One of the many concerns with genetic engineering is the opportunity it presents in changing a pre-implantation embryo from a subject to an object.[21] The barrier is broken down between being a person and a thing. Where embryos and people are objects to be tinkered with and mastered.[22] Allowing the genomes to be freely altered opens up Pandora's box. It raises a litany of ethical issues that we are currently incapable of attempting to answer. For instance, would it be permissible for a child to hold their parents responsible for unwanted consequences as a result of altering the child's genome?[23] The potential opportunities genetic engineering can introduce can radically change our species. The ability to foresee or even predict how genetic modifications will interact is beyond our current capabilities.[24]

Genetic engineering is central to a goal of transhumanism: morphological freedom. Morphological freedom is the right to self-modification according to one's desires.[25] This is a foundational right to the transhumanist movement. There is an emphasis on individual freedom so individuals can choose their enhancements because of various conceptions of the "good life".[26] The emphasis on liberty is parallel to the way societies throughout the world emphasize liberty today. This emphasis on liberty decentralizes authority within the movement and empowers individuals to make decisions for themselves. The positive implications of this are clear. There is no central authority that decides what advancements or enhancements are acceptable. It gives people full reign over their lives and allows their imaginations to govern how they live. The downsides of this are clear as well. Immediately, there is tension between Persson and Savulescu's proposal of making people more altruistic and morphological freedom. For instance, someone can alter that part of themselves so they no longer act altruistically.

Materialism is central to transhumanism and its mission. Subscribing to a materialist worldview brings about exciting possibilities like molecular nanotechnology. This technology would give individuals the ability to transform materials at the atomic level into new objects. Rearranging atoms can be done in a variety of ways, from chemical reactions to an assembler the size of a nanometer. [27] The image of an atomic assembler used to be a work of science fiction, but now seems to be on the horizon. The ability to manipulate matter at the atomic level is truly groundbreaking and has countless implications for every profession. This technology builds off materialism because, as transhumanists claim, there is nothing inherently special about how objects are formed or their natures. Researchers have manipulated the nature of other organisms, and doing the same to humans is no different or should be seen as controversial.[28]

Making decisions with a little help from our friends

We have data showing people do not make decisions that always benefit themselves. This epistemic base is used in a variety of social sciences. Operationalizing this knowledge presents ethical challenges because it walks the line between respecting a person's autonomy and acting paternalistically. Navigating this area is an unenviable task. There is a way, however, to help people make better choices by nudging them. Nudge theory is a concept that's been popularized by behavioral scientists. These social scientists conducted experiments that found people weren't good decision-makers and they sought to address this disparity. To aid in decision-making, nudges encourage a particular choice in a variety of ways. Encouraging a particular choice is a foundational concept in nudge theory. This section will explain what nudge theory is, and how nudges are used in the real world.

What is a Nudge?

A nudge is altering a person's behavior in a way that does not limit options or alter economic motivations.[29] The line between nudging someone towards a choice and heavily persuading them is thin. Nudging someone towards an option does not mean another option is banned, or taxes are levied on people for choosing the alternative. Individuals must maintain their freedom of choice when they are being nudged. Nudging's starting point is rooted in the belief that people do not act optimally when making decisions. Claiming people act irrationally is too harsh, saying they do not act optimally gives people space to be human.[30] The fine line between autonomy and paternalism is crucial to understanding how much a person can be nudged. Walking this line can be difficult in healthcare because of the epistemic advantage doctors and other healthcare professionals have over their patients. Fundamental to nudging a person to make a better choice is creating an environment that aids better decision-making. This role is aptly named "Choice architect". This involves asking oneself what a reasonable and thoughtful person would want a product to do.[31] Of course, defining reasonable and thoughtful is a task within itself, but the choice architect would have a good sense of their target population's needs.

Choice architecture is a key to nudging, but it is not without criticism. Critics claim choice architecture ignores the larger factors that contribute to an individual's decision.[32] The larger factors that contribute to a person's decision-making pose philosophical questions: like whether we have the power to freely choose. Opponents of choice architecture lay blame further up the chain than private companies or research teams. They claim it allows the government to shirk its responsibility to govern and not use more effective techniques.[33] Further, nudges at a governmental level have negative implications for democracy. Nudges were shown to reduce opportunities for public deliberation in favor of using technocratic tools.[34] In order for nudges to not be an obstacle to deliberation and decision-making they must account for the structural factors that influence a person's decision-making process, and recognize that nudges are meant to help people make certain decisions.[35] More will be said about the latter part later on in the chapter.

The belief and empirical support that people generally act irrationally is fundamental to nudges. The goal of a nudge is to help them make better decisions that will improve their current situation.[36] Nudges can be interpreted by people who claim they are manipulating people or acting paternalistically. Manipulation has been defined in several ways. For instance, it can be defined as influencing someone's beliefs, wants, or emotions in a way that is contra to that person's self-interest.[37] Manipulating a person's beliefs, wants or desires can take two forms: intricate and blunt. Intricately manipulating means using known facts, emotions, and feelings of a person to influence their decision-making. While blunt manipulation can be thought of as using general knowledge about people to influence their decision-making.[38] Paternalism can be defined in numerous ways, as well. For our purposes, it can be thought of as an instance in which a person, B, interferes with another person, T, to promote T's own good.[39] The distinctions between manipulation, paternalism, and nudges are crucial for our conversation on what constitutes a nudge, and how to justify their use in a healthcare setting.

The theory that supports a nudge is libertarian paternalism. This form of paternalism is considered to be weak and nonintrusive.[40] Being weak and nonintrusive allows policymakers, or in this case, healthcare professionals to recommend or nudge people in a direction that will potentially make their lives better. Conflicts arise when nudges can be interpreted as interfering with autonomy. Before the discussion about how these two concepts interact, it is crucial to define each. Autonomy is a combination of the Greek word *autos* and *nomos,* meaning self-rule.[41] Autonomy is fundamental to Western society and allows individuals to act freely in the manner they decide. There is a broad conception of autonomy where it is seen as an end in itself.[42] This perspective of autonomy is taken for granted in Western society. Using this perspective of autonomy in a healthcare setting is challenging because of how broad its nature is. There are limits on autonomy, which usually extend to when it is interfered with by another person. Various academic disciplines have tried to contextualize and pontificate about the limits and value of autonomy. These discussions are not pertinent to this chapter.

Nudges exist on a spectrum. If we start with libertarian paternalism's core approach, the nudges progressively get more targeted as guiding choices through default options.[43] As the nudges become more targeted, the use of empirical support is needed to justify why that option is better. For instance, generic prescription rates increased from 75% to 98% after an intervention in a patient's electronic health record. The intervention proposed by the researchers in this particular study was not groundbreaking but rather was a simple change in the order the prescription drugs appeared. The study's implications are clear, generic prescriptions have been associated with higher adherence and better health outcomes.[44] This would be considered an instance of changing defaults. Changing defaults works well when both clinicians and patients do not have strong preferences for the options.[45] The effectiveness of nudges can be enhanced or inhibited if shareholders, clinicians, and administrative leadership personnel are not engaged. Successfully implementing nudges requires a buy-in from all relevant parties. However, implementing nudges can be challenging because of the complex power relations within institutions. There are various political and social beliefs held by the employees and patients.

As shown, nudges can be a pathway to improve health outcomes and can be successfully implemented in the workplace. However, nudges are not without criticism. Opponents posit nudges reinforce individualism because of their emphasis on an individual's choice, instead of creating better institutions.[46] The structural argument against nudges seems to be persuasive, but nudges operate within the structures. Nudges are a way for a preferred policy outcome to be accomplished without overhauling the entire system. Opponents may argue this approach reinforces existing inequalities, or that nudges do not account for the power of external factors, like wealth and social status, to influence decision-making.[47] They argue we ought to focus on creating a more just society through principles like solidarity rather than tackling individual behavior.

Why Should We Use Nudges?

Nudging people to make better choices, especially in healthcare, is a worthwhile venture. In the domain of healthcare, the justifications for the nudges oscillate between promoting a good or preventing harm. While the difference between promoting a good or preventing harm may be negligible in other fields, it matters in healthcare where the stakes can be high. The solutions to a healthier population can be structural, like addressing the social determinants of health or empowering people to make better choices.

The practicality and simplicity of nudges make them worthwhile tools for accomplishing the desired goal. For instance, it is difficult to get a majority of the population vaccinated against influenza. A simple change in the electronic health record of a patient proved to be an effective solution to increasing the percentage of patients that received a flu vaccine.[48] The integration of technology and medicine is proving to be an effective way to improve patient care. The authors of this study used active choice as a nudge in the flu study. This meant physicians or their assistants would have to "accept" or "cancel" an order for the flu vaccine.[49] The results of this study suggest active choice within electronic health records can be expanded to other forms of preventive medicine. Implementing active choice resulted in a 37.3% increase in influenza vaccinations.[50] This nudge is cost-effective and straightforward to implement. Given our current and future reliance on vaccines, it is clear this nudge may become a permanent feature in electronic health records.

The discussion of nudges usually revolves around helping individuals make better choices for themselves, but nudges can also be used to benefit future people. For instance, nudges ought to be used to improve the decision-making of pregnant women.[51] The implications of nudges for pregnant women are tremendous. However, these implications are not unqualified. When deciding whether a given nudge should be implemented, we ought to ask ourselves if the benefits outweigh the burdens that will be put on pregnant women.[52] There needs to be a stronger justification for a nudge during all stages of pregnancy. To do this, we can look to the harm principle. Nudging pregnant women to make better choices for themselves and their babies walks the fine line between invading privacy and promoting their good. The privacy-good calculus conducted by the nudgers and nudgees would have to be conducted transparently to ease concerns about paternalism and manipulation. Further, this process could lead to building trust between pregnant women, doctors, and social scientists. The effects are not exclusive to pregnant women and their future babies. A nudge like this one serves as a reminder that we have obligations to future people.

As previously mentioned, there's pushback against nudges because of their focus on individual decision-making. Opponents argue we ought to create better institutions and address structural issues rather than burdening individuals. These tasks are not mutually exclusive. We can, and ought to, address unjust structures while nudging people to make better decisions related to their healthcare. We can use data to empower people to better decisions about their healthcare choices and address structural inequities that put people at risk for poor health. Further, even if we have more just institutions and address social determinants of health that does not mean people will make good decisions about their health. Engaging individuals on a personal basis is crucial to ensure the proposed nudge will have a reasonable chance of succeeding. This will allow future nudges to be more efficient. But it is also important to engage relevant stakeholders, like other employees in the health system and leadership.[53] For nudges to be effective, it is important to have top-down engagement in addition to patient engagement. Engaging all relevant stakeholders democratizes the nudging process and allows people to provide feedback, potentially making the nudge better.

Nudges are an effective tool in prescribing medications and ensuring people participate in preventive medicine. But, nudges in a clinical setting, where doctors provide information about a patient's diagnosis and prognosis, have not been explored. Doctors have an epistemic advantage over the patients they treat. This knowledge gap places doctors in a precarious position where doctors can frame information about their patient's conditions in a way that impacts decision-making. For instance, there is no standard way to deliver a cancer diagnosis to a patient. Doctors can alter how they deliver information to the patient, opting to frame prognoses or treatments in a different light.[54] A paradigmatic case of a clinical nudge is a situation where a patient opts for a cancer treatment that will not benefit her, even though the doctor informed her about the unlikelihood of chemotherapy helping. This is an instance where a patient's autonomy and medical paternalism interact. Here, the doctor can rely on their epistemic advantage and decide not to offer the cancer treatment as an option before speaking with her. A situation like this becomes more complicated when other principles like beneficence and justice enter into this scenario. However, that conversation is beyond the scope of this chapter.

Implementing nudges is a worthwhile endeavor, however, it is not without skepticism. The reasons that justify the nudges ought to be scrutinized. This evaluative step brings us to question the purpose of the nudge, and what goal we want to accomplish. The purpose and goal of the nudge can be further unpacked, and various conclusions can be drawn based on the framework used to evaluate the nudge. Before nudges are implemented, the theoretical grounding must be shored up. The justifications for the proposed nudge should avoid a presupposed conception of good health. This is where nudges and paternalism diverge once again. Nudges do not limit the choice options for people, whereas paternalism may ban or prohibit certain options. The evaluation of a nudge is conducted by the nudgee, where the agent seeks to reach their desired end and not someone else's.[55] Further, the gulf between paternalism and nudges widens when people's preferences are accounted for. In the context of health, only people who prioritize being healthy would be nudged.[56] Ensuring only the people who want to be nudged are nudged, can be tricky. In healthcare, the beneficent duty of a doctor to their patient ends when the patient exercises their autonomy. Doctors may believe their patients ought to care more about their health and try to frame information in a way to startle the patient or caution them about their lifestyle. But, as shown, this blurs the line between nudging and manipulation.

Nudging toward Transhumanism: A Perfect Union?

As shown, transhumanism is an ambitious movement that seeks to eradicate and ultimately overcome ailments that have plagued people for thousands of years. For instance, the opportunity to rid our civilization of general suffering is enticing, but will everybody opt to forgo suffering? Will physical, emotional, and psychological hardships we face in life maintain their value? The answer to the latter is beyond the scope of the chapter. However, through nudging, we may be able to make transhumanism more popular among people. The second half of this section raises questions about how transhumanism expands the role of medicine, whether a nudge addresses a fundamental issue or a superficial one, and how we should view potential conclusions once we test transhumanist technologies on humans.

Is it permissible to Nudge someone toward Transhumanism?

Modern technology at large changes our abilities to act.[57] Nudging individuals to become transhuman is crucial to our future success as a civilization. The marquee technologies of transhumanism are ways away, but once they are here, transhumanists believe everybody should welcome them because of the benefits they will offer. The benefit-burden calculus individuals will perform is unique to themselves. Before discussing the question of whether it is permissible to nudge someone toward transhumanism, it is important to say something about the kinds of enhancements that will be discussed in this section. This section will discuss a sort of sibling to Persson and Savulescu's proposal about moral enhancement, and then discuss the biological enhancement Bostrom and others talk about.

The idea of morally enhancing people via genetic engineering raises ethical questions. If people carry genes that make them predisposed to erratic behavior, and this trait is linked to criminality, how responsible is someone for a crime they commit?[58] Identifying genes linked to specific behavioral issues, like violent behavior, is not science fiction.[59] Identifying so-called 'criminal' genes reinforces attitudes that uphold genetic determinism. Where a person's biology in itself is ethically controversial, and open to politicization because certain traits are deemed risky or abnormal.

Persson and Savulescu's argument about moral enhancement is not as far-fetched as some may believe it to be. Even if we are years away from genetically engineering people to be more altruistic, there are pharmaceuticals already on the market that can unintentionally alter moral decision-making.[60] While this is a step back from genetically engineering altruism, it represents a real-world case of impacting moral decision-making because these pharmaceuticals are widely prescribed and used. For example, propranolol, a drug prescribed for hypertension, angina, migraines, and posttraumatic stress disorder (PTSD), has been shown to reduce implicit racial bias.[61] Propranolol has a wide range of neural effects, but its main purpose is to block the effects of adrenaline and noradrenaline on the adrenoceptors.[62] There's only been one study to test this hypothesis, but if it is replicable, then the implications are clear. Propranolol's unexpected benefits are beneficial from a moral standpoint. Developing methods to tame implicit racial bias causes headaches because they are implicit, our brains make assumptions about what we see subconsciously. This conclusion may be significant because of propranolol's wide use. However, more studies will need to be conducted to see if this same phenomenon occurs outside of the laboratory.[63]

The popularly prescribed selective serotonin reuptake inhibitors (SSRIs) also have unintentional consequences when it comes to moral decision-making. SSRIs block the reabsorption of serotonin in the brain. They're used to alleviate depression, mood disorders, and excessive anxiety.[64] One study found SSRIs seem to make people more cooperative and less critical of one another.[65] Another study split participants into pairs and had them play the Dictator game. Here, the 'dictator' decides how much money is split between the dictator and another participant. The study found the participants who took an SSRI divided the sum of money more fairly than the control group.[66] Another study found SSRIs increased a person's aversion to harming others.[67] This study's implications complicate the positive side of SSRIs. If people taking SSRIs are in a position of power, they may be less likely to inflict harm. This may complicate a judge's role in society. They may be more likely to give a defendant a lighter sentence because of the SSRI. Further, as the 'dictator' game showed, those that take SSRIs and find themselves in a bargaining scenario may acquiesce to a worse deal. They may open themselves up to being exploited because of how they assess what is considered fair while on an SSRI. Whether SSRIs cause morally better or worse behavior is up for debate because of a variety of factors, like the person's predisposition to cooperation before taking SSRIs.[68] The general conclusions drawn by these studies show that SSRIs may improve moral decision-making. They can affect moral decision-making, but on the whole, it is premature to say whether they are positive or negative.

Nudging someone toward an SSRI or propranolol is hardly controversial, however, further disclosures must be made by doctors and clinicians about their moral side effects. The moral side effects align with making people better, morally speaking. Nudging patients toward these two prescriptions utilizes the knowledge created after studying the side effects. Whether the patients have access to the kinds of unintentional side effects the medications produce is a separate issue.

While Persson and Savulescu's proposal to genetically modify humans to make them morally better is far-off, products exist on the market that are unintentionally effective in impacting moral decision-making. The more pie-in-the-sky technologies are those that would become a part of our physical bodies. These technologies face numerous regulatory hurdles that prevent testing on human subjects. As things stand currently, physical technologies would not make us act in a more moral sense in the same way prescription drugs or altruistic genetic engineering would.

If we begin with morphological freedom, a fundamental tenet of transhumanism, it seems as though it is permissible to nudge individuals toward transhumanism. Morphological freedom, as previously defined is the right to modify one's body according to their desires.[69] The right to modify our bodies is supported by our age-old desire for self-creation.[70] Throughout human history, we chose new ways to modify our personalities via virtues, though self-creation through adding technology to biology is new. This form of freedom is not new, it's grounded in and an extension of the language of autonomy. This freedom is a negative right, meaning it can be exercised but others are not obliged to support its exercise.[71] Morphological freedom goes hand-in-hand with how autonomy operates and is viewed within bioethics. Where it gives individuals the right to consent to treatments, research trials, and enhancements. A prominent proponent of morphological freedom frames it in a way that suggests we ought to nudge people to embrace morphological freedom.[72] Morphological freedom enhances an individual's autonomy. They can choose to inhabit a physical body or exist in a virtual form.

If we merge the nanotechnology transhumanism espouses with morphological freedom, we see an almost perfect theoretical marriage. This partnership is a complete manifestation of a materialistic worldview. Where organisms and objects can be rearranged at the atomic level. The physical technologies transhumanism strives to merge with our biology have not been tried on human subjects. Transhumanism's desire to merge biology and technology presents unique ethical questions, in their eyes, we will become new beings through this union. The challenges to human experimentation ought to be looked at, especially given the abuses that took place throughout history. The technologies that would be implanted within humans would raise new and old ethical concerns. If we cognitively enhance the brain via a computer chip, we have to be sure it is a secure chip and cannot be hacked. Further, these technologies may differ in marked ways from current medical treatments. Where the technologies may be centered on cosmetic surgery, and not provide any medical benefit.[73] It seems as though decision-making will be up to individuals themselves because the technologies would walk the line between reducing morbidity and cosmetically enhancing a person's appearance. The proposed advancements of molecular nanotechnology, genetic engineering, and radical life extension allow people to let their imaginations take over and create the life they want to live. It's a manifestation of expressive individualism. These technologies will give individuals the ability to alter themselves, however, whether the unmitigated opportunity to do so is good, remains to be discussed.

One potential issue that will likely be ameliorated as time goes by is the number of transhumanist options. For nudging to be nudging, all available options must be available. If there are only two choices for physical technologies being implanted in the body, where one is opting to have a technological implant and the other is opting not to. Then we run into the issue of defining what "all available options" is. This phrase remains undefined in two ways: the first is whether more than one technological option is available and the other is whether the nudgee knows if all available options are presented. The latter question is more interesting because of the obligation to inform on the part of the nudger. There is also a level of trust and truth-telling that must be reached for the nudgee to know the nudger is being transparent and not trying to manipulate the nudgee's choice. Trust and truth-telling are integral components of medicine and patient care. The stakes of trust and truth-telling rise as procedures become riskier. For trust and truth-telling to exist in nudging someone toward transhumanism, they must exist within society at large.[74] The institutional values are a mirror of what is practiced in society. These virtues will make their way into healthcare if they are practiced by the public. The normative guidance of trust and truth-telling cannot exist in isolation within society, they must be embodied for them to actually exist.[75] Explaining how common morality ought to be practiced is beyond the scope of the chapter.

How Transhumanism Challenges our Ability to Evaluate Nudges

The prospects of both transhumanism and nudging seem to push people to make decisions that will better their lives. The decisions will make them healthier, happier, and possibly live longer. These outcomes may be true, but health, happiness, and longevity are not always unqualified goods. The components of health, happiness, and longevity ought to be scrutinized. In other words, what goes into defining health can become an ethical issue.

The intersection of nudges and transhumanism brings together a great deal of information. Information the nudgers possess that they want to inform the nudgees about. Transhumanists and their advocates possess information about what a person can accomplish with their new bodies or minds. These two groups of people assume a duty they have to the informed and uninformed. This obligation is an other-directed epistemic obligation, where experts are obliged to ensure other people are informed about new information about a given topic.[76] This obligation to inform others can become troublesome because of a lack of trust between parties. Even if the information is communicated in an easily understandable way, that does not mean it will overcome a lack of trust. Trust is a fundamental virtue that requires confidence in another person in that they will act for the right reasons by broader moral norms.[77] Building trust between and among people is crucial to ensure cohesion. Trust is even more crucial when it comes to decision-making in healthcare. For individuals to be nudged toward transhumanism, they have to trust the evidence behind the nudge and the nudger themselves.

On its face, nudging a person towards a particular choice seems to be an unqualified good. However, this may not always be the case. The option which the person is meant to select is supported by empirical evidence they might not have access to. This complicates the issue of autonomy and respecting the person as a knower. The knowledge used to justify nudges is not something found or observed, but rather it is created. This body of knowledge is the product of a process that is shaped by social and political forces. These forces influence how people design experiments, and how the data is interpreted.[78] The knowledge produced is constantly being interpreted in a variety of ways. The data is under a continuous hermeneutical exercise conducted by various researchers. They may arrive at different conclusions after interpreting the same data. How scientists interpret the data is influenced by non-epistemic values, like their cultural values, social values, and personal values.[79] Science, broadly speaking, has worked to mitigate these values so they do not influence research or impact results, and studies. The creation of the double-blind test and random controlled trials are seen as the gold standard of clinical tests. Their efficacies have been questioned for several reasons, one being the methods employed are malleable.[80] The study design, implementation, analysis, and publication of trials require the researchers to make choices about the data and how it is interpreted and published.

The knowledge created helps fill in the image of what a good life might look like. Where everybody is happy because they have an ideal body and brain. But defining an ideal body and brain is troublesome because of what is excluded from the definition of ideal. Nudging people towards an ideal choice that could make their lives better is loaded with ideological presuppositions. Transhumanists are not shy about their ideal body; they want to transcend physical and mental limitations. Transhumanist thinkers write about the need to overcome death or make their physical condition better. Every time transhumanist thinkers write about the ideal body, in their minds, they reject the undesired bodies.[81] The undesired bodies are the disabled bodies. Moreover, the definition of the desired body is fluctuating. Just like knowledge, it is open to the various social and political forces that define it. The body is opened up for further politicization through defining "ideal". Social rules, like what the ideal body is, or being transhuman, are regulated by the same fields of knowledge used to support nudging. Transhumanism assumes the authority of defining what a life worth living is. It is a life free from suffering, pain, disease, and ultimately death.[82] Transhumanism assumes this authority by declaring acceptable social norms, it is not necessarily supported by government authority.

When we evaluate nudges, we must ask ourselves a question about the nudge itself: is this going to adequately address the disparity we want to address? Some argue the nudge does not address the underlying issue of why the suboptimal behavior takes place or poor decision-making. It is a quick fix, short-term solution.[83] The proposed transhumanist interventions may treat superficial insecurities individuals have while leaving the true cause of the insecurity untouched. Where the true cause may be a bigger issue, like loneliness, or self-esteem. In other words, the proposed nudge provides a moment of security and comfort without adequately addressing why the individual feels that way in the first place. Justice and solidarity theorists call for us to address the root causes of inequalities and problems. The larger discourse surrounding justice and solidarity is a prominent feature in virtue ethics, where our interdependence is highlighted and built into our relationships and institutions. Nudging someone towards transhumanism is an inflection point for virtue ethics. If nudges are the end-all-be-all for addressing the inequalities that exist throughout society, we will lose sight of what we owe to one another. We are meant to learn how to reason from one another and evaluate what options are good for us. Through this acknowledged dependency, we can step back and evaluate the end we are pursuing.[84] Assisting one another through nudging is good, but it cannot be the only option to help people make better decisions for themselves.

While nudging reminds us of our obligations to one another, it might rob us of something greater. Nudges isolate the individual's decision-making without questioning whether the nudge is right for them. Nudges might remove the opportunity to understand one another more personally and develop friendships. The nudge towards transhumanism might be bolstered by quantitative data, where instead the individuals need to make decisions based on qualitative data. Qualitative data may alter the nudge entirely and change the person's mind, or exclude the individual from the nudge because it is not what they need. Rather, they need to address a more fundamental issue the nudge does not address. Here, we can learn about personal histories and understand the good each other is pursuing.[85] When we learn one another's personal histories, we learn to speak for the other. We learn what good they are pursuing, and we help them flourish.[86] This interaction builds trust among and between people and communities. This relationship allows people to work together to achieve one another's respective ends. Whereas nudges further isolate the individual in their decision-making. Nudges do not account for the web of relationships that aids them in decision-making. The nudge to transhumanism can be rushed without understanding the end the individual is pursuing.

Merging nudging and transhumanism raises questions about the ontology of medicine and health. Health as a general concept can be bifurcated into neutralism and positive health. Where neutralism is when the body is disease-free, and positive health involves something beyond the freedom from disease.[87] Transhumanism adopts a positive view of health. Where individuals are not merely free from disease, but fundamentally enhanced in some way. This can be thought of as being "well", or striving for wellness. The nature of health can be further split into naturalism and normativism. Naturalism posits that health is contingent on biological facts, and normativism argues that health depends on normative considerations.[88] Normative considerations, like certain capacities, make health a value-laden concept. This definition and conception align with how transhumanism views the human body. Where we can be healthier if we overcome our biological limitations and disease-caused suffering. This value-laden conception of health is open to endless interpretations because there is an endless dispute over the types of values. For instance, because of the expanded values in health, doctors may be inclined to nudge their patients to adopt a heart-healthier diet.[89] This expanded role for medicine may be good, but it gives doctors and clinicians more latitude in control over their patients' lives. A patient's daily life is further medicalized by doctors.[90] A person may not perceive themselves to be unhealthy, but the doctor believes certain aspects of their daily life are unhealthy. This falls in line with transhumanism and nudging because transhumanists see the human condition can always be improved.

The idea that health is seen in a positive light, where there is something beyond freedom from disease, ought to give us pause. It entertains the idea that health is more value-laden than we first considered. The nature and justification of the values in conceiving a positive view of health lead medicine and its goals to be socially constructed. This conception of health is not wrong, but it does have shortcomings. It removes the ends of medicine; they would be made extrinsic to the discipline.[91] An important end that might go by the wayside if medicine, as broadly defined, is entirely socially constructed is the patient-physician relationship. The relationship might be sacrificed and become centered around a physician-society-based relationship.[92] This view of medicine would leave it to be defined by the whims of society. It would be removed from its *telos.* A socially-constructed view of medicine asks basic questions about who decides what the goal of medicine actually is, whether is it medicine anymore, and whether the socially-constructed values are universally applicable. Transhumanism and its proponents seem to eschew questions about the nature of medicine and adopt a socially-constructed positive view of health.

The hopes that transhumanism offers are promising and inspire hope for everybody, but we ought to be cautious when it comes to interpreting and acting upon the data that shows the technologies are efficacious. Testing neural implants and other transhumanist technologies on humans won't happen in the immediate future. But when it does happen and results are published, we ought to scrutinize the results and evidence. There is evidence that false findings are published in a variety of social science disciplines.[93] The issue of non-replication produces confusion about the legitimate conclusion of the study. Good science is predicated upon the ability to reproduce the findings, in the same manner, repeatedly. Researchers and policymakers ought to be suspicious of test results, given the stakes of transhumanist technologies and the possibility of nudging people toward them. Even reaching a definitive conclusion in a given study might be impossible given the ambiguity of identifying the truth in any research question.[94] Further, if we examine who is publishing the research, we can sometimes see pharmaceutical companies funding and publishing the research.[95] This is an obstacle to accessing the real results. The truth, in science may be a Platonic form where we never truly experience it. It may be the case that as time progresses, the regulations for human trials will improve along with the methodologies used in scientific research.

Transhumanism is marketed as an intervention. However, there seems to be a theoretical disparity about what it treats. Contemporary medical interventions treat high blood pressure, depression, blood clots, etc. These medical issues are supported by various theories of diseases. Diseases generally have two constituent parts: they must create some sort of dysfunction, and that dysfunction must be deemed harmful.[96] Transhumanism, at its core, challenges this theory of disease. An ideal, healthy body is problematized because, in its natural state, it's deemed unhealthy. Creating and marketing transhumanistic medical interventions challenges our conceptions of health. Further, nudging people who perceive themselves to be healthy toward this raises ethical questions. Caution ought to be exercised in how these truly awesome technologies are marketed and designed. The idea of overcoming suffering is alluring, it might not be alluring while risking our self-perception.

We are only starting to understand the social determinants of health and how they impact our well-being and overall health.[97] Social determinants are forces outside of medicine and one's biology that influences how healthy a person is. This is troubling for transhumanists because the social determinants represent forces outside of a person's control. They are forces that someone is born and raised in. Addressing social determinants is also out of the purview of nudging. To adequately address the social determinants, structural changes would have to be made. Nudging an individual towards transhumanism ignores the structural factors that contribute to someone's unhealthy state.[98] The opportunity to enact structural change and alter individual decision-making may present itself. But addressing the structural factors might change an individual's mind about choosing transhumanism. Addressing structural inequities that lead to poor health outcomes means ending the status quo, like austere policies and neoliberal thinking.[99] Here, we see another virtue ethic against nudging someone towards transhumanism, nudges can be an obstacle to solidarity. Nudges can ignore the larger collective responsibility we have toward one another and, instead, reinforce the individual ontology that prevails in the West.[100]

Conclusion

Transhumanism presents a groundbreaking opportunity to overcome mental and physical limitations. It's an ambitious movement that forces us to grapple with our own physical and mental limitations. The mental and physical limitations people have are difficult to face, they're our weaknesses. Nudging provides an opportunity to aid people in overcoming their shortcomings. Libertarian paternalism ensures the nudges do not breach the line of paternalism or manipulation. This is an ideological offspring of the libertarian vein of American society, where individual agents are largely left alone to make decisions about their lives. Nudges are a powerful tool in healthcare and ought to be further explored. They are a cost-effective and simple measure for improving decision-making and a person's health. It is important to be mindful of how nudges and autonomy interact. The line between paternalism, nudging, and manipulation must be respected. Simply rearranging choices likely does not violate autonomy, but the use of subliminal messaging or framing information in a particular way might violate autonomy. Further as previously stated, nudges serve as a reminder that we have obligations to future generations. Nudging can be used to ensure people are more moral. If appropriate, people can be nudged towards particular SSRIs. The unintentional consequences are well-documented. However, when constructing and implementing nudges, we must be mindful of the direction we are nudging people toward. The direction we want to nudge people is loaded with presuppositions and faulty assumptions. Where people may not need the nudge, but actually need assistance dealing with an underlying issue. Also, we have to consider what this nudge may say about what we, as a society, value. This is problematized by transhumanism excluding the disabled body in favor of a being without it. When healthcare professionals nudge people toward an option, they must be mindful of what the nudge is actually doing. The effects of the nudge extend beyond the clinical encounter and into a person's life. Where certain choices are considered suboptimal. The nudges that may rob a person of the opportunity to undergo an exercise is self-interpretation. Where they decide what is ultimately best for them and their needs.

Notes

[1] Bostrom, Nick "A History of Transhumanist Thought," *Journal of Evolution and Technology.* 14 (1) April 2005 ISSN: 1541-0099, 2
[2] Ibid. 3
[3] Ibid., 3-4.
[4] Ibid., 6.
[5] Friedrich Nietzsche, *Thus Spoke Zarathustra.* Cambridge University Press, 2006.
[6] Nick Bostrom, ("Transhumanist FAQ" https://nickbostrom.com/views/transhumanist.pdf, 2003. Accessed 16 September 2022, 5.
[7] Francesca Ferrando, *Philosophical Posthumanism.* Bloomsbury, 2019, 28.
[8] Bostrom, "FAQ," 13-17
[9] Ibid., 17-19
[10] Ingmar Perssonand Julian Savulescu, "Moral Transhumanism," *Journal of Medicine and Philosophy.* 35 (2010): 656-669.
[11] Ibid., 666
[12] Julian Savulescu, "Procreative Beneficence: Why We Should Select The Best Children", *Bioethics,* 15 No. 5/6(2001): 413-426
[13] Julian Savulescu, "Why Genetic Testing for genes for criminally is morally required", *Princeton Journal of Bioethics.* 4 (2001), 79-97, 86.
[14] Melinda C. Hall, *The Bioethics of Enhancement: Transhumanism, Disability, and Biopolitics.* Lexington Books. 2017, 9.
[15] Savulescu, "Procreative Beneficence," 423.
[16] Hall, 2.
[17] John K. Davis, *New Methuselahs: The Ethics of Life Extension,* Massachusetts Institute of Technology, 2018, 8-14.
[18] Davis, 10.
[19] Bostrom, "FAQ," 7-8.
[20] Jurgen Habermas, *The Future of Human Nature, Polity Press,* 2003: 23
[21] Ibid.
[22] Ibid., 13
[23] Ibid., 13
[24] Eric Lander, Francoise Baylis, Feng Zhang, et. al, "Adopt a moratorium on heritable genome editing." *Nature.* 2019, 166.

[25] Anders Sandberg, *The Transhumanist Reader,* Edited by Max More and Natasha Vita-More, Wiley-Blackwell. 2013, 56-64.
[26] Nick Bostrom, "Transhumanist Values", *Journal of Philosophical Research.* 30 (Supplement): 3-14 (2005). https://doi.org/10.5840/jpr 2005_26 11
[27] Bostrom, "FAQ," 9.
[28] Bostrom, "FAQ," 35.
[29] Richard H. Thaler, Cass R. Sunstein, *Nudge: The Final Edition.* Penguin Books. 2021: 8.
[30] Ibid., 27.
[31] Ibid., 16-17.
[32] Barbara Prainsack, "The value of healthcare data: to nudge, or not?," *Policy Studies* 41, no. 5 (2020), https://doi.org/10.1080/01442872.2020. 1723517., 553
[33] Prainsack, 553.
[34] Christopher McCrudden, and Jeff King, "The Dark Side of Nudging: The Ethics, Political Economy, and Law of Libertarian Paternalism," In *Choice Architecture in Democracies, Exploring the Legitimacy of Nudging,* edited by Alexandra Kemmerer, Christoph Mollers, Maximilian Steinbeis and Gerhard Wagner, 75-140.
[35] Prainsack, 553
[36] Thaler and Sunstein, 7
[37] Anne Barnhill, "What is Manipulation?" *Manipulation: Theory and Practice.* Eds: Christian Coons and Michael Weber, Oxford University Press, 2014: 51-72.
[38] Barnhill, 58.
[39] Donald VanDeVeer,*Paternalistic Interventions: The Moral Bounds of Benevolence.* Princeton University Press, 1986: 12.
[40] Thaler, Sunstein, 7
[41] Tom L Beauchamp, and James F Childress. *Principles of Biomedical Ethics, Eighth Edition.* Oxford University Press, 2019: 99
[42] Cass R. Sunstein, *Why Nudge? The Politics of Libertarian Paternalism,* Yale University Press, 2014: 127.
[43] Joseph D. Harrison and Mitesh S. Patel, "Designing Nudges for Success in Health Care," *AMA Journal of Ethics* 22, no. 9(2020): 796-801, 797.
[44] Mitesh S Patel, Day, Susan C., et. al., "Generic Medication Prescription Rates After Health System-wide redesign of default options within the electronic health record." *JAMA Internal Med.* 176, no. 6(2016): 847-848.
[45] Harrison and Patel, 798.
[46] Prainsack, 549.
[47] Ibid., 552
[48] Mitesh Patel, Kevin G.Volpp, Dylan S. Small, et. al., "Using Active Choice Within the Electronic Health Record to Increase Influenza Vaccination Rates," *Journal of General Internal Medicine,* 32, no. 7(2017): 790-795.
[49] Ibid., 791.
[50] Ibid., 792.
[51] Hafez Ismaili M'hamdi, Medard Hilhorst, Eric A P.Steegers, et. al., "Nudge me, help my baby: on other-regarding nudges," *Journal of Medical Ethics,* 43(2017):702-706.
[52] Ibid., 705.

[53] Mitesh Patel, Kevin G. Volpp, David A. Asch, "Nudge Units to Improve the Delivery of Health Care," *N Engl J Med.* 378, no. 3(2018): 214-216.
[54] Ajay Aggarwal, Joanna Davies, Richard Sullivan, "'Nudge' in the clinical consultation-an acceptable form of medical paternalism?" *BMC Medication Ethics,* 15, no. 31(2014): 1-6.
[55] Thaler and Sunstein, 7.
[56] Bart Engelen, "Ethical Criteria for Health-Promoting Nudges: A Case-by-Case Analysis," *The American Journal of Bioethics* 19, no. 5(2019): 48-59, 52.
[57] Hans Jonas, *The Imperative of Responsibility,* University of Chicago Press. 1984.
[58] Savulescu, 79-97.
[59] H G Brunner, et al., "X-linked borderline mental retardation with prominent behavioral disturbance: phenotype, genetic localization, and evidence for disturbed monoamine metabolism," *American Journal of Human Genetics* vol. 52, no. 6 (1993): 1032-9.
[60] Neil Levy, Thomas Douglas, Guy Kahane, Guy, et. al., "Are you Morally Modified? The Moral Effects of Widely Used Pharmaceuticals," *Philos Psychiatr Psychol.* 21, no. 2(2014): 111-125.
[61] Ibid., 116.
[62] Ibid., 113.
[63] Sylvia Terbeck, Guy Kahane, Sarah McTavish, et. al., "Propranolol reduces implicit negative racial bias," *Psychopharmacology.* 222, (2012): 419-424, 424.
[64] Levy, Douglas, Kahane, et. al., 116.
[65] Brian Knutson, Owen M.Wolkowitz, Steve W. Cole, "Selective Alteration of Personality and Social Behavior by Serotonergic Intervention," T*he American Journal of Psychiatry* 155, no. 3 (1998): 373-9.
[66] Wai S.Tse, and Alyson J. Bond, "Serotonergic intervention affects both social dominance and affiliative behaviour," *Psychopharmacology* 161 (2002): 324-330.
[67] Molly J. Crockett, Luke Clark, Marc D. Hauser, et al., "Serotonin selectively influences moral judgment and behavior through effects on harm aversion," *Proceedings of the National Academy of Sciences of the United States of America* 107, no. 40 (2010): 17433-8.
[68] Levy, Douglas, Kahane, et. al., 118.
[69] Sandberg, 56.
[70] Ibid., 59
[71] Ibid., 57
[72] Ibid., 59
[73] Nick Bostrom and Anders Sandberg, "Cognitive Enhancements: Methods, Ethics, Regulatory Challenges," *Sci Eng Ethics.* 15(2009): 311-341, 323.
[74] Beauchamp and Childress, 458.
[75] Quill Kukla, "Living with Pirates: Common Morality and Embodied Practice" *Cambridge quarterly of healthcare ethics: CQ: the international journal of healthcare ethics committees* 23, no. 1 (2014): 75-85, 81.
[76] Stephen John,"Artificial Ignorance, Epistemic Paternalism, and Epistemic Obligations," *Epistemic Paternalism: Conceptions, Justifications and Implications,.* Amiel Bernal and Guy Axtell, eds. 17-27, 18.

[77] Beauchamp and Childress, 40.
[78] Michel Foucault, *Power,* James D. Faubion, ed., The New Press. 2001, 8.
[79] Phyllis Rooney,"On Values in Science: Is the Epistemic/Non-Epistemic Distinction Useful," *PSA: Proceedings of the Biennial Meeting of the Philosophy of Science Association* (1992): 13–22.
[80] Jacob Stegenga,*Medical Nihilism,* Oxford University Press 2018: 12.
[81] Hall, 72.
[82] Melinda Hall,"Vile Sovereigns in Bioethical Debate," *Disability Studies Quarterly,* 33, no. 4(2013): 1-23, 13.
[83] Danielle Meehan,"Epistemic Vices and Epistemic Nudges: A Solution?" *Epistemic Paternalism: Conceptions, Justifications and Implications,* Amiel Bernal and Guy Axtell eds., 249-261, 252.
[84] Alasdair MacIntyre, *Dependent Rational Animals,* Carus Publishing Company, 1999: 83.
[85] Ibid., 149.
[86] Ibid., 150.
[87] Jacob Stegenga,*Care & Cure: An Introduction to Philosophy of Medicine,* University of Chicago Press, 2018: 7.
[88] Ibid., 7, 15
[89] Irving Kenneth Zola,"Medicine as an institution of social control," *Ekistics,* 41, no. 245(1976): 210-214.
[90] Ibid., 210
[91] Edmund Pellegrino, "Philosophy of Medicine: Should it be Telelogically or Socially Constructed?" *Kennedy Institute of Ethics Journal,* 11, no. 2(2001): 169-180.
[92] Ibid., 178.
[93] John PA Ioannidis, "Why most published research findings are false," *PLoS Med* 2, no. 8(2005): e124, 0696-0701.
[94] Ibid.
[95] Joel Lexchin, et al., "Pharmaceutical industry sponsorship and research outcome and quality: systematic review," *BMJ (Clinical research ed.)* 326, no. 7400 (2003): 1167-70.
[96] Jacob Stegenga, "Effectiveness of medical interventions," *Studies in history and philosophy of biological and biomedical sciences* 54 (2015): 34-44, 37.
[97] P. Bravemanand L. Gottlieb, "The Social Determinants of Health: It's Time to Consider the Causes of the Causes," *Public Health Reports*. 2014;129 (1_suppl2): 19-31, 20.
[98] Prainsack, 551
[99] Ibid.
[100] Ibid., 552

Literature

Aggarwal, Ajay, Davies, Joanna, Sullivan, Richard. 2014. "'Nudge' in the clinical consultation-an acceptable form of medical paternalism?" *BMC Medication Ethics.* 15, no. 31: 1-6.

Barnhill, Anne. 2014. "What is Manipulation?" *Manipulation: Theory and Practice.* Eds: Christian Coons and Michael Weber. Oxford University Press, 51-72.

Beauchamp, Tom L and Childress, James F. 2019. *Principles of Biomedical Ethics. Eighth Edition.* Oxford University Press.

Bostrom, Nick. 2005. "A History of Transhumanist Thought." *Journal of Evolution and Technology.* 14, no. 1. April.

Bostrom, Nick 2003. "Transhumanist FAQ" https://nickbostrom.com/views/transhumanist.pdf, Accessed 16 September 2022

Bostrom, Nick. 2005. "Transhumanist Values", *Journal of Philosophical Research.* 30 (Supplement): 3-14.

Bostrom, Nick and Sandberg, Anders. 2009. "Cognitive Enhancements: Methods, Ethics, Regulatory Challenges" *Sci Eng Ethics.* 15: 311-341, 323.

Braveman P, Gottlieb L. 2014. "The Social Determinants of Health: It's Time to Consider the Causes of the Causes." *Public Health Reports.* 2014;129(1_suppl2):19-31.

Brunner, H G, et al. 1993. "X-linked borderline mental retardation with prominent behavioral disturbance: phenotype, genetic localization, and evidence for disturbed monoamine metabolism." *American Journal of Human Genetics* vol. 52, no. 6: 1032-9.

Crockett, Molly J, Clark, Luke, Hauser, Marc D., et al. 2010. "Serotonin selectively influences moral judgment and behavior through effects on harm aversion." *Proceedings of the National Academy of Sciences of the United States of America* 107, no. 40: 17433-8.

Davis, John K. 2018. *New Methuselahs: The Ethics of Life Extension.* Massachusetts Institute of Technology.

Engelen, Bart. "Ethical Criteria for Health-Promoting Nudges: A Case-by-Case Analysis." *The American Journal of Bioethics.* 19(5), 2019. 48-59

Ferrando, Francesca. *Philosophical Posthumanism*. 2019. Bloomsbury.

Foucault, Michel. 2001. *Power.* Edited by James D. Faubion. The New Press.

Habermas, Jurgen. 2003. *The Future of Human Nature. Polity Press.*

Hall, Melinda. 2013. "Vile Sovereigns in Bioethical Debate." *Disability Studies Quarterly*, 33, no. 4, 1-23.

Hall, Melinda C. 2017. *The Bioethics of Enhancement: Transhumanism, Disability, and Biopolitics*. Lexington Books.

Harrison, Joseph D. and Patel, Mitesh S. 2020. "Designing Nudges for Success in Health Care" *AMA Journal of Ethics.* 22, no. 9: 796-801.

Ioannidis, John PA. 2005. "Why most published research findings are false." *PLoS Med* 2, no. 8: e124.

John, Stephen. 2020. "Artificial Ignorance, Epistemic Paternalism, and Epistemic Obligations." *Epistemic Paternalism: Conceptions, Justifications and Implications*. Amiel Bernal and Guy Axtell, eds. 17-27.

Jonas, Hans. 1984. *The Imperative of Responsibility.* University of Chicago Press.

Knutson, Brian, Wolkowitz, Owen M., Cole, Steve W. 1998. "Selective Alteration of Personality and Social Behavior by Serotonergic Intervention" T*he American journal of psychiatry* vol. 155, no. 3: 373-9.

Kukla, Quill. 2014. "Living with Pirates: Common Morality and Embodied Practice" *Cambridge quarterly of healthcare ethics: CQ: the international journal of healthcare ethics committees* 23, no. 1: 75-85.

Lander, Eric, Baylis, Francoise, Zhang, Feng, et. al, 2019. "Adopt a moratorium on heritable genome editing." *Nature.*

Levy, Neil, Douglas, Thomas, Kahane, Guy, et. al.,2014. "Are you Morally Modified? The Moral Effects of Widely Used Pharmaceuticals." *Philos Psychiatr Psychol.* 21, no. 2: 111-125.

Lexchin, Joel, et al., 2003. "Pharmaceutical industry sponsorship and research outcome and quality: systematic review." *BMJ (Clinical research ed.)* 326, no. 7400: 1167-70.

MacIntyre, Alasdair. 1999. *Dependent Rational Animals.* Carus Publishing Company.

McCrudden, Christopher, and King, Jeff. 2017. "The Dark Side of Nudging: The Ethics, Political Economy, and Law of Libertarian Paternalism." In *Choice Architecture in Democracies, Exploring the Legitimacy of Nudging,* edited by Alexandra Kemmerer, Christoph Mollers, Maximilian Steinbeis and Gerhard Wagner. 75-140.

Meehan, Danielle. 2020. "Epistemic Vices and Epistemic Nudges: A Solution?" *Epistemic Paternalism: Conceptions, Justifications and Implications.* Amiel Bernal and Guy Axtell, eds. 249-261.

M'hamdi, Hafez Ismaili, Hilhorst, Medard, Steegers, Eric A P. et. al., 2017. "Nudge me, help my baby: on other-regarding nudges" *Journal of Medical Ethics*. 43, 702-706.

Nietzsche, Friedrich. 2006. *Thus Spoke Zarathustra*. Cambridge University Press.

Patel, Mitesh S, Day, Susan C., et. al. 2016. "Generic Medication Prescription Rates After Health System-wide redesign of default options within the electronic health record." *JAMA Internal Med.* 176, no. 6 847-848.

Patel, Mitesh, Volpp, Kevin G., Small, Dylan S., et. al.,2017. "Using Active Choice Within the Electronic Health Record to Increase Influenza Vaccination Rates." *Journal of General Internal Medicine.* 32, no. 7:. 790-795.

Patel, Mitesh, Volpp, Kevin G., Asch, David A. 2018. "Nudge Units to Improve the Delivery of Health Care." *N Engl J Med.* 378(3):. 214-216.

Pellegrino, Edmund. 2001. "Philosophy of Medicine: Should it be Telelogically or Socially Constructed?" *Kennedy Institute of Ethics Journal.* 11(2). 2001, 169-180.

Persson, Ingmar and Savulescu, Julian. 2010. "Moral Transhumanism." *Journal of Medicine and Philosophy*, 35: 656-669.

Prainsack, Barbara. 2020. "The value of healthcare data: to nudge, or not?," *Policy Studies* 41, no. 5, https://doi.org/10.1080/01442872.2020.1723517.

Rooney, Phyllis. 1992. "On Values in Science: Is the Epistemic/Non-Epistemic Distinction Useful" *PSA: Proceedings of the Biennial Meeting of the Philosophy of Science Association* 1992: 13–22.

Sandberg, Anders. 2013. *The Transhumanist Reader.* Max More and Natasha Vita-More, eds. Wiley-Blackwell. 56-64.

Savulescu, Julian. 2001. "Why Genetic Testing for genes for criminally is morally required." *Princeton Journal of Bioethics*. 4: 79-97.

Savulescu, Julian. "Procreative Beneficence: Why We Should Select The Best Children", *Bioethics.* 15 (5/6) 2001. 413-426

Stegenga, Jacob. 2018. *Care & Cure: An Introduction to Philosophy of Medicine.* University of Chicago Press.

Stegenga, Jacob. 2018. *Medical Nihilism.* Oxford University Press.

Stegenga, Jacob. 2015. "Effectiveness of medical interventions." *Studies in history and philosophy of biological and biomedical sciences,* 54: 34-44.

Sunstein, Cass R. 2014. *Why Nudge? The Politics of Libertarian Paternalism.* Yale University Press.

Terbeck, Sylvia, Kahane, Guy, McTavish, Sarah, et. al. 2012. “Propranolol reduces implicit negative racial bias” *Psychopharmacology.* 222: 419-424.

Thaler, Richard H., Sunstein, Cass R. 2021. *Nudge: The Final Edition.* Penguin Books.

Tse, Wai S., Bond, Alyson J. 2002. “Serotonergic intervention affects both social dominance and affiliative behaviour”. *Psychopharmacology.* 161: 324-330.

VanDeVeer, Donald. 1986. *Paternalistic Interventions: The Moral Bounds of Benevolence.* Princeton University Press.

Zola, Irving Kenneth. 1976. “Medicine as an institution of social control” *Ekistics.* 41, no. 245: 210-214.

CHAPTER FOURTEEN

RESILIENCE IN EMBRYO INTERVENTION TO RESOLVE SERIOUS GENETIC DISEASE

BRIANNE HELFRICH

Introduction

Currently, in developed countries, one-third of all pediatric hospitalizations and a large cause of childhood deaths are due to genetically determined disorders.[1] In recent years, the field of gene and cell therapy has approved its first therapies following decades of effort.[2] Human gene therapy has quickly become one of the newest and most ethically concerning advances in our modern world, as it serves to bridge the gap between science and medicine.[3] Gene therapy is a treatment of disease. It transfers genetic material, either DNA or RNA, into the cells of the patients.[4] These therapies treat diverse clinical indications and tissue targets, such as neuromuscular disease, cancer, and inherited blindness. While the potential of these therapies' successes to be extrapolated to various conditions and patient populations is exciting, new advancements in technology are dramatically expanding the impact of these medicines on treating human disease. Although these approved therapies are life-changing for the affected patients, they provide an even broader impact on society. These therapies lay the foundation on which treatments for many other conditions can be developed.[5] Within the concept of gene therapy, many discussions circulate surrounding ethical considerations concerning somatic versus germline treatments. Despite raising ethical, legal, and social concerns, somatic gene therapy is suggested to be less challenging to the current existing ethical and legal frameworks set in place. In contrast, however, the notion of altering the germline or embryo such that alterations are passed on to future generations raises a substantial amount of ethical and social concerns. Currently, it is debated as to whether it should even be allowed within the context of basic research. At the moment, germline and embryo gene therapy are not being conducted for clinical purposes and are prohibited in several jurisdictions.[6] Ex vivo therapy is widely considered. It is a gene therapy technique where genetic manipulation of the cell is done outside of the body, increasing its level of safety.[7] The Human Genome Project and other related advances in molecular biology promise progress for the long-term curative treatment of many severe genetic disorders. However, the current approach for controlling these disorders remains prevention. This includes the accepted application of preimplantation genetic testing.[8] Regardless of the great advancements made, the field of gene therapy has been marked by mistakes, failures, and ethical scandals, which, in turn, has led to intense periods of public scrutiny. Gene therapy has been resilient in offering hope for the reduction of human suffering. This remains appealing to researchers, health providers, and the public, despite the evident ethical and technical issues surrounding the concept.[9]

Since the 1950s, ethical debates have surrounded what is now coined as human gene editing.[10] Originally in a 1999 United States clinical trial and a European study a few years later, gene therapy was the culprit of a very small number of deaths. However, since then, a series of successful clinical trials have taken place over the past few decades. This has created a sufficient amount of confidence in the advancement and forward movement of gene therapy treatments.[11] In developed countries, roughly 3.5% of live births suffer from some form of birth defect. The demand for new therapeutic treatments has been heard, as the social and economic impacts of these disorders are extremely important. More importantly, the emotional and physical distress these disorders cause for young patients and their families creates a larger desire for our society to adopt gene therapy interventions. The incredible advancements in fundamental research have resulted in the identification of genes and factors involved in genetic diseases. They have created a better understanding of the mechanisms responsible for pathological manifestations of diseases.[12] For many young pediatric patients who are affected by rare genetic diseases, such as spinal muscular atrophy, gene therapies bring about a sense of resiliency to the patient and their families. It provides an opportunity for the child to live a relatively "normal" life.However, there are still fears regarding genetic intervention efficacy and the potential risks associated.[13]

Ex Vivo Gene Therapy

The first gene therapy trials entailed the transplantation of autologous cells whose underlying genetic defect was corrected using an ex vivo gene delivery method. In ex vivo therapy, the patient's cells are collected, cultured, modified, and transplanted back to the patient.[14] This method is preferable due to the fact it does not require a histocompatibility donor and avoids graft-versus-host disease.[15] Ex vivo gene therapies are based on their ability to isolate progenitor, stem, or differentiated cells from a patient or healthy donor and expand them. This method of therapy can be done with or without modification, then implanted within the patient to establish a transient or stable graft of the infused cells and their progeny. Ex vivo gene therapy may be used for several purposes. This intervention can replace genes in genetic diseases, destroy malignant cells in neoplastic diseases and modify patient immune responses. It can also provide trophic molecules and immunize against infectious diseases.[16] Compared to in vivo therapies, ex vivo has a higher efficacy and safety rate. There is no exposure of the patient to the gene transfer vector and the target cells of transduction can be more effectively selected. The main disadvantage of ex vivo gene therapy is the possibility that the ex vivo culture may decrease the reconstitution potential of the infused cells, as well as, the risk of malignancy triggered by vector insertion. However, one could argue that the risks of ex vivo therapy are greatly undermined by the benefits of said therapy. The benefit of eradicating life-threatening diseases outweighs the risks associated with this treatment. Society's most current knowledge regarding ex vivo therapies stem from information obtained from hematopoietic cells, either lymphocytes or stem and progenitor cells. As society advances and our technology aids in discovery, the knowledge gained from these cells will eventually expand into and instruct emerging studies.[17]

Preimplantation Genetic Testing

Preimplantation genetic testing is the practice of obtaining a cellular biopsy sample from a developing embryo or oocyte which was acquired from a cycle of in vitro fertilization. This practice evaluates the genetic composition of the sample and uses this information to determine which embryo is optimal for transfer into the maternal uterus.[18] The genetic testing of preimplantation embryos is most used for preimplantation genetic diagnosis.[19] Preimplantation genetic diagnosis (PGD) was developed as an alternative form of prenatal diagnosis. It is carried out on embryos and has most frequently been employed in assisted reproduction. PGD detects chromosome aneuploidy or structural chromosome rearrangements. However, despite misconceptions, PGD is not a therapeutic procedure for embryos as there are no changes to the DNA or any other genetic structures. It is solely a diagnostic procedure that can identify whether a specific embryo carries a single gene disorder. The information provided by the PGD is then used by the couple and their healthcare provider to make decisions on which embryo is best suited for implantation.[20] The amount of extra information that PGD yields will inevitably prove to be beneficial in terms of selecting embryos for healthy growth, however, will raise various ethical concerns.[21] It is argued that many who object to abortion, also object to preimplantation genetic diagnosis. Although PGD does not destroy a fetus and no embryos are harmed, many argue that embryos are created that will never be used. Those who view human life as beginning at fertilization, view these two concepts in conjunction with each other.[22] However, PGD allows for a lower possibility of parents passing on specific genetic diseases to their children conceived by IVF, which overall can improve the quality of life.

The Importance of Somatic Cell Therapy Versus Germ Line Therapy

Somatic cells are those present in the tissues of the body with the exclusion of sperm, egg cells, and their precursors. Somatic cell therapy involves the removal of specific cells, modifying them, and replacing them within the patient.[23] This therapy can cure certain diseases; however, it will not prevent the transmission of disease from parent to offspring. The use of somatic cells within the genetic intervention is more ethically accepted by society. As a result, over two dozen gene therapies have been approved for clinical use by the drug regulatory agencies of various countries. More are in the process of clinical trials.[24] Somatic cell therapies have been tested and implemented for many more years than germline therapy.[25] Although editing of germline cells has been achieved in animals, germline therapy in humans poses many more risks. There are many major technical challenges to be addressed in developing a therapy safe and effective for humans.[26] Germ line therapy is applied to gametes or preimplantation embryos which holds the promise of potentially preventing genetic disease to be passed not only to the patient's offspring but to future generations. A common concern surrounds the topic of mosaicism which is the presence of unrepaired and repaired cells within the same embryo. To achieve the benefit of preventing the passing of disease to the offspring, the offspring must take the risk of receiving undesirable mutations and mosaicism. When looking at this risk-benefit ratio, it is deemed rather low for benefit as there are several potential alternative therapies available.[27] Germ line therapy is highly contentious due to the fact the resulting genetic changes could be inherited by future generations. Therefore, the technology would cross a line many view as ethically untouchable.[28] More serious objections to germline therapy stem from the unacceptability of using a person solely as a vehicle for creating genetic change in their offspring. It is argued that it is morally unacceptable to use the promise of future benefits to experiment on embryos or fetuses. Especially, when in today's society, more effective technologies exist to aid parents in producing healthy children.[29] The possibility of making germline heritable changes moves the conversation away from individual-level concerns. This now becomes a more significant and complex technical, religious, and social concern. The appropriateness of the degree of intervention within nature and the potential effects of such changes on the acceptance of children born with disabilities become ethical concerns due to germline therapy.[30]

Ethical Considerations for Ex Vivo Intervention

Genetic intervention has been a source of ethical conversation for many decades. There are major concerns in the application of gene therapy as a form of human medicine. However, a core set of agreed-upon ethical principles have been established for governing gene therapies. Although these principles were slow to emerge, in comparison to the speed of technical and scientific developments, they provide guidelines for the conduction of gene therapy.[31] Despite a widespread agreement that it would be ethical to use somatic cell gene therapy to correct grave diseases, the public remains with an air of uneasiness surrounding this topic. It is observed that the basis for this concern lies less with the procedure's clinical risks, but with the fear that human nature will change as a result of genetic intervention. The potential for misuse of gene therapy technology is a legitimate concern. This, therefore, justifies a moral line to be drawn regarding the allowance and terms of gene therapy intervention.[32] In recent years, there is a renewed interest in biomedical ethics to inform the public surrounding the ethical quandaries of gene therapy. This helps researchers make informed ethical decisions through education, as it provides a framework for broader societal discussions of the risks, benefits, safety, and equity of genetic intervention.[33]

Safety and Efficacy

The concept of safety surrounding gene therapy has attracted a lot of attention since the very beginning. Many are worried about the possibility that gene therapy might fail to provide benefits or cause some unexpected side effects.[34] However, novel therapy involves a calculation that the expected benefits outweigh the possible risks. Since such a calculation can never be precisely quantified, certain ethical decisions must be made at various levels to maintain a high level of safety.[35] When reviewing clinical protocols for human gene therapy, safety should be the primary issue evaluated.[36] For proposed clinical trials regarding somatic gene therapy, Institutional Review Boards must analyze the protocol and written consent. For the protection of human subjects to be a top priority, human trials must be preceded by appropriate animal trials. The experiment must be designed based on the results of animal experimentation. There should also be a knowledge of the natural history of the disease under study, causing the anticipated results to justify the performance of the experiment.[37] Ex vivo gene therapy is a technique used where genetic manipulation of cells is done outside of the body.[38] It is argued that ex vivo genetic intervention is the more safe method due to the success of the editing can be verified before the cells are reimplanted into the patient.[39] Ex vivo targets a specific organ rather than treating a whole organism, causing a more precise and accurate therapy.[40] Safety and efficacy are of utmost importance. Hence, due to a lack of knowledge, germline therapy is not a standard practice. In the context of ongoing debates surrounding genetic intervention, clinical practitioners and researchers must conduct careful scientific research and disseminate the results to build a solid evidence base.[41]

Informed Consent and Patient Autonomy

In modern medical ethics, patient autonomy is considered one of the core principles in decision-making about an individual's health.[42] This principle is applied to medical therapeutics and medical research.[43] The informed consent process regarding human studies educates and shares the purpose of said gene therapy trial. This process should be collaborative between the potential research subjects as well as the researcher. The risks, benefits, and what is involved should be discussed. This process needs to provide enough information regarding the trial so the participant can assert their autonomy regarding their contribution or not.[44] To obtain proper informed consent, the researcher must be transparent. They must abide by the principles of honesty and open communication, to share information in ways that are accessible and understandable to stakeholders. To adhere to these principles, one must make a commitment to fully disclose information possible in a timely manner. Also, one must encourage meaningful public input into the policymaking process related to human gene therapy and other related novel technologies. Obtaining a high level of respect for patient autonomy stems from the ideology that one must have respect for persons. This requires recognition of the personal dignity of all individuals, respect towards individual decisions, and the acknowledgment of the centrality of personal choice. Within the context of gene therapy, it must be recognized that all people have equal moral value, regardless of their genetic qualities. Those involved must adhere to the responsibility of committing to the prevention of recurring abusive forms of eugenics that were practiced in the past and destigmatizing disability.[45]

Diverse Gene Therapy Regulations Among Varying Geographical Locations

In 2007, over 1340 gene therapy clinical trials have been completed, are ongoing, or have been approved for practice worldwide.[46] In recent years, there has been a rapid expansion of the development of gene therapy and its treatment possibilities. Gene therapies created a sense of hope for those suffering from a broad range of genetic diseases worldwide. These therapies aim to treat the underlying causes of the disease and not solely relieve the symptoms.[47] Although gene therapies can create great benefits for society, the unique nature of these therapies poses challenges regarding regulation within traditional frameworks. While some countries around the world have established frameworks for advanced gene therapies, all countries may experience differences between them.[48] The United States has accepted the use of somatic gene therapy to help fight and resolve genetic diseases affecting many. The use of gene therapy within Europe has become a more accepted and common practice. The United Kingdom has increasing numbers of clinical trials regarding somatic gene therapy with many therapies now being placed in clinical settings.

Regulation of Gene Therapy in the United States

Currently, in the United States, only somatic gene therapy is permitted. The regulation of gene therapy interventions is conducted by the US Food and Drug Administration (FDA) and the National Institute of Health (NIH).[49] The United States Food and Drug Administration regulates and oversees a wide range of products entering the United States market, including gene and cell therapies. The regulatory approach of genetic intervention therapies is similar to that of other medical products and requires a multitiered framework of statutes, regulations, and guidance documents.[50] The FDA creates a strong emphasis on safety and the control measures associated with gene therapy. Their approval is required for all gene therapy protocols to be placed into practice. The NIH focuses heavily on the regulation of gene therapy research and trials. The Recombinant DNA Advisory Committee (RAC), a subcommittee of the NIH, reviews the gene therapy protocols. This committee is composed of professionals in the science, medical, and law fields.[51] When conducting gene therapy research using human subjects, the United States possesses a special process for review and approval. All protocols that seek to be carried out in institutions receiving federal funding, must submit to both the FDA and the NIH Office of Biotechnology Activities. Currently, the FDA has jurisdiction over human cells, tissues, and tissue-based products intended for human use. The Office of Biotechnology Activities submission allows for a determination as to whether the protocol requires a public review by the Recombinant DNA Advisory Committee (RAC). In the United States, a review by the RAC is carried out if there are novel characteristics of the protocol that are deserving of public discussion.[52] As the United States solely allows somatic gene cell therapy, the need for government regulation has been limited. However, as technology and society advance, government regulation will become increasingly important as new developments arise for germline therapy. Although germline therapy currently poses risks due to a lack of knowledge and intervention methods, it could provide potential benefits for the eradication of future diseases. Future generations will be forever affected by current decisions; therefore, one must proceed cautiously.[53]

The Acceptance of Gene Therapy Treatment in Europe

Europe, more specifically the United Kingdom, has become recognized as a world leader in the discovery and advancement of gene therapies.[54] Clinical trials within the United Kingdom, including somatic gene therapies, are increasing in number and branching out from specialist academic measures toward societal experimental medicine.[55] This country has begun to develop the capacity to operate at a level required for manufacturing and provide these gene therapies to patients in need.[56] Gene therapy within the United Kingdom encompasses a broad range of studies with antisense and small inhibitor RNA, as well as DNA and viruses.[57] During the time between 1993 and 2012, the Gene Therapy Advisory Committee established a working relationship with scientists and clinicians in the United Kingdom. This relationship was established to build a regulatory infrastructure to support the early development of genetic intervention therapy. For the Gene Therapy Advisory Committee, it was important to receive collaboration from members distinguished with knowledge of scientific detail, medicine regulation, and an extensive background surrounding public healthcare ethics. This was proven to be extremely beneficial as it provided scientific evidence to inform ethical decisions in the emerging field of somatic gene therapy. This committee ensures that scientific progress is being made by approving only trials which are designed to answer meaningful and impactful clinical questions. In gene therapy trials, scrutiny of the procedures is deemed appropriate as there is a large public concern regarding developments involving genetic manipulation of human subjects.[58] To ensure patient safety and ethical standards within clinical trials of gene therapy throughout Europe, the 2001 EU directive entitled "Good Clinical Practice 2001/EC" regulates and reviews the risk-benefit analysis of the study. In this directive, the review period for "standard" medicinal products is roughly 60 days. However, for clinical gene therapy intervention, the directive allows an additional 30 days for review. This additional review was set in place due to the infancy of gene therapy at the beginning and the lack of knowledge surrounding the therapy hindered a proper assessment of the risks and benefits.[59] Patient benefit is of the utmost importance, therefore in efforts to improve safety and treatment efficacy, it is critical and an ethical obligation for novel therapies to be accessible in a timely manner.[60]

The Advantages of Somatic Genetic Intervention

Some of the illnesses and diseases found throughout our society today are direct reflections of our genetics. Many severe genetic disorders that currently have no therapeutic options could potentially be treated by gene therapy. Gene therapy uses genetic information to correct a mutation and provide new functions for the infected human cells. Diseases such as severe combined immunodeficiency and spinal muscular atrophy affect the lives of infants and before gene therapy resulted in many deaths. For the past two decades, gene therapy has strived to fix genetic defects presented in diseased cells of the body and create a large benefit to society.[61] To effectively maximize the benefits of gene therapy, it requires low toxicity, sustained therapeutic gene expression in target cells, and a high profile of safety.[62]

Severe Combined Immune Deficiencies (SCID) Immune System Reconstruction

Human Severe Combined Immunodeficiency is a prenatal disorder that greatly affects the development of T lymphocytes. Over the last few years, more knowledge has been collected regarding the genetic defects causing SCID and the characterization of the molecular and immunological mechanisms have developed.[63] The estimated frequency of SCIDS is between 1:50,000 and 1:100,000 live births. However, this may be a slight underestimate due to the possibility of early death preventing a correct diagnosis.[64] SCID is most commonly overserved and diagnosed in the first few months of life, around months 4-7. Although diagnosis is sometimes not presented at birth, human SCID is already present even if clinically silent in most affected newborns. The knowledge of the genetic basis of SCID is essential for the diagnosis and treatment of the disease.[65] Studies of the patterns of inheritance, immunological characteristics, and genotypes have led to the identification of 11 distinct SCID conditions. The clinical presentation of different SCID conditions is characterized by the early onset of infections, typically presenting in the gut and respiratory tract. SCID is considered a medical emergency that if not treated, leads to death within the first year of life, as its clinical symptoms are so severe.[66] ADA-SCID, a rare inherited genetic disorder, was one of the earliest targets for gene therapy.[67] Adenosine deaminase (ADA) deficiency leads to an increased number of toxic purine degradation by-products which affects lymphocytes, causing adenosine deaminase-deficient severe combined immunodeficiency (ADA-SCID).[68] 25 years after the first gene therapy attempt in humans, gene therapy for ADA-SCID received approval. Due to the present expression of the ADA enzyme under normal conditions and the rate of survival of ADA-expressing cells, ADA-SCID was considered an ideal candidate for somatic gene therapy.[69] In May 2016, Strimvelis was the first ex vivo gene therapy approved for clinical practice by the European Medicine Agency. Strimvelis contains retroviral transduced autologous CD34+ cells which express ADA.[70] This gene therapy is used for the treatment of patients who have no suitable matched-related stem cell donor available. Due to the advancements in medical research and technology, the European Society for Immunodeficiencies and the European Society for Blood and Marrow Transplantation have updated their joint guidelines to recommend gene therapy as the first treatment option for those infected with ADA-SCID and no acceptable stem cell donor. The approval of gene therapy for ADA-SCID represents a public healthcare benefit and steps forward in the management of this rare genetic disease affecting infants.[71]

Spinal Muscular Atrophy

Spinal Muscular Atrophy (SMA) is an autosomal recessive disorder.[72] It is an inherited genetic disease characterized by the degeneration of alpha motor neurons in the spinal cord, as the SMN1 gene is deleted or mutated.[73] Due to this, those with this neuromuscular disorder experience progressive muscle weakness, resulting in paralysis.[74] Before treatment developments, it was the most common genetic cause of child mortality and had an incidence of 1 in 10,000 live births.[75] Recently, disease-modifying therapies have been introduced into clinical practice to alleviate the suffering of these patients.[76] Onasemnogene Abeparvovec, which is marketed as Zolgensma®, is a gene therapy medication that the FDA approved in May 2019 as a sufficient treatment for spinal muscular atrophy.[77] Zolgensma® was developed by the Swiss drug maker Novartis.[78] It uses the adeno-associated virus 9 (AAV9) vector to deliver the SMN1 gene to the infected patients. During the phase 1 and phase 3 clinical trials of this drug, it showed that a single administration resulted in the improvement of motor functions for the vast majority of infants with spinal muscular atrophy. Currently, phase 3 clinical trials are being conducted for infants diagnosed with SMA who are pre-symptomatic.[79] As mentioned, in the United States, Zolgensma® was approved for medical practice in children under the age of two who are diagnosed with SMA through genetic testing in 2019, and approved in Japan and the European Union in 2020.[80] The approval of this gene therapy has provided society with real-life data that can develop and modify future gene therapies.[81]

Conclusion

Gene therapy is an emerging treatment option for a small but growing number of diseases within our society. Although the concept faced scientific and ethical uncertainty at the beginning, its resiliency and foundation of approach have been proven to provide benefits.[82] For many years, gene therapy has been governed by ethical and moral norms. It has been subject to regulatory oversight; however, this experience offers new guidance to better-suited norms and mechanisms for gene therapy of somatic cells.[83] Gene therapy is understood as the ability to improve genetic disease through the correction of altered genes. This therapy became possible with the advances in genetics and bioengineering.

Currently, viral and non-viral methods of gene therapy are used either in vivo or in vitro/ex vivo.[84] However, only somatic gene therapy is being conducted due to the risks associated with inherited genetic changes. As our understanding of genes grows, so does the ability to manipulate and modify them.[85] Although several protocols have been successful, gene therapy remains a complex concept and many techniques will need new developments in the future.[86] With more than 1700 approved clinical trials worldwide, society has not only gained new information and knowledge regarding gene therapy but has learned to understand the concern that surrounds it.[87]

Somatic gene therapy is gaining more and more recognition worldwide as countries begin to adopt these therapeutic practice standards. Gene therapy has been proven to aid those suffering from genetic diseases and is deemed effective as a treatment option. Those pediatric patients suffering from severe combined immunodeficiency and spinal muscular atrophy are being treated with gene therapies that intervene in the underlying cause. Various regulations are currently in place for varying geographical locations. As society continues to evolve and alter our ethical norms of medical practice, gene therapies have the potential to become revolutionary within the health science field. Although there are only a limited number of gene therapies that are currently being used, the number of clinical trials continues to increase.[88]

Notes

[1] E. Kanavakis and J. Traeger-Synodinos, "Preimplantation Genetic Diagnosis in Clinical Practice," *Journal of Medical Genetics* 39, no.1 (January 1, 2002): 6-11.
[2] Karen Bulaklak and Charles A. Gersbach, "The Once and Future Gene Therapy," Nature Communications 11, no. 1 (November 16, 2020): 5820. https://doi.org/10.1038/s41467-020-19505-2.
[3] "Gene Therapy Regulation: A Proactive Approach," Accessed March 9, 2022. https://www.ndsu.edu/pubweb/~mcclean/plsc431/students/ross. htm.
[4] Ran Tang and Zhigang Xu, "Gene Therapy: A Double-Edged Sword with Great Powers," *Molecular and Cellular Biochemistry* 474, no. 1–2 (November 2020): 73–81. https://doi.org/10.1007/s11010-020-03834-3.
[5] Bulaklak and Gersbach.
[6] Heidi C. Howard, Carla G. van El, Francesca Forzano, Dragica Radojkovic, Emmanuelle Rial-Sebbag, Guido de Wert, Pascal Borry, and Martina C. Cornel, "One Small Edit for Humans, One Giant Edit for Humankind? Points and Questions to Consider for a Responsible Way Forward for Gene Editing in Humans," *European Journal of Human Genetics* 26, no. 1 (January 2018): 1–11. https://doi.org/ 10.1038/s41431-017-0024-z.

[7] Kevin Gregory-Evans, A. M. A. Emran Bashar, and Malcolm Tan, "Ex Vivo Gene Therapy and Vision," *Current Gene Therapy* 12, no. 2 (April 1, 2012): 103–15. https://doi.org/10.2174/156652312800099 607.
[8] Kanavakis and Traeger-Synodinos.
[9] Kateryna Kratzer, Landon J. Getz, Thibaut Peterlini, Jean-Yves Masson, and Graham Dellaire, "Addressing the Dark Matter of Gene Therapy: Technical and Ethical Barriers to Clinical Application," *Human Genetics*, April 8, 2021. https://doi.org/10.1007/s00439-021-02272-5.
[10] John H. Evans, "Setting Ethical Limits on Human Gene Editing after the Fall of the Somatic/Germline Barrier," *Proceedings of the National Academy of Sciences* 118, no. 22 (June 2021): e2004837117. https://doi.org/10.1073/pnas.2004837117.
[11] "Gene Therapy Regulation: A Proactive Approach."
[12] Marie-Claude Senut, and Fred H. Gage, "Prenatal Gene Therapy: Can the Technical Hurdles Be Overcome?" *Molecular Medicine Today* 5, no. 4 (April 1, 1999): 152–56. https://doi.org/10.1016/S1357-4310(99)01441 -0.
[13] Eric Bender, "Regulating the Gene-Therapy Revolution," *Nature* 564, no. 7735 (December 12, 2018): S20–22. https://doi.org/10.1038/d41586-018-07641-1.
[14] Brian R. Davis and Nicole L. Prokopishyn, "Chapter 65 - Stem Cell Gene Therapy." In *Essentials of Stem Cell Biology* (Second Edition), edited by Robert Lanza, John Gearhart, Brigid Hogan, Douglas Melton, Roger Pedersen, E. Donnall Thomas, James Thomson, and Ian Wilmut, 591–97. San Diego: Academic Press, 2009. https://doi.org/10.1016/ B978-0-12-374729-7.00065-2.
[15] Tang and Xu.
[16] G. Karpati and H. Lochmüller, "The Scope of Gene Therapy in Humans: Scientific, Safety and Ethical Considerations," *Neuromuscular Disorders*: NMD 7, no. 5 (July 1997): 273–76. https://doi.org/10. 1016/s0960-8966(97)00051-5.
[17] Luigi Naldini, "Ex Vivo Gene Transfer and Correction for Cell-Based Therapies," *Nature Reviews Genetics* 12, no. 5 (May 2011): 301–15. https://doi.org/10.1038/nrg2985.
[18] P. R. Brezina, D. S. Brezina, and W. G. Kearns, "Preimplantation Genetic Testing," *BMJ* 345, no. sep18 2 (September 18, 2012): e5908–e5908. https://doi.org/10.1136/bmj.e5908.
[19] Kristien Hens, Wybo Dondorp, Alan H. Handyside, Joyce Harper, Ainsley J. Newson, Guido Pennings, Christoph Rehmann-Sutter, and Guido de Wert, "Dynamics and Ethics of Comprehensive Preimplantation Genetic Testing: A Review of the Challenges," *Human Reproduction Update* 19, no. 4 (July 1, 2013): 366–75. https://doi.org/10.1093/humupd/dmt009.
[20] Stern, "Preimplantation Genetic Diagnosis."
[21] Hens et al.
[22] Faith Lagay, "Preimplantation Genetic Diagnosis," *AMA Journal of Ethics* 3, no. 8 (August 1, 2001): 265–68. https://doi.org/10.1001/ virtualmentor.2001.3.8.gnth1-0108.
[23] Committee on Human Gene Editing: Scientific, Medical, and Ethical Considerations et al., *Human Genome Editing*.
[24] Tang and Xu.
[25] "Why Treat Gene Editing Differently in Two Types of Human Cells?"

[26] Committee on Human Gene Editing: Scientific, Medical, and Ethical Considerations et al., *Human Genome Editing.*
[27] Tang and Xu.
[28] Committee on Human Gene Editing: Scientific, Medical, and Ethical Considerations et al., *Human Genome Editing.*
[29] M. Lappé, "Ethical Issues in Manipulating the Human Germ Line," *The Journal of Medicine and Philosophy* 16, no. 6 (December 1991): 621–39. https://doi.org/10.1093/jmp/16.6.621.
[30] Committee on Human Gene Editing: Scientific, Medical, and Ethical Considerations et al., *Human Genome Editing.*
[31] Kratzer et al.
[32] W. French Anderson, "Human Gene Therapy: Why Draw a Line?" *The Journal of Medicine and Philosophy: A Forum for Bioethics and Philosophy of Medicine* 14, no. 6 (December 1, 1989): 681–93. https://doi.org/10.1093/jmp/14.6.681.
[33] Kratzer et al.
[34] Tang and Xu.
[35] Allen R. Dyer, "The Ethics of Human Genetic Intervention: A Postmodern Perspective," *Experimental Neurology* 144, no. 1 (March 1997): 168–72. https://doi.org/10.1006/exnr.1996.6404.
[36] Howard M.Temin, "Safety Considerations in Somatic Gene Therapy of Human Disease with Retrovirus Vectors," *Human Gene Therapy* 1, no. 2 (July 1, 1990): 111–23. https://doi.org/10.1089/hum.1990.1.2-111
[37] Dyer.
[38] Gregory-Evans, Bashar, and Tan.
[39] Committee on Human Gene Editing: Scientific, Medical, and Ethical Considerations et al., *Human Genome Editing.*
[40] Gregory-Evans, Bashar, and Tan.
[41] Howard et al.
[42] Alireza Parsapoor, Mohammad-Bagher Parsapoor, Nima Rezaei, and Fariba Asghari, "Autonomy of Children and Adolescents in Consent to Treatment: Ethical, Jurisprudential and Legal Considerations," *Iranian Journal of Pediatrics* 24, no. 3 (June 2014): 241–48.
[43] Dyer.
[44] "Gene Therapy."
[45] Committee on Human Gene Editing: Scientific, Medical, and Ethical Considerations et al., *Human Genome Editing.*
[46] J. Gene Med, Michael L. Edelstein, Mohammad R. Abedi, and Jo Wixon, "Gene Therapy Clinical Trials Worldwide to 2007 - An Update," 2007.
[47] Eve Hanna, Cécile Rémuzat, Pascal Auquier, and Mondher Toumi, "Gene Therapies Development: Slow Progress and Promising Prospect," *Journal of Market Access & Health Policy* 5, no. 1 (January 3, 2017): 1265293. https://doi.org/10.1080/20016689.2017.1265293.
[48] Daniela Drago, Betsy Foss-Campbell, Keith Wonnacott, David Barrett, and Adora Ndu, "Global Regulatory Progress in Delivering on the Promise of Gene Therapies for Unmet Medical Needs," *Molecular Therapy. Methods & Clinical Development* 21 (April 5, 2021): 524–29.

https://doi.org/10.1016/j.omtm.2021.04.001.
[49] "Gene Therapy Regulation: A Proactive Approach."
[50] Alexander M. Bailey, Judith Arcidiacono, Kimberly A. Benton, Zenobia Taraporewala, and Steve Winitsky, "United States Food and Drug Administration Regulation of Gene and Cell Therapies." In *Regulatory Aspects of Gene Therapy and Cell Therapy Products: A Global Perspective*, edited by Maria Cristina Galli and Mercedes Serabian, 1–29. Advances in Experimental Medicine and Biology. Cham: Springer International Publishing, 2015. https://doi.org/10.1007/978-3-319-18618-4_1.
[51] "Gene Therapy Regulation: A Proactive Approach."
[52] Carson Strong, "Regulatory and Ethical Issues for Phase I In Utero Gene Transfer Studies," *Human Gene Therapy* 22, no. 11 (November 2011): 1323–30. https://doi.org/10.1089/hum.2011.062.
[53] "Gene Therapy Regulation: A Proactive Approach."
[54] UK Bioindustry Association, "A Guide to Cell and Gene Therapy and UK Excellence in the Field."
[55] Emma Morris, Martin Gore, Andrew Baker, and Adrian J Thrasher, "Moving Forward on Shifting Sands: Ethical Regulation of Gene Therapy Clinical Trials in the United Kingdom," *Molecular Therapy* 21, no. 4 (April 2013): 715–16. https://doi.org/10.1038/mt.2013.43
[56] UK Bioindustry Association, "A Guide to Cell and Gene Therapy and UK Excellence in the Field."
[57] Leonard W. Seymour, "The Future of Gene Therapy in the UK," *Trends in Biotechnology* 24, no. 8 (August 1, 2006): 347–49.
https://doi.org/10.1016/j.tibtech.2006.06.002.
[58] Morris et al.
[59] Marcel J.H. Kenter, "Is It Time to Reform Oversight of Clinical Gene Therapy in the EU?" *British Journal of Clinical Pharmacology* 85, no. 1 (January 2019): 8–10. https://doi.org/10.1111/bcp.13795.
[60] Morris et al.
[61] Fulvio Mavilio and Giuliana Ferrari, "Genetic Modification of Somatic Stem Cells," *EMBO Reports* 9, no. Suppl 1 (July 2008): S64–69.
https://doi.org/10.1038/embor.2008.81.
[62] M, B Baradaran Jafarlou, T A Saedi, V Jafarlou, D Shanehbandi, M Maralani, and F Othman, "An Overview of the History, Applications, Advantages, Disadvantages and Prospects of Gene Therapy," *Journal of Biological Regulators and Homeostatic Agents* 30, no. 2 (2016): 315-321.
[63] Fausto Cossu, "Genetics of SCID," *Italian Journal of Pediatrics* 36, no. 1 (November 15, 2010): 76. https://doi.org/10.1186/1824-7288-36-76.
[64] Marina Cavazzana-Calvo and Alain Fischer, "Gene Therapy for Severe Combined Immunodeficiency: Are We There Yet?" *Journal of Clinical Investigation* 117, no. 6 (June 1, 2007): 1456–65.
https://doi.org/10.1172/JCI30953.
[65] Cossu.
[66] Cavazzana-Calvo and Fischer.

[67] Heide Stirnadel-Farrant, Mahesh Kudari, Nadia Garman, Jessica Imrie, Bikramjit Chopra, Stefania Giannelli, Michela Gabaldo, et al., "Gene Therapy in Rare Diseases: The Benefits and Challenges of Developing a Patient-Centric Registry for Strimvelis in ADA-SCID," *Orphanet Journal of Rare Diseases* 13, no. 1 (April 6, 2018): 49. https://doi.org/10.1186/s13023-018-0791-9.
[68] Aisling M. Flinn and Andrew R. Gennery, "Adenosine Deaminase Deficiency: A Review," *Orphanet Journal of Rare Diseases* 13 (April 24, 2018): 65. https://doi.org/10.1186/s13023-018-0807-5.
[69] A Aiuti, MG Roncarolo, L. Naldini, "Gene therapy for ADA-SCID, the first marketing approval of an ex vivo gene therapy in Europe: paving the road for the next generation of advanced therapy medicinal products," *EMBO Mol Med.* 2017;9(6):737-740. doi:10.15252/emmm.201707573
[70] Emily South, Edward Cox, Nick Meader, Nerys Woolacott, and Susan Griffin, "Strimvelis® for Treating Severe Combined Immunodeficiency Caused by Adenosine Deaminase Deficiency: An Evidence Review Group Perspective of a NICE Highly Specialised Technology Evaluation," *PharmacoEconomics - Open* 3, no. 2 (June 2019): 151–61. https://doi.org/10.1007/s41669-018-0102-3.
[71] Stirnadel-Farrant et al.
[72] Hossamaldein Gaber Ali, Khalid Ibrahim, Mahmoud Fawzi Elsaid, Reem Babiker Mohamed, Mahmoud I. A. Abeidah, Azhar Othman Al Rawwas, Khaled Elshafey, et al., "Gene Therapy for Spinal Muscular Atrophy: The Qatari Experience," *Gene Therapy* 28, no. 10 (November 2021): 676–80. https://doi.org/10.1038/s41434-021-00273-7.
[73] Tamara Dangouloff and Laurent Servais, "Clinical Evidence Supporting Early Treatment Of Patients With Spinal Muscular Atrophy: Current Perspectives," *Therapeutics and Clinical Risk Management* 15 (October 2, 2019): 1153–61. https://doi.org/10.2147/TCRM.S172291.
[74] Ali et al.
[75] Dangouloff and Servais.
[76] Katarzyna Kotulska, Aviva Fattal-Valevski, and Jana Haberlova, "Recombinant Adeno-Associated Virus Serotype 9 Gene Therapy in Spinal Muscular Atrophy," *Frontiers in Neurology* 12 (October 1, 2021). https://doi.org/10.3389/fneur.2021.726468.
[77] Ali et al.
[78] Mahajan, "Onasemnogene Abeparvovec for Spinal Muscular Atrophy."
[79] Kotulska, Fattal-Valevski, and Haberlova.
[80] Rajiv Mahajan, "Onasemnogene Abeparvovec for Spinal Muscular Atrophy: The Costlier Drug Ever," *International Journal of Applied and Basic Medical Research* 9, no. 3 (2019): 127–28. https://doi.org/10.4103/ijabmr.IJABMR_190_19.
[81] Kotulska, Fattal-Valevski, and Haberlova.
[82] Esther Landhuis, "The Gene Fix," *Scientific American* 325, no. 5 (November 2021): S3–5.
[83] Committee on Human Gene Editing: Scientific, Medical, and Ethical Considerations et al., *Human Genome Editing*.
[84] Christoph F. Rochlitz, "Gene Therapy of Cancer," n.d.

[85] LeRoy Walters and Julie G. Palmer. *The Ethics of Human Gene Therapy*. Oxford University Press, 1997.
[86] Giulliana Augusta Rangel Gonçalves and Raquel de Melo Alves Paiva, "Gene Therapy: Advances, Challenges and Perspectives," *Einstein* 15, no. 3 (2017): 369–75. https://doi.org/10.1590/S167945082017RB4024.
[87] Thomas Wirth, Nigel Parker, and Seppo Ylä-Herttuala, "History of Gene Therapy," *Gene* 525, no. 2 (August 2013): 162–69. https://doi.org/10.1016/j.gene.2013.03.137.
[88] Hanna et al.

Literature

Aiuti A, Roncarolo MG, Naldini L. 2017. "Gene therapy for ADA-SCID, the first marketing approval of an ex vivo gene therapy in Europe: paving the road for the next generation of advanced therapy medicinal products." *EMBO Mol Med*. 9(6):737-740. doi:10.15252/emmm.201707573

Anderson, W. French. 1989. "Human Gene Therapy: Why Draw a Line?" *The Journal of Medicine and Philosophy: A Forum for Bioethics and Philosophy of Medicine* 14, no. 6 (December 1): 681–93. https://doi.org/10.1093/jmp/14.6.681.

Ali, Hossamaldein Gaber, Khalid Ibrahim, Mahmoud Fawzi Elsaid, Reem Babiker Mohamed, Mahmoud I. A. Abeidah, Azhar Othman Al Rawwas, Khaled Elshafey, et al. 2021. "Gene Therapy for Spinal Muscular Atrophy: The Qatari Experience." *Gene Therapy* 28, no. 10 (November): 676–80. https://doi.org/10.1038/s41434-021-00273-7.

Bailey, Alexander M., Judith Arcidiacono, Kimberly A. Benton, Zenobia Taraporewala, and Steve Winitsky. 2015. "United States Food and Drug Administration Regulation of Gene and Cell Therapies." In *Regulatory Aspects of Gene Therapy and Cell Therapy Products: A Global Perspective*, edited by Maria Cristina Galli and Mercedes Serabian, 1–29. Advances in Experimental Medicine and Biology. Cham: Springer International Publishing. https://doi.org/10.1007/978-3-319-18618-4_1.

Bender, Eric. 2018. "Regulating the Gene-Therapy Revolution." *Nature* 564, no. 7735 (December 12): S20–22. https://doi.org/10.1038/d41586-018-07641-1.

Brezina, P. R., D. S. Brezina, and W. G. Kearns. 2012. "Preimplantation Genetic Testing." *BMJ* 345, no. sep18 2 (September 18): e5908–e5908. https://doi.org/10.1136/bmj.e5908.

Bulaklak, Karen, and Charles A. Gersbach. 2020. "The Once and Future Gene Therapy." *Nature Communications* 11, no. 1 (November 16): 5820. https://doi.org/10.1038/s41467-020-19505-2.

Cavazzana-Calvo, Marina, and Alain Fischer. 2007. "Gene Therapy for Severe Combined Immunodeficiency: Are We There Yet?" *Journal of Clinical Investigation* 117, no. 6 (June 1): 1456–65. https://doi.org/10.1172/JCI30953.

Committee on Human Gene Editing: Scientific, Medical, and Ethical Considerations, National Academy of Sciences, National Academy of Medicine, and National Academies of Sciences, Engineering, and Medicine. 2017. *Human Genome Editing: Science, Ethics, and Governance*. Washington, D.C.: National Academies Press. https://doi.org/ 10.17226/24623.

Cossu, Fausto. 2010. "Genetics of SCID." *Italian Journal of Pediatrics* 36, no. 1 (November 15): 76. https://doi.org/10.1186/1824-7288-36-76.

Dangouloff, Tamara, and Laurent Servais.2019. "Clinical Evidence Supporting Early Treatment Of Patients With Spinal Muscular Atrophy: Current Perspectives." *Therapeutics and Clinical Risk Management* 15 (October 2): 1153–61. https://doi.org/10.2147/TCRM.S172291.

Davis, Brian R., and Nicole L. Prokopishyn. 2009. "Chapter 65 - Stem Cell Gene Therapy." In *Essentials of Stem Cell Biology (Second Edition)*, edited by Robert Lanza, John Gearhart, Brigid Hogan, Douglas Melton, Roger Pedersen, E. Donnall Thomas, James Thomson, and Ian Wilmut, 591–97. San Diego: Academic Press. https://doi.org/10.1016/ B978-0-12-374729-7.00065-2.

Drago, Daniela, Betsy Foss-Campbell, Keith Wonnacott, David Barrett, and Adora Ndu. 2021. "Global Regulatory Progress in Delivering on the Promise of Gene Therapies for Unmet Medical Needs." *Molecular Therapy. Methods & Clinical Development* 21 (April 5): 524–29. https://doi.org/10.1016/j.omtm.2021.04.001.

Dyer, Allen R. 1997. "The Ethics of Human Genetic Intervention: A Postmodern Perspective." *Experimental Neurology* 144, no. 1 (March): 168–72. https://doi.org/10.1006/exnr.1996.6404.

Evans, John H. 2021. "Setting Ethical Limits on Human Gene Editing after the Fall of the Somatic/Germline Barrier." *Proceedings of the National Academy of Sciences* 118, no. 22 (June): e2004837117. https://doi.org/10.1073/pnas.2004837117.

Flinn, Aisling M., and Andrew R. Gennery. 2018. "Adenosine Deaminase Deficiency: A Review." *Orphanet Journal of Rare Diseases* 13 (April 24): 65. https://doi.org/10.1186/s13023-018-0807-5.

"Gene Therapy." N. d. Accessed January 24, 2022. https://eds-p-ebscohost-com.authenticate.library.duq.edu/eds/ebookviewer/ebook/bmxlYmtfXzYwOTE4OF9fQU41?sid=f22a596b-ff96-41d6-9907-ec092311f64f@redis&vid=4&format=EB&rid=1.

"Gene Therapy Regulation: A Proactive Approach." Accessed March 9, 2022. https://www.ndsu.edu/pubweb/~mcclean/plsc431/students/ross.htm.

Gonçalves, Giulliana Augusta Rangel, and Raquel de Melo Alves Paiva. 2017. "Gene Therapy: Advances, Challenges and Perspectives." *Einstein* 15, no. 3: 369–75. https://doi.org/10.1590/S167945082017RB4024.

Gregory-Evans, Kevin, A. M. A. Emran Bashar, and Malcolm Tan. 2012. "Ex Vivo Gene Therapy and Vision." *Current Gene Therapy* 12, no. 2 (April 1): 103–15. https://doi.org/10.2174/156652312800099 607.

Hanna, Eve, Cécile Rémuzat, Pascal Auquier, and Mondher Toumi. 2017. "Gene Therapies Development: Slow Progress and Promising Prospect." *Journal of Market Access & Health Policy* 5, no. 1 (January 3): 1265293. https://doi.org/10.1080/20016689.2017.1265293.

Hens, Kristien, Wybo Dondorp, Alan H. Handyside, Joyce Harper, Ainsley J. Newson, Guido Pennings, Christoph Rehmann-Sutter, and Guido de Wert. 2013. "Dynamics and Ethics of Comprehensive Preimplantation Genetic Testing: A Review of the Challenges." *Human Reproduction Update* 19, no. 4 (July 1): 366–75. https://doi.org/10.1093/humupd/dmt009.

Howard, Heidi C., Carla G. van El, Francesca Forzano, Dragica Radojkovic, Emmanuelle Rial-Sebbag, Guido de Wert, Pascal Borry, and Martina C. Cornel. 2018. "One Small Edit for Humans, One Giant Edit for Humankind? Points and Questions to Consider for a Responsible Way Forward for Gene Editing in Humans." *European Journal of Human Genetics* 26, no. 1 (January): 1–11. https://doi.org/10.1038/s41431-017-0024-z.

Jafarlou, M, B Baradaran, T A Saedi, V Jafarlou, D Shanehbandi, M Maralani, and F Othman. 2016. "An overview of the history, applications, advantages, disadvantages and prospects of gene therapy." Journal of Biological Regulators and Homeostatic Agents 30, no. 2: 315-321.

Kanavakis, E., and J. Traeger-Synodinos. 2002. "Preimplantation Genetic Diagnosis in Clinical Practice." *Journal of Medical Genetics* 39, no. 1 (January 1): 6–11. https://doi.org/10.1136/jmg.39.1.6.

Karpati, G., and H. Lochmüller. 1997. “The Scope of Gene Therapy in Humans: Scientific, Safety and Ethical Considerations.” *Neuromuscular Disorders: NMD* 7, no. 5 (July): 273–76. https://doi.org/10.1016/s0960-8966(97)00051-5.

Kenter, Marcel J.H. 2019. “Is It Time to Reform Oversight of Clinical Gene Therapy in the EU?” *British Journal of Clinical Pharmacology* 85, no. 1 (January): 8–10. https://doi.org/10.1111/bcp.13795.

Kotulska, Katarzyna, Aviva Fattal-Valevski, and Jana Haberlova. 2021. “Recombinant Adeno-Associated Virus Serotype 9 Gene Therapy in Spinal Muscular Atrophy.” *Frontiers in Neurology* 12 (October 1). https://doi.org/10.3389/fneur.2021.726468.

Kratzer, Kateryna, Landon J. Getz, Thibaut Peterlini, Jean-Yves Masson, and Graham Dellaire. 2021. “Addressing the Dark Matter of Gene Therapy: Technical and Ethical Barriers to Clinical Application.” *Human Genetics*, April 8. https://doi.org/10.1007/s00439-021-02272-5.

Lagay, Faith. 2001. “Preimplantation Genetic Diagnosis.” *AMA Journal of Ethics* 3, no. 8 (August 1): 265–68. https://doi.org/10.1001/virtualmentor.2001.3.8.gnth1-0108.

Landhuis, Esther. 2021. “The Gene Fix.” *Scientific American* 325, no. 5 (November): S3–5.

Lappé, M. 1991. “Ethical Issues in Manipulating the Human Germ Line.” *The Journal of Medicine and Philosophy* 16, no. 6 (December): 621–39. https://doi.org/10.1093/jmp/16.6.621.

Mahajan, Rajiv. 2019. “Onasemnogene Abeparvovec for Spinal Muscular Atrophy: The Costlier Drug Ever.” *International Journal of Applied and Basic Medical Research* 9, no. 3: 127–28. https://doi.org/10.4103/ijabmr.IJABMR_190_19.

Mavilio, Fulvio, and Giuliana Ferrari. 2008. “Genetic Modification of Somatic Stem Cells.” *EMBO Reports* 9, no. Suppl 1 (July): S64–69. https://doi.org/10.1038/embor.2008.81.

Med, J. Gene, Michael L. Edelstein, Mohammad R. Abedi, and Jo Wixon.2007. “Gene Therapy Clinical Trials Worldwide to 2007 - An Update.”.

Morris, Emma, Martin Gore, Andrew Baker, and Adrian J Thrasher. 2013. “Moving Forward on Shifting Sands: Ethical Regulation of Gene Therapy Clinical Trials in the United Kingdom.” *Molecular Therapy* 21, no. 4 (April): 715–16. https://doi.org/10.1038/mt.2013.43.

Naldini, Luigi. 2011. “Ex Vivo Gene Transfer and Correction for Cell-Based Therapies.” *Nature Reviews Genetics* 12, no. 5 (May): 301–15. https://doi.org/10.1038/nrg2985.

Parsapoor, Alireza, Mohammad-Bagher Parsapoor, Nima Rezaei, and Fariba Asghari. 2014. "Autonomy of Children and Adolescents in Consent to Treatment: Ethical, Jurisprudential and Legal Considerations." *Iranian Journal of Pediatrics* 24, no. 3 (June): 241–48.

Rochlitz, Christoph F. N.d. "Gene Therapy of Cancer,"

Senut, Marie-Claude, and Fred H. Gage. 1999. "Prenatal Gene Therapy: Can the Technical Hurdles Be Overcome?" *Molecular Medicine Today* 5, no. 4 (April 1): 152–56. https://doi.org/10.1016/S1357-4310(99)01441 -0.

Seymour, Leonard W. 2006. "The Future of Gene Therapy in the UK." *Trends in Biotechnology* 24, no. 8 (August 1): 347–49. https://doi.org/10.1016/j.tibtech.2006.06.002.

South, Emily, Edward Cox, Nick Meader, Nerys Woolacott, and Susan Griffin. 2019. "Strimvelis® for Treating Severe Combined Immunodeficiency Caused by Adenosine Deaminase Deficiency: An Evidence Review Group Perspective of a NICE Highly Specialised Technology Evaluation." *PharmacoEconomics - Open* 3, no. 2 (June): 151–61. https://doi.org/10.1007/s41669-018-0102-3.

Stern, Harvey J. 2014. "Preimplantation Genetic Diagnosis: Prenatal Testing for Embryos Finally Achieving Its Potential." *Journal of Clinical Medicine* 3, no. 1 (March): 280. https://doi.org/10.3390/jcm 3010280.

Stirnadel-Farrant, Heide, Mahesh Kudari, Nadia Garman, Jessica Imrie, Bikramjit Chopra, Stefania Giannelli, Michela Gabaldo, et al. 2018. "Gene Therapy in Rare Diseases: The Benefits and Challenges of Developing a Patient-Centric Registry for Strimvelis in ADA-SCID." *Orphanet Journal of Rare Diseases* 13, no. 1 (April 6): 49. https://doi.org/10.1186/s13023-018-0791-9.

Strong, Carson. 2011. "Regulatory and Ethical Issues for Phase I In Utero Gene Transfer Studies." *Human Gene Therapy* 22, no. 11 (November): 1323–30. https://doi.org/10.1089/hum.2011.062.

Tang, Ran, and Zhigang Xu. 2020. "Gene Therapy: A Double-Edged Sword with Great Powers." *Molecular and Cellular Biochemistry* 474, no. 1–2 (November): 73–81. https://doi.org/10.1007/s11010-020-03834-3.

Temin, Howard M. 1990. "Safety Considerations in Somatic Gene Therapy of Human Disease with Retrovirus Vectors." *Human Gene Therapy* 1, no. 2 (July 1): 111–23. https://doi.org/10.1089/hum.1990.1.2-111

UK Bioindustry Association. 2018. "A Guide to Cell and Gene Therapy and UK Excellence in the Field," October.

Walters, LeRoy and Julie G. Palmer. 1997.*The Ethics of Human Gene Therapy*. Oxford University Press.

"Why Treat Gene Editing Differently in Two Types of Human Cells? | SciTech Connect." 2015. Accessed February 6, 2022. https://scitechconnect.elsevier.com/treat-gene-editing-differently-human-cells/, https://scitechconnect.elsevier.com/treat-gene-editing-differently-human-cells/.

Wirth, Thomas, Nigel Parker, and Seppo Ylä-Herttuala. 2013. "History of Gene Therapy." *Gene* 525, no. 2 (August): 162–69. https://doi.org/10.1016/j.gene.2013.03.137.

CHAPTER FIFTEEN

THE RESILIENCE OF GENETIC TESTING TO PREVENT DISEASE, EUTHANASIA, AND ASSISTED SUICIDE & THE PURSUIT OF SUSTAINABLE HEALTHCARE

SCOTT DYER

Introduction

The scientific breakthroughs that have led to advancements in genetic testing have also forced a newfound grappling of the matter in a normative ethical lens. The use of PGD, or pre-implantation genetic diagnosis, allows scientists to determine whether an embryo has a genetic predisposition to a range of diseases after undergoing in-vitro fertilization (IVF). These diseases range from anything that could be fatal to the fetus, to conditions that may not affect the child until they reach adulthood. PGD is also strongly encouraged for women over the age of 40 years old due to the significantly increased risk of having embryos with a chromosomal abnormality, which can lead to various ailments for the baby or even a miscarriage.[1 2]

If the embryo is found to have the genes for future disease, the embryo can then be discarded and not be implanted in a womb to be brought to term. Further, PGD and artificial reproductive technology (ART) allow for the manipulation of the embryo's genes to intentionally introduce genetic variants that will result in a given condition. For example, it is, while controversial, possible for Deaf parents to choose to give their future child deafness.[3] Other attributes are also possible to choose, such as the sex of the child or even their eye color.[4 5] While such practices are illegal in most countries, PGD and ART are largely unregulated in the United States. Thus, reproductive tourism has sprouted from the quickly growing bud of genetic engineering. Reproductive tourism comes with its own unique set of challenges that will be explored further in this chapter.

Aside from PGD to test for future diseases, an alternative exists in the form of the newborn heel-prick. This is done within 24 hours of birth, so obviously the infant cannot be simply "discarded" as is the case with embryonic testing. Contemporary utilization of genetic testing has resulted in a two-pronged use of the practice—one scientific, one capitalistic. While the sequencing of the human genome was a significant step forward for science, some companies have capitalized off this technology to sell people information about their own genes. Direct-to-consumer (DTC) companies like 23andMe and AncestryDNA allow people to submit a sample of their DNA, usually via saliva, and pay for a synopsis of their genetic origins and, often for a higher price, any disease-predisposition they may have. Genetic testing as a commodity is also largely unregulated, and the companies that have such extensive amounts of genetic information have an ethical duty to properly conduct their business in a way that protects their customers' genetic information and material.

While the bulk of this chapter will focus on the ethics surrounding genetic testing, parallels will be drawn between healthcare and the environment. While preventing future diseases in humans via PGD does reduce the carbon footprint of the healthcare industry, other obstacles remain. For example, excessive utilization of single-use plastics, specific anesthetic gases that are detrimental to the ozone layer, and a lack of renewable energy, all contribute to the greenhouse gas emissions put forth by hospitals and healthcare systems around the world.

Lastly, this chapter will delve into the nature of genetic testing to prevent the need for future use of euthanasia or assisted suicide. By discovering an embryo has a genetic predisposition to future debilitating disease, and then not implanting that embryo and avoiding gestation, human suffering will be avoided, and billions of dollars will be saved.[6] [7] The embryos do not have to be discarded in vain, however. They could be utilized for general scientific research and not, as one might say, "wasted." This will be explored using the normative ethical framework of utilitarianism and deontology, as well as the four principles of bioethics, namely, autonomy, beneficence, nonmaleficence, and justice.

Preimplantation Genetic Diagnosis

The utilization of preimplantation genetic diagnosis (PGD) in an in-vitro fertilization pregnancy was first done in humans in 1989, which yielded a healthy baby in 1990.[8] PGD gained momentum in the late 20th century because it was considered to be less ethically problematic and not as invasive as embryo biopsies.[9] While PGD was initially intended to screen for cystic fibrosis, the procedure has evolved immensely. Now, non-medical manipulation of the embryo, such as selecting the sex of the baby, is also an option. While this practice of sex selection is illegal across all of Europe, it is allowed in the United States, where about nine percent of all PGD procedures are carried out for this reason.[10] This lack of regulation in the U.S. has opened the door to other controversial PGD pursuits, which will be examined later in this chapter.

History and Brief Overview of Embryonic Testing and Manipulation

The origins of PGD can be traced back to 1890 when embryos were transferred between different rabbits by Walter Heape.[11] Tests on animal embryos continued over the next hundred or so years and led to the first IVF pregnancy occurring in Australia in 1973. While this aforementioned IVF itself was successful, the pregnancy ultimately resulted in a miscarriage.[12] Political discourse in the United Kingdom surrounding PGD in the 1980s was largely responsible for the scientific advancements surrounding the procedure. UK Parliament members wanted proof of the efficacy of PGD before passing legislation approving the practice. This prompted a significant increase in research on the matter.[13]

Today, PGD is largely unregulated in the United States. Across the European Union, however, regulation exists to varying degrees. Some countries strictly prohibit the procedure while others allow it with close supervision and oversight. This has resulted in global reproductive tourism, wherein potential parents travel to foreign countries where they can legally carry out PGD that is otherwise criminalized in their home country.[14] This cross-border reproductive care (CBRC) brings with it myriad ethical issues that bioethics in general has yet to fully overcome but will be briefly examined later on in the text.[15]

Ethical Dilemmas Involved in PGD

The number of ethical issues attached to pre-implantation genetic diagnosis is multiplied when considered in a global lens. This is mainly due to the CBRC that came about as a result of differing laws and regulations around the world. American laws have remained mostly untouched in the realm of PGD, giving prospective parents free rein to effectively "design" their baby. The controversial uses of PGD range from choosing the sex or eye color of the baby to purposely giving offspring a gene for deafness, or achondroplasia, which causes dwarfism.[16] [17] Because PGD of this magnitude is not allowed across Europe, something known as "reproductive tourism" now occurs. This is when people travel to a given place outside of their home state or country (typically to the United States) to participate in PGD that would otherwise be illegal for them.[18] The ethical quandaries that come from this are numerous. Once back home, medical personnel could be wary of treating the mother after she underwent an illegal procedure, putting both the mother and the fetus in jeopardy. Additionally, the home country of the parents could hypothetically deny the legal status of parenthood if it was determined that a pregnancy was illegally obtained.[19] Further legal troubles could arise if a country's laws strictly prohibit one from crossing borders to obtain PGD, which could endanger all parties involved.[20]

Other ethical issues stem from parents who pursue PGD to choose the sex of their child. Proponents of non-medical sex selection say that allowing parents to choose the sex of their baby lies in their right as autonomous individuals. Further, it can help with "family balancing," or the desire for specific numbers of males and females within a given family. One of the most significant reasons critics of non-medical sex selection offer against it is the inherent sexism they say is intertwined with the process. Parents who utilize this part of PGD are, they say, putting a grossly extreme value on the gender of their offspring before knowing anything else about them – parents may be motivated by stereotypical ideas of gender, such as a father wanting a son to share in traditionally masculine activities.[21]

Aside from the legal risks and ethical challenges, the use of PGD is expensive. Therefore, due to the lack of socialized medicine in the United States, only wealthy parents would be able to afford the procedure, disproportionally affecting those with lower incomes.[22] Financially speaking, the use of PGD has been estimated to save tens of billions of dollars, due to the avoidance of medical costs for patients who would have been born with a lifelong disease, such as cystic fibrosis.[23] Thus there is a case to be made for government subsidies for those undergoing PGD, especially in the name of autonomy, nonmaleficence, and beneficence. When one or both parents carry a gene for a given disease, I argue that it is even more important that the utilization of PGD be made financially possible for them.

If PGD can prevent the onset of an illness, it could be argued that a government is morally obligated to ensure that its future citizen can be born as healthy as possible. This shows respect for the autonomy of both the mother and the child. A mother who wishes to utilize PGD to ensure a healthy child should not be prevented from doing so. Autonomy is also upheld in the child because, if born healthy, the child can live their life to the fullest without fear of future disease. Meanwhile, using PGD to prevent disease also upholds the principles of beneficence and nonmaleficence by taking preventative measures to ensure someone is healthy before they are born.[24] While there are several ethical dilemmas surrounding PGD, the use of it to *prevent* a child from being born with an otherwise genetically carried disease is not only ethical but could be considered unethical if not carried out. This is especially true if both parents have the same genetic predisposition, making it significantly more likely the disease will be passed on.

Due to the costs of undergoing PGD easily reaching tens of thousands of dollars, providing PGD to everyone is simply not, at the moment, a feasible idea.[25] That said, couples or individuals who are known to carry specific diseases should be able to undergo the procedure for free or extremely discounted rates. Until the procedure is significantly more cost-effective so that anyone can be included in a socialized program, PGD should be made more readily available for those who qualify under predetermined guidelines. Those guidelines should, I argue, be conducive enough so those who carry genes for the most debilitating and expensive diseases qualify. A disease such as Spinal Muscular Atrophy 1 (SMA1) carries annual costs of care around $75,000 for moderate cases and over $600,000 for severe cases and typically results in death within three years of birth.[26] Other forms of the disease, known as SMA2, SMA3, and SMA4, also carry expensive price tags, however, are milder cases and those who suffer from them can still lead relatively normal functioning lives.[27]

Arguably more controversial is the use of PGD to *incur* a given disease, disorder, or condition in the offspring. Parents purposely giving their child dwarfism or deafness is, I argue, inherently unethical. Those who support the choice of, say, deafness using PGD do so for several reasons, including the nature of the underlying culture of the Deaf community. It is understood in Deaf circles as a "minority cultural experience based on shared language," and not fundamentally a disability.[28] While the minority cultural experience may be true, this fails to understand deafness for what it is – lacking an important human function. Thus, purposely giving someone Deafness effectively prevents them from being able to experience the world in a given way, i.e., through sound. Other advocates then say that the onus is on the parents to ensure a proper upbringing of a deaf child that was purposefully given the deaf gene.[29] This fails to consider several possibilities. For example, the parents could tragically die in an accident, leaving the child to be brought up by hearing-enabled people. While this, admittedly, is an extreme example, choosing to make your child unable to hear is also an extreme measure. Additionally, all the above can also be applied to the conversation surrounding the ethics of parents with dwarfism who purposely choose to have children that also suffer from the same condition.

Newborn Testing & The Neonatal Heel-Prick

A much less controversial way of testing for genetically carried diseases or future ailments is the neonatal heel prick. About 24 hours after birth, a medical professional will take a blood sample from the newborn via the baby's heel. The blood is then tested for several diseases, many of which, while rare, could also be quite detrimental to the child.[30] One of the main differences when compared to preimplantation genetic diagnosis is the fact that this test is done once the baby is born, whereas PGD is done before the embryo is even implanted in the mother's womb. Ethical quandaries arise but are not limited to when medical establishments save these blood samples for extended periods. These blood samples, while used for medical research, have been saved without the consent of the parents, sometimes for several decades.[31] [32] The following section will give a brief overview of the neonatal heel prick, as well as an analysis of any ethical implications.

Usage & Efficacy Especially Compared to PGD

The testing of a blood sample from a newborn, also known as The Guthrie Test, began in 1962 after being invented by Dr. Robert Guthrie to screen newborns for phenylketonuria, a congenital birth defect.[33] Since its inception, the heel prick has evolved to now test for nearly a dozen diseases or congenital conditions.[34] Both PGD and the heel-prick test for many of the same ailments.[35] [36] Thus, parents vying for reproduction must carry out preimplantation genetic diagnosis in the hopes of not even allowing a given condition to come to fruition in the first place. This is not easy, given the current financial strains that the high costs of PGD put on prospective parents, so there is necessarily an important conversation to be had in the political realm of bioethical policymaking, i.e., if PGD can prevent diseases from even being brought into the world in the first place, it should be offered for anyone trying to have children via IVF as a matter of national health and individual healthcare.

Ethical Dilemmas in Contemporary Usage

The use of the Guthrie Test is not as controversial as PGD, mainly because it does not include the scrapping of embryos. However, there are still ethical concerns that come from the neonatal heel prick. Firstly, the fact that it objectively hurts the baby is a genuine concern for some, and the withdrawal of the blood can leave temporary bruises or swelling.[37] Studies are being done, however, to test alternative, less painful methods of taking the blood sample. A study published in *Pediatrics International* in 2020 found that using an automatic lancet device was less painful than the traditional needle for the heel-prick.[38]

Second, the parents can undergo stress related to the test for multiple reasons. False positives could lead to undue stress, or a true positive could determine that one of the parents was a carrier for a given disease and passed it onto their child, which could very well cause inter-familial issues or acute stress on the individual parent. Third, a newborn that is found to have one of the diseases tested for in the Guthrie Test will have to become a frequent visitor of medical facilities. This could result in additional illness due to proximity to other sick individuals. However, it is now commonly accepted that despite all the above concerns, the benefits of the neonatal heel prick far outweigh the costs of withholding the procedure.[39]

Another concern that comes along with the implementation of the Guthrie Test is the fact that many states in America have an opt-out option, where a child's blood sample is automatically taken, tested, and stored. If the parents do not want this to occur, then they must make it clear to the hospital and opt out of the procedure.[40] This is done because several facilities across the country (and the world) store these samples, sometimes indefinitely, for various medical and public health studies, the pursuit of scientific advancement, and, among other things, larger genome understanding. Some critics say that this fails to allow proper informed consent by the parents. Without providing explicit written consent to keep their baby's blood sample, they say the state does not have the right to collect, store, and study the blood.[41]

On a larger scale, the ethical issue of Guthrie Test blood specimens being stored for long periods is prevalent across both the U.S. and the globe. Lawsuits have been brought against state governments in recent years due to the retaining of these blood samples without proper informed consent from the parents. The Minnesota Supreme Court dismissed a lawsuit that claimed the blood samples being kept violated the Genetic Privacy Act because the Court determined that the act did not apply to neonatal testing.[42] In Texas, the state's Department of Health Services was sued for using blood samples for medical research without the clear approval of the parents.[43] Afterward, Texas law was altered to allow the retention of blood samples while the state also agreed on the destruction of five million blood samples that had been saved without the consent of the parents before the new law was passed.[44]

Contrary to U.S. law, Denmark, Sweden, and Norway have laws that allow the retention of these blood samples and their use in medical studies for up to 1,000 years.[45] This also largely differs from Ireland, where the general policy is to keep the blood for a minimum of six months unless the parents provide a written request for the facility to destroy the sample before the six months are up.[46] Current Irish laws surrounding this practice stem from a 2009 complaint from an Irish citizen claiming their blood sample was being held without their consent. This led to much public debate, especially after the minister of health made the decision to destroy the entirety of Ireland's blood samples taken via the Guthrie Test. Such a decision was deemed by many in the medical community and otherwise to be a grave error, one "based on a very narrow and legalistic perspective that does not take the medical and societal value of the cards into account."[47] Those in the field of genetics say that without these blood samples, the evolution of genetic understanding, science, and technology would have been significantly more difficult, if not impossible. The retention of blood samples seems to be the largest ethical issue facing many communities in this regard. Here I am cautious to side with those wishing to destroy the blood samples. Instead, a proper, independent, ethical oversight of the blood samples should be established to ensure they are used to pursue medical science and the greater good.

Medicinal and Commercial Uses of Contemporary Genetic Testing

Modern use of genetic testing is done in two realms: medicine and commodity. On the one hand, scientists use DNA, algorithms, and genetic testing techniques in attempts at, among other things, curing diseases and assisting with reproductive efforts. On the other hand, a curious citizen can send their DNA to a company to test it and find the geographical lineage of their ancestors and learn whether or not they may have a genetic predisposition to various ailments (for a [usually higher] price, of course). Both can be beneficial but also come with their own unique pros and cons.

One of the most impactful projects that allowed genetic science, and the greater body of science at large, to get to where it is today, was the completion of the Human Genome Project, i.e., the sequencing of the human genome. The human body and our DNA are now understood at molecular levels that were once unfathomable. This has helped further the evolution of assisted reproductive technology (ART) as well. ART has given prospective parents newfound freedom regarding family planning. Thanks to aspects of ART such as in-vitro fertilization and preimplantation genetic diagnosis, people who previously could not have children are now able to both reproduce and have much higher chances of ensuring healthy offspring. However, dangerous possibilities arise from this evolution in medical science. Thanks to these advancements parents can now, as mentioned above, choose various ailments to purposely impart on their child.

DNA science has also been commodified on an international level. Commercial DTC organizations such as 23andMe and AncestryDNA can now extract and test one's DNA sent through the mail. These tests reveal the geographical origins of the DNA, as well as if the sender is predisposed to various health conditions. The company then sends the results back to the owner of said DNA, where they can pore through the results themselves. While certainly fascinating, these tests also come with their ethical quandaries. Those who have access to large amounts of the population's DNA, and what they do with it, should be closely regulated. Further, companies like this can reveal deep family secrets that some may not have wanted to be made known.

Genetic testing is largely done using algorithms. If a scientist knows the algorithm within the DNA that creates a certain result, they can alter the DNA in a way that results in a specifically chosen outcome. This is done for everything from experimental treatments to combat a disease to choosing specific traits for offspring in IVF. However, this, too, comes with its ethical challenges. Innate discrimination is a significant possibility in some DNA sequencing, as is stress and anxiety due to newfound awareness of disease predisposition. Subsequent sections below will delve into all the above, with an emphasis on the ethical implications in each aspect.

Medicinal Use

The application of DNA and genetic science in medicine is vast. Upon the completion of the Human Genome Project in April of 2003, the blueprints of a human being were finally fully mapped out.[48] This largely singlehandedly drove significant advancements in understanding the health and disease of the human body.[49] One of the initial changes to our understanding of human disease thanks to the complete mapping of the genome was the introduction of genomics. Genomics, or the understanding of the relationship between given genes and environmental factors they exist in gave scientists a deeper insight into how one's genetic makeup can affect the general health and well-being of a person when considered in conjunction with their immediate environment.[50] This deeper understanding gives us a more comprehensive look at why, genetically speaking, we are the way we are.

The ability of scientists to determine the probability of someone having a specific genetic disease has drastically changed prognostic medicine. Genetic testing for diseases, especially rare ones, has become a much more widespread practice since the completion of the genome project. This entails scientists cloning the gene or genes that work in conjunction with a certain disease, then they create a DNA-based test that seeks out distorted genes that are predisposed to the disease.[51] While this is certainly beneficial for individuals and society at large, it can be difficult for someone who may have a rare disease to find testing. There are two underlying financial reasons for this. First, it can be quite costly for someone to get extensive genetic testing done for rare diseases, meaning those with lower socioeconomic status may likely refrain from doing so.[52] The other financial reason lies in the fact that clinical laboratories may often forego offering tests for rare genetic diseases due to not being cost-effective. If the demand is low, there is no money to be made.[53]

Due to the rare nature of a given disease, it can cost the healthcare system over $10,000 per patient on their journey to find a proper diagnosis. Marshall et al. termed this journey a patient's "diagnostic odyssey," due to the time and effort it takes to produce any semblance of a diagnosis for some rare diseases. Often the patient may experience a "prolonged and expensive diagnostic odyssey culminating in a delayed diagnosis or…no diagnosis at all."[54] Thus, ethical issues arise when considering rare disease testing. What may seem impractical and expensive could also save lives and money.

Aside from the advancements of scientific understanding in the way of genomics and genome sequencing, assisted reproductive technology has also blossomed thanks to genetic testing. ART, according to the Centers for Disease Control and Prevention (CDC), consists of any "fertility treatments in which either eggs or embryos are handled."[55] For example, PGD, as mentioned above, allows potential parents to ensure the embryos they choose to bring to term in IVF will not be carriers of specific diseases.[56] This genetic testing of embryos has been estimated to save tens of billions of dollars, due to the avoidance of medical costs for patients who would have been born with a lifelong disease, such as cystic fibrosis.[57] However, the financial concerns previously outlined remain relevant here; PGD is an expensive procedure, and not widely available to the average middle or lower-class couple or individual. The ethical dilemmas attached to the monetary aspects of genetic testing are not isolated to ART or PGD but rather are intertwined throughout all of healthcare. These issues must be dealt with on the national stage where policy changes are implemented to ensure affordability in healthcare to the masses.

Genetic Testing as a Commodity

The public's interest in genetic testing has skyrocketed over the last decade.[58] [59] Services provided by DTC genetic testing companies allow people to trace their ancestral lineage to see their roots in the world. Further, and for a higher price, these companies also offer additional genetic testing and analysis that can give the buyer deeper insight into the very structure of their DNA. In other words, they can purchase a report outlining possible genetic diseases they may have or be predisposed to.[60] [61] While this piqued the interest of the general public, many health experts have warned to proceed cautiously.[62]

Various public health issues can arise with the sudden onslaught of this direct-to-consumer technology. Bypassing medical facilities and utilizing services such as these raises several questions. First, no test is 100 percent accurate, and discrepancies may occur in one's results. Secondly, these companies have databases containing millions of people's DNA. The fact that this data is owned by commercial companies as opposed to healthcare institutions raises concerns about privacy, safety, and security. Without significant oversight or governmental regulations in place, these companies largely decide how this data is stored, who has access to it, and what is done with it.[63]

Aside from the large-scale logistical aspects of data storage and privacy, more intimate problems may also come out of this new technology. On the surface, these technologies work in such a way that biologically related individuals are connected via their DNA matches. This poses the risk of someone discovering their previously understood notions of familial relatedness may be wrong. Family secrets can be unraveled in a matter of seconds after receiving one's results. Children may discover their mother or father is not their biological parent, or vice versa. This grappling with Truth can be considered in two lights: the right to know versus the right not to know. Either way, the companies providing these services have a duty to inform their customers of the possibility of such unanticipated results, and how to navigate those waters if one finds themselves in them.[64]

Aside from familial issues, legal ownership of one's genetic material is also compromised when utilizing one of these services, which number in the hundreds worldwide.[65] A company's acquisition of a person's DNA during one of these transactions works like anything else being bought or sold in a free market. Once the company doing the testing receives the genetic material, it is theirs to do with what they please. Oftentimes, unbeknownst to the original owner of the DNA, the company will then sell the material to a third party, with all the profits going to the company. Paradoxically, many of these companies assert that one has the right to "unrestricted access to their genetic data" due to the fact it is an inherent part of their person.[66] Meanwhile, these companies also sell that genetic information and collect the entirety of the profit without issue.[67]

Racial discrimination is another underlying problem with many of these services. This lies in the fact that a given genealogy provided by one of these companies often provides more detailed information for those with white, European backgrounds. This racial bias stems from their databases that connect everyone who uses the service. In the beginning, these tests were available only in the United States. Poor funding and infrastructure for underrepresented groups such as Africans, Asians, and indigenous peoples have caused these companies to struggle to collect data from them. Therefore, an African American individual will very likely not get as thorough a report as their European American counterparts.[68]

Where these services are working to remove the financial burden that traditional genetic testing puts on individuals, they still come with their own modern ethical issues. Regulation and oversight of the storage of genetic materials is of the utmost importance. DNA tells the story of humankind at large, but also the backgrounds and possible futures of individuals, and thus must be protected. In terms of testing for genetic diseases, an ethical question of duty naturally arises. While there is no legal obligation for these companies to tell someone about a genetic predisposition to a disease, one could argue there is a moral one. The ethical duty to warn someone of a potentially serious future ailment should override the pursuit of financial gain. Lastly, steps should be taken by these companies to ensure two things: that the customer knows the possible ramifications of DNA testing (discovering previously unknown familial facts), and proper representation of currently underrepresented populations.

Algorithms, DNA, Genetics

An algorithm can be defined, broadly, as "a step-by-step procedure for solving a problem or accomplishing some end."[69] Thus, scientists have utilized algorithms to map out genes and find discrepancies in otherwise seemingly normal DNA. This allows science to pinpoint genetic variations that could cause disease. This use of DNA, however, also comes with its own unique set of ethical challenges that must be recognized, confronted, and overcome.

Use of Algorithms in DNA Science

Scientists have begun to utilize genetic algorithms (GA) as a metaheuristic – using a higher-level heuristic or learning mechanism to problem solve – a method of finding solutions to several problems throughout medical science.[70] On a larger scale, the mapping of the human genome was done with the help of complex algorithms. These algorithms sought out overlapping DNA sequences to recreate the human genome.[71] Further, the use of algorithms has begun to advance our understanding of genetic diseases.

For example, GAs are prevalent in diagnosing Parkinson's disease. It is now widely accepted within the scientific community that mitochondrial dysfunction is a significant factor in the onset of Parkinson's. So, a team of researchers created a GA that identifies patterns of mitochondrial mutations in those suffering from Parkinson's disease. The result was an algorithm that could, with perfect accuracy, analyze those mutations in a patient's DNA to diagnose Parkinson's.[72] This is one example of many where a GA has been used in healthcare to predict or help overcome a genetic disease.

While the use of GAs in medicine can significantly increase the efficacy of medical care, one major concern lies in the disconnect between physicians and scientists creating some of these algorithms. Some computer scientists may not understand the complexities of medicine and medical care. Similarly, physicians may not be able to comprehend the intricate math that goes into the creation of the GA. A multidisciplinary interconnectedness must be established to overcome this obstacle. Computer scientists, physicians, engineers, and others must work together to improve their interactions in solving problems.[73]

Ethical Implications

Like any other medical technology, especially new and emerging ones, ethical implications abound in the use of algorithms and GAs in DNA and medical science. On a macro level, no one should have unfettered access to this type of data. It should be protected and strictly regulated. On a micro-level, specific individuals' data should be carefully guarded. When it comes to widespread genetic testing, there are two trains of thought. One side says testing should be limited to those who have family members that are known carriers of rare genetic diseases. The other side says anyone and everyone should be tested for the sake of maintaining both individual and public health.[74] However, some also argue that the data being used is not technically using human *subjects*. Thus, how the data, its storage, and handling are carried out need not be extensively regulated and may not even fall under traditional guidelines for using human subjects in clinical settings.[75] Certainly, significant oversight for scientific and medical institutions storing this data may not be necessary, but the truth may not be the same for corporate entities, like direct-to-consumer companies offering genetic testing. This naturally begs the question of to what extent the medical and scientific community should be regulated in this regard compared to what the oversight should look like for those DTC organizations.

Revelations from the implementation of genetic testing can cause stress and anxiety among those who discover a predisposition to a genetic disease.[76] Additionally, insurance companies could add predatory surcharges to people's bills who are otherwise perfectly healthy due to the outcome of a genetic test should they learn of a customer's results. However, laws throughout the U.S. have been found to prevent this type of discrimination by insurance companies.[77] Nonetheless, because these genetic tests can offer so much insight into a person's medical future, caution must be taken in how the data is stored and disseminated, if at all.

Further, genetic testing done by prospective parents could reveal certain diseases they may be predisposed to and could pass on to their children. This could very well lead to acute psychological distress. Many experts opine that genetic testing should come with genetic counseling so people can properly navigate these new and difficult emotions.[78] The entire range of ethical issues stemming from genetic testing is beyond the scope of this chapter. However, some of the more relevant and important ones have been outlined here.

Testing Today for a Healthy Tomorrow

The most common diseases that result in euthanasia or assisted dying requests, after cancer, include multiple sclerosis, dementia, Parkinson's, and Huntington's disease.[79] Some of these neurodegenerative disorders can now be detected in an embryo's DNA via preimplantation genetic diagnosis.[80] Other debilitating diseases, such as SMA1 discussed earlier, can also be detected via preimplantation genetic diagnosis.[81] Moving forward, it is important for prospective parents to strongly consider discarding those embryos to avoid future human suffering and, subsequently, pursuits of euthanasia or assisted death. Until scientists can safely delete disease-carrying genes in the embryo and replace them with healthy ones, physicians and reproductive scientists should encourage patients who have embryos like those described here to seriously consider discarding them.

CRISPR-Cas9 and Gene Manipulation

Thanks to genetic testing, as has been discussed at length above, we can quite literally predict the future of a human's health by using PGD. If the embryo that will one day become a human is going to have a disease, such as SMA1, then we can further predict, with a substantial amount of accuracy, that that human will, at the very most, live to only three years old. Further, those three years will be spent in pain, physically deteriorating, in and out of hospitals, and costing their parents, caregivers, and the healthcare system hundreds of hours. Financially, tens, if not hundreds of thousands of dollars will be spent caring for these children who live to simply suffer. Unless, of course, that embryo was never implanted and allowed to grow into a person. Had that been the case, all of that pain and suffering could have been avoided. Furthermore, all that time and money could have gone toward other resources.

Another approach is also in the making and already widely utilized: CRISPR and, more specifically, CRISPR-Cas9. CRISPR, which is an acronym for Clustered Regularly Interspaced Short Palindromic Repeats, is, in sum, how bacteria defend themselves. It works as a memory bank of genetic code that "remembers" past viruses that infiltrated it. If those same virus cells return, the bacteria cells remember them and move to eradicate them. The Cas9 addition refers to a protein within the bacteria cell that scientists can assign to seek, destroy, and edit a particular part of a DNA strand.[82]

Aside from not implanting embryos found to have disease predisposition via PGD, scientists can use CRISPR-Cas9 to edit the DNA of an embryo to remove some inheritable diseases. The Cas9 protein is used to target a specific part of the DNA that it then cuts out of the strand. The cell responds by repairing its DNA. In doing so, parts of the DNA get added or taken off the ends, thus disabling a specific disease gene.[83]

When this was tested in a laboratory setting, the goal was to get the cell to repair its DNA with a template put forth by the scientists. Specifically, they used sperm from a donor with a gene mutation for a disease and injected it into an egg from a healthy woman without the disease. The sperm also had the Cas9 instructions to seek and destroy the specific part of the DNA that carried the mutation. Additionally, the template containing a healthy version of the gene was included in the Cas9 instructions in the hopes the cell would utilize it when fixing its DNA. The Cas9 successfully found and removed the sperm cell's disease-causing elements. However, it utilized the mother's normal gene as a model for replacement, and not the one the scientists introduced. The experiment worked, but not in the way it was anticipated to.[84]

Some unintended side effects of this type of gene manipulation are possible. The embryo undergoes what is called "large deletion," where the DNA strand now has a normal gene but also, thanks to the cutting of the strand, a missing one. While the DNA strand is now missing the disease gene, some scientists critiquing this process say the embryo is by no means healthy due to the random DNA that replaces what Cas9 removes. When mice embryos were tested with Cas9 editing, large deletions were found to be prevalent, thus hindering this use in human embryos.[85]

Genetic Testing's Correlation to Future Health, Euthanasia, and Assisted Suicide

The utilization of algorithms can help to avoid debilitating diseases and immense human suffering. These algorithms used in genetic testing are essentially tools that can be used to bypass future pain and suffering, the need for a future abortion, or even euthanasia or assisted suicide years down the line. Additionally, embryonic genetic testing before implantation of the embryo in the mother can prevent future harm and suffering for those who are amongst the living. Chromosomal abnormalities, which can be detected by PGD, are responsible for 50 percent or more of all miscarriages that occur in the first trimester.[86] Thus, a woman should be able to choose to not implant an embryo found to have a chromosomal abnormality.

Similarly, diseases that can cause future suffering such as single-cell gene disorders like cystic fibrosis, hemochromatosis, Tay-Sachs, and sickle cell anemia can all be avoided so long as the embryo does not get implanted.[87] However, if the choice is made to move forward with pregnancy despite such awareness of predisposition, then the simple awareness of those genes in the child can provide a head start in their healthcare; steps can be taken early on to delay the onset and deal with the disease earlier than what would have been previously possible.

The algorithms used in detecting those previously mentioned diseases within the genome of an embryo directly allow for one to discard that embryo. However, genetic testing can also help adults either circumvent the need for future assisted dying or allow for better-advanced care planning. For example, recent improvements in science have allowed for better detection of genetic predisposition to Parkinson's disease.[88] However, there are currently over 500 different clinical genetic tests for Parkinson's, with specific genes tested numbering between five and 62. Thus, there is a dire need for additional research into which genes should be focused on for more effective genetic testing of Parkinson's.[89]

Dementia is similar to Parkinson's in that it is difficult to perform genetic testing for the disease. However, advancements in medical science are slowly allowing for this to become a reality. Mendelian risk genes simply increase the chances that one will get dementia. Deterministic genes, however, tell one that they are certain to one day have it.[90] Additionally, those who have been diagnosed with dementia can undergo multigene testing that can provide higher accuracy diagnosis. In turn, this can identify specific dementia-causing genes that the patient has, which can help the patient's family members in their own diagnostic and genetic testing endeavors.[91]

In sum, genetic testing can drastically reduce future suffering. An embryo found to have genes for a debilitating disease can be discarded, directly preventing suffering. If an adult is found to have a genetic predisposition to a neurodegenerative disease like dementia or Parkinson's, steps can be taken to set goals of care before they lose decision-making capacity. If the patient is in a location where the practice is legal, a request for euthanasia or assisted suicide can be implemented. The procedure could be arranged for a later date while the patient is still of sound mind and deemed to have the proper decision-making capabilities. This would prevent not only their own suffering but also avoid the mental anguish and financial strains such diseases so often impose on loved ones caring for such patients.

Ethical Rationalization of Genetic Testing

One of the main points of genetic testing in both embryos and humans is to avoid undue suffering. Embryos with genetic diseases can be destroyed and humans can sooner establish better goals of care and advance directives if found to have an imminent or higher chance of disease. Further, we can gain knowledge and understanding of the human genome in terms of genomics as well as medical and genetic science. The use of genetic testing to avoid future disease and suffering and, thus, the need for euthanasia or assisted suicide is the focal point of this chapter. However, various normative ethical principles can be applied here to bolster this argument.

Thinking through ideas of utilitarianism and deontology is necessary for a larger understanding of how medical ethics applies to this conversation. The utilitarian approach says what is best for the most number of people is what should be done – it is a society-based approach. Utilitarianism does, however, allow for the harming of a few individuals, as long as the greater good is upheld. Deontology, on the other hand, says that one must act insofar as no harm to an individual occurs. Contrary to the acceptable nature of individuals being harmed in utilitarianism, society at large may be harmed by the implementation of a deontological framework.[92] On a broader level, deontology deals with duty, and what one should do in a given situation independently of how it may affect others. This non-consequentialist approach ignores what types of consequences may occur outside of the decision made for the individual.[93]

Autonomy, beneficence, nonmaleficence, and justice, otherwise known as the four main principles of bioethics, are undoubtedly intertwined in this. In autonomy, respect for individual freedom and one's decision-making regarding their own body must always be upheld. Beneficence says that what is best for the patient's well-being must always be sought while nonmaleficence says harm must always be avoided. Lastly, justice is the pursuit of equality and fairness both within society and individual cases.[94] All four principles apply to varying degrees in different circumstances in the general scope of healthcare ethics, but all apply here.

Utilitarianism & Deontology

The utilitarian nature of genetic testing is widespread. It can be considered in multiple ways, from reducing general suffering to large-scale, expensive financial considerations. It has been theorized that larger healthcare systems in general work as utilitarian organizations while individual clinicians work in a deontological manner.[95] That said, there are specific parts of genetic testing in healthcare that apply to a larger utilitarian ideology.

When an embryo is created in IVF, one of two things occurs. On one hand, the embryo can be deemed "healthy" wherein it falls under the category of "wanted" by the prospective parent(s). On the other hand, the embryo may be found to have some type of abnormality that leads the prospective parent(s) to deem the embryo *not* wanted. When this happens, the embryo can either be promptly discarded, or utilized for research. If the embryo is immediately destroyed, there is a lost opportunity for research.

The unwanted embryo can and should be utilized for research purposes. The overall benefit of doing so may help advance medical science in ways that were once not possible. Technology can evolve, thanks to embryonic research, in ways that may allow people to live longer, better lives due to advancements in therapies. Future suffering may be eased thanks to therapeutic medicines resulting from second-hand discoveries of this research. More specifically, embryonic stem cells are hoped to help scientists discover ways to induce in vitro organ and tissue growth for transplantation purposes.[96] Conversely, the utilization of unwanted embryos for research purposes will help avoid conducting possibly harmful experiments on living people or animals. Finally, some argue, it could even be considered appropriate to purposefully create embryos for such research.[97] Doing so would allow for faster advancements in medical science that could lead to the saving of countless lives and immeasurable suffering.

Financially speaking, genetic testing can save billions, if not trillions, of dollars. Lifetime costs for some diseases can be exorbitantly high, causing financial burdens for patients, their families, insurance companies, healthcare institutions, and society in general. For example, the average hemophilia patient can easily spend over $300,000 a year on treatment. More severe cases can reach over $600,000 annually.[98] The average life expectancy for someone with hemophilia is 77 years old.[99] This results in 20 to 46 million dollars for a lifetime of treatment, not to mention the seizures, possible paralysis, and chronic pain that comes along with a hemophilia diagnosis.

According to the CDC, about 400 babies are diagnosed with hemophilia in the U.S. each year.[100] That carries with it an approximately 13.2-billion-dollar price tag for the total cost of lifetime care required for each year of new cases. A 2020 publication reported about 30,000 people in the U.S. were estimated to have hemophilia.[101] This means that at that moment, the cost of care for every American citizen with hemophilia combined is between nine and 18 billion dollars for the year. Moving forward, if genetic testing, specifically PGD, is utilized for sufferers of hemophilia who are trying to procreate and choose to discard any embryos found to have a predisposition to the disease, trillions of dollars will be saved over time. The example of hemophilia is just one of many extremely expensive, painful, and now, *preventable* diseases. Others include but are not limited to, cystic fibrosis, Tay Sachs, SMA, and muscular dystrophy.

Several reasons in the deontological realm of genetic testing justification are quite similar to those found in the utilitarian camp. For instance, when undergoing PGD, if there are multiple possible embryos to choose from, there is a moral obligation for the hopeful parents to choose the embryo which offers the future child the best life. The embryos with the least amount of "genetic endowment" should not be implanted in the womb.[102] This idea touches on the overarching deontological idea that establishes precedence in upholding the sanctity of a human life.[103] By discarding all embryos except for the one with the objectively healthiest future life, a human life is thereby protected, and sanctity upheld. Allowing a damaged or physically inferior embryo to gestate is unethical. It would be doing a disservice to the sanctity of life and that future human, who would be purposely subjected to disease and suffering. The choosing of the embryo with the best genetic endowment is both a deontological and utilitarian endeavor in that one would be saving more than just a future human from undue suffering. Aside from the general trouble of parents dealing with a child with a disease, it would save money and time for healthcare workers and the healthcare system at large. Finally, it also serves as another step in the eradication of a given disease, which serves as both a utilitarian and deontological benefit.

Autonomy, Beneficence, Nonmaleficence, Justice

The four principles of bioethics are pillars upholding normative ethics. To ignore them in a chapter such as this would be to ignore the fundamental nature of healthcare ethics. Autonomy, beneficence, nonmaleficence, and justice all apply broadly to genetic testing. Some fit better than others in different respects, but all are relevant.

Autonomy is perhaps the most immediately relevant of the four principles. In the lens of PGD and prenatal genetic testing in general, a patient has the right to decide their own care, including foregoing such testing, despite what their physician may think or even if the patient has a genetic marker for a given ailment. When facing the possibility of undergoing PGD or prenatal testing of a fetus, to uphold autonomy, a physician must ensure that proper informed consent is given. For example, the severity of a given disease being tested for, how, if at all, treatable the future child's condition will be, and if abortion will have to be considered, should all be made clear to the potential parent(s) or mother undergoing PGD or fetal testing.[104] This underscores the right to know, and how all available information should be given to ensure the patient is making the best decision they see fit. In a more general sense, autonomy applies in that it allows parents to choose the best possible life for their child.

Beneficence and nonmaleficence simultaneously apply to both the mother and the potential fetus and/or future child, should a child be brought to term. Beneficence can be upheld when a woman can choose what happens to her embryo or fetus if it is deemed to have something substantially wrong with it, i.e., a chromosomal abnormality or a future debilitating disease. As was discussed above, such an abnormality often leads to miscarriage. A woman choosing to not go forth with embryo implantation, in this case, may well save herself from future pain and suffering that a miscarriage would entail. Regarding the future life, if an embryo is found to have a probability of an extremely debilitating disease, the discarding of that embryo also saves the future life from pain and suffering. All the above also work in conjunction with nonmaleficence. Avoiding harm is of the utmost importance in medicine, and genetic testing now allows us to actively avoid *future* harm.

In his text on global bioethics, Henk ten Have suggests a few theoretical ways justice can be approached in healthcare ethics. For our purposes here, two stand out. First, the principle of equality of opportunity, states "everybody should have the same chance to realize a healthy life through satisfying basic needs."[105] If an embryo is purposely gestated despite the awareness of an incapacitating ailment, justice is not being upheld – a chance at a healthy life was not given. Genetic testing allows us to make sure that does not happen by making sure a future human has the same chance as other healthy individuals of obtaining as healthy a life as possible.

Secondly, ten Have's capabilities approach, which states "policies should promote and sustain capabilities that are necessary for human flourishing," is relevant.[106] In this context, it would be an injustice to institute any policies that do not allow for proper genetic testing and procedures that follow. Because genetic testing allows one to choose from a set of embryos with the best genetic endowment, the future flourishing of that potential human is recognized. Similarly, being able to discard or utilize research embryos that were deemed unfit for the best possible life is also upholding justice. It would be wrong to force someone to be born who would be known to have some type of disease that significantly impedes their flourishing that could have been avoided; using such an embryo for research and the advancement of medical science would contribute to the flourishing of future humans.

Genetic testing, in sum, can largely circumvent human suffering and the need for euthanasia and assisted suicide in the future. By discarding embryos with genetic markers for debilitating diseases, such as Huntington's, which is one of the leading causes of euthanasia and assisted suicide requests, a future human would be saved from having to make this choice. This upholds both a utilitarian approach as well as a deontological one. The avoidance of diseases leading to euthanasia or assisted dying allows society at large to focus time and money on alternative healthcare efforts, while the well-being of individuals is upheld.

Sustainable Healthcare Practices

On a global level, the healthcare system contributes approximately 4.5% of the world's total greenhouse gas emissions. In the United States, about 8.5% of the overall greenhouse gas emissions come from hospitals and healthcare, which is about ¼ of all global greenhouse gas emissions.[107] This is mainly due to three factors: first, gases emitted from hospitals and healthcare institutions from things like boilers, exhaust systems, and medical gas releases; second, indirect impacts through electricity purchased for the building(s); and third, greenhouse gases that are emitted throughout the supply chain during the production of energy, products, goods, and services for a given institution. Furthermore, the use of disposable and single-use products, plastics, and tools significantly contributes to medical waste and landfills.[108]

Hospitals and Healthcare Systems

A healthcare system is an organization that entails various hospitals, people, and resources in the name of healthcare. Such systems generally have policies that are centralized and enacted by a governing body or board that establishes rules that apply to the whole. Thus, general standards can be enacted system-wide that work to lessen the carbon footprint of hospitals. From utilizing solar panels to limiting plastics and single-use instruments and opting for sterilization, to choosing specific gases and materials to use within the hospital, there is a moral obligation to enact green and planet-friendly norms.

Some shocking and concerning numbers emerge when looking at specific gases emitted from healthcare facilities—for our purposes we will examine the difference between desflurane and propofol, two commonly used anesthetics. In recent years, there has been a growing movement encouraging doctors to refrain from using desflurane. The reason is simple—desflurane is hundreds of times worse for the environment than propofol. Desflurane lasts in the atmosphere for over 20 years. Desflurane has a gCO2e, or grams of carbon dioxide equivalent, of about 32,000, which, compared to propofol's 12 gCO2e, is about 2,600 times worse for the environment.[109] In simpler terms, using desflurane for one hour has the greenhouse effect of driving over 200 miles, whereas propofol has the greenhouse effect of driving less than one mile.[110] [111]

There are several other ways hospitals can reduce their carbon footprint. Energy efficiency can be maximized via the utilization of solar energy and solar panels. Electric vehicles (ambulances, etc.) can reduce both long-term financial costs and greenhouse gas emissions of traditional gasoline-powered vehicles. Sustainably grown food in hospital cafeterias can reduce the hospital's impact on the environment. General recycling practices as well as policies that ensure water and energy conservation should be implemented across the board.

Moral Obligation

Hospitals and healthcare systems have a duty to enact regulations that benefit not only patients but also the world at large. Moving towards renewable energies and net-zero carbon emissions is not a cheap endeavor. The financial difficulties posed by an environmentally friendly overhaul of entire hospitals and health systems can be overcome through several means. The billions, if not trillions, of dollars saved over time from eliminating preventable diseases outlined above (hemophilia, etc.) can be reallocated to this cause. The reduction of preventable diseases through the utilization of PGD and IVF will allow hospital funds to, over time, be redistributed to lessen carbon footprints.

Additionally, hospitals have an ethical duty to start budgeting time, money, and effort toward some of these solutions. Especially as 24-hour, seven-day-a-week operations, it is of the utmost importance that they become more environmentally conscious. A system whose goal is to heal people should not also harm the world we live in. When it does, when any type of system harms the air we breathe, taints the water supply, and disrupts food growth, humanity is harmed on both individual and collective levels.

Conclusion

Genetic testing has evolved significantly over the last few decades. In doing so, the capabilities of science and medicine have evolved alongside it. Scientists can now, to an extreme degree of accuracy, tell the future of embryos. By analyzing DNA, we can predict whether or not a future life will have a given disease. Moreover, the completion of the Human Genome Project has effectively given us access to the blueprints of a human being. Additionally, the commercialization of DNA testing and tracing now allows people to discover their ancestral genealogy and whether or not they are predisposed to some genetic diseases. The medicinal, scientific, and ethical implications surrounding this are vast, and the data should be treated with respect and dignity.

All of this has been made possible by the use of algorithms. If a scientist knows the algorithm that creates a certain outcome, they can alter the DNA to create an outcome they implement themselves. Algorithms can not only predict disease but also prevent it. If a scientist wanted to knock out a certain gene to combat a given disease, an algorithm can do this. However, if it becomes known that a specific person is likely to have a future disease, they could face harassment or discrimination.

This is largely why some argue that genetic tests should have their own specific class and be more stringently regulated than other medical information. In an article in the New England Journal of Medicine, Roche and Annas say DNA itself can be understood "as a coded probabilistic medical record, that makes genetic privacy unique and differentiates it from the privacy of medical records."[112] This effectively puts DNA in its own exclusive category, one that should be carefully maintained and regulated to ensure privacy is upheld. DNA does, after all, tell us things about ourselves we may not even want to know, let alone allow others to know.

Further, PGD offers a novel way for us to help future generations before they are even born. However, left unfettered, the use of PGD to alter someone's genes can be utilized as a means for selfish and inherently unethical ends. Purposely giving one's child deafness or dwarfism is problematic and never justifiable, thus laws should be enacted to end and prohibit this practice in the United States. Using PGD to determine which embryos to utilize in IVF, or discard those that would otherwise give way to someone with painful, lifelong disease is the only overtly ethical use of the technology thus far; it saves the child and family from unnecessary pain and suffering while also saving a significant amount of money for the state, healthcare at large, and the families of the potential offspring.

Further, there is a case to be made for the government to cover healthcare costs of PGD, if it can ensure the safety and well-being of their future citizens. Choosing the sex or eye color of a baby for non-medical purposes is a much more nuanced conversation that goes beyond the scope of this chapter, but could, on rare occasions, be ethically justified. If one day we can easily choose the hair color, eye pigment, sexual orientation, and height of the child, this conversation of designer babies will be forced back to the forefront of ethics in genetics.

Genetic testing today allows for a healthier tomorrow – for both society at large and individuals. It can eradicate disease by ensuring only embryos without genes for certain diseases are gestated. On a large enough timeline, this will remove these diseases from the human genome. While doing such would likely result in minimal overall impact on the human gene pool, there are some concerns with potential significant alterations that may come from the complete removal of a gene.[113] While this generally happens naturally in humans through evolution, doing so on small scales in short periods could result in unforeseen consequences. Besides health maintenance, financial stability is also considered. The exorbitant amounts of money that can be saved from genetic testing of embryos can go towards myriad other healthcare initiatives.

If the United States healthcare system were itself a country, it would be the 13th highest contributor to greenhouse gas emissions worldwide.[114] The highest percentage of these emissions come from electricity, heating, and/or cooling of facilities.[115] The inherent nature of the duty of healthcare is compromised when the system itself is detrimental to the health of the planet. To foster resiliency in healthcare, medical institutions have a duty to uphold and pursue beneficence and nonmaleficence. If, while healing, the world at large is also harmed, then there is an obligation to reduce that harm as much as possible. Enacting regulations for more environmentally friendly practices, like the installation of solar panels and the switch from desflurane to propofol, works to ease the burden of healthcare on the environment. Additionally, the onus is on the healthcare systems themselves, the corporations, the executives, the board and directors who run hospitals and health systems, to act in a utilitarian nature, to do what is best for the most amount of people, as opposed to what is best for their wallets. A nurse recycling their water bottle is a drop in the bucket compared to the change that could be enacted by those at the top of the administrative hierarchy. If doctors pledge to do no harm, then that should extend from their individual patients to the impact of their practice and hospitals on the world at large.

Finally, the deeper, philosophical implications of genetic testing are clear. In a utilitarian lens, society would be better off with fewer people carrying genes for diseases. Deontologically speaking, individuals would be saved from pain, suffering, and potential social stigma. Autonomy, beneficence, nonmaleficence, and justice are all also upheld. Autonomy here is twofold. First, individual freedom is maintained when someone is allowed to choose the embryo with the most promising future in terms of health. Second, that future human would be able to have the most autonomy possible if they are free of diseases. Beneficence and nonmaleficence work hand in hand here. The maintaining of patient health in regard to the mother who would be carrying the child, as well as the future child being free of disease, makes sure the best possible outcome is pursued and that undue pain and suffering are held at bay. As for justice, genetic testing allows for human flourishing to be not only considered but ensured, while all future humans are given the best chance of survival and pursuits of happy, healthy lives.

Notes

[1] "Preimplantation Genetic Testing," accessed February 6, 2022, https://nyulangone.org/locations/fertility-center/preimplantation-genetic-testing.

[2] "Miscarriage - Prenatal Genetics," https://www.gbmc.org, accessed February 6, 2022, https://www.gbmc.org/miscarriage.

[3] Jacqueline Mae Wallis, "Is It Ever Morally Permissible to Select for Deafness in One's Child?," *Medicine, Health Care, and Philosophy* 23, no. 1 (March 2020): 3–15, https://doi.org/10.1007/s11019-019-09922-6.

[4] Robert Ranisch, "Germline Genome Editing versus Preimplantation Genetic Diagnosis: Is There a Case in Favour of Germline Interventions?," *Bioethics* 34, no. 1 (January 2020): 60–69, https://doi.org/10.1111/bioe. 12635.

[5] "Choose Your Baby's Eye Color | The Fertility Institutes Offers Eye Color Screening and Selection," accessed April 25, 2022, http://www.fertility-docs.com/programs-and-services/pgd-screening/choose-your-babys-eye-color.php.

[6] I. Tur-Kaspa, et al., "PGD for All Cystic Fibrosis Carrier Couples: Novel Strategy for Preventive Medicine and Cost Analysis," *Reproductive Biomedicine Online* 21, no. 2 (August 2010): 186–95,
https://doi.org/10.1016/j.rbmo.2010.04.031.

[7] Crystal S. Balnkenship, "To Manage Costs of Hemophilia, Patients Need More Than Clotting Factor," *Biotechnology Healthcare* 5, no. 4 (2008): 37–40.

[8] A. H. Handyside, et al., "Birth of a Normal Girl after in Vitro Fertilization and Preimplantation Diagnostic Testing for Cystic Fibrosis," *The New England Journal of Medicine* 327, no. 13 (September 24, 1992): 905–9,
https://doi.org/10.1056/NEJM199209243271301.

[9] Anastasia A. Theodosiou and Martin H. Johnson, "The Politics of Human Embryo Research and the Motivation to Achieve PGD," *Reproductive BioMedicine Online* 22, no. 5 (May 1, 2011): 457–71, https://doi.org/10.1016/j.rbmo.2011.01.008.
[10] Michelle J. Bayefsky, "Comparative Preimplantation Genetic Diagnosis Policy in Europe and the USA and Its Implications for Reproductive Tourism," *Reproductive Biomedicine & Society Online* 3 (December 2016): 41–47, https://doi.org/10.1016/j.rbms.2017.01.001.
[11] Walter Heape and Michael Foster, "III. Preliminary Note on the Transplantation and Growth of Mammalian Ova within a Uterine Foster-Mother," *Proceedings of the Royal Society of London* 48, no. 292–295 (January 1, 1891): 457–58, https://doi.org/10.1098/rspl.1890.0053.
[12] "History of the Monash IVF Research Program from 1971 to 1989 | The Embryo Project Encyclopedia," accessed February 6, 2022, https://embryo.asu.edu/pages/history-monash-ivf-research-program-1971-1989.
[13] Theodosiou and Johnson.
[14] S. Soini, "Preimplantation Genetic Diagnosis (PGD) in Europe: Diversity of Legislation a Challenge to the Community and Its Citizens," *Medicine and Law* 26, no. 2 (June 2007): 309–23.
[15] Bayefsky.
[16] Wallis.
[17] Susannah Baruch, David Kaufman, and Kathy L. Hudson, "Genetic Testing of Embryos: Practices and Perspectives of US in Vitro Fertilization Clinics," *Fertility and Sterility* 89, no. 5 (May 1, 2008): 1053–58, https://doi.org/10.1016/j.fertnstert.2007.05.048.
[18] Bayefsky.
[19] Richard F. Storrow, "Assisted Reproduction on Treacherous Terrain: The Legal Hazards of Cross-Border Reproductive Travel," *Reproductive BioMedicine Online*, Symposium: Cross-Border Reproductive Care – Ethical, Legal, and Socio-Cultural Perspectives, 23, no. 5 (November 1, 2011): 538–45, https://doi.org/10.1016/j.rbmo.2011.07.008.
[20] Storrow.
[21] H. Strange and R. Chadwick, "The Ethics of Nonmedical Sex Selection," *Health Care Analysis: An International Journal of Health Care Philosophy and Policy* 18, no. 3 (September 1, 2010): 252–66, https ://doi.org/10.1007/s10728-009-0135-y.
[22] Bayefsky.
[23] Tur-Kaspa et al.
[24] James M Kemper, Christopher Gyngell, and Julian Savulescu, "Subsidizing PGD: The Moral Case for Funding Genetic Selection," *Journal of Bioethical Inquiry* 16, no. 3 (September 2019): 405–14, https: //doi.org/10.1007/s11673-019-09932-2.
[25] "PGD and IVF Costs – What Is the Cost for Preimplantation Genetic Diagnosis – PGD and PGS, Preimplantation Genetic Screening? – Advanced Fertility Center of Chicago," accessed April 25, 2022, https://advancedfertility.com/fertility-treatment/affording-care/pgd-cost/.

[26] Tamara Dangouloff et al., "Systematic Literature Review of the Economic Burden of Spinal Muscular Atrophy and Economic Evaluations of Treatments," *Orphanet Journal of Rare Diseases* 16, no. 1 (January 23, 2021): 47, https://doi.org/10.1186/s13023-021-01695-7.
[27] "Spinal Muscular Atrophy (SMA): Types, Symptoms & Treatment," Cleveland Clinic, accessed April 25, 2022, https://my.clevelandclinic.org/health/diseases/14505-spinal-muscular-atrophy-sma.
[28] Simo Vehmas and Tom Shakespeare, "Disability, Harm, and the Origins of Limited Opportunities," *Cambridge Quarterly of Healthcare Ethics* 23, no. 1 (January 2014): 41–47, https://doi.org/10.1017/S0963180113000 42X.
[29] Wallis.
[30] "Why Do They Prick Your Newborn Baby's Heel?," Cleveland Clinic, October 6, 2020, https://health.clevelandclinic.org/why-do-they-prick-your-newborn-babys-heel/.
[31] Lauran Neergaard, "Ethics Debate over Blood from Newborn Safety Tests," *Boston.Com*, February 8, 2010, http://archive.boston.com/news/health/articles/2010/02/08/ethics_debate_over_blood_from_newborn_safety_tests/.
[32] John Crowe, "Plan to Destroy Heelprick Cards Is a Serious Mistake," The Irish Times, accessed February 6, 2022, https://www.irishtimes.com/ news/health/plan-to-destroy-heelprick-cards-is-a-serious-mistake-1.1330334.
[33] Michele Caggana et al., "Newborn Screening: From Guthrie to Whole Genome Sequencing," *Public Health Reports* 128, no. Suppl 2 (2013): 14–19.
[34] "Default - Stanford Children's Health," accessed February 7, 2022, https://www.stanfordchildrens.org/en/topic/default?id=newborn-screening-tests-90-P01967.
[35] Harvey J. Stern, "Preimplantation Genetic Diagnosis: Prenatal Testing for Embryos Finally Achieving Its Potential," *Journal of Clinical Medicine* 3, no. 1 (March 17, 2014): 280–309, https://doi.org/10.3390/ jcm3010280.
[36] James J Pitt, "Newborn Screening," *The Clinical Biochemist Reviews* 31, no. 2 (May 2010): 57–68.
[37] "Default - Stanford Children's Health."
[38] Tatenobu Goto et al., "Neonatal Pain Response to Automatic Lancet versus Needle Heel-prick Blood Sampling: A Prospective Randomized Controlled Clinical Trial," *Pediatrics International* 62, no. 3 (March 2020): 357–62.
[39] Dana-Teodora Anton-Paduraru, "OC-24 Newborn Screening for Cystic Fibrosis: To Screen or Not to Screen (Pros and Cons)," *Archives of Disease in Childhood* 102, no. Suppl 2 (June 1, 2017): 24, https://doi.org/ 10.1136/archdischild-2017-313273.24.
[40] Michelle H. Lewis et al., "State Laws Regarding the Retention and Use of Residual Newborn Screening Blood Samples," *Pediatrics* 127, no. 4 (April 2011): 703–12, https://doi.org/10.1542/peds.2010-1468.
[41] Bradford L. Therrell, "Ethical, Legal and Social Issues in Newborn Screening in the United States," *The Southeast Asian Journal of Tropical Medicine and Public Health* 34 Suppl 3 (2003): 52–58.
[42] "Bearder v. State," Justia Law, accessed February 7, 2022,

https://law.justia.com/cases/minnesota/supreme-court/2011/a10-101.html.
[43] "Beleno et al v. Texas Department of State Health Services et Al," Justia Dockets & Filings, accessed February 7, 2022, https://dockets.justia.com/docket/texas/txwdce/5:2009cv00188/353950.
[44] Lewis et al.
[45] Luciano Vittozzi et al., "Evaluation of Population Newborn Screening Practices for Rare Disorders in Member States of the European Union," *Orphanet Journal of Rare Diseases* 5, no. S1 (October 2010): P26, 1750-1172-5-S1-P26, https://doi.org/10.1186/1750-1172-5-S1-P26.
[46] "A Practical Guide to Newborn Bloodspot Screening in Ireland," n.d., 68.
[47] Crowe.
[48] "The Human Genome Project," Genome.gov, accessed March 12, 2022, https://www.genome.gov/human-genome-project.
[49] W. Gregory Feero, Alan E. Guttmacher, and Francis S. Collins, "Genomic Medicine--an Updated Primer," *The New England Journal of Medicine* 362, no. 21 (May 27, 2010): 2001–11, https://doi.org/10. 1056/NEJMra0907175.
[50] Fiona M Walter and Jon D Emery, "Genetic Advances in Medicine: Has the Promise Been Fulfilled in General Practice?," *The British Journal of General Practice* 62, no. 596 (March 2012): 120–21, https://doi.org/10.3399/bjgp12X629955.
[51] M. M. Gottesman and F. S. Collins, "The Role of the Human Genome Project in Disease Prevention," *Preventive Medicine* 23, no. 5 (September 1994): 591–94, https://doi.org/10.1006/pmed.1994.1094.
[52] Judith Kruse et al., "Genetic Testing for Rare Diseases: A Systematic Review of Ethical Aspects," *Frontiers in Genetics* 13 (January 26, 2022): 1–14.
[53] Leslie G. Biesecker, "Orphan Tests," *Cambridge Quarterly of Healthcare Ethics* 5, no. 2 (ed 1996): 300–306, https://doi.org/10. 1017/S0963180100007076.
[54] Deborah A. Marshall, et al., "The Value of Diagnostic Testing for Parents of Children with Rare Genetic Diseases," *Genetics in Medicine* 21, no. 12 (December 1, 2019): 2798–2806, https://doi.org/10.1038/s41436-019-0583-1.
[55] "What Is Assisted Reproductive Technology? | Reproductive Health | CDC," November 9, 2020, https://www.cdc.gov/art/whatis.html.
[56] Romualdo Sciorio, Raffaele Aiello, and Alfonso Maria Irollo, "Review: Preimplantation Genetic Diagnosis (PGD) as a Reproductive Option in Patients with Neurodegenerative Disorders," *Reproductive Biology* 21, no. 1 (March 2021): 100468, https://doi.org/10.1016/j.repbio.2020.100468.
[57] Tur-Kaspa et al.
[58] "Special Report: Genetic Testing Goes Mainstream," *Science News* (blog), May 22, 2018, https://www.sciencenews.org/article/consumer-genetic-testing-ancestry-dna.
[59] "2017 Was the Year Consumer DNA Testing Blew Up," MIT Technology Review, accessed March 13, 2022, https://www.technologyreview.com/2018/02/12/145676/2017-was-the-year-consumer-dna-testing-blew-up/.
[60] 23andMe, "DNA Genetic Testing For Health, Ancestry And More - 23andMe," accessed March 13, 2022, https://www.23andme.com/.

[61] "AncestryDNA® | DNA Tests for Ethnicity & Genealogy DNA Test," accessed March 13, 2022, https://www.ancestry.com/dna/.
[62] "Consumer Genetic Testing Is Booming: But What Are the Benefits and Harms to Individuals and Populations? | Blogs | CDC," accessed March 13, 2022, https://blogs.cdc.gov/genomics/2018/06/12/consumer-genetic-testing/.
[63] The Regulatory Review, "Privacy Problems in the Genetic Testing Industry | The Regulatory Review," January 23, 2021, https://www.theregreview.org/2021/01/23/saturday-seminar-privacy-problems-genetic-testing/.
[64] Maarten H. D. Larmuseau, "Growth of Ancestry DNA Testing Risks Huge Increase in Paternity Issues," *Nature Human Behaviour* 3, no. 1 (January 2019): 5–5, https://doi.org/10.1038/s41562-018-0499-9.
[65] Alexander Nill and Gene Laczniak, "Direct-to-Consumer Genetic Testing and Its Marketing: Emergent Ethical and Public Policy Implications," *Journal of Business Ethics* 175, no. 4 (February 1, 2022): 669, https://doi.org/10.1007/s10551-020-04632-z.
[66] M Popovsky, "Exaggerated Benefts and Underestimated Harms: The Direct-to-Consumer Genetic Test Market and How to Manage It Going Forward," *Dartmouth Law Journal* 8 (2010): 65–87.
[67] Nill and Laczniak.
[68] Dieter Holger, "Dna Testing for Ancestry Is More Detailed for White People," *PCWorld* 37, no. 1 (January 2019): 96–103.
[69] "Definition of ALGORITHM," accessed March 13, 2022, https://www.merriam-webster.com/dictionary/algorithm.
[70] Ali Ghaheri et al., "The Applications of Genetic Algorithms in Medicine," *Oman Medical Journal* 30, no. 6 (November 2015): 406–16, https://doi.org/10.5001/omj.2015.82.
[71] "Human Genome Project: Sequencing the Human Genome | Learn Science at Scitable," accessed March 13, 2022, http://www.nature.com/scitable/topicpage/dna-sequencing-technologies-key-to-the-human-828.
[72] Rafal Smigrodzki et al., "Genetic Algorithm for Analysis of Mutations in Parkinson's Disease," *Artificial Intelligence in Medicine* 35, no. 3 (November 2005): 227–41, https://doi.org/10.1016/j.artmed.2004.11.006.
[73] Ghaheri et al.
[74] Cgc Janice Berliner MS, *Ethical Dilemmas in Genetics and Genetic Counseling: Principles Through Case Scenarios* (Oxford [England]: Oxford University Press, 2015), https://authenticate.library.duq.edu/login?url=https://search.ebscohost.com/login.aspx?direct=true&db=nlebk&AN=833180&site=eds-live&scope=site.
[75] Kärt Pormeister, "Genetic Research and Consent: On the Crossroads of Human and Data Research," *Bioethics* 33, no. 3 (March 2019): 347–56, https://doi.org/10.1111/bioe.12475.
[76] "What Are the Risks and Limitations of Genetic Testing?: MedlinePlus Genetics," accessed March 13, 2022, https://medlineplus.gov/genetics/understanding/testing/riskslimitations/.

[77] Mark A. Hall and Stephen S. Rich, "Laws Restricting Health Insurers' Use of Genetic Information: Impact on Genetic Discrimination," *The American Journal of Human Genetics* 66, no. 1 (January 1, 2000): 293–307, https://doi.org/10.1086/302714.
[78] Leyla Dinc and Fusun Terzioglu, "The Psychological Impact of Genetic Testing on Parents," *Journal of Clinical Nursing* 15, no. 1 (January 2006): 45–51, https://doi.org/10.1111/j.1365-2702.2005.01228.x.
[79] J. M. Trejo-Gabriel-Galán, "Euthanasia and Assisted Suicide in Neurological Diseases: A Systematic Review," *Neurologia (Barcelona, Spain)*, June 2, 2021, S0213-4853(21)00090-6, https://doi.org/10. 1016/j.nrl.2021.04.016.
[80] Chun-Hua Liao et al., "Preimplantation Genetic Diagnosis of Neurodegenerative Diseases: Review of Methodologies and Report of Our Experience as a Regional Reference Laboratory," *Diagnostics* 9, no. 2 (April 23, 2019): 44, https://doi.org/10.3390/diagnostics9020044.
[81] L. Fallon et al., "Preimplantation Genetic Diagnosis for Spinal Muscular Atrophy Type I," *Neurology* 53, no. 5 (September 22, 1999): 1087–90, https://doi.org/10.1212/wnl.53.5.1087.
[82] "Questions and Answers about CRISPR," Broad Institute, December 17, 2014, https://www.broadinstitute.org/what-broad/areas-focus/project-spotlight/questions-and-answers-about-crispr.
[83] Michael Le Page, "What Is Crispr?," *New Scientist* 240, no. 3207 (December 8, 2018): 7–7.
[84] Anna Groves, "Repairing the Future. (Cover Story)," *Discover* 40, no. 4 (May 2019): 28–33.
[85] Michael Kosicki, Kärt Tomberg, and Allan Bradley, "Repair of Double-Strand Breaks Induced by CRISPR-Cas9 Leads to Large Deletions and Complex Rearrangements," *Nature Biotechnology* 36, no. 8 (September 2018): 765–71, https://doi.org/10.1038/nbt.4192.
[86] Philip John Hardy and Kathy Hardy, "Chromosomal Instability in First Trimester Miscarriage: A Common Cause of Pregnancy Loss?," *Translational Pediatrics* 7, no. 3 (July 2018): 211–18, https://doi.org/ 10.21037/tp.2018.03.02.
[87] Genetic Alliance and District of Columbia Department of Health, *Single-Gene Disorders*, *Understanding Genetics: A District of Columbia Guide for Patients and Health Professionals* (Genetic Alliance, 2010), http://www.ncbi.nlm.nih.gov/books/NBK132154/.
[88] Fangzhi Jia, Avi Fellner, and Kishore Raj Kumar, "Monogenic Parkinson's Disease: Genotype, Phenotype, Pathophysiology, and Genetic Testing," *Genes* 13, no. 3 (March 2022): 471-N.PAG, https://doi.org/ 10.3390/genes13030471.
[89] Lola Cook, et al., "The Commercial Genetic Testing Landscape for Parkinson's Disease," *Parkinsonism & Related Disorders* 92 (November 2021): 107–11, https://doi.org/10.1016/j.parkreldis.2021.10.001.
[90] "The Role of Genes in Your Alzheimer's Risk," Mayo Clinic, accessed April 26, 2022, https://www.mayoclinic.org/diseases-conditions/alzheimers-disease/in-depth/alzheimers-genes/art-20046552.

[91] Aamira J Huq et al., "Genetic Testing in Dementia-A Medical Genetics Perspective," *International Journal of Geriatric Psychiatry* 36, no. 8 (August 2021): 1158–70, https://doi.org/10.1002/gps.5535.
[92] Jharna Mandal, Dinoop Korol Ponnambath, and Subhash Chandra Parija, "Utilitarian and Deontological Ethics in Medicine," *Tropical Parasitology* 6, no. 1 (2016): 5–7, https://doi.org/10.4103/2229-5070 .175024.
[93] Gerard Garbutt and Peter Davies, "Should the Practice of Medicine Be a Deontological or Utilitarian Enterprise?," *Journal of Medical Ethics* 37, no. 5 (May 2011): 267–70, https://doi.org/10.1136/jme.2010.036111.
[94] Tom L. Beauchamp and James F. Childress, *Principles of Biomedical Ethics*, Eighth edition. (Oxford University Press, 2019).
[95] Garbutt and Davies.
[96] Davor Solter et al., *Embryo Research in Pluralistic Europe* (Springer Science & Business Media, 2003).
[97] Thomas Douglas and Julian Savulescu, "Destroying Unwanted Embryos in Research. Talking Point on Morality and Human Embryo Research," *EMBO Reports* 10, no. 4 (April 2009): 307–12, https://doi.org/10. 1038/embor.2009.54.
[98] Tyler W. Buckner et al., "Health Care Resource Utilization and Cost Burden of Hemophilia B in the United States," *Blood Advances* 5, no. 7 (April 8, 2021): 1954–62, https://doi.org/10.1182/bloodadvances. 2020003424.
[99] Shermarke Hassan et al., "Mortality, Life Expectancy, and Causes of Death of Persons with Hemophilia in the Netherlands 2001–2018," *Journal of Thrombosis and Haemostasis* 19, no. 3 (March 2021): 645–53, https://doi.org/10.1111/jth.15182.
[100] J. M. Soucie, B. Evatt, and D. Jackson, "Occurrence of Hemophilia in the United States. The Hemophilia Surveillance System Project Investigators," *American Journal of Hematology* 59, no. 4 (December 1998): 288–94, https://doi.org/10.1002/(sici)1096-8652(199812)59:4<288 ::aid-ajh4>3.0.co;2-i.
[101] John Michael Soucie et al., "Occurrence Rates of Haemophilia among Males in the United States Based on Surveillance Conducted in Specialized Haemophilia Treatment Centres," *Haemophilia: The Official Journal of the World Federation of Hemophilia* 26, no. 3 (May 2020): 487–93, https://doi.org/10.1111/hae.13998.
[102] Julian Savulescu and Guy Kahane, "The Moral Obligation to Create Children with the Best Chance of the Best Life," *Bioethics* 23, no. 5 (2009): 275, https://doi.org/10.1111/j.1467-8519.2008.00687.x.
[103] Dieter Birnbacher, "Prenatal Diagnosis Yes, Preimplantation Genetic Diagnosis No: A Contradictory Stance?," *Reproductive BioMedicine Online* 14 (January 1, 2007): 109–13, https://doi.org/10.1016/S1472-6483(10)60741-9.
[104] Amanda Courtright-Lim and Mary Drago, "Ethics of Genetic Testing," *Medicine* 48, no. 10 (October 1, 2020): 675–79, https://doi.org/10.1016/j.mpmed.2020.07.014.
[105] Henk ten Have, *Global Bioethics: An Introduction* (Taylor & Francis Group, 2016), 221.
[106] Ibid., 222.

[107] Matthew J. Eckelman et al., "Health Care Pollution And Public Health Damage In The United States: An Update," *Health Affairs* 39, no. 12 (December 2020): 2071–79, https://doi.org/10.1377/hlthaff.2020.01247.
[108] Maggie Z. X. Xiao et al., "A Roadmap for Environmental Sustainability of Plastic Use in Anesthesia and the Perioperative Arena," *Anesthesiology* 135, no. 4 (October 1, 2021): 729–37, https://doi.org/10. 1097/ALN.0000000000003845.
[109] Jeffrey S. Yasny and Jennifer White, "Environmental Implications of Anesthetic Gases," *Anesthesia Progress* 59, no. 4 (2012): 154–58, https://doi.org/10.2344/0003-3006-59.4.154.
[110] "Doctors Pledge to Do No Harm. The Entire Health Care Sector Should Do the Same by Battling Climate Change," *STAT* (blog), August 11, 2021, https://www.statnews.com/2021/08/11/doctors-health-care-sector-battling-climate-change/.
[111] OAR US EPA, "Greenhouse Gas Emissions from a Typical Passenger Vehicle," Overviews and Factsheets, January 12, 2016, https://www.epa.gov/greenvehicles/greenhouse-gas-emissions-typical-passenger-vehicle.
[112] Patricia A. Roche and George J. Annas, "DNA Testing, Banking, and Genetic Privacy," *The New England Journal of Medicine* 355, no. 6 (August 10, 2006): 545–46, https://doi.org/10.1056/NEJMp068136.
[113] *Human Genome Editing: Science, Ethics, and Governance* (The National Academies Press, 2017), 117,118.
[114] Diane Gordon, "Sustainability in the Operating Room: Reducing Our Impact on the Planet," *Anesthesiology Clinics* 38, no. 3 (September 1, 2020): 679–92, https://doi.org/10.1016/j.anclin.2020.06.006.
[115] "Healthcare's Climate Footprint," accessed March 29, 2022, https://www.arup.com/perspectives/publications/research/section/healthcares-climate-footprint.

Literature

23andMe. N. d. "DNA Genetic Testing For Health, Ancestry And More - 23andMe." Accessed March 13, 2022. https://www.23andme.com/.

"A Practical Guide to Newborn Bloodspot Screening in Ireland." N.d., 68.

Alliance, Genetic, and District of Columbia Department of Health. 2010. *Single-Gene Disorders. Understanding Genetics: A District of Columbia Guide for Patients and Health Professionals*. Genetic Alliance. http://www.ncbi.nlm.nih.gov/books/NBK132154/.

"AncestryDNA® | DNA Tests for Ethnicity & Genealogy DNA Test." N. d.Accessed March 13, 2022. https://www.ancestry.com/dna/.

Anton-Paduraru, Dana-Teodora. 2017. "OC-24 Newborn Screening for Cystic Fibrosis: To Screen or Not to Screen (Pros and Cons)." *Archives of Disease in Childhood* 102, no. Suppl 2 (June 1): A9–A9. https://doi.org/10.1136/archdischild-2017-313273.24.

Baruch, Susannah, David Kaufman, and Kathy L. Hudson. 2008. "Genetic Testing of Embryos: Practices and Perspectives of US in Vitro Fertilization Clinics." *Fertility and Sterility* 89, no. 5 (May 1): 1053–58. https://doi.org/10.1016/j.fertnstert.2007.05.048.

Bayefsky, Michelle J. 2016. "Comparative Preimplantation Genetic Diagnosis Policy in Europe and the USA and Its Implications for Reproductive Tourism." *Reproductive Biomedicine & Society Online* 3 (December): 41–47. https://doi.org/10.1016/j.rbms.2017.01.001.

Beauchamp, Tom L., and James F. Childress. 2019. *Principles of Biomedical Ethics*. Eighth edition. Oxford University Press.

Biesecker, Leslie G. 1996. "Orphan Tests." *Cambridge Quarterly of Healthcare Ethics* 5, no. 2 (ed): 300–306. https://doi.org/10.1017/S0963180100007076.

Birnbacher, Dieter. 2007. "Prenatal Diagnosis Yes, Preimplantation Genetic Diagnosis No: A Contradictory Stance?" *Reproductive BioMedicine Online* 14 (January 1): 109–13. https://doi.org/10.1016/S1472-6483 (10)60741-9.

Blankenship, Crystal S. 2008. "To Manage Costs of Hemophilia, Patients Need More Than Clotting Factor." *Biotechnology Healthcare* 5, no. 4 (2008): 37–40.

Broad Institute. 2014. "Questions and Answers about CRISPR," December 17. https://www.broadinstitute.org/what-broad/areas-focus/project-spotlight/questions-and-answers-about-crispr.

Buckner, Tyler W., Iryna Bocharova, Kaitlin Hagan, Arielle G. Bensimon, Hongbo Yang, Eric Q. Wu, Eileen K. Sawyer, and Nanxin Li. 2021. "Health Care Resource Utilization and Cost Burden of Hemophilia B in the United States." *Blood Advances* 5, no. 7 (April 8): 1954–62. https://doi.org/ 10.1182/bloodadvances.2020003424.

Caggana, Michele, Elizabeth A. Jones, S.I. Shahied, Susan Tanksley, Cheryl A. Hermerath, and Ira M. Lubin. 2013. "Newborn Screening: From Guthrie to Whole Genome Sequencing." *Public Health Reports* 128, no. Suppl 2: 14–19.

"Choose Your Baby's Eye Color | The Fertility Institutes Offers Eye Color Screening and Selection." N. d. Accessed April 25, 2022. http://www.fertility-docs.com/programs-and-services/pgd-screening/choose-your-babys-eye-color.php.

Cleveland Clinic. N. d. "Spinal Muscular Atrophy (SMA): Types, Symptoms & Treatment." Accessed April 25, 2022. https://my.clevelandclinic.org/health/diseases/14505-spinal-muscular-atrophy-sma.

"Consumer Genetic Testing Is Booming: But What Are the Benefits and Harms to Individuals and Populations? | Blogs | CDC." N. d. Accessed March 13, 2022. https://blogs.cdc.gov/genomics/2018/06/12/consumer-genetic-testing/.

Cook, Lola, Jeanine Schulze, Jennifer Verbrugge, James C Beck, Karen S Marder, Rachel Saunders-Pullman, Christine Klein, Anna Naito, and Roy N Alcalay. 2021. "The Commercial Genetic Testing Landscape for Parkinson's Disease." *Parkinsonism & Related Disorders* 92 (November): 107–11. https://doi.org/10.1016/j.parkreldis.2021.10.001.

Courtright-Lim, Amanda, and Mary Drago. 2020. "Ethics of Genetic Testing." *Medicine* 48, no. 10 (October 1): 675–79. https://doi.org/10.1016/j.mpmed.2020.07.014.

Crowe, John. N. d. "Plan to Destroy Heelprick Cards Is a Serious Mistake." The Irish Times. Accessed February 6, 2022. https://www.irishtimes.com/news/health/plan-to-destroy-heelprick-cards-is-a-serious-mistake-1.1330334.

Dangouloff, Tamara, Camille Botty, Charlotte Beaudart, Laurent Servais, and Mickaël Hiligsmann. 2021. "Systematic Literature Review of the Economic Burden of Spinal Muscular Atrophy and Economic Evaluations of Treatments." *Orphanet Journal of Rare Diseases* 16, no. 1 (January 23): 47. https://doi.org/10.1186/s13023-021-01695-7.

"Default - Stanford Children's Health." N. d. Accessed February 7, 2022. https://www.stanfordchildrens.org/en/topic/default?id=newborn-screening-tests-90-P01967.

"Definition of ALGORITHM." N. d. Accessed March 13, 2022. https://www. merriam-webster.com/dictionary/algorithm.

Dinc, Leyla, and Fusun Terzioglu. 2006. "The Psychological Impact of Genetic Testing on Parents." *Journal of Clinical Nursing* 15, no. 1 (January): 45–51. https://doi.org/10.1111/j.1365-2702.2005.01228.x.

Douglas, Thomas, and Julian Savulescu. 2009. "Destroying Unwanted Embryos in Research. Talking Point on Morality and Human Embryo Research." *EMBO Reports* 10, no. 4 (April): 307–12. https://doi .org/10.1038 /embor.2009.54.

Eckelman, Matthew J., Kaixin Huang, Robert Lagasse, Emily Senay, Robert Dubrow, and Jodi D. Sherman. 2020. "Health Care Pollution And Public Health Damage In The United States: An Update." *Health Affairs* 39, no. 12 (December): 2071–79. https://doi.org/10.1377/hlthaff.2020.01247.

Fallon, L., G. L. Harton, M. E. Sisson, E. Rodriguez, L. K. Field, E. F. Fugger, M. Geltinger, et al. 1999. "Preimplantation Genetic Diagnosis for Spinal Muscular Atrophy Type I." *Neurology* 53, no. 5 (September 22): 1087–90. https://doi.org/10.1212/wnl.53.5.1087.

Feero, W. Gregory, Alan E. Guttmacher, and Francis S. Collins. 2010. "Genomic Medicine--an Updated Primer." *The New England Journal of Medicine* 362, no. 21 (May 27): 2001–11. https://doi.org/10.1056/NEJMra 0907175.

Garbutt, Gerard, and Peter Davies. 2011. "Should the Practice of Medicine Be a Deontological or Utilitarian Enterprise?" *Journal of Medical Ethics* 37, no. 5 (May): 267–70. https://doi.org/10.1136/jme.2010.036111.

Ghaheri, Ali, Saeed Shoar, Mohammad Naderan, and Sayed Shahabuddin Hoseini. 2015. "The Applications of Genetic Algorithms in Medicine." *Oman Medical Journal* 30, no. 6 (November): 406–16. https://doi .org/10. 5001/omj.2015.82.

Gordon, Diane. 2020. "Sustainability in the Operating Room: Reducing Our Impact on the Planet." *Anesthesiology Clinics* 38, no. 3 (September 1): 679–92. https://doi.org/10.1016/j.anclin.2020.06.006.

Goto, Tatenobu, Takeshi Inoue, Chinami Kamiya, Hiroyuki Kawabe, Machiko Higuchi, Megumi Suyama, Tomoki Goto, et al. 2020. "Neonatal Pain Response to Automatic Lancet versus Needle Heel-prick Blood Sampling: A Prospective Randomized Controlled Clinical Trial." *Pediatrics International* 62, no. 3 (March): 357–62.

Gottesman, M. M., and F. S. Collins. 1994. "The Role of the Human Genome Project in Disease Prevention." *Preventive Medicine* 23, no. 5 (September): 591–94. https://doi.org/10.1006/pmed.1994.1094.

Groves, Anna. 2019. "Repairing the Future. (Cover Story)." *Discover* 40, no. 4 (May): 28–33.

Hall, Mark A., and Stephen S. Rich. 2000. "Laws Restricting Health Insurers' Use of Genetic Information: Impact on Genetic Discrimination." *The American Journal of Human Genetics* 66, no. 1 (January 1): 293–307. https:// doi.org/10.1086/302714.

Handyside, A. H., J. G. Lesko, J. J. Tarín, R. M. Winston, and M. R. Hughes. 1992. "Birth of a Normal Girl after in Vitro Fertilization and Preimplantation Diagnostic Testing for Cystic Fibrosis." *The New England Journal of Medicine* 327, no. 13 (September 24): 905–9. https://doi.org/10. 1056/NEJM199209243271301.

Hardy, Philip John, and Kathy Hardy. 2018. "Chromosomal Instability in First Trimester Miscarriage: A Common Cause of Pregnancy Loss?" *Translational Pediatrics* 7, no. 3 (July): 211–18. https://doi.org/10.21037/tp.2018.03.02.

Hassan, Shermarke, Rory C. Monahan, Evelien P. Mauser-Bunschoten, Lize F. D. van Vulpen, Jeroen Eikenboom, Erik A. M. Beckers, Louise Hooimeijer, et al. 2021. "Mortality, Life Expectancy, and Causes of Death of Persons with Hemophilia in the Netherlands 2001–2018." *Journal of Thrombosis and Haemostasis* 19, no. 3 (March): 645–53. https://doi. org/10.1111/jth.15182.

Have, Henk ten. 2016. *Global Bioethics: An Introduction*. Taylor & Francis Group.

"Healthcare's Climate Footprint." N. d. Accessed March 29, 2022. https://www.arup.com/perspectives/publications/research/section/healthcares-climate-footprint.

Heape, Walter, and Michael Foster. 1891. "III. Preliminary Note on the Transplantation and Growth of Mammalian Ova within a Uterine Foster-Mother." *Proceedings of the Royal Society of London* 48, no. 292–295 (January 1): 457–58. https://doi.org/10.1098/rspl.1890 .0053.

"History of the Monash IVF Research Program from 1971 to 1989 | The Embryo Project Encyclopedia." N. d. Accessed February 6, 2022. https://embryo.asu.edu/pages/history-monash-ivf-research-program-1971-1989.

Holger, Dieter. 2019. "Dna Testing for Ancestry Is More Detailed for White People." *PCWorld* 37, no. 1 (January): 96–103.

Human Genome Editing: Science, Ethics, and Governance. The National Academies Press, 2017.

"Human Genome Project: Sequencing the Human Genome | Learn Science at Scitable." N. d. Accessed March 13, 2022. http://www.nature.com/scitable/topicpage/dna-sequencing-technologies-key-to-the-human-828.

Huq, Aamira J, Adrienne Sexton, Paul Lacaze, Colin L Masters, Elsdon Storey, Dennis Velakoulis, Paul A James, and Ingrid M Winship. 2021. "Genetic Testing in Dementia-A Medical Genetics Perspective." *International Journal of Geriatric Psychiatry* 36, no. 8 (August): 1158–70. https://doi.org/10.1002/gps.5535.

Janice Berliner MS, Cgc. 2015. *Ethical Dilemmas in Genetics and Genetic Counseling: Principles Through Case Scenarios*. Oxford [England]: Oxford University Press. https://authenticate.library.duq.edu/ login?url=https://search.ebscohost.com/login.aspx?direct=true&db=nlebk&AN=833180&site=eds-live&scope=site.

Jia, Fangzhi, Avi Fellner, and Kishore Raj Kumar. 2022. "Monogenic Parkinson's Disease: Genotype, Phenotype, Pathophysiology, and Genetic Testing." *Genes* 13, no. 3 (March 2022): 471-N.PAG. https://doi.org/10.3390/genes 13030471.

Justia Dockets & Filings. N. d. "Beleno et al v. Texas Department of State Health Services et Al." Accessed February 7, 2022. https://dockets.justia.com/ docket/texas/txwdce/5:2009cv00188/353950.

Justia Law. "Bearder v. State." N. d. Accessed February 7, 2022. https://law.justia.com/cases/minnesota/supreme-court/2011/a10-101.html.

Kemper, James M, Christopher Gyngell, and Julian Savulescu. 2019. "Subsidizing PGD: The Moral Case for Funding Genetic Selection." *Journal of Bioethical Inquiry* 16, no. 3 (September): 405–14. https://doi.org/10 .1007/s11673-019-09932-2.

Kosicki, Michael, Kärt Tomberg, and Allan Bradley. 2018. "Repair of Double-Strand Breaks Induced by CRISPR-Cas9 Leads to Large Deletions and Complex Rearrangements." *Nature Biotechnology* 36, no. 8 (September): 765–71. https://doi.org/10.1038/nbt.4192.

Kruse, Judith, Regina Mueller, Ali A. Aghdassi, Markus M. Lerch, and Sabine Salloch. 2022. "Genetic Testing for Rare Diseases: A Systematic Review of Ethical Aspects." *Frontiers in Genetics* 13 (January 26): 1–14.

Larmuseau, Maarten H. D. 2019. "Growth of Ancestry DNA Testing Risks Huge Increase in Paternity Issues." *Nature Human Behaviour* 3, no. 1 (January): 5–5. https://doi.org/10.1038/s41562-018-0499-9.

Le Page, Michael. 2018. "What Is Crispr?" *New Scientist* 240, no. 3207 (December 8): 7–7.

Lewis, Michelle H., Aaron Goldenberg, Rebecca Anderson, Erin Rothwell, and Jeffrey Botkin. 2011. "State Laws Regarding the Retention and Use of Residual Newborn Screening Blood Samples." *Pediatrics* 127, no. 4 (April): 703–12. https://doi.org/10.1542/peds.2010-1468.

Liao, Chun-Hua, Ming-Yuh Chang, Gwo-Chin Ma, Shun-Ping Chang, Chi-Fang Lin, Wen-Hsiang Lin, Hsin-Fu Chen, et al. 2019. "Preimplantation Genetic Diagnosis of Neurodegenerative Diseases: Review of Methodologies and Report of Our Experience as a Regional Reference Laboratory." *Diagnostics* 9, no. 2 (April 23): 44. https://doi.org /10.3390/ diagnostics9020044.

Mandal, Jharna, Dinoop Korol Ponnambath, and Subhash Chandra Parija. 2016. "Utilitarian and Deontological Ethics in Medicine." *Tropical Parasitology* 6, no. 1: 5–7. https://doi.org/10.4103/2229-5070 .175024.

Marshall, Deborah A., Karen V. MacDonald, Sebastian Heidenreich, Taila Hartley, Francois P. Bernier, Meredith K. Gillespie, Brenda McInnes, A. Micheil Innes, Christine M. Armour, and Kym M. Boycott. 2019. "The Value of Diagnostic Testing for Parents of Children with Rare

Genetic Diseases." *Genetics in Medicine* 21, no. 12 (December 1): 2798–2806. https://doi.org/10.1038/s41436-019-0583-1.

"Miscarriage - Prenatal Genetics." N. d. https://www.gbmc.org. Accessed February 6, 2022. https://www.gbmc.org/miscarriage.

MIT Technology Review. N. d. "2017 Was the Year Consumer DNA Testing Blew Up." Accessed March 13, 2022. https://www.technologyreview .com2018/02/12/145676/2017-was-the-year-consumer-dna-testing-blew-up/.

Neergaard, Lauran. N. d. "Ethics Debate over Blood from Newborn Safety Tests." *Boston.Com*, February 8, 2010. http://archive.boston.com/news /health/articles/2010/02/08/ethics_debate_over_blood_from_newborn_ safety_tests/.

Nill, Alexander, and Gene Laczniak. 2022. "Direct-to-Consumer Genetic Testing and Its Marketing: Emergent Ethical and Public Policy Implications." *Journal of Business Ethics* 175, no. 4 (February 1): 669–88. https://doi.org/10.1007/s10551-020-04632-z.

"PGD and IVF Costs – What Is the Cost for Preimplantation Genetic Diagnosis – PGD and PGS, Preimplantation Genetic Screening? – Advanced Fertility Center of Chicago." N. d. Accessed April 25, 2022. https://advancedfertility.com/fertility-treatment/affording-care/pgd-cost/.

Pitt, James J. 2010. "Newborn Screening." *The Clinical Biochemist Reviews* 31, no. 2 (May): 57–68.

Popovsky, M. 2010. "Exaggerated Benefts and Underestimated Harms: The Direct-to-Consumer Genetic Test Market and How to Manage It Going Forward." *Dartmouth Law Journal* 8: 65–87.

Pormeister, Kärt. 2019. "Genetic Research and Consent: On the Crossroads of Human and Data Research." *Bioethics* 33, no. 3 (March): 347–56. https://doi.org/10.1111/bioe.12475.

"Preimplantation Genetic Testing." N. d. Accessed February 6, 2022. https://nyulangone.org/locations/fertility-center/preimplantation-genetic-testing.

Ranisch, Robert. 2020. "Germline Genome Editing versus Preimplantation Genetic Diagnosis: Is There a Case in Favour of Germline Interventions?" *Bioethics* 34, no. 1 (January): 60–69. https://doi.org/10.1111/bioe. 12635.

Review, The Regulatory. 2021. "Privacy Problems in the Genetic Testing Industry | The Regulatory Review," January 23. https://www.theregreview.org/2021/01/23/saturday-seminar-privacy-problems-genetic-testing/.

Roche, Patricia A., and George J. Annas. 2006. "DNA Testing, Banking, and Genetic Privacy." *The New England Journal of Medicine* 355, no. 6 (August 10): 545–46. https://doi.org/10.1056/NEJMp068136.

Savulescu, Julian, and Guy Kahane. 2009. "The Moral Obligation to Create Children with the Best Chance of the Best Life." *Bioethics* 23, no. 5: 274–90. https://doi.org/10.1111/j.1467-8519.2008.00687.x.

Sciorio, Romualdo, Raffaele Aiello, and Alfonso Maria Irollo. 2021. "Review: Preimplantation Genetic Diagnosis (PGD) as a Reproductive Option in Patients with Neurodegenerative Disorders." *Reproductive Biology* 21, no. 1 (March): 100468. https://doi.org/10.1016/j.repbio.2020. 100468.

Smigrodzki, Rafal, Ben Goertzel, Cassio Pennachin, Lucio Coelho, Francisco Prosdocimi, and W. Davis Parker. 2005. "Genetic Algorithm for Analysis of Mutations in Parkinson's Disease." *Artificial Intelligence in Medicine* 35, no. 3 (November): 227–41. https://doi.org/10 .1016/j.artmed.2004.11. 006.

Soini, S. 2007. "Preimplantation Genetic Diagnosis (PGD) in Europe: Diversity of Legislation a Challenge to the Community and Its Citizens." *Medicine and Law* 26, no. 2 (June): 309–23.

Solter, Davor, D. Beyleveld, M. B. Friele, J. Holówka, H. Lilie, R. Lovell-Badge, C. Mandla, U. Martin, and R. Pardo Avellaneda. 2003. *Embryo Research in Pluralistic Europe*. Springer Science & Business Media.

Soucie, J. M., B. Evatt, and D. Jackson. 1998. "Occurrence of Hemophilia in the United States. The Hemophilia Surveillance System Project Investigators." *American Journal of Hematology* 59, no. 4 (December): 288–94. https://doi.org/10.1002/(sici)1096-8652(199812)59:4<288::aid-ajh4>3.0.co;2-i.

Soucie, John Michael, Connie H. Miller, Brandi Dupervil, Binh Le, and Tyler W. Buckner. 2020. "Occurrence Rates of Haemophilia among Males in the United States Based on Surveillance Conducted in Specialized Haemophilia Treatment Centres." *Haemophilia: The Official Journal of the World Federation of Hemophilia* 26, no. 3 (May): 487–93. https://doi.org/10.1111/hae.13998.

Science News. 2018. "Special Report: Genetic Testing Goes Mainstream," May 22. https://www.sciencenews.org/article/consumer-genetic-testing-ancestry-dna.

STAT. 2021. "Doctors Pledge to Do No Harm. The Entire Health Care Sector Should Do the Same by Battling Climate Change," August 11.

https://www.statnews.com/2021/08/11/doctors-health-care-sector-battling-climate-change/.

Stern, Harvey J. 2014. "Preimplantation Genetic Diagnosis: Prenatal Testing for Embryos Finally Achieving Its Potential." *Journal of Clinical Medicine* 3, no. 1 (March 17): 280–309. https://doi.org/10.3390 /jcm3010280.

Storrow, Richard F. 2011. "Assisted Reproduction on Treacherous Terrain: The Legal Hazards of Cross-Border Reproductive Travel." *Reproductive BioMedicine Online*, Symposium: Cross-Border Reproductive Care – Ethical, Legal, and Socio-Cultural Perspectives, 23, no. 5 (November 1): 538–45. https://doi.org/10.1016/j.rbmo.2011.07.008.

Strange, H., and R. Chadwick. 2010. "The Ethics of Nonmedical Sex Selection." *Health Care Analysis: An International Journal of Health Care Philosophy and Policy* 18, no. 3 (September 1): 252–66. https://doi. org/10.1007/s10728-009-0135-y.

Genome.gov. N. d. "The Human Genome Project." Accessed March 12, 2022. https://www.genome.gov/human-genome-project.

Mayo Clinic. 2022. "The Role of Genes in Your Alzheimer's Risk." Accessed April 26. https://www.mayoclinic.org/diseases-conditions /alzheimers-disease/in-depth/alzheimers-genes/art-20046552.

Theodosiou, Anastasia A., and Martin H. Johnson. 2011. "The Politics of Human Embryo Research and the Motivation to Achieve PGD." *Reproductive BioMedicine Online* 22, no. 5 (May 1): 457–71. https://doi.org/10.1016/j.rbmo.2011.01.008.

Therrell, Bradford L. 2003. "Ethical, Legal and Social Issues in Newborn Screening in the United States." *The Southeast Asian Journal of Tropical Medicine and Public Health* 34 Suppl 3: 52–58.

Trejo-Gabriel-Galán, J. M. 2021. "Euthanasia and Assisted Suicide in Neurological Diseases: A Systematic Review." *Neurologia (Barcelona, Spain)*, June 2, S0213-4853(21)00090-6. https://doi.org/10.1016/j.nrl.2021.04.016.

Tur-Kaspa, I., G. Aljadeff, S. Rechitsky, H. E. Grotjan, and Y. Verlinsky. 2010. "PGD for All Cystic Fibrosis Carrier Couples: Novel Strategy for Preventive Medicine and Cost Analysis." *Reproductive Biomedicine Online* 21, no. 2 (August): 186–95. https://doi.org/10.1016/j. rbmo.2010.04.031.

US EPA, OAR. 2016. "Greenhouse Gas Emissions from a Typical Passenger Vehicle." Overviews and Factsheets, January 12. https://www. epa.gov/greenvehicles/greenhouse-gas-emissions-typical-passenger-vehicle.

Vehmas, Simo, and Tom Shakespeare. 2014. "Disability, Harm, and the Origins of Limited Opportunities." *Cambridge Quarterly of Healthcare Ethics* 23, no. 1 (January): 41–47. https://doi.org/10.1017/S096318011300042X.

Vittozzi, Luciano, Georg F Hoffmann, Martina Cornel, and Gerard Loeber. 2010. "Evaluation of Population Newborn Screening Practices for Rare Disorders in Member States of the European Union." *Orphanet Journal of Rare Diseases* 5, no. S1 (October): P26, 1750-1172-5-S1-P26. https://doi.org/10.1186/1750-1172-5-S1-P26.

Wallis, Jacqueline Mae. 2020. "Is It Ever Morally Permissible to Select for Deafness in One's Child?" *Medicine, Health Care, and Philosophy* 23, no. 1 (March): 3–15. https://doi.org/10.1007/s11019-019-09922-6.

Walter, Fiona M, and Jon D Emery. 2012. "Genetic Advances in Medicine: Has the Promise Been Fulfilled in General Practice?" *The British Journal of General Practice* 62, no. 596 (March): 120–21. https://doi.org /10. 3399/bjgp12X629955.

"What Are the Risks and Limitations of Genetic Testing?: MedlinePlus Genetics." N. d. Accessed March 13, 2022. https://medlineplus.gov/genetics/understanding/testing/riskslimitations/.

"What Is Assisted Reproductive Technology? | Reproductive Health | CDC,"2020. https://www.cdc.gov/art/whatis.html.

Cleveland Clinic. 2020. "Why Do They Prick Your Newborn Baby's Heel?," October 6. https://health.clevelandclinic.org/why-do-they-prick-your -newborn-babys-heel/.

Xiao, Maggie Z. X., Syed Ali Akbar Abbass, Lisa Bahrey, Edward Rubinstein, and Vincent W. S. Chan. 2021. "A Roadmap for Environmental Sustainability of Plastic Use in Anesthesia and the Perioperative Arena." *Anesthesiology* 135, no. 4 (October 1): 729–37. https:// doi.org/10. 1097/ALN.0000000000003845.

Yasny, Jeffrey S., and Jennifer White. 2012. "Environmental Implications of Anesthetic Gases." *Anesthesia Progress* 59, no. 4: 154–58. https://doi.org/10.2344/0003-3006-59.4.154.

CHAPTER SIXTEEN

THE RESILIENT USE OF HYPERSPECTRAL IMAGING IN THE RESEARCH OF CANCER IMAGING

ETHAN HERSTEK

Introduction

Medical imaging began, in the simplest of ways, with the discovery of X-rays in 1895 by a German professor of physics, Wilhelm Rontgen. Over the years, many different imaging modalities have been introduced into the medical field to diagnose diseases and progress the medical industry. We as scientists, are always trying to push the boundaries of medicine as far as possible to ensure current and future generations are privileged with the best technology that mankind has to offer. Each modality, from CT scans to multispectral and fluorescence imaging, has been engineered to address a specific issue. Each modality uses different technological advances, and methods to see into the human body allowing us to create images to address issues within human anatomy correctly and accurately. This chapter will detail the inner workings of hyperspectral imaging in the medical industry. We will then focus on the applications of this imaging modality, in terms of cancer imaging. Lastly, we will study how bioethics plays a part in the adoption of hyperspectral imaging in the medical field of cancer research and imaging.

Background

Before we can examine how hyperspectral imaging works and how it is a beneficial modality over other current technologies that are in use, we must learn about some crucial basics. Current imaging modalities capture image data throughout the electromagnetic spectrum, not just the visible light spectrum. But what exactly does that mean? To understand exactly how this works, we must first understand how we as humans interpret objects around us in different colors, as well as understand how everyday cameras work. Why is one shade of green different from another? Why is that apple red? How does our camera replicate the surroundings we are seeing with our eyes into an image? These sound like simple questions, but asking these scientific questions allows us to understand the basics of imaging, which is necessary when various imaging modalities will be studied.

Every color that we see on a day-to-day basis is incorporated into what is referred to as the visible light spectrum. The visible light spectrum is the segment of the electromagnetic spectrum humans can see. Look at a rainbow, for instance. We have all heard of ROYGBIV. These are the basic colors that make up this visible light. Diving deeper into this concept, we must understand white light. White light is the combination of all colors on the visible light spectrum. White light is the light given off by the sun, as well as most of our household lightbulbs. This light consists of the visible light spectrum. Each color involved in this spectrum has a different frequency and wavelength that is either absorbed or reflected differently by objects in contact with white light. Back to the question of why the apple you put on your teacher's desk is red. When the white light given off by the sun hits the apple, the apple has properties that absorb each color except for the color we see, red, which is reflected. This applies to every object that you see from green blades of grass to the blue color of the sky to the brown color of school desks.[1]

Now, how do our everyday cameras like our phones work? The cameras that we are familiar with capture visible light mainly in three colors, red, blue, and green. Our cameras consist of sensing elements (sensels) covered by filters. The filters correspond to their specific colored sensel, red, blue, or green. If red light strikes a green filter and sensel, it will not be able to see it, as only red light will be able to penetrate this filter and be captured by this specific sensel. This is also true for the red and blue components as well. When you take a picture, each of these sensels is filled with the corresponding brightness of each color. These sensels in turn are combined to make pixels. Each pixel corresponds to one singular color. This process is repeated in each pixel that ends up creating the pictures we take with our cameras.[2]

Hyperspectral Imaging

Now that we understand the basics of imaging and light, we can dive deeper into hyperspectral imaging. Hyperspectral imaging is an imaging modality that is widely used throughout the medical and agricultural industries as well as being used by the military and NASA to image space. In addition to the above-explained visible light spectrum, and white light, what makes multispectral imaging special is that it allows us to obtain images by being able to see wavelengths that the human eye, as well as other imaging modalities, cannot.

Hyperspectral imaging divides the regions of the electromagnetic spectrum that we cannot see into many more bands than just RGB and creates visible images from them. As we already know, our basic cameras use the primary color bands of red, green, and blue filters to make images, and each image is comprised of thousands of pixels in which each pixel only relates to one specific color. In hyperspectral imaging, we use at times hundreds to thousands of different bands and filters with different widths to acquire image data in each pixel. In a typical RGB camera, there is no depth to the pictures taken. With hyperspectral imaging, since there is a continuous number of spectral bands, we now have depth to our images. A hyperspectral image is made up of many different images, but we can see colors in these images that we can't with the normal eye because each of those individual pictures is comprised of individual pixels that incorporate an entire spectra. And just to reiterate, an RGB camera's pixels only consider one spectra (red, green, or blue) in each pixel.[3]

Hyperspectral imaging takes advantage of various filters as well. Using various bandpass filters in the system, when necessary, we can select which part of an object's wavelength we want to image. We can use these bandpass filters at specific points in the wavelengths of substances or objects to be able to differentiate between colors that a normal camera or the human eye cannot. This is especially extremely useful in imaging colors that are very similar. For example, take three substances that you can think of that are similar in color, such as sugar, flour, and calcium. To our eyes, we see each substance as white, as we are seeing them in the visible light spectrum. In reality, these substances do not have the same chemical properties, and in turn, reflect different electromagnetic radiation outside the visible light spectrum. These substances give off the same color at the lower wavelength of their spectrum, but they give off different colors at their higher wavelengths. In turn, if we add a bandpass filter at the higher wavelength of these substances, the colors separate. Adding this bandpass filter allows us to only image these substances at their higher wavelengths which give off different colors and therefore we can differentiate between these substances. To more easily understand this example, take a look at your hand and spread out your fingers as far as possible. At the point where your fingers come out of your palm, they are close together, but towards the tips of your fingers, they are more separated. This is similar to how these three substances' wavelengths are configured. The bandpass filter that we are adding in this example will only take into account the tips of your finger which are more distinguishable at their tips than they are at the palm.

We can summarize hyperspectral imaging with an example. If a plane were to fly over a landscape and take a picture with an RGB camera, it would only be able to capture the light that is reflected off the Earth's surface within the visible light spectrum and result in a picture that we are accustomed to seeing. But remember, the sun gives off light from the entire electromagnetic spectrum, and visible light is only a small part of that. If the same plane flies overhead taking the same picture with a Hyperspectral camera, the camera will be able to capture almost all the light that is being reflected from the Earth's surface and would result in a very different picture. In reality, they are reflecting much more. Depending on the application needed from this picture, using certain filters, we would be able to differentiate the topography of the landscape much more clearly than with an RGB camera, as the chemical makeup of each material on the ground reflects a different color which normal cameras and our eyes just can not differentiate.

Hyperspectral Imaging in Cancer Detection

The wavelengths of light used in HSI can penetrate different depths into biological tissues and because of this, in recent times, hyperspectral imaging has been at the forefront of research in cancer detection, imaging, and image-guided surgery. Hyperspectral imaging provides an effective method in the finding of tumors and can give information about surrounding healthy and infected tissues that current imaging modalities are unable to do. Hyperspectral imaging gathers spatial and spectral information from tissues simultaneously, giving images depth. This modality collects information in three dimensions—vertical, horizontal, and one spectral—resulting in what was previously described as a data cube. This modality is based on using the previously explained principle that everything, including tissues, has different chemical properties with different spectral reflectance responses, giving them each a unique "fingerprint", that current imaging systems are unable to obtain as efficiently as hyperspectral imaging. The development of diseases is followed by changes in the optical and pathological properties of the tissue that is infected.[4]

Our tissue structure and the chromophores in our tissue cause the light from the hyperspectral system to undergo certain property changes. The light will be subject to scattering, and absorption when traveling through the tissues. This scattering and absorption act differently in healthy tissue than it does in healthy tissues. The absorption spectrum will also reflect a certain degree coinciding with angiogenesis and metabolism which reveals the scatter and absorption information that the hyperspectral imaging system can detect and turn into its data for imaging. These changes in chemical composition, as well as the measuring of the absorption, fluorescence, emission, or reflectance spectrum, are easily picked up by hyperspectral imaging systems and in turn, can be turned into data for diagnostic information.[5]

It is because of these unique fingerprints that this modality has shown to be a more effective and accurate imaging modality in terms of finding cancerous tumors and separating them from healthy tissues, compared to current modalities such as CT scans, MRI, and X-rays. HSI captures images under different spectral bands (visible light, infrared, ultraviolet) while being a non-invasive way to image the body. Overall, HSI has been shown to give a doctor or surgeon a high-resolution image of a patient's body and detect cancer growth earlier than current imaging modalities. One advantage that hyperspectral imaging has over current imaging modalities in cancer imaging is the incorporation of spatial information can show an exact size, shape, and location of targeted objects. This spatial information also exposes composition-specific data under hundred of wavelengths. This is important as the most effective way to remove cancerous tumors is to remove one layer of healthy tissue after the tumor is completely removed. Current imaging modalities struggle to differentiate this last layer of infected tissue from the healthy tissue underneath.[6]

There are also various disadvantages that other imaging systems have that hyperspectral imaging does not. CT scans implement radiation exposure and have poor soft tissue contrast. PET exposes the patient to radiation, has low spatial resolution and long acquisition time, and is one of the most expensive imaging modalities. MRI has a long acquisition time and limited sensitivity. Ultrasound is unable to be used for a full body image. Hyperspectral imaging does not expose the patient to any negative radiation, has a high spatial resolution, and can be used for a full body image. It is for these reasons as well as previously explained reasons that hyperspectral imaging is the lead imaging modality in the imaging of tumors and cancer detection.

Hyperspectral imaging is also able to detect oxygen saturation in the body. This enables the creation of 2D maps of the blood oxygenation of tissues. This allows the expansion of hyperspectral imaging into other areas of the medical detection field such as mapping of brain activity, monitoring of wound healing, ophthalmological diagnoses of eye and brain diseases, and detection of blood traces.[7] Hyperspectral imaging can take this data based on healthy tissue and blood activity and contrast it with other patients for a possibility of detecting cancers earlier than ever before.[8]

Future of Hyperspectral Imaging

Hyperspectral imaging is a technology that we have not yet been able to reach the full potential of as we have with other imaging modalities. Because of this, the future of hyperspectral imaging is bright, not just in medicinal applications, but in agriculture, space exploration, food quality, and mineral mapping to just name a few.

One aspect that will improve upon hyperspectral imaging in the future is its use in combination with artificial intelligence. Hyperspectral imaging has been and continues to be an expensive imaging modality, as well as having difficult processing techniques to produce data because of its complexity. This has prevented it from being incorporated into various industries as fast as some would like. This is not a surprise as this is usually the case with new technologies, especially in the medical field. Combining this technology with new image processing and artificial intelligence, the hope is that this technology can be incorporated more thoroughly throughout various industries which it is currently breaking into. Artificial intelligence will be able to easier compress the large hyperspectral data cube into the desirable data, as well as better and easier data handling. Making these improvements will enable the technology to be more widespread. As the technology becomes more widespread, then the cost of obtaining a system will decrease. Also, the more widespread this technology can become, the more it can be incorporated into research for various purposes outside the cancer imaging this chapter has talked about.[9]

Also, another benefit of incorporating hyperspectral imaging with artificial intelligence is to ensure that the images collected from the scans allow for 100% diagnostic accuracy. In scientific experiments, something that always must be accounted for is human error. Incorporating artificial intelligence when imaging potential cancerous areas ensures that human error does not need to be accounted for. Based on the imaging results of the potentially affected area, instead of having a human interpret these images, artificial intelligence can be used along with specific programming where it would make it impossible to misinterpret an area and lead to a false diagnosis.[10]

All research that has been conducted on the use of hyperspectral imaging in the research of cancer imaging has been promising. This technology is only growing every day and continuously proves that it should be at the forefront of cancer research and imaging.

Introduction of Ethics in HSI

For the final piece of this chapter, we will look at the key bioethics when adopting hyperspectral imaging into the medical community. This includes but is not limited to a patient's autonomy, the requirement and obtaining of consent, and justice when dealing with fair access to this technology. We will also look at beneficence and non-maleficence and how these two pillars of bioethics link with hyperspectral imaging. Along with ensuring that this imaging modality can accurately detect cancer at early stages, and being a dependable imaging modality overall, there are essential pieces that have been mentioned minimally throughout this chapter. We need to ensure that our cancer detection methods are as non-invasive as possible, guaranteeing that our techniques do not negatively impact the health and well-being of our patients. We need to ensure that this technology is accessible to people of all ages, races, and classes. We also need to ensure the patient has sufficient knowledge to be able to give consent. Multiple imaging modalities are currently used that can potentially harm a patient's health which should be avoided at all costs. As hyperspectral imaging is slowly making its transition from research and development to being used on patients in a medical setting, we need to ensure that at every point of this transition, we are following the correct bioethics.[11]

Beneficence and non-maleficence

Beneficence and non-maleficence are two pillars of bioethics that should be talked about when studying the ethics of hyperspectral imaging in cancer detection. Beneficence and non-maleficence are two interrelated concepts whose basic premise is to not harm others. Although they share the same basic premise, there is one major difference that separates beneficence and non-maleficence. Non-maleficence comes from the Latin maxim *primum non nocere* meaning "first, do no harm."[12] This is a pillar of bioethics that states that a medical practitioner has a duty to do no harm or allow harm to be caused to a patient in any way shape or form.

Beneficence refers to actions that promote the well-being of others. This involves helping to prevent or remove harm or to improve the situation of others. Non-maleficence simply involves not doing any harmful action. A simple example can sum up the difference between these two important pillars of bioethics for easy interpretation. Beneficent actions involve rescuing a person from danger, getting a homeless person food, or helping an alcoholic quit drinking. Non-maleficent actions involve not giving a person a cigarette, or not putting someone in a place that could cause them harm or danger.[13]

Both pillars explained above are involved when adopting this technology into cancer research. Hyperspectral imaging is intertwined with beneficence as it is an action that is performed on a patient that promotes the well-being of the patient. Beneficence ensures that medical practitioners have a moral duty to promote the course of action they believe is in the best interest of a specific patient's health. Every patient and their perspective health issue are individual, so, therefore, a doctor needs to make sure that hyperspectral imaging is the best course of action for that specific patient. What is best for one patient may not be the best for another patient. This means that when hyperspectral imaging is intertwined into this field, there need to be experts in this technology that can promise that this technology will best suit that patient.[14]

Non-maleficence acts as a threshold for treatment. This means that if the treatment a patient receives does not provide more good than harm, then it should not be provided. Comparing this to beneficence where all valid treatment options are presented and then the best is chosen. Non-maleficence is a constant in clinical practice. Hyperspectral imaging does not do any harm to a patient's body. Using this technology does not go against any aspect of non-maleficence, and therefore is a better option when comparing it to current imaging modalities that can cause bodily harm.[15]

Justice

Justice, as it exists in society today, is defined by the principles of equality and fairness. In the field of bioethics, justice is more narrowly defined by the allocation of innovations equally throughout society. The integration of hyperspectral imaging does not follow this pillar of ethics. This technology is only available to the wealthy for their potential treatments and research of their cancer. This technology has the opportunity to benefit millions of people but will most likely never be able to do so. For example, X-rays, which have been around for and incorporated into the medical community for a very long time, can still cost thousands of dollars for one scan. Low-income areas throughout the world will not be able to have access to this emerging technology and in turn, this technology will not be able to be equally distributed throughout society. Because of the demands of justice to address such an imbalance of access, efforts should be undertaken by governments working with corporations to enhance such access globally.[16]

Autonomy

Autonomy is a central pillar of bioethics and needs to be considered at every stage of incorporating new technology into the medical field in which tests will be run on patients. According to Beauchamp and Childress, autonomy is, at minimum self-rule that is free from both controlling interference by others and limitations of knowledge. Autonomy explains that a patient must have a certain level of understanding of the technology that will be used. This is important as a person cannot give consent unless they have an adequate understanding of procedures involving their own body. If a patient does not understand the test that they will be subject to, they are unable to give consent.

To ensure that patient autonomy is practiced when using this imaging modality, a couple of steps need to be exercised. Professionals with expertise in hyperspectral imaging should be present before each scan of the patient. These professionals should routinely visit patients before procedures and thoroughly explain the process that they will be undergoing. Hyperspectral imaging is lucky in that there are no negative effects this imaging modality can have on the body. This minimizes some problems that could arise when explaining this technology to a patient. This is a clear advantage when contrasting this technology with other existing imaging modalities. As long as a professional who completely understands the inner workings of hyperspectral imaging is present to talk to the patient, then autonomy should be no issue when using this technology.[17]

Conclusion

This chapter has detailed the inner workings of a new technology that is hoped to take over the field of cancer imaging, diagnoses, and treatment. The basic information as to how this technology works has been detailed. We have talked about why this imaging modality can be such an addition to the medical field, and the improvements that will inevitably happen in the future. If we as scientists want to help as many patients as we can, then this technology should be focused on as being the future of care in this field. In the field of bioethics, the most important area in which cancer and bioethics intersect is in the care of dying. Cancer is one of the leading causes of death in the United States, and worldwide. Hyperspectral imaging intends to mitigate these numbers as much as the technology will let us. Adopting hyperspectral imaging over current imaging modalities can lead to positives in terms of patient ethics. HSI is one of the most non-invasive types of imaging modalities for cancer detection methods. Current imaging modalities such as X-ray, CT, and PET scans use ionizing radiation which can be detrimental to patient health. These more prominent imaging modalities can be even more detrimental to health if a patient needs to undergo multiple scans along their journey to treat their specific cancer. The previous information presented can be used as guidelines as we incorporate hyperspectral imaging more and more into the medical field. If these guidelines are followed than I believe that we will be able to incorporate this growing technology successfully and safely.

Notes

[1] Vedantu, "White Light," *VEDANTU, Vedantu,* 29 Nov. 2022, https://www.vedantu .com/physics/white-light.

[2] Tim Brown, et al. How Camera Sensors Work. (a) in a Typical Imaging System, a Lens," https://www.researchgate.net/figure/How-camera-sensors-work-a-In-a-typical-imaging-system-a-lens-concentrates-light-onto_fig2_296618875. In "Using phenocams to monitor our changing Earth: toward a global phenocam network," *Frontiers in Ecology and the Environment* 14, no. 2 (2016).

[3] "What Is Hyperspectral Imaging? How Does It Work?" *HAIP Solutions,* 28 June 2021, https://www.haip-solutions.com/what-is-hyperspectral-imaging/.

[4] Y Zhang, X Wu, L He, C Meng, S Du, J. Bao, Y Zheng, "Applications of Hyperspectral Imaging in the Detection and Diagnosis of Solid Tumors," *Translational Cancer Research*, U.S. National Library of Medicine, https://pubmed.ncbi.nlm.nih.gov/35117471/.

[5] Ibid.

[6] Ibid.

[7] "Medical Hyperspectral Imaging," *Imec*, https://www.imechyperspectral.com/en/applications/medical-hyperspectral-imaging.

[8] "Hyperspectral Imaging [AMP] Cancer Detection." Hyperspectral Imaging and Cancer Detection, http://www.ott.emory.edu/industry/featured/hyperspectral.html.

[9] Gemma Church, "The Future of Hyperspectral Imaging," The Future of Hyperspectral Imaging | *Electro Optics*, 26 July 2021,

https://www.electrooptics.com/viewpoint/future-hyperspectral-imaging.
[10] Stefano Selci, "Article Versions Notes," *Diagnostics* | Free Full-Text | Cerebrospinal Fluid Biomarkers in INPH: A Narrative Review | Notes, 2019, https://www.mdpi.com/2075-4418/12/12/2976/notes.
[11] V, Mallardi. "[The Origin of Informed Consent]." *Acta Otorhinolaryngologica Italica : Organo Ufficiale Della Societa Italiana Di Otorinolaringologia e Chirurgia Cervico-Facciale*, U.S. National Library of Medicine, 2005, https://pubmed.ncbi.nlm.nih.gov/16602332/.
[12] "What Is Nonmaleficence - Meaning and Definition," *Pallipedia*, 2016, https://pallipedia.org/nonmaleficence/.
[13] "Medical Ethics: Non-Maleficence," *The Medic Portal*, 1 Feb. 2021, https://www.themedicportal.com/application-guide/medical-school-interview/medical-ethics/medical-ethics-non-maleficence/.
[14] "Medical Ethics: Non-Maleficence."
[15] "Medical Ethics: Non-Maleficence."
[16] Matti Häyry, "Roles of Justice in Bioethics," *Cambridge Core*, Cambridge University Press, https://www.cambridge.org/core/elements/roles-of-justice-in-bioethics/2292775B3C024A98E20584153145637A.
[17] Mallardi.

Literature

Brown, Tim, et al. .2016. How Camera Sensors Work. (a) in a Typical Imaging System, a Lens." https://www.researchgate.net/figure/How-camera-sensors-work-a-In-a-typical-imaging-system-a-lens-concentrates-light-onto_fig2_296618875. In "Using phenocams to monitor our changing Earth: toward a global phenocam network," *Frontiers in Ecology and the Environment* 14, no. 2.

Church, Gemma. 2021. "The Future of Hyperspectral Imaging." The Future of Hyperspectral Imaging | *Electro Optics*, 26 July. https://www.electrooptics.com/viewpoint/future-hyperspectral-imaging.

Häyry, Matti. "Roles of Justice in Bioethics." *Cambridge Core*, Cambridge University Press. https://www.cambridge.org/core/elements/roles-of-justice-in-bioethics/2292775B3C024A98E20584153145637A.

"Hyperspectral Imaging." N. d. Hyperspectral Imaging - an Overview | ScienceDirect Topics. https://www.sciencedirect.com/topics/computer-science/hyperspectral-imaging.

"Hyperspectral Imaging [AMP] Cancer Detection." N. d. *Hyperspectral Imaging and Cancer Detection*. http://www.ott.emory.edu/industry/featured/hyperspectral.html.

"Journal of Imaging." N. d. An Open Access Journal from MDPI. https://www.mdpi.com/journal/jimaging.
Mallardi, V, 2005. "[The Origin of Informed Consent]." *Acta Otorhinolaryngologica Italica : Organo Ufficiale Della Societa Italiana Di Otorinolaringologia e Chirurgia Cervico-Facciale.* U.S. National Library of Medicine. https://pubmed.ncbi.nlm.nih.gov/16602332/.
"Medical Ethics: Non-Maleficence." 2021. *The Medic Portal*, 1 Feb,
Medical Hyperspectral Imaging." N. d. *Imec.* https://www.imechyperspectral.com/en/applications/medical-hyperspectral-imaging.
Selci, Stefano. "Article Versions Notes." *Diagnostics* | Free Full-Text | Cerebrospinal Fluid Biomarkers in INPH: A Narrative Review | Notes, 2019. https://www.mdpi.com/2075-4418/12/12/2976/notes.
Sivasubramanian, Maharajan, Hsia, Yu and Lo, Leu-Wei. 2014. "Advantages and Disadvantages of Imaging Modalities Used with Targeted Nanoparticles," in "Nanoparticle-facilitated Functional and Molecular Imaging for the Early Detection of Cancer." *Frontiers in Molecular Biosciences.* https://www.researchgate.net/figure/Advantages-and-disadvantages-of-imaging-modalities-used-with-targeted-or-non-targeted_tbl1_273515519.
Vedantu. 2022. "White Light," *Vedantu*, 29 Nov. https://www.vedantu .com/physics/white-light.
"What Is Hyperspectral Imaging? How Does It Work?" 2021. HAIP Solutions, 28 June. https://www.haip-solutions.com/what-is-hyperspectral-imaging/.
"What Is Nonmaleficence - Meaning and Definition." 2016. *Pallipedia.* https://pallipedia.org/nonmaleficence/.
Zhang, Y, X Wu X, L He, C Meng, S Du, J. Bao, Y Zheng, "Applications of Hyperspectral Imaging in the Detection and Diagnosis of Solid Tumors." *Translational Cancer Research.* U.S. National Library of Medicine. https://pubmed.ncbi.nlm.nih.gov/35117471/.

CHAPTER SEVENTEEN

THE RESILIENT USE OF PHOTOACOUSTIC FLOW CYTOMETRY FOR ANALYZING PSEUDOMONAS AERUGINOSA WITH BACTERIOPHAGE TAGS

JENNIFER SCHINKE

Antibiotic-Resistant Threat

The CDC explains that antibiotic resistance is currently one of the greatest global health challenges. Antibiotic resistance occurs whenever bacteria within a patient have the ability to destroy the drug prescribed to kill it.[1] Antibiotics became popular in the 1940s and have evolved as a treatment for infection. These medications have saved millions of lives and have been used in a wide range of patient treatments. Bacterial resistance to antibiotics (antibiotic resistance) can happen naturally, through genetic mutations, or when encountering other bacteria whose resistance genes are contagious upon bacterial mating.[2] Throughout the years, antibiotics have been overprescribed, causing a decrease in antibiotic effectiveness due to bacteria growing accustomed (resistant) to these medications.[3] Doctors occasionally prescribe antibiotics before receiving evidence of infection if they believe that not prescribing them causes greater risk.[4] This is one of the many contributors to the antibiotic resistance threat. If diagnostic technologies produced faster results, doctors would no longer need to prematurely prescribe antibiotics.

Nearly 30% of all antibiotics prescribed are misdiagnosed, accounting for about 47 million patients that took them unnecessarily. When a patient is prescribed an antibiotic that is not meant for them, it can cause an adverse drug effect such as toxicity or an allergic reaction. Antibiotics have both positive and negative attributes. They may kill the infection at hand but, they can also hurt the body as antibiotics also kill the "good" bacteria that protects the patient's microbiome.[5] The CDC stresses that society needs updated diagnostic tools for the identification of bacterial infections as the current technologies are outdated. Many of these tools are time-consuming, expensive, and do not determine if the infection is bacterial, fungal, or viral.[6]

This poses the importance of swift identification of bacterial infection. Identification will be achieved by detecting *Pseudomonas aeruginosa* using photoacoustic flow cytometry, a newly developed optical method for fast detection of any desired analyte. *Pseudomonas aeruginosa* is a species of bacteria that is multidrug-resistant and accounted for an estimation of nearly 3,000 fatalities in 2017.[7] It commonly infects surgical sites and the bloodstream, and is known to cause infections such as pneumonia and urinary tract infections. Successful identification of *Pseudomonas aeruginosa* will provide the foundation for an updated diagnostic tool to aid in faster bacterial identification. Faster diagnostics will lead to more precise treatment plans and could decrease the number of unnecessary antibiotics prescribed.

Introduction to *Pseudomonas*

Pseudomonas is a species of rod-shaped, gram-negative bacterium reaching a maximum size of 1 by 3 microns.[8] Gram-negative and gram-positive bacterium are classified by results of gram stain testing, a diagnostic tool that became popular in 1882. Gram staining is useful for the identification of bacterial infection but has limitations as it cannot diagnose the species of bacteria present.[9] The resulting stain color classifies bacteria between gram-negative or gram-positive. The cell wall of gram-positive bacterium contains a thick layer of peptidoglycan, the polymer responsible for purple staining. Gram-negative bacterium differs from this, as their cell walls only contain a thin layer of peptidoglycan resulting in a light pink stain rather than purple. Another defining difference between the two is that gram-negative bacterium has an outer membrane containing lipopolysaccharide (LPS). LPS is an endotoxin that protects bacteria and can cause serious infection in patients if released into the bloodstream.[10]

Pseudomonas is commonly found in water, soil, and vegetation. It generally favors moist environments and tends to colonize sinks, mops, medical equipment, etc. within the hospital setting. Hospitals are the most common place patients become exposed and it is commonly transmitted through direct contact. *Pseudomonas* does not pose a severe threat to healthy people but can be extremely dangerous to any hospitalized patient, especially those that are immunocompromised.[11] *Pseudomonas* is made up of more than 140 different strains, but this research will only be focusing on the most common strain, *Pseudomonas aeruginosa*. This strain is known to cause localized infection in the lungs, blood, or other areas of post-surgical patients and has a fatality rate of 50 percent.[12] Having a faster method of identifying bacterial infections such as those caused by *Pseudomonas aeruginosa*, could lead to proactive treatment. Finding a solution for faster bacterial identification could lower fatality rates and aid in the antibiotic resistance threat.

Diagnosing *Pseudomonas*

Diagnosis of *Pseudomonas aeruginosa* is currently achieved by culturing and is identified by the use of biochemical media. This diagnosis process could take up to 4 days before any conclusive results are available.[13] Depending on the severity of the patient's symptoms, antibiotics may be distributed before positive bacterial identification and an uncertain diagnosis may occur. Uncertain diagnoses are when a doctor believes that not prescribing an antibiotic could cause more harm than waiting for the results, this vastly contributes to the antibiotic resistance threat and must be addressed. Faster identification of bacteria could eliminate this issue and lead to proper diagnosis in a timely manner.[14] The goal of this research is to change the "gold standard" of lab culturing diagnosis and instead use a newly developed technique called photoacoustic flow cytometry for efficient identification of *Pseudomonas aeruginosa*. Faster bacterial identification will allow for timely treatment plans and should eliminate the overuse of antibiotics.

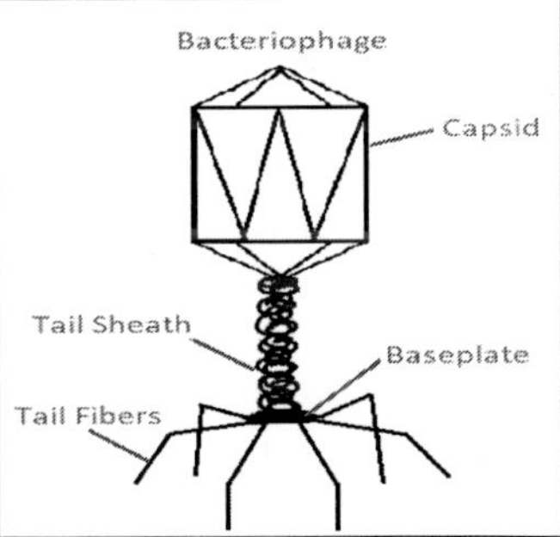

Figure 1: Diagram of bacteriophage

Introduction to Bacteriophage

Bacteriophages, also referred to as phages, are viruses that infect and kill their singular target host cell. The phage used throughout this research was acquired from the Baylor College of Medicine and was created using cell line PA01. This tells us that *Pseudomonas aeruginosa* will be the host cell phage is targeting. Phage, as shown in Figure 1, is made up of its capsid, tail sheath, tail fibers, and baseplate. The capsid contains the DNA/RNA necessary for the replication of the virus. The tail sheath is utilized as a passageway for sending the genetic material to the tail fibers. Once the baseplate opens, the tail fiber is then responsible for puncturing the bacteria to determine host cell recognition. If recognized, the fiber tail will then inject its genetic material inside the bacterial cell membrane and the virus will continue replicating within the bacterial cell until it reaches the final stage of the lytic life cycle.[15]

Lytic Life Cycle

There are 6 different stages of the lytic life cycle and those include the following: absorption, penetration, replication, maturation, release, and reinfection. Absorption occurs when receptor sites from the bacterial host are absorbed by the attachment sites from the phage. Contingent on the phage being able to absorb the bacteria, the phage will then penetrate through the bacterial wall and the genome will be injected into the cytoplasm. Next, phage hijacks the bacterial metabolic system and rapid replication of the genome occurs. For about 40 minutes, the genome will begin to mature and the cell will grow as it becomes more concentrated. Once the cell has reached its threshold of growth, osmotic lysis occurs and phage is released from the cell. Once the cell has burst, there is a possibility that 50 to 200 phages are produced, causing reinfection within the remaining bacteria.[16]

Application

Since it's known that phage only targets its host, we will be utilizing this knowledge to tag *Pseudomonas aeruginosa* with phage for detection via the method of photoacoustic flow cytometry. The phages will be dyed with Direct Red and mixed with the bacterial cells. Attachments will happen almost instantaneously resulting in a highly concentrated red cell. As long as this experiment is completed before the release stage of the lytic life cycle, the laser will excite the highly red-concentrated cell and generate a photoacoustic wave that will be detected by the system. This process will determine if *Pseudomonas aeruginosa* is present within a sample because if no bacteria are present, then the cell and phage mixture will not have a reaction and this will result in null detections. This method allows for the identification of bacteria in under an hour, compared to a current wait time of nearly 4 days.[17] The difference of an hour compared to four days could drastically improve healthcare treatment plans and aid in decreasing antibiotic distribution.

Photoacoustic Flow Cytometry

Photoacoustic flow cytometry (PAFC) is a newly developed optical method for the detection of pathologic analytes such as cancerous and bacterial cells. This process is completed using a pump that distributes a sample at 30 μL/min through the flow chamber. There is an optical fiber placed perpendicular to the flow channel within the chamber that delivers 532nm laser light with 5-ns pulses for absorption in the targeted cells. For single-cell detection, the laser is calibrated to emit 4 mJ of energy.[18] At the base of the chamber, there is a transducer placed containing a focal length of 0.8in. The transducer is placed accordingly so that the distance between the sample absorption area and transducer is optimal. Ultrasound gel is the coupling medium used that fills the space between the flow channel and transducer to ensure that the photoacoustic waves generated do not get lost when the laser excites a positive sample.[19] Once the signal has been received via the transducer, it is then amplified and sent to a program that counts each instance a cell has been detected. Every shot the laser delivers has the potential to cause a true or false detection within the PAFC system. False detections can occur from air bubbles within the ultrasound gel or past samples stuck in the flow channel. Although false detections may occur, they are distinguishable as the waves generated are much smaller than those of true detections. It is known that running a positive control sample of red-dyed distilled water through the PAFC system will always result in true detections. This is how we determine if detections are valid when testing flow chambers. Theoretically, coating bacterial cells with red-dyed phage should cause the same photoacoustic response as the positive control sample.

Novel Flow Chamber

Photoacoustic flow chambers have been used in prior research involving bacterial identification. Working firsthand with the original flow chamber model made it easy to realize that an update was necessary. The original flow chamber is made up of 10-μm-thick quartz tubing that passes through a 3-D printed flow chamber.[20] The quartz tubing acts as an optically transparent channel for the sample being tested but has limitations. Prior to PAFC testing, air bubble removal within the flow chamber is crucial as these bubbles can cause false detections. It has been found that using a syringe is the best way to remove air bubbles but that can be problematic. Due to the delicateness of the quartz, it often breaks if the channel is bumped/touched slightly by a syringe during PAFC system preparation. Additionally, assuming the channel is unbothered, the chambers can only withstand a minimal number of experiments before breaking naturally. This causes an immense amount of waste since the chamber cannot be fixed or repurposed.

A novel flow chamber has been created to eliminate the use of quartz tubing as the flow channel. This is achieved by using polydimethylsiloxane (PDMS) and piano wire to make a hollow channel within the chamber. PDMS was chosen due to its ideal properties. It is acoustically matched with water; thus, both will allow for efficient transmission of acoustic energy through the boundary of PDMS and the cell suspension. Acoustic transparency allows for the reduction of reflections and leads to minimal energy loss whenever the laser excites a sample.[21] This is the characteristic that allows for robust detections within the PAFC.

The chamber was first designed in Fusion360 and then 3D-printed. PEEK tubing was attached to both sides of the chamber with fingertight fittings and piano wire with a diameter of 0.457mm was threaded through both fittings before filling with polydimethylsiloxane. Polydimethylsiloxane (PDMS) will set after 48 hours at room temperature. Next, the piano wire is then removed and distilled water is pushed through the channel to ensure proper flow. The hollow channel created is unbreakable leading to a more reliable chamber that can withstand many more uses than the prior model. This novel flow chamber eliminates the delicate aspect of the prior model and thus saves time, money, and resources. The importance of having a reliable chamber is crucial for the detection of cells using PAFC.

Purpose

The purpose of this research is to use photoacoustic flow cytometry to efficiently identify bacteria and this will be demonstrated using *Pseudomonas aeruginosa* tagged with dyed bacteriophage. A new method of bacterial identification diagnostics is necessary as the current waiting period is a key contributor to the overuse of antibiotics. Faster bacterial identification can allow for proactive treatment potentially lowering the cases of fatality and the amount of overused antibiotics. It will also allow for more tests to be run within a laboratory saving time and making it possible to obtain bacterial identification results in a single day. This could vastly increase the number of results that labs provide to the healthcare system and could evolve diagnostics as a whole for bacterial identification.

A novel flow chamber was created to allow for a structurally improved flow channel. Detections will be accomplished from the attachment of red-dyed bacteriophages to the bacterial host, *Pseudomonas aeruginosa*. The red-dyed phage will infect the bacteria resulting in detectable, red-coated bacterial cells within minutes for speedy testing. This is a valid form of detecting if *Pseudomonas* is present in a sample because if *Pseudomonas* is not present, phages will not have a host to infect, and thus, no detections will be made. This research will prove that using PAFC is an easy and efficient way for bacterial identification. Similar research has previously been completed with *Staphylococcus* and it was found that using modified bacteriophages as optical tags is a valid form of bacterial detection.[22] This method of bacterial identification could be a new way for the medical field to urgently determine if bacteria are present within a patient.

Emerging Developments

Healthcare is an emerging field of study and must be improved upon to keep up with medical demands. The once "gold standard" for laboratory diagnosis is no longer as successful as it once was. The need for updated diagnostic tools is greater than ever, and photoacoustic flow cytometry may be the solution for a faster method of bacterial identification. Theoretically, the method of using bacteriophages as optical tags paired with PAFC could be demonstrated with any strain of bacteria. Granted that the appropriate accompanying phage is present to test each possible bacterium, this could then be the new standard for all bacterial identification. The idea of using optical tags to detect samples with PAFC has proven quite successful but additional studies are required to prove the range of bacterium this can be tested on. *Staphylococcus* and *Salmonella* are currently the only other bacterial species that have been successfully detected, but more studies are expected to come. Once proven valid, this method could completely evolve bacterial diagnostics and aid in improved patient care.

Future Directions

Photoacoustic flow cytometry can be used to detect not only bacterial cells but cancerous cells as well. The concept of detecting circulating tumor cells (CTCs) is similar to the methodology followed for bacterial identification, with the biggest difference being that phage is not the optical tag. Phage only attacks bacterial cells, so preparing optical tags for cancerous cells requires a slightly different process. For bacterial cells, tagging with red-dyed phage is straightforward as they will immediately attach to the cell wall leading to a red-coated, detectable cell. For cancerous cells, microspheres are the optical tag used to cause detections in PAFC and require little preparation to aid in cell attachment. Black-dyed microspheres are first coated in MART-1, a protein responsible for recognizing tumor-specific antigens. Once the microspheres are coated, they can then easily attach to CTCs making them highly detectable using PAFC.[23] In the future, photoacoustic flow cytometry could serve as the lead diagnostic tool for the identification of both bacterial and cancerous cells within a patient.

Bioethical Analysis

After diving into the background, purpose, and future directions of this research, it's important to analyze the bioethical dilemmas faced. Biomedical ethics is described as the education of moral theory and principles.[24] All healthcare professionals must understand the foundation of bioethics as it provides the ethical guidelines to follow in any biomedical situation. Ethical dilemmas may be faced during bacterial identification using PAFC, so knowledge of biomedical ethics is required. The study of biomedical ethics has been in effect for over 40 years and has paved the way for the ethical education of medical professionals. Some main ideas of bioethics that can be applied to the above research are as follows: respecting patient autonomy, demonstrating acts of beneficence, and providing a just culture in healthcare. Understanding both the morals and principles that define bioethics is particularly crucial in delivering fair and ethical healthcare to all.

Autonomy

The Greek words "autos" and "nomos" translate to "self" and "law." These are the two words autonomy is derived from in the Greek language. The original definition of autonomy represented self-governing independent city-states; this word later evolved to have one of the most important definitions in bioethics. The meaning of this word eventually extended to define an individual's self-law to act accordingly with one's self-chosen ideas.[25] These theories of autonomy postulate that being autonomous means you must possess self-reasoning, self-deliberating, and self-managing. Three conditions must be fulfilled to be considered fully autonomous and those include the following: intentionality, understanding, and non-control.[26] Intentionality is an action based on the response to a particular question at hand. When a person has intentions, they have a planned outcome; this plan may not always come to fruition, but the intent is constant. The second condition of autonomy is understanding. An act cannot be autonomous if the person acting does not fully understand what they are doing. Illness and immaturity can tend to block this condition, hence, blocking full autonomy if afflicted by either. The last condition that needs to be fulfilled in addition to intentionality and understanding, is non-control. Non-control guarantees no outside influence such as another person's ideas, opinions, etc.[27] Being fully autonomous means that one must be in control of themselves. They must have a clear understanding of their full intentions while not being swayed by any outside sources.

Bioethics teaches healthcare professionals how to promote autonomy by respecting patient boundaries. Some of these boundaries include providing doctor-patient confidentiality, informed consent, and not tampering with the autonomy of their patients.[28] Informed consent is described as the process of communicating patient diagnosis, treatment, and care options. Patients should be fully aware of their medical situation before giving consent for care.[29] Autonomy defines the roles of both the patient and the professional. The role of a healthcare provider is to educate patients about their options and give their professional advice, leaving the decision solely for the patient based on their own beliefs, morals, etc. Patient consent for medical testing is crucial in respecting one's autonomy. This is not only ethically emphasized but legally too. The law of consent legally enforces autonomy and protects patients' rights.[30] This shows the ethical and legal importance of understanding autonomy and can be applied to the above research involving bacterial identification. Patient samples of *Pseudomonas* have been obtained both ethically and legally from the ATCC. All patients fully understood and had the intention of releasing their specimens for research purposes. It is clear that autonomy is never violated during research involving bacterial identification using PAFC.

Beneficence

The means of this research is to demonstrate beneficence while respecting patient autonomy to provide efficient diagnostic tools in healthcare. The principle of beneficence is defined as a moral obligation to act for the well-being and benefit of others.[31] Beneficence demonstrates acts of kindness and mercy for the greater good while having no ill intent. Fields pertaining to biomedical research, medicine, and public health all demonstrate beneficence by working towards a common goal of wanting to help people by improving healthcare.[32] The principle of positive beneficence explains that all people have an obligation to help others under certain circumstances. Five conditions must be met to insist on this moral obligation. 1) The person (or thing – research in this case) must be at risk of damaging health, life, etc. 2) Action is necessary to prevent condition one. 3) Action could probably prevent condition one. 4) Action to prevent condition one will not cause any financial or physical burden nor cause danger to others. 5) Benefits gained after preventing condition one hold stronger precedence over any situation suffered during that prevention.[33] This model can be applied to the above research by defining condition one as outdated bacterial identification diagnostic tools which cause great risk in healthcare. Outdated diagnostic tools have led to the abuse of antibiotics which has contributed to the antibiotic resistance threat that's currently happening in the US. Conditions two through five are all met in accordance with condition one, hence, it is shown that under the principle of positive beneficence, moral obligation is necessary. This research will likely lead to new diagnostic methods that could help millions of people and save many lives. The above research has been centered around the principle of beneficence as this work serves as a foundation aiding in the development of improved diagnostic tools. Faster methods of bacterial identification are necessary for efficient patient care thus, this research is carried out with the intent to help others and evolve the healthcare system.

Justice

Justice revolves around the ideas of equality and human rights.[34] With this in mind, equal access to healthcare is crucial in leading a just medical field. Justice ensures that human qualities such as age, gender, race, etc. are incomparable to others. This means that a person cannot be considered more or less "important" than another person based on the characteristics that he or she may possess. Justice is modeled by a system of fair treatment to everyone due to the disregard for all personal achievements.[35] For example, consider the following two patients: one being a poor man and the other being rich and famous. To be just, both men must be treated the same way; bioethics explains that medical professionals are expected to overlook any special qualities that define a person and view every patient as equal.

Aristotle defined the principle of justice as, "Equals must be treated equally, and unequals must be treated unequally."[36] As the first part of this principle is without a doubt indisputable, the same cannot be said for the second half referring to "unequals". This principle is now considered the formal principle due to its lack of explanation to decipher between an equal and unequal person.[37] Bioethicists have speculated upon this and determined that material principles may be used to aid in providing additional theories of justice. Material principles also can classify patient relevancy for some biomedical distribution.[38] Since this ethical dilemma is not covered in Aristotle's principle of justice, this poses the importance of material principles in bioethics. An example of these principles would include the principle of need. The principle of need has been used to further elaborate the principle of justice in certain situations. Take, for instance, dealing with the distribution of medical services, devices, supplies, etc. Unfortunately, healthcare does not have unlimited access to doctors, medicine, and medical devices. So, this poses the importance of determining the need for distribution while remaining just. The formal principle of justice does not provide enough input on how to act justly in such a dilemma. Thus, the principle of need is deemed valid as a material principle of justice to aid in the ethical understanding of acting just.[39] Understanding justice in bioethics is crucial as it outlines the actions necessary to promote fair medical attention. The goal of the above research is to ensure equal access to bacterial identification diagnostic tools for anyone with access to healthcare.

Notes

[1] CDC, "Pseudomonas aeruginosa infection. Centers for Disease Control and Prevention," retrieved December 5, 2022, from https://www.cdc.gov/hai/organisms/pseudomonas.html 2019.

[2] NYU Langone Health. (n.d.). *Types of antibiotic-resistant infections*. Types of Antibiotic-Resistant Infections. Retrieved December 1, 2022, from https://nyulangone.org/conditions/antibiotic-resistant-infections/types#:~:text=Types%20of%20Antibiotic-Resistant%20Infections%201%20Methicillin-Resistant%20Staphylococcus%20Aureus,a%20lung%20infection.%20...%203%20Carbapenem Resistant%20Enterobacteriaceae%20

[3] NYU Langone Health.

[4] D. Hyun, "Why doctors prescribe antibiotics-even when they shouldn't," *The Pew Charitable Trusts* Jone 30, 2017. Retrieved October 26, 2022, from https://www.pewtrusts.org/en/research-and-analysis/articles/2017/06/30/why-doctors-prescribe-antibiotics-even-when-they-shouldnt

[5] CDC, 2019.

[6] CDC, 2019.

[7] CDC, 2019.

[8] B. H. Iglewski, Chapter *27 Pseudomonas*, National Library of Medicine, 1996 Retrieved October 7, 2022, https://www.ncbi.nlm.nih.gov/ books/NBK8326/

[9] Cleveland Clinic. (n.d.). *Gram stain: What it is, purpose, procedure & results*. Cleveland Clinic. Retrieved December 8, 2022, from https://my.clevelandclinic.org/health/diagnostics/22612-gram-stain

[10] R. Bailey, *Gram positive vs. Gram negative bacteria*, ThoughtCo., February 6, 2020. Retrieved December 5, 2022, from https://www.thoughtco.com/gram-positive-gram-negative-bacteria-4174239#:~:text=The%20differences%20between%20Gram%20positive%20and%20Gram%20negative,murein.%20These%20bacteria%20stain%20purple%20after%20Gram%20staining.
[11] Iglewski.
[12] CDC, 2019.
[13] Iglewski.
[14] Hyun.
[15] P. G. Leiman, F. Arisaka, M. J. Van Raaij, V. A. Kostyuchenko, A. A. Aksyuk, S. Kanamaru, and M. G. Rossmann, "Morphogenesis of the T4 tail and tail fibers," *Virology Journal* 7, (2010): 355. https://doi.org/ 10.1186/1743-422X-7-355
[16] G. Kaiser, *The lytic life cycle of bacteriophages*, Biology LibreTexts, April 9, 2022. Retrieved November 2, 2022, from https://bio.libretexts.org/Bookshelves/Microbiology/Book%3A_Microbiology_(Kaiser)/Unit_4%3A_Eukaryotic_Microorganisms_and_Viruses/10%3A_Viruses/10.07%3A_Bacteriophage_Life_Cycles%3A_An_Overview/10.7A%3A_The_Lytic_Life_Cycle_of_Bacteriophages
[17] R. H. Edgar, J. Cook, C. Noel, A. Minard, A. Sajewski, M. Fitzpatrick, R. Fernandez, J. D. Hempel, J. A. Kellum, and J. A. Viator, "Bacteriophage-mediated identification of bacteria using photoacoustic flow cytometry," *Journal of Biomedical Optics* 24, no. 11 (2019):1-7.
[18] Ibid.
[19] J. A.Viator, et al., "Photoacoustic detection of circulating melanoma cells in late stage patients," *Journal of innovative optical health sciences*, November 2020. Retrieved October 10, 2022, from https://www.ncbi.nlm.nih.gov/pmc/articles/PMC8218985/ Publication: J. A. Viator, M. Hazur, A. Sajewski, A. Tarhini, M. E. Sanders, and R. H. Edgar, "Photoacoustic detection of circulating melanoma cells in late stage patients," *Journal of innovative optical health sciences* 13, no. 6, (2020).
[20] R. H. Edgar, et al.
[21] Y. Fu, J. Fischer, K. Pan, G. H. Yeoh, & Z. Peng, "Underwater sound absorption properties of polydimethylsiloxane/ carbon nanotube composites with steel plate backing," *Applied Acoustics, September 26, 2020*. Retrieved November 7, 2022, from https://www.sciencedirect.com/science/article/pii/S0003682X20307726#:~:text=Also%2C%20PDMS%20is%20acoustically%20transparent%2C%20which%20means%20that,studies%20have%20chosen%20PDMS%20as%20the%20matrix%20material.
[22] Edgar, R., Samson, A.-P., Cook, J., Dougals, M., Urish, K., Kellum, J., Hempel, J., & Viator, J. *Photoacoustic discrimination of antibiotic-resistant and sensitive staphylococcus aureus isolates*. Lasers in surgery and medicine. Retrieved November 8, 2022, from https://pubmed.ncbi. nlm.nih.gov/34940986/ 7
[23] T. Kocsis, "Labeling Melanoma Cells With Black Microspheres for Improved Sensitivity in Detection via Photoacoustic Flow Cytometry," (MA Thesis, Duquesne University, 2022).

[24] T. L. Beauchamp, and J. F. Childress, *Principles of Biomedical Ethics,* ed. 8th (Oxford: Oxford University Press, 2019).
[25] Ibid.
[26] Ibid.
[27] Ibid.
[28] A. W. Partin, "Ethics and Informed consent," *ScienceDirect Topics*,2021. Retrieved December 14, 2022, from https://www.sciencedirect.com/topics/medicine-and-dentistry/informed-consent
[29] K. Nunez, "What You Need to Know About Informed Consent," *Healthline*, 2019. Retrieved December 15, 2022, from https://www.healthline.com/health/informed-consent
[30] S. McLean, *Autonomy, consent, and the law*. (Oxfordshire, UK: Routledge-Cavendish, 2009).
[31] Beauchamp and Childress.
[32] Ibid.
[33] Ibid.
[34] Regis College, "What are examples of social justice in Health Care?" *Regis College Online* 2021. Retrieved December 15, 2022, from https://online.regiscollege.edu/blog/social-justice-in-health-care/
[35] Beauchamp and Childress.
[36] Ibid.
[37] Ibid.
[38] Ibid.
[39] Ibid.

Literature

Bailey, R. 2020. *Gram positive vs. Gram negative bacteria*. ThoughtCo. Retrieved December 5, 2022, from https://www. thoughtco.com/gram-positive-gram-negative-bacteria-4174239#:~:text =The%20differences %20between%20Gram%20positive%20and%20Gram%20negative,mu rein.%20These%20bacteria%20stain%20purple%20after%20Gram%2 0staining.

Beauchamp, T. L., and Childress, J. F. 2019. *Principles of Biomedical Ethics.* Ed. Eight. Oxford: Oxford University Press.

Centers for Disease Control and Prevention (CDC). 2021. "2019 Antibiotic Resistance Threats Report." *Centers for Disease Control and Prevention*. Centers for Disease Control and Prevention 23 Nov. https://www. cdc.gov/drugresistance/biggest-threats.html.

Centers for Disease Control and Prevention (CDC). 2019. *Pseudomonas aeruginosa infection*. Centers for Disease Control and Prevention. Retrieved December 5, 2022, from https://www.cdc.gov/hai/organisms/pseudomonas.html

Cleveland Clinic. (n.d.). *Gram stain: What it is, purpose, procedure & results*. Cleveland Clinic. Retrieved December 8, 2022, from https://my.clevelandclinic.org/health/diagnostics/22612-gram-stain

Edgar, R. H., Cook, J., Noel, C., Minard, A., Sajewski, A., Fitzpatrick, M., Fernandez, R., Hempel, J. D., Kellum, J. A., and Viator, J. A. 2019. "Bacteriophage-mediated identification of bacteria using photoacoustic flow cytometry." *Journal of biomedical optics* 24, no. 11: 1–7. https://doi.org/10.1117/1.JBO.24.11.115003

Edgar, R., Samson, A.-P., Cook, J., Dougals, M., Urish, K., Kellum, J., Hempel, J., & Viator, J. 2021. *Photoacoustic discrimination of antibiotic-resistant and sensitive staphylococcus aureus isolates*. Lasers in surgery and medicine. Retrieved November 8, 2022, from https://pubmed.ncbi.nlm.nih.gov/34940986/

Fu, Y., Fischer, J., Pan, K., Yeoh, G. H., & Peng, Z. 2020. "Underwater sound absorption properties of polydimethylsiloxane/ carbon nanotube composites with steel plate backing." *Applied Acoustics*. Retrieved November 7, 2022, from https://www.sciencedirect.com/science/article/pii/S0003682X20307726#:~:text=Also%2C%20PDMS%20is%20acoustically%20transparent%2C%20which%20means%20that,studies%20have%20chosen%20PDMS%20as%20the%20matrix%20material

Hyun, D. 2017. "Why doctors prescribe antibiotics-even when they shouldn't." The Pew Charitable Trusts. Retrieved October 26, 2022, from https://www.pewtrusts.org/en/research-and-analysis/articles/2017/06/30/why-doctors-prescribe-antibiotics-even-when-they-shouldnt

Iglewski, B. H. 1996. Chapter *27 Pseudomonas*. National Library of Medicine. Retrieved October 7, 2022, https://www.ncbi.nlm.nih.gov/books/NBK8326/

Kaiser, G. 2022. "The lytic life cycle of bacteriophages." *Biology LibreTexts*. Retrieved November 2, 2022, from https://bio.libretexts.org/Bookshelves/Microbiology/Book%3A_Microbiology_(Kaiser)/Unit_4%3A_Eukaryotic_Microorganisms_and_Viruses/10%3A_Viruses/10.07%3A_Bacteriophage_Life_Cycles%3A_An_Overview/10.7A%3A_The_Lytic_Life_Cycle_of_Bacteriophages

Kocsis, T. 2022. "Labeling Melanoma Cells With Black Microspheres for Improved Sensitivity in Detection via Photoacoustic Flow Cytometry." MA Thesis., Duquesne University.

Leiman, P. G., Arisaka, F., Van Raaij, M. J., Kostyuchenko, V. A., Aksyuk, A. A., Kanamaru, S., & Rossmann, M. G. 2010. "Morphogenesis of the T4 tail and tail fibers." *Virology Journal* 7 [355].

https://doi.org/10.1186/1743-422X-7-355
McLean, S. 2009. *Autonomy, consent, and the law*. Oxfordshire, UK: Routledge-Cavendish.
Nunez, K. 2019. "What You Need to Know About Informed Consent." *Healthline*. Retrieved December 15, 2022, from https://www.healthline.com/health/informed-consent
NYU Langone Health. (n.d.). *Types of antibiotic-resistant infections*. Types of Antibiotic-Resistant Infections. Retrieved December 1, 2022, from https://nyulangone.org/conditions/antibiotic-resistant-infections/types#:~:text=Types%20of%20Antibiotic-Resistant%20Infections%201%20Methicillin-Resistant%20Staphylococcus%20Aureus,a%20lung%20infection.%20...%203%20Carbapenem-Resistant%20Enterobacteriaceae%20
Partin, A. W. 2021. "Ethics and Informed consent." *ScienceDirect Topics*. Retrieved December 14, 2022, from https://www.sciencedirect.com/topics/medicine-and-dentistry/informed-consent
Regis College. 2021. "What are examples of social justice in Health Care?" *Regis College Online*. Retrieved December 15, 2022, from https://online.regiscollege.edu/blog/social-justice-in-health-care/
Viator, J. A., Hazur, M., Sajewski, A., Tarhini, A., Sanders, M. E., & Edgar, R. H. 2020. "Photoacoustic detection of circulating melanoma cells in late stage patients." *Journal of innovative optical health sciences*. Retrieved October 10, 2022, from https://www.ncbi.nlm.nih.gov/pmc/articles/PMC8218985/

IV. GENERATIONAL CHALLENGES

ON ECOLOGY AND CLIMATE

CHAPTER EIGHTEEN

LAUDATO SÍ AND THE ECOLOGY OF CREATION

JAMES SWINDAL

Pope Francis illustrates a radical ecology of creation in his 2015 encyclical letter, *Laudato Sí*: *On Care for our Home*, written to all of God's people.[1]

The Inspiration of St. Francis of Assisi

The Pope celebrates St. Francis of Assisi, who is considered the patron saint of ecology who linked ecology to justice for the poor and considered all creatures "brothers and sisters." He was mainly concerned for the poor within God's creation. He was generously self-giving to the less fortunate and to nature. He lived in harmony with God's creation and shows us the reality of concern for justice for all of nature. Pope Francis chose St. Francis for inspiration for this encyclical because he was constantly faithful to the gospel and invites us to see nature as a magnificent book where God speaks to us personally, and illustrates his love for us by giving us glimpses of who He is as dwelling in creation.

The Pope mentions that a philosopher who is prophetic about our relationship with the natural world is St. Thomas Aquinas. In his Five Ways of demonstrating God's existence, Aquinas elaborates on the importance of nature. By nature, we can learn God's love for us by appreciating the beauty it provides, Additionally, nature manifests God's existence by showing us realities that are effects of God's causality. Through nature, we can see how not man but God creates and gives beauty and life to all creation. St. Francis emphasized that if we do not treat God's creation with care, we abuse the gift that He gives us.

Three Examples of the Effects of Modern Anthropocentrism

Pope Francis believes that once certain environmental resources have been depleted, war will inevitably break out. He warns that there will be deadly competition for the scarce remaining resources and that this will be exacerbated by the threat of nuclear and biological weapons. Pope Francis states that war has extremely detrimental effects on the environment, and future wars utilizing chemical, biological, or nuclear warfare will have the potential to "alter the balance of nature." He believes that everyone must try to have a positive impact on the environment whenever possible. Specifically, he stresses the importance of political action and intervention to prevent warfare and further environmental deterioration from occurring.

Environmental deterioration is similarly linked to human and ethical degradation. The most problematic is the pursuit of financial gain, which fails to consider the effects on human dignity and the natural environment. Many people claim that they are innocent because distractions dull our consciousness and make us numb to how finite and limited our world is. As a result, the environment is fragile and defenseless to the threats of human-caused problems in our world. Environmental deterioration affected by modern anthropocentrism has a bigger impact than our human minds can comprehend.

The Pope says that in times of crisis, such as our current environmental one, people tend to feign ignorance about it. Superficial ecology is the mindset that nothing is imminently wrong with the environment. He says that this allows people to justify continuing their daily routines despite their environmental harms. He is suggesting that we take responsibility for our impact on the environment and not be complacent in allowing further destruction. The philosopher David Hume also discusses how we humans use our reason to "anthropocentrize" in order to explain the unfamiliar and mysterious world by using only our self-interests.

Many problematic effects are linked to modern anthropocentrism according to Pope Francis. Among them are wars, pollution, and deterioration. These often lead to a constantly rising temperature, the killing of species, and bacteriological and biological warfare. A philosopher that can be connected to this topic is Albert Camus. He was a philosopher who wrote *The Plague* about how people went back to normal right after a pandemic was over by pretending it never happened. It is common that if something doesn't affect people directly, they choose to pretend it doesn't exist. Pope Francis applies this to modern anthropocentrism, stating that because these problems may not affect persons directly, they put them off and make them somebody else's problem. Yet the environmental problems that we have solved, "show that men and women are still capable of intervening positively" (58). Pope Francis knows we can do better, it is just a matter of whether or not we want to.

The Co-Dependence of All Creatures

According to the Pope, we are all creatures of God. He explains that all life is a manifestation of God, and not one creature can live without the other. He says,

> the spectacle of their countless diversities and inequalities tells us that no creature is self-sufficient. Creatures exist only in dependence on each other, to complete each other, in the service of each other (86).

He makes this point as a way to defend why humans should care about the environment. When we neglect our environment, plants and animals of all kinds are affected in some way. And since all creatures depend on each other, we depend on the very animals and plants that we are injuring. Since God is in every creature, when we ignore his creations we also ignore Him. But Francis says that once we have realized how God is connected to every living thing, then we will truly begin to respect the ecological value of each.

In his Fifth Way of the demonstration of God's existence in the *Summa Theologica*, Aquinas concludes that "there is a being with intelligence who orders all the things of nature to their ends, and we call this being God" (45). Aquinas enriches the Pope's claim that God is in every creature and connects all living things.

****

These are just a few ecological insights from this remarkable, timely, and powerful encyclical letter to benefit all humankind and indeed all creatures.

Note

[1] This chapter resulted from an integrated class project, in PHIL 160C, *Reality and Illusion*, Fall 2021, written by the students and coordinated by the faculty instructor Professor James Swindal.

CHAPTER NINETEEN

CLIMATE CHANGE: MITIGATION AND ADAPTATION THROUGH A JUST TRANSITION

CALLI CAHILL

All efforts to mitigate climate change are essential to the success of future generations. However, one aspect of the fight against global warming is seldom discussed and even less understood: adaptation and resilience. They seem to fall into a category separate from other mitigation efforts, like alternative energy. In our current state of emergency, it is necessary to mitigate and adapt to all climate-related changes. The challenge is that all adaptations must reflect the sustainability of the future. In other words, the adaptations cannot worsen or contribute to the climate problem but be an active approach to protect life during the mitigation transitions. Even with all the positive impacts of energy reform, warming will increase, and conditions will worsen. What we do now will determine *how much* worse. So, adaptation is a crucial part of climate remission because, no matter what, we will be living life differently in the future.

This chapter explores one of the most encompassing plans to adapt and encourage resilience: the Green New Deal (GND). This "just transition" sets up a means to protect the most vulnerable during the change to an emission-free society: racial minority groups, the poor, First Nations Peoples, and those living in developing nations. Through the use of narratives, this discussion aims to give voice to those most affected by climate change and offer equitable solutions.

The GND begins by outlining the current state of affairs. Temperatures at or above 2 °C beyond pre-industrial levels will lead to trillions of dollars in economic loss, mass migration, the death of 99% of coral reefs, severe storms, and wildfires.[1] Unless emissions are down 40% before 2030 and with zero emissions by 2050, economies and ecosystems are in dire circumstances. Because the United States is the main contributor of greenhouse gases (20% by the end of 2021[2]) and has the technological capabilities to limit its footprint, it must spearhead this transition. However, during any kind of sensitive phase or change, vulnerable communities are the ones who suffer the most. Currently, these communities are already feeling the effects of a disintegrating system.

Many First Nations and developing countries live a traditional lifestyle, emphasizing family farming, small business, and spirituality, involving little to no GHG emissions. This lifestyle requires daily adaptation and resilience that comes with generational knowledge of the land's resources and their complex interplay. Global warming causes the deterioration of fragile ecosystems, putting many indigenous communities at risk. For example, the Nyngatom of southwest Ethiopia face grim prospects as one of the many native groups struggling to survive in a warming world.[3] Their generational knowledge is no longer relevant to their current situation because it is unprecedented. The wet season is no longer as wet, and temperatures are rising, leading to the disappearance of various wildlife. It is not only a threat to physical well-being, but it is also a spiritual and social matter as well:

> Examples of this degradation include formerly cattle-rich pastoralists becoming poor, women becoming more dependent on their husbands, leather skirts — attributes of clan affiliation and family status — being replaced by cotton, and seasonal ceremonies falling out of sync as a result of changes in the timing of natural indicators. This […] will require new institutional settings and the accordant shifts in societal hierarchies and power.[4]

The Nyngatom have had their cultural and spiritual identities stripped from them in a way that will alter their lived experience for generations to come. As a result of Western industrialization, climate change has become another mode of colonization. Like the Nyngatom, much of the developing world lives closer to the equator, which will experience the brunt of increased temperatures and variability with extreme hot and cold swings. The global trend is that the most marginalized people are the most affected by global warming but the least able to mitigate it. Already, the climate has disrupted communities near and far: climate refugees- those who have to flee their homeland due to climate change consequences- have emerged.

In 2016, the first of the refugees in the United States came from a coastal community in Louisiana. The Isle de Jean Charles is home to several indigenous communities and is sinking into an over-exploited and increasingly rising ocean. Their resettlement plan was the first, at the time, to allocate money to a "climate resilience" grant that would evaluate how to move people out of dangerous zones effectively.[5] Still, this does not quell the deep spiritual ties to the land itself. Chief Albert Naquin of the Biloxi-Chitimacha-Choctaw tribe laments, "We're going to lose all our heritage, all our culture. It's all going to be history." For over one hundred years, the Biloxi-Chitimacha-Choctaw people fished, hunted, and cultivated that land until, in 1955, loggers and oil companies cut deep channels into the beaches. Now, only 10% of their land remains intact. Further ice sheet melts and oceanic thermal expansion will continue to engulf it. For the Biloxi-Chitimacha-Choctaw tribe, it is a spiritual and cultural loss where the chance for adaptation is already in the rearview mirror. Eventually, they will be forced to abandon their ancestral lands entirely.

In some cases, adaptation and mitigation are possible, but climate change has further increased the inequities that already exist. Urban neighborhoods with low-income and minority communities are heating up, and individuals suffer as temperatures increase. Concrete, cheap materials, and the lack of trees all contribute to the poor conditions of inner-city areas.[6] Shakira Franklin lives in West Baltimore, a site 6 degrees warmer than Baltimore's coolest neighborhoods. "I can actually feel me riding out of the heat. When I get to a certain place when I'm on my way, I'll turn off my air and I'll roll my windows down." It is no coincidence that the hottest ZIP codes correspond to the poorest communities, like West Baltimore, while the wealthiest remain the coolest. "This means that as the planet warms, the urban poor in dozens of large U.S. cities will actually experience more heat than the wealthy, simply by virtue of where they live."[7]

The narratives above are representative of entire nations struggling with climate change. The authors of the GND recognize climate issues occurring now and seek to legislate ethical innovations. Foremost, the GND seeks to end current inequities in the shift to a sustainable future:

> [T]o promote justice and equity by stopping current, preventing future, and repairing historic oppression of indigenous peoples, communities of color, migrant communities, deindustrialized communities, depopulated rural communities, the poor, low-income workers, women, the elderly, the unhoused, people with disabilities, and youth (referred to in this resolution as ''frontline and vulnerable communities''). [8]

"Frontline and vulnerable communities" describes current marginalized populations who will continue to endure discrimination if the transition occurs without the GND's promise of protection. Basic economic freedoms are among the pledges: education, universal healthcare, job transition training, and new industries (among other such protections). Instead of sticking to the status quo, the GND wishes to rebuild the economy, fostering new growth opportunities for those historically and socially barred from such experiences.

During the Conference of Parties at the UN Framework Convention on Climate Change, the International Indigenous Peoples Forum on Climate Change (IIPFCC) stated:

> [W]e reiterate the need for recognition of our traditional knowledge, which we have sustainably used and practiced for generations; and the need to integrate such knowledge in global, national and sub-national efforts. This knowledge is our vital contribution to climate change adaptation and mitigation.[9]

Indigenous communities have been advocating for adopting their sustainable lifestyle for centuries. In honoring their stewardship over the land and in offering reparations, the GND seeks to "[obtain] the free, prior, and informed consent of indigenous peoples for all decisions that affect indigenous peoples and their traditional territories, honoring all treaties and agreements with indigenous peoples, and protecting and enforcing the sovereignty and land rights of indigenous peoples;"[10] The purpose of the proposal by the GND is to *minimize* impact, advocate for sustainability, and honor autonomous First Nation governments to protect and preserve their livelihood and heritage.

Another issue facing a just transition lies in the housing crisis. The United States lacks affordable housing and vacant space to accommodate low-income individuals and families. Low-income neighborhoods are built from inexpensive materials that tend to absorb heat. As part of the resiliency plan, the GND promises to "[upgrade] all existing buildings in the United States and building new buildings to achieve maximum energy efficiency, water efficiency, safety, affordability, comfort, and durability, including through electrification;"[11]

While the GND focuses on the United States, it is not limited to a national cause. The beauty of the resolution is that it is broad enough to apply to various countries and cultures. "As the GND brand spreads, traditional advocacy and policy groups are likely to break off more manageable chunks under the same rubric. After all, the GND is so sweeping that virtually any climate policy can claim to be part of it."[12]

In short, climate change transition begins with social justice. In moving away from a consumerist approach to climate change, the developed world will come closer to building social equity. Achieving social equity in the future requires a radical rearrangement of hierarchies and superstructures.

This would mean that developing countries would educate women, uphold indigenous and worker's rights, and enforce anti-racist and anti-classist policies in their mission towards climate adaptation. Furthermore, removing the current socioeconomic barriers to clean power, housing, and nutrition is paramount.

As social equity increases, generational knowledge and the latest technology will lead to adaptive advancement. These include architectural innovation, building heat-resistant homes in the Middle East, and erecting sea walls in the Indonesian archipelago to prepare for impending floods. Fracking, drilling, and mining will need to cease worldwide while, at the same time, initiating a new green economy that guarantees jobs, training, and livable wages. Furthermore, the GND will support and invest in family farming and agriculture that promotes land health and sustainable planting. These practices are essential for developing nations as those confronted with food and water shortages grow closer to day zero. The impact of the GND is global, even if implementation occurs little by little.

The global society has historically placed marginalized groups in positions where few ethical choices are left: Those in developing nations do not have the resources available to make informed decisions about their future; indigenous people by their colonizer's disruption to their physical and spiritual ecosystems; and urban and low-income populations are forced into inner-city hothouses where they experience chronic illness just because of their zip code. The first step is recognizing the problem through education. Mitigation is important, but so are people. In the transition to zero emissions, we must protect those who have been systematically disadvantaged. The GND seeks to create a realistic plan for how the world might first mitigate and then adapt. This chapter does not endorse a particular political ideology but demonstrates how the GND details a just and accessible initiative. As such, it is only as a global community that we can come together to solve the greatest threat to humanity: global warming.

Notes

[1] H.R., "Recognizing the duty of the Federal Government to create a Green New Deal" 116th Cong., H. Res 109 (2019).

[2] Simon Evans, "Analysis: Which countries are historically responsible for climate change?" *Carbon Brief* (2021).

[3] Kirsty Galloway McLean, "Land use, climate change adaptation and indigenous peoples," *United Nations University* (2012).

[4] Ibid.

[5] Coral Davenport & Campbell Robertson, "Resettling the first American 'climate refugees,'" *The New York Times* (2016).
[6] Meg Anderson & Sean McMinn, "As Rising Heat Bakes U.S. Cities, The Poor Often Feel It Most," *NPR* (2019).
[7] Ibid.
[8] H.R., "Recognizing."
[9] Galloway McLean.
[10] H.R., "Recognizing." (2019)
[11] Ibid.
[12] David Roberts, "The Green New Deal, Explained," *Vox* (2019).

Literature

Anderson, Meg & McMinn, Sean. 2019. "As Rising Heat Bakes U.S. Cities, The Poor Often Feel It Most." *NPR*.

Davenport, Coral & Campbell, Robertson. 2016. "Resettling the First American 'Climate Refugees.'" *The New York Times*.

Evans, Simon. 2021. "Analysis: Which countries are historically responsible for climate change?" *Carbon Brief*.

Galloway McLean, Kirsty. 2012. "Land Use, Climate Change Adaptation and Indigenous Peoples." *United Nations University*.

H.R. 2019. "Recognizing the duty of the Federal Government to create a Green New Deal." 116th Cong., H. Res 109.

Roberts, David. 2019. "The Green New Deal, Explained." *Vox*.

CHAPTER TWENTY

ANXIETY AND CLIMATE CHANGE

AMANDA KOLSON

Over the recent decades, environmental scientists have noted detrimental changes within ecosystems that have resulted from human activities. In the prologue to Donald Vandeveer's and Christine Pierce's *The Environmental Ethics and Policy Book*, these environmental issues are analyzed. This includes ozone depletion, pollution, over-fishing, loss of soil productivity, deforestation, endangered species, and human overpopulation.[1] The authors end this article with a call to action, that for the world to have sustainable resources and ecosystems, there needs to be a change at the international level. However, climate change is not just a threat to the environment, but the current and future generations' mental health. Recent psychological research notes that anxiety about climate change and other environmental issues has begun to rise. The environmental issues that are outlined above are influencing individuals' mental health in a harmful way.

Background

In the field of psychology, anxiety has been defined as, "the central nervous system's physiological and emotional response to a vague sense of threat or danger."[2] Anxiety is a universal emotional response and is normal to experience whenever stimuli in one's environment pose a risk. However, anxiety levels can be disproportionate to environmental stimuli and could rise to the point of disorder, in which one's ability to go about daily activities is inhibited. Psychologists explain this disorder as Generalized Anxiety Disorder (GAD) which is defined as, "a disorder marked by persistent and excessive feelings of anxiety and worry about numerous events and activities."[3] Throughout this chapter, when the term "anxiety" is used, it is referencing the emotional response of anxiety, not GAD. While individuals with GAD may also experience symptoms of anxiety that can relate to climate change, the focus will be on the general population.

Evidence

In previous studies, researchers have analyzed whose mental health is directly affected by environmental dangers. The most apparent instance is the emotional response to natural disasters. Susan Clayton in an article titled, "Climate Anxiety: Psychological Responses to Climate Change," states:

> Climate change is associated with increased frequency and severity of extreme weather events, and the impacts of discrete events such as natural disasters on mental health have been demonstrated through decades of research showing increased levels of PTSD, depression, anxiety, substance abuse, and even domestic violence following the experience of storms.[4]

These natural disasters induced by climate change include flooding, hurricanes, and wildfires. These often result in a loss of livable space, which leads to migration and its associated psychological stressors. However, more gradual environmental changes have also been noted to negatively affect mental health. Rising temperatures have been correlated to higher numbers of suicides and hospitalizations for mental health issues.[5] While a direct cause for this correlation has yet to be determined, many researchers have found this directional relationship to be present. Also noted, air pollutants have been found to affect cognitive abilities in the elderly and impulsivity/attention issues in children.[6]

Mental health and climate change are more recent topics of research. This could be because many scientists and civilians seem to be most focused on the physical dangers of the environment. After all, the environment itself is a physical living being. However, now mental health professionals and policymakers need to focus on the unforeseen risks that climate change poses to mental health, specifically anxiety. Ulrich Beck in the article titled, "The Role of Technology and the Natural Sciences in the Risk Society," discusses how new technological advancements in society can have unintended risks that harm humans and the environment only after it is too late to make a significant difference.[7] An example of this would be how scientists can theorize or do calculations to determine the nuclear fallout of an atomic bomb, but it is only after the bomb is detonated that individuals truly see the entirety of the damage.[8] Psychologists are experiencing this to some degree regarding mental health as a new risk due to climate change. It can significantly inhibit one's life and may cause certain populations to need treatment.

On a broader scale, climate change can affect anyone's mental health, regardless of whether or not the above conditions have been met. Clayton explains, "this anxiety response is important to understand in part because of the range of potential sufferers: anyone who knows about climate change – in other words, given the reach of communications technology, almost everyone – could be affected by climate anxiety regardless of their own vulnerability or relative safety."[9] The very nature of climate change is causing many individuals to have this type of anxiety in which there is a present and ongoing threat to one's own and future generations. In addition, there is a feeling of uncertainty about the future because stopping the progression of these environmental dangers may not be possible. Overall, there is a distinct link between climate change and mental health issues, specifically related to anxiety.

Therapy Approaches

In terms of mental health, therapy is usually sought out whenever an individual is having mental or emotional struggles that inhibit their daily lives and need to seek professional help. Not only has climate change led to an overall feeling of anxiety in general populations, but those at risk of developing anxiety-related disorders may need therapy approaches that are structured to address climate change anxieties. Taylor explains this difference, stating, "adaptive anxiety can motivate climate activism … maladaptive anxiety can take the form of anxious passivity, where the person feels anxious but incapable of addressing the problem of climate change and may take the form of an anxiety disorder triggered or exacerbated by climatic stressors."[10] In this section, maladaptive forms of anxiety will be examined as well as the accompanying therapeutic approaches.

As noted by many researchers across varying fields, climate change disproportionately affects marginalized communities. Pope Francis in the *Laudato Si'* explains how the poor and homeless communities have limited access to resources such as clean water, which is further explained as the human right to have clean and accessible water.[11] Taylor explains that populations such as the poor, indigenous peoples, children, and those living in volatile lands are most at risk for anxiety psychopathology due to the risk of environmental dangers (i.e., natural disasters) and lack of community support.[12] These populations are more likely to develop a maladaptive form of anxiety that could be diagnosed as a disorder but may not have the means to seek treatment.

In an article titled, "Anxiety Disorders, Climate Change, and the Challenges Ahead: Introduction to the Special Issue," Taylor introduces three ways in which therapy can be altered to assist those experiencing symptoms of anxiety about environmental dangers. These are outlined to be, "(1) programs that motivate people to overcome anxious passivity and thereby take action to mitigate the effects of climate change, (2) treatment programs that address anxiety associated with exposure to climatic stressors, and (3) programs that build resilience at an individual and community level, to help people better cope with the challenges ahead."[13] The first therapeutic recommendation is to help the client get out of a passive mindset when it comes to climate change, so that way they feel as though they have a form of control. One way to do this is to promote sustainable practices, and for news outlets to include ways to practice climate activism in addition to having information about environmental crises. The second recommendation addresses individuals who have had first-hand experience with environmental dangers. Therapy should be accessible to these individuals by any means, which includes in-person and online options. The third recommendation is to build programs that help to build adaption and resiliency skills with clients that are struggling to do so. This can be done on an individual basis and community level.

Callicott in the article "Animal Liberation: A Triangular Affair," explains the notion of a community that could be applied to therapeutic approaches. Callicott explains that humans are one part of the environment and therefore have a duty to protect the whole.[14] Doing active steps at protecting the whole, or the environment can help with getting a patient out of a passive mindset. This could be done by practicing small sustainable acts every day, like recycling. Furthermore, Callicott elaborates stating that the communities to which an individual is connected are similar to the rings in a tree.[15] The communities that matter most to that individual and therefore have the most influence are those closest to the center and vice versa. Therefore, humans have a duty to their communities and the environment. This is called the biotic community. By being a part of this biotic community, humans become deeply connected to the environment. Using this mindset in therapy may help clients to see that their acts do matter regarding climate change, no matter the scale, and that they can be part of a community that can make a positive change.

Conclusion

Climate change is real and is not only damaging the environment but is becoming an ongoing threat to individuals' mental health. Marginalized communities are disproportionately affected by climate change, and children are at risk of developing climate change-induced anxiety because they will have to learn how to live in a world with constant environmental dangers. These findings have caused the scientific and psychological communities to face new challenges of not only combating climate change but also providing accurate therapeutic responses to new anxiety symptoms. Future directions should focus on how to implement environmental ethical practices at a systematic level so that environmental dangers can be lessened, and therefore anxiety can be less prominent. Overall, there is still much work to be done for both climate change and mental health.

Notes

[1] Union of Concerned Scientists, "Prologue: World Scientists' Warning to Humanity," In D. Vandeveer, & C. Pierce, *The Environmental Ethics and Policy Book* (Belmont: Wadsworth/Thomson Learning, 2003), xxvii-xxx.
[2] R.J. Comer & J.S. Comer, *Abnormal Psychology* (New York: Worth Publishers, 2018), G-1.
[3] Ibid., G-7.
[4] Susan Clayton, "Climate Anxiety: Psychological Responses to Climate Change," *Journal of Anxiety Disorders* (2020): 1.
[5] Ibid.
[6] Ibid.
[7] Ulrich Beck, "The Role of Technology and the Natural Sciences in the Risk Society," *Theory, Culture, and Society* (1992): 249-253.
[8] Ibid., 250.
[9] Clayton, 2.
[10] S. Taylor, "Anxiety Disorders, Climate Change, and the Challenges Ahead: Introduction to the Special Issue," *Journal of Anxiety Disorders* (2020): 1.
[11] Pope Francis, Encyclical Letter *Laudato Si'* of the Holy Father Francis on Care for Our Common Home, Vatican City: Libraria Editrice Vaticana, 2015, 23.
[12] Taylor, 1.
[13] S. Taylor, "Anxiety Disorders," 1.
[14] J.B. Calicott, "Animal Liberation: A Triangular Affair," *Environmental Ethics* 2, no. 4: 115-132.
[15] J.B. Calicott, "Second Thoughts on 'A Triangular Affair'," *Conceptual Foundations of the Land Ethic* (1987): 134.

Literature

Beck, U. 1992. "The Role of Technology and the Natural Sciences in the Risk Society." *Theory, Culture, and Society*: 249-253.

Callicott, J. B. 1980. "Animal Liberation: A Triangular Affair." *Environmental Ethics*: 115-132.

Callicott, J. B. 1987. "Second Thoughts on 'A Triangular Affair.'" *The Conceptual Foundations of the Land Ethic*: 133-135.

Clayton, S. 2020. "Climate Anxiety: Psychological Responses to Climate Change." *Journal of Anxiety Disorders* 74. https://doi.org/10.1016/j.janxdis.2020.102263

Comer, R. J., & Comer, J. S. 2018. *Abnormal Psychology.* New York: Worth Publishers.

Pope Francis. 2015. Encyclical Letter Laudato Si' of the Holy Father Francis on Care for Our Common Home. Vatican City: Libraria Editrice Vaticana.

Taylor, S. 2020. "Anxiety Disorders, Climate Change, and the Challenges Ahead: Introduction to the Special Issue." *Journal of Anxiety Disorders* 74. doi: 10.1016/j.janxdis.2020.102313

Union of Concerned Scientists. 2003. "Prologue: World Scientists' Warning to Humanity." In D. Vandeveer, & C. Pierce, *The Environmental Ethics and Policy Book* (Belmont: Wadsworth/Thomson Learning), xxvii-xxx.

Chapter Twenty-One

Democracy's Impact on Climate Change

Lei' asha Battle

Problem

Climate change is a dilemma that needs to be addressed not just within Pittsburgh, Pennsylvania, but across the world. Many people do not believe in climate change due to it being more of a long-term than a short-term effect. However, alarming evidence of human influence on the globe has shown significant changes in the earth's climate over even a short period of time. The events that have arisen from climate change are rising sea levels, deforestation, rising greenhouse gases, and even concentrations of the earth's core temperature have been rising steadily over the past few years. These negative effects have been accumulating leading the earth to a costly future. On top of that people often react slowly towards the situation at hand which is not good for society. All of this makes climate change a "defining issue of our time"[1] that needs to be addressed sooner rather than later before it is too late.

Vital Factor of Democracy

To fight climate change democracy plays an important factor in making change policy globally. The influence of a democratic government where the authority lies with the people can be the determinant of improving the environment. To put it simply, democracy impacts decisions towards actions for climate change. Action can take place through the collective will of a people in a democracy. These can be such things as rules and regulations as well as shifts towards policies that promote renewable energy. Whether a person may be rich or poor, democracy is effective for both groups to be an influence in making a difference for climate change because in a democracy everyone has a voice. According to Li and Reuveny, democracy allows for a two-fold way to confront environmental needs. The first is by raising "public awareness"[2] and the second is "responsiveness to the environmental needs of the public."[3]

Democracy awareness is demonstrated by "information flows"[4] such as reports to the media that inform the people of environmental problems. Members of society are more likely to be aware and know of situations with climate change that they then, in turn, can change, especially through democratic action. Informational topics on environmental issues are addressed more freely and encouraged through the openness of democracy. Democracy such as in the United States guarantees political rights such as freedom of speech and assembly that promote climate change action. Moreover, because climate change affects us all, peoples' rights to such things as life and the pursuit of happiness are directly affected. Hence climate change policy is a cause that needs to be heard by everyone and addressed by their respective political leaders.

The last reason democracy is a vital factor is its responsiveness to the environmental needs of the public. Democracies are more likely to react or listen to environmentalists who "stand a greater chance of affecting policy-making…than they do in an autocracy."[5] The more democratic governments there are helps improve the more environmentally friendly outcomes that are better for all local health. Democratic change eliminates or reduces the chance of human activity being a direct factor in climate change.

Examples

Now that it has been shown that, "democratic governments have the greatest number of effects on CO2 emissions followed by terrestrial protected areas and improved water sources"[6] and therefore are a great catalyst for change, portraying examples is essential. Democracy may be a determinant, but the determinant cannot be a determinant without the people; the two are reciprocal. Collaboration between people allows democracy to be truly impactful for climate change. Although it takes everyone, perhaps the most notable examples are those indigenous peoples who live closest to nature. Their voices ought to be heard and democracies are the best chance for this.

As an example, climate change affects water and land quality, but democracy improves the quality of both for the better. With democracy "sustainable development policies" result in "improved water sources"[7] for all individuals. Drinking water sources that are safe ("potable") are important for households for survival. Clean water that is affordable and sanitary throughout the United States showcases how democracy plays a role in implementing clean water standards. The government protecting water resources simply reduces risks to public health.

Besides water quality, land quality, and natural resources are protected by democracy as well. Climate change is taking a huge toll on ecosystems around the world. Many water sources are drying out, forests are being cleared at an alarming pace, and sea levels are rising. Many of these natural ecosystems and resources are not only being devastated but for them to recover takes years. With democracy, "terrestrial protected areas" become key for countries to have high-quality land along with protection for the ecosystems. National parks across the United States are prime examples of landmarks that are under the protection of democracy. Grassroots action within a democracy can stop actions such as deforestation, land degradation, and much more.

Communities all around the world are fighting for environmental justice. People in a healthy democracy collaborate to address the climate crisis both within their neighborhoods as well as around the world. Communities that are most vulnerable, for example, due to poverty, tend to speak up more on the subject of fighting climate change. This refers especially to minorities who live in communities less likely to get the kind of resources needed to mitigate climate change. Yet, these communities (Black, Latino, Asian), "overwhelmingly support government action on climate change and environmental protection"[8] compared to the nationwide and white averages. Democracy enacts a voice for the will of the people fighting against climate change to have a better sustainable environment for all.

Finally, an example of people collaborating is the Earth Summit held by the "United Nations Framework Convention on Climate Change (UNFCC)" which is a crucial environmental meeting by the 185 member states in addressing the next steps for climate change. Their goal is, "to move from consciousness-raising to agenda-setting to agreement on action by Member States to follow-up on conference commitments and to effective assistance for the countries that need help in realizing their commitments."[9] Key to this goal is listening to those voices that are closest to nature yet often marginalized such as indigenous communities.

Future Generations

Through democratic collaborations with people today as well as keeping future generations in mind, people can use adaptive measures to fight climate change. Countries across the globe are acknowledging the need to adapt their ways to protect biodiversity along with ecosystems.

Small businesses and firms are adapting their businesses to be environmentally friendly. There is now much pressure to get large firms to do the same. Future generations are more prepared to have safety and contingency measures for situations such as rising sea levels, harsh weather, and especially higher temperatures just to name a few. Conference of the Parties (COPs) also known as the Climate Summit meets every year to discuss new pledges for change. Their perspectives are toward the future becoming more adaptable to have a better "green economy." World leaders are collectively agreeing they need to act together to adapt to the inevitable impacts of climate change for future generations. From here on a child of the next generation should not have to pay the price of our (in)actions today. The climate crisis needs to be diminished because human beings ought not to have to choose between migrating or starvation. This is why the voices of people, through democratic action, are essential for providing rules and regulations to fight climate change now and for future generations. Democracy is a great catalyst for improved, qualitative environmental outcomes.

Conclusion

All in all, democracy brings the drastic action needed to make future impacts on climate change at an unprecedented scale. As read above, democracy is a vital factor in bringing change now and in the future. Policy decisions can give protection for many things including rainforest protection, stopping deforestation, and is good overall in reducing pollution within our neighborhoods. Democracy overall is a twofold source of communication between the voices in local communities and between people around the world. This communication leads people to come together to collaborate against climate change. Environmental factors take the lives of people every day, and humanity knows climate change is the biggest health threat to organized life in the world.[10] However, people know that with democracy change will come. Future generations will adapt to a better lifestyle than this generation not only for survival but to flourish which is the end goal of all political communities. Hopefully, future generations will not make the same mistakes as this and past generations and the best hope for this is a robust democracy where all voices are heard.

Notes

[1] United Nations, *United Nations*, (2019).
[2] Quan Li, and Rafael Reuveny, "Democracy and Environmental Degradation," *International Studies Quarterly*, vol. 50, no. 4 (2006): 936.
[3] Ibid., 937.
[4] Ibid.
[5] Ibid.
[6] Yoonsoo Nam, "Do Democratic Governments Improve Environmental Quality?" *Graduate Student Theses, Dissertations, & Professional Papers*, 4500 (2015): 36.
[7] Ibid., 17.
[8] Tim Lau, "Why the Climate Crisis Demands Democracy Reform," Brennan Center for Justice, (2021).
9 United Nations, "Climate Change," United Nations (2019).
[10] Nuclear weapons, too.

Literature

Lau, Tim. 2021. "Why the Climate Crisis Demands Democracy Reform." *Brennan Center for Justice*. https://www.brennancenter.org/our-work/analysis-opinion/why-climate-crisis-demands-democracy- reform

Li, Quan, and Reuveny, Rafael. 2006. "Democracy and Environmental Degradation," *International Studies Quarterly*, vol. 50, no. 4: 935–56. http://www.jstor.org/stable/4092786.

Nam, Yoonsoo. 2015. "Do Democratic Governments Improve Environmental Quality?" *Graduate Student Theses, Dissertations, & Professional Papers*, 4500. https://scholarworks.umt.edu/cgi/viewcontent.cgi?article=5506&context=etd.

United Nations. 2019. "Climate Change." *United Nations*. https://sustainabledevelopment.un.org/milestones/unced.

CHAPTER TWENTY-TWO

CLIMATE CHANGE AND DEMOCRACY

JOHN SPARBER

Throughout the entirety of planet Earth's existence, there has been a constant change in the global climate. This is something that has happened naturally for millions of years and will continue to happen for millions of years after humans are gone. Although the Earth naturally cycles through warming and cooling periods, the problem that we currently face is global warming due to human beings who have contributed to the acceleration of this process. The way in which this has happened and continues to happen is through the lack of care we have shown to the environment. This lack of care can take many forms such as pollution, deforestation, and the burning of fossil fuels such as coal, gas, and oil. Through these acts, we humans have added to the problem and have sped up the global warming crisis to the point that the cycles of heating and cooling are no longer natural. In this chapter, I will be addressing two issues. First, the problems faced by the world and what could contribute to the demise of organized human life on Earth. Second, I will be exploring some governments worldwide and what they have done and plan to do in the future to combat the ever-growing threat of climate change. Through exploring these points, we will be able to better understand the situation that we currently face and what lies in front of us so that we may reverse the damage that we have caused.

The issue that will be addressed in this chapter is that of climate change and the effects carried by it, I'll start by giving a brief background on climate change and the issues that come with it. In the broad scheme of things, climate change is a rather recent issue as the first mention of it came around 180 years ago when there was documented warming globally and an increase in global ocean levels. Recently it has become a much larger issue as the severity of the situation has been identified by global governments. The reason for global recognition came from the realization that pollution, which is produced by all nations albeit at disproportionate levels, was contributing to the increase in greenhouse gasses. When the global amount of greenhouse gasses increase it affects the atmosphere by trapping heat and therefore warming the earth. Currently, we can see a global increase in temperature which has resulted in the rise in ocean water levels and the loss of many species. In terms of global water levels the, "sea level has risen about 8–9 inches (21–24 centimeters) since 1880, with about a third of that coming in just the last two and a half decades. The rising water level is mostly due to a combination of meltwater from glaciers and ice sheets and thermal expansion of seawater as it warms."[1] As a result, there is the fear that due to warmer climates, food shortages will become more abundant, as farmland that used to be useful to produce certain crops now can't produce them as the climate is too hot for the crop to survive or there is lack of water to irrigate farmland.

When thinking about climate change it is very easy to distance yourself from the issue. Considering the amount of data that is now available concerning climate change, we can easily view this as something abstract and therefore brush it off as not directly affecting us. This can be seen through how we view certain issues for example a forest fire in a faraway foreign country or the deforestation occurring in South America. Although we may not be locally involved, it doesn't mean these things don't affect us. Indeed, because climate change is an Earth-affecting phenomenon, it links us all and affects us as one global, "local" community.

One of the larger global events that impact the environment is deforestation. Currently, this is one of the most detrimental activities happening worldwide as everyone is scrambling to clear forests to develop into real estate or create more farmable land. As stated in *National Geographic*, "About 17 percent of the Amazonian rainforest has been destroyed over the past 50 years, and losses recently have been on the rise."[2] By doing so we are clearing one of the planet's best resources in the process. The reason that this is such a concerning issue is because the Amazon accounts for an eighth of all trees in the world and the destruction of them means that there is a significantly smaller amount of carbon dioxide being turned into oxygen. The consequence is that "Removing trees deprives the forest of portions of its canopy, which blocks the sun's rays during the day and retains heat at night. That disruption leads to more extreme temperature swings that can be harmful to plants and animals."[3] Adding to the problem is the fact that a majority of the wood taken from the forest is then burned contributing to the total amount of carbon dioxide in the air.

Something that can be discussed now that I have identified a few key issues that climate change possess is to offer ways in which we can combat climate change. Starting on a small scale we can first take a look at the things we do as individuals that affect the environment and be more mindful of these activities. This can mean many things but can be as simple as changing from gas to electric (such as an automobile) and using solar or wind power to power our electric appliances and even whole households. On a larger scale, we can view more generalized issues that would require change but could be implemented now. Things that fall under this category would be changing the diet of populations to institute more plant-based dietary options. This would help the ever-growing demand for beef and beef products and the realization that not only is it an unsustainable diet for the future but livestock such as cows account for a sizeable percentage of the carbon dioxide and methane in the Earth's atmosphere. In the United States alone this accounts for 4% of all greenhouse gasses produced.

When looking toward the future there are also a few things that can be implemented that would contribute to the reduction of human influence on climate change. To start, when developing new housing they should be built in a way that utilizes resources such as solar panels and geothermal heating and cooling which would reduce the amount of gas needed to maintain a comfortable household. Further, and on a much larger scale, there should be a global consensus on the problem at hand and the cooperation of other countries would contribute to benefiting the environment. But this requires that people push for change through their respective political regimes, most notably democracies so that those political leaders who understand the urgency of this issue implement the changes that are needed.

Revisiting the issue of deforestation, there is strong evidence that mature, robust democracies offer safeguards against deforestation. Using what is called a "Polity Index" political scientists use this to rank the impact of regimes, especially democracy, with deforestation rates. The result is, "a dome-shaped curve suggesting that, in general, more democracy is better for the trees. Young or struggling democracies tend to have the worst rates of deforestation, while more-democratic countries are losing less forest."[4] It seems imperative that not only democracies should be encouraged, but strong robust ones where political leaders are held to account by the people. This is especially true for places where deforestation has a detrimental impact on indigenous peoples, for example in Africa and Brazil where deforestation is a central issue.

Another aspect to consider when looking at something on the global level is that of world governments and their willingness to participate in the fight against climate change. Currently, some policies have been put into place to safeguard the environment though they are hard-pressed to keep countries accountable to the terms they agree on. Under much pressure from their respective constituencies, in 2015 arguably the largest policy was put in place on a global scale with the introduction of the Paris Climate Conference. This agreement included 196 countries agreeing to the restrictions and requests that the agreement offered. In response, the Paris Climate Conference was the largest global consensus on climate-related issues this century. What they hope to accomplish through the agreement is to reach peak pollution levels as soon as possible to reverse the amount of pollution being released. They hope that as a result, the Climate Conference will aim the world in a direction where it will be climate neutral by the end of the century.

When looking into where the government has failed in this aspect, we can take a direct look at large manufacturing companies. Many of these companies have lobbyists who can control the voting interest in environmental issues in order to sway it in their favor. This is absurd as the government allows these companies to slip in restrictions due to political loopholes, harming our environment even further. A way that would increase effectiveness and the speed at which we address these issues is if we chose to make them an issue of our own. The reason that this would be effective is because it may begin with one person's actions, but can quickly turn to democratic, collective action that affects the future outcome of our planet.

Climate change is something that is not going to go away anytime soon. It is something that will continue to occur naturally and will continue to advance at an exponential rate if we choose to do nothing. The consequences are already being seen today as the generations prior did not care enough to do anything to address the issue. Consequently, we are all now seeing the repercussions as we are living in ever-warming climates. Though we can personally change things in our lives to benefit the environment, this is something that is a global problem and therefore demands collective, global action. If we can't all work together then we won't be able to achieve our goal of climate neutrality. How we need to change is by finding ways that we don't have to rely on natural resources as they not only are bound to run out but are also key in contributing to climate change. If we can find ways in which to adjust our lifestyles we will be able to give to our planet rather than take from it. This will not only allow us to live in a better world but it will save it for future generations to explore and enjoy. But if we do nothing they will be in a much worse world than the one we are currently living in. This is simply unacceptable.

Notes

[1] Lindsey, Rebecca, "Climate Change: Global Sea Level Rise," *Climate Change: Global Sea Level | NOAA Climate.gov* (2022), https://www.climate.gov/news-features/understanding-climate/climate-change-global-sea-level.
[2] Christina Nunez, "Deforestation and Its Effects on the Planet," *Deforestation Facts and Information, National Geographic,* (2019), https://www.nationalgeographic.com/environment/article/deforestation
[3] Ibid.
[4] M. Buitenzorgy, and A. P. J. Mol, "Does democracy lead to a better environment? Deforestation and the democratic transition peak," *Environmental and Resource Economics* 48 (2011): 59–70.

Literature

Lindsey, Rebecca. 2020. "Climate Change: Global Sea Level." *Climate Change: Global Sea Level | NOAA Climate.gov*.

Nunez, Christina. 2022. "Deforestation and Its Effect on the Planet." *Environment National Geographic*. https://www.nationalgeographic.com/environment/article/deforestation#:~:text=About%2017%20percent%20of%20the,have%20been%20on%20the%20rise

"Global Climate Change Policies," 2020. *European Environment Agency*, 23 Nov. https://www.eea.europa.eu/themes/climate-change-adaptation/global-climate-change-policies.

ON ENVIRONMENT AND CONSERVATION

CHAPTER TWENTY-THREE

TECHNICAL PROGRESS AND SOCIETY: THE SELF IN THE MIDDLE

ALEXANDER MOLINARO

Life in the modern age has drastically changed how people relate to the environment and one another. In the wake of industrial capitalism, human society became alienated in its connections between community and nature. Previously, humans worked with the environment and saw themselves as a part of their community, however, they have become distanced. In one aspect, the world has never been more connected in all of human history. With the rise of technology and social media, in theory, people can now be more connected with others and relate to the world in new ways. However, with an influx of technological development, the inverse has happened as people feel lonelier and more isolated in today's world. According to a Cigna study from NPR, more than three out of every five American feels isolated, misunderstood, and lack meaningful relationships.[1] Within modern life, daily activities have become more convenient, but this convenience has inadvertently isolated people as what social theorist Max Weber might call an "unintended consequence" of the "progress" of modernization. It has removed people from their natural, place-bound connections to community and the natural environment while redefining modern humans primarily as consumers. With all the technological advancements that have distinguished modern life, these points of consumer "progress" are at the cost of the environment. Cars, industry, airplanes, and computers have affected the environment in disastrous ways. One of the social-psychological effects of these technological advancements is that people have become more alienated from their environments: both social/communal and natural.

French sociologist Jacques Ellul talks much about "Technical Progress" in modern society. He discusses how technical progress alienates the individual from their surroundings. By making life more convenient, the individual becomes a potential consumer in their capitalistic culture. Life, itself, then becomes a commodity, and it does so at the expense of the environment. Ellul does not claim that all technical progress is inherently bad. On the contrary, he claims that there are a lot of benefits to some forms of technical advances. Medical progress can help people live longer and fight off diseases. People can communicate with others all around the world like never before. Ellul does offer a warning and demonstrates how the negative effects of technical progress are already happening. Modern humans are increasingly isolated, and their existence is increasingly commodified. The world around them is shown in front of them with a price tag. People are more disconnected from nature than ever. The natural tendencies to group as a community and live with the environment have been disconnected. Thus, the individual is left to retreat to isolation. The complexities of modern life force the individual to be alone, but, at the same time, social media offers a panacea to loneliness. Now, an individual can see a whole new digital world. They interact with people across the world. The possibilities are limitless for new connections. However, in doing so, the individual cuts themselves off from their community to retreat to the digital world. Within the digital world, people can now create personas that are not authentic. They offer up an ideal version of themselves–further commodifying and objectifying themselves. Ellul states, "One of the great designs of advertising is to create needs; but this is possible only if these needs correspond to an ideal life that man accepts."[2] The individual then becomes alienated even from themselves. While technical progress has made impressive advancements, it has come at the expense of environmental well-being and the integral development of interaction in the community and with nature.

For most of human history, individuals had to live within the realm of their communities and environment. Humans, like every other animal, thought of themselves as part of a larger community and of the natural world. However, in the wake of capitalism and the Industrial Revolution, the sanctified bond between humans and the natural world resulted in a divorce. The earth was seen as a commodity, and its resources were open to exploitation for profit. This ethical, psychological, cultural, and economic reorientation was a product of alienation. Cities were then built to distinguish the division between "human" and "nature": the human is where *people* live and this is cast in opposition to the threatening outside environment. Communities and individuals were cut off from nature and the environment. With the advance of capitalism, the Earth and its "resources" were seen as any other object that was open for economic exploitation. Urban and industrial cities were built which have had severe consequences to the natural world. Industries that included steel mills and coal mines were seen as necessary for sustaining life within the city and for capitalistic motives, but, in effect, greatly harmed and polluted the environment. In recent years, globalization has become an important force for the spread of capitalism. Industries now can go around the world and exploit resources and labor from developing countries for profit. From the University of Sussex, Peter Newell goes over the effects of global capitalism in his book, *Globalization and the Environment: Capitalism, Ecology, and Power*. He goes over the current "state" of the planet including the depletion of half of the tropical forests, human activity increasing atmospheric carbon dioxide, and the persistent presence of toxic chemicals.[3] Newell also states that "the fate of the planet's ecology is bound up with the fate of contemporary capitalism."[4] As humans have further separated themselves from the natural world, the Earth does not have this luxury. In fact, the earth is dependent on the works of humans. As modern society has become more and more advanced, the further it moves away from the natural world and the greater it affects the earth. Through pollution, exploitation, and globalization, humans have commodified the Earth as an object for human development and a container for pollution.

As humans began to view the environment as a commodity, they began to physically separate themselves from the natural world as well. This separation rendered the environment "dangerous," "wild," "inhospitable," and "unsafe to live in." The need for urban cities and communities can be seen as fortresses that separate individuals and communities from the natural world. The modern world was then consumed by industrial cities as this was seen as the prime example of modern, developed life. Trees and grass were replaced by skyscrapers and pavement. Human interaction was cut off from the outside environment. Man-made life became the new natural world. Matthew Thomas Clement of Texas State University discusses the removal of human interaction and the environment where human society is supposed to evolve passed the need to live within the realm of the environment.[5] A perfect community in a post-modern industrial society should live independently from the environment with limited interaction with it. This can even be reflected in how sociologists have talked about environmentalism in the past. From the start of sociology, sociologists discussed humans living in a post-industrial revolution, modern world. There was little dialogue connecting human nature and the natural world. Humanity's destiny has therefore been taken in context to industrial life and capitalism. Clement comments on this through the modern use of recycling. He says, "Compared with earlier stages of modernization, by the 1990s, there were more people participating, more materials being recycled, more curbside recycling programs, and more materials recovery facilities."[6] The uptake in recycling in the latter half of the twentieth century gave evidence to the point that it was easier to recycle used goods than to cut down on new materials being bought. More materials being bought increases profits but harm the environment. However, to keep up with capitalistic profit margins, cities began promoting recycling to avoid decreasing consumption. The knowledge to actively help the planet was known, but the execution was for economic purposes. Since the advance of modern capitalism, the motivations of humans in modern society can only be seen through the mode of profit/economic gain and not in relation to the environment as an intrinsic good.

Capitalistic and industrial institutions (systems) not only affected the environment around them but also the people who rely on these institutions to live. Likewise, these institutions made it so that humans had to rely on man-made industrial establishments (capitalism and capital establishments and ideas of labor and value) instead of the natural world. In this respect, individuals living within capitalistic, economic systems are forced to live and abide by these systems that are destroying the environment and social connectedness. In other words, economic depressions/recessions and environmental disasters occur often, and, the people subjugated to these systems, pay the price of environmental strain and social alienation from these modes of economics. Even still, the reliance on man-made institutions alienates the individual from the natural world and further isolates themselves. James Quane and William Julius Wilson comment on how relying on urban institutions affects individuals; especially racial minorities who are disproportionately affected by capitalism and poverty. They comment saying, "The limitations caused by this disconnect from capital-enhancing opportunities deny them the ability to think strategically [..] their lives are predominately influenced by more constraining institutions that infuse the social and economic rationalities under which they operate."[7] By living in a consumerist, capitalist environment, people lose a sense of agency that they once had. For example, racial and ethnic minorities are forced to live under urbanized institutions created by a white majority in power. The white majority, unable to properly understand the situation of racial minorities, cannot effectively create opportunities and systems that accurately address the needs for these individuals. Even though the system is not designed to help racial minorities, the racialized individual is left alienated and to blame. The individual within the system struggles to work within these systems and are then limited in their ability to express themselves and are isolated. This tension then develops between the individual and the capitalistic institutions where they are then forced to become consumers. Further, the resentment extends first to individuals of institutions, then between people and people—creating a purchasing battle that has a strain on the environment and relationships. They are designed to participate in a capitalistic culture that sees people primarily as consumers. This then turns the community into a potential buying market. To achieve social success, one must acquire enough material goods to compete with the social status of the rest of the community. When an individual is unable to meet consumer expectations, a sense of shame and hostility emerges from this sense of lack. Because an individual cannot succeed in purchasing material and consumer goods, they are then seen as a failure within capitalistic institutions. Similarly, because the self is tied to material consumption, one's "self" is embarrassed when it does not have the financial means to contribute to capitalist consumption. The self is then in conflict with its designed society and community. Not to mention, either, that the more material consumption there is, the more resources are extracted from the environment, the more waste there is, and the more the environment is affected. Thus, a successful sense of the self is linked to consumption and economic output; when these areas of social status are not meant, a successful sense of the self is depleted, social agency is stressed, and communal connectedness is strained, as well. These institutions are supposed to help the well-being of the community. Instead, they play into the economic sphere of capitalism in a societal system that has already pushed out and objectified the natural world.

The main purpose for the development of cities was to separate people from the environment; likewise, other attempts to "reconnect" with nature have had similar alienating effects. To combat urbanization, the twentieth century saw a rise in suburban neighborhoods. These areas were designed to be a mix of safety from the outside world as well as some sort of reconnection to the natural landscape. In a journal article titled, "From Alienated to Aleatory: Nature in the Post-Post Modern World," Christine Cerqueria Gaspar comments on how people relate to nature today. Especially with the popularity of suburbia and its proximity to nature, Gaspar comments about how there is still a disconnect between nature and the community while also isolating members in suburbia. She states, "The development of American suburbia stems from the search for a middle landscape between this newly tamed nature and the traditionally sheltering urban environment [...] The result, many have argued, is a placeless middle ground."[8] She argues her point by saying that even within suburban communities, nature is driven farther and farther out till it again becomes the "outside wildness" that is seen within urbanization. Nature, once again, is seen outside of the lived experience. It has become a controlled commodity that can be interacted with when desired. If one wants to engage with nature, one can go on a nature trail outside their living space. However, one does not have to rely on nature to live. For example, it could be the middle of winter in Pittsburgh, Pennsylvania and one could still buy strawberries or a pineapple from their local Walmart (which may be closer to their residency than a nature trail). Suburban communities still uphold the same "nature-less" aspects that a city suffers from. Even still, within suburban communities, people are still isolated from one another. Not only are they still seen as consumers (this time perhaps with more spending power) but are even physically distanced from their community. It is no longer necessary to interact with other community members. The result is an alienated community separated from others and the environment.

A growing concern in the modern world is isolation which comes from a lack of a communal spirit and connection to one's environment. Increasingly, more and more Americans are saying that they feel isolated and lack strong connections to others or even the environment they come from. Social bonds continue to become weaker and material goods are becoming a centerpiece for "happiness." Researchers Paolo Parigi and Warner Henson II chronicle the idea of social isolation in America through different theories about modernity throughout the years. Discussions regarding social isolation in modernity can even be traced back to Emile Durkheim. Durkheim was greatly concerned with the changes in socialization in the wake of the Industrial Revolution and his landmark case study on suicide. Durkheim notes how the agents of socialization were changing, as traditional means of socializing were becoming increasingly weaker.[9] Alexis de Tocqueville noted how Americans were isolated from their families and community in the nineteenth century, as well.[10]This trend of modern American isolationism continued into the late twentieth century. However, the authors note that today, the problem of isolation occurs due to the lack of meaning in a relationship. They say, "An isolate is no longer (simply) a person without connections but is now also somebody who creates connections that carry little meaning. This is the contemporary face of isolation.[11]" Friendships and relationships have become cheap and vapid. Relationships can be created just as easily as they fall apart. The strength of long-lasting and meaningful relationships is what is making people isolated; a sense of not receiving strong, challenging, and authentic relationships. Everyone may know one another, but there is a generalized antipathy. People may even choose to be with people that they may not even enjoy; all for the sake of being around someone. The desire to be with someone and communicate with them overcomes the fact that they may not even enjoy being with the person they are with. The lack of truly meaningful and fulfilling relationships leads people to try to establish any sort of relationship that they can.

Relationships have become weaker in a post-modern world as social media creates a space where people can meet other new people, but the personas people create are an ideal, alienated (or at least constructed/manipulated) version of themselves. As capitalism and industrialization progressed, people have become disconnected from the environment, their community, and eventually themselves. This is nowhere more apparent than in the world of social media. In a study to determine the strength of online and real-life relationships, Gustavo Mesch and Ilan Talmud created a survey to compare the two types of relationships. They found that "online friends tend to be perceived as less close than face-to-face friends [...] they are as less integrated in the daily life of the individuals and as more distant."[12] In an increasingly online world, more people are spending time virtually than ever. With the Covid-19 pandemic, online social gatherings, classrooms, and workplace environments became normal. However, this has affected the relationships people maintain. As the study shows, online relationships are less meaningful and weaker overall. This can be attributed to how people compose themselves online. Online, people can present themselves in a new and personalized way. When making an online profile, people can present the ideal form of themselves. They can show off what they want to be shown off and hide away all their insecurities. When other people see their online persona, they accept this to be their real self, and, when starting a new relationship, people play up to their ideal online self–leaving out their true self in the meantime. Therefore, online relationships are composed of idealized, yet fake, presentations of a person. The individual then becomes inauthentic, and an alienated form. Instead of accepting their flaws and shortcomings, people can throw them away and highlight what they are confident in. Yet, it is in understanding other people's flaws and imperfections that true, meaningful relationships can flourish. By removing this aspect, people lose a part of themselves, and, in turn, suffer to establish meaningful relationships. With a lack of genuine and meaningful social relationships, the self then becomes consumed and absorbed by itself. The self then gains meaning through consumption—whether it be material possessions or through virtual means. Simon Gottschalk, a sociologist from the University of Nevada, Las Vegas, discusses how an individual navigates their sense of self in online spaces. In it, he defines an age of the "Hypermodern" as a new mode of consumption, and that mode is "hyperconsumption." Now more than ever, the commodities people buy can give them a fulfilling satisfaction on their own, and, through this, people have become more individualistic. This is apparent within virtual spaces, where people can now show themselves in any way they want, socialize without the stress of physical interaction, and distract themselves through online media and engagement. Gottschalk explains,

> Moreover, as television, radio, the internet, and other enterprises increasingly operate on a 24/7/365 schedule, consumption also becomes 'turboconsumption,' and the pampered impatient turboconsumer expects to have 'what I want, when I want, and where I want it'(p. 102; see also Gottschalk 2009). In the hypermodern moment, therefore, the expectation of instant gratification is fueled by the impulse of a bulimic constant gratification as well.[13]

The self, therefore, becomes crucially linked to the material it consumes. Overt capitalistic and virtual consumption become the main identifier of the hypermodern self. Rather than seek outside relationships, work towards starting or maintaining them, or become connected with the outside environment, individuals retreat into themselves and their virtual spaces for gratification. Now that gratification becomes easier, the need to be gratified increases while the strength decreases. The self gives up on keeping social and environmental relationships and, instead, links itself within virtual spaces and media for "social" gratification, and, in doing so, consumes media constantly. The desire for these easier ways of consumption affects personal relationships as when one looks at virtual spaces, the relationships they create are not as strong and fulfilling as before. These virtual forms of gratification and relationships are just distractions from the lack of meaningful relationships with their community and environment. While more people can connect online, the "curated" self is not reflective of their true identity and hinders the ability to establish strong and meaningful relationships.

In all, modern society has created a more convenient and online world. However, this world disconnects the individual from their community and environment. It then replaces these natural forms of connections with consumerism and meaningless relationships. Modern isolation and lack of environmental connections have been discussed for decades now. However, the problem is growing more and more. As the world becomes more digital, people can only become more and more isolated and alienated. Capitalism has bred a population full of potential consumers. Old ways of connecting with others are becoming harder and human's integral relationship to the natural world has been replaced by man-made institutions. Modern, industrial society has then created a nature-less and community-less world for isolated and lonely individuals.

Notes

[1] Elena Renken, "Most Americans Are Lonely, and Our Workplace Culture May Not Be Helping," *National Public Radio* (January 23, 2020): https://www.npr.org/sections/health-shots/2020/01/23/798676465/most-americans-are-lonely-and-our-workplace-culture-may-not-be-helping
[2] Jacques Ellul, *The Technological Society* (New York: Vintage, 1965), 407.
[3] Peter Newell, *Globalization and the Environment: Capitalism, Ecology and Power*. Cambridge: Polity, 2012.
[4] Ibid., 1933.
[5] Matthew Thomas Clement, "Urbanization and the Natural Environment: An Environmental Sociological Review and Synthesis," *Organization & Environment*, 23:3 (2010): 291–314.
[6] Ibid., 303.
[7] James M.Quane and William Julius Wilson, "Critical Commentar. Making the Connection between the Socialisation and the Social Isolation of the Inner-City Poor," *Urban Studies* 49:14 (2012): 2982.
[8] Christine Cerqueira Gaspar, "From Alienated to Aleatory: Nature in the Post-Post-Modern World," *Thresholds* 26 (2003): 23.
[9] Paolo Parigi, and Warner Henson II, "Social Isolation in America., *Annual Review of Sociology* 40 (2014): 160.
[10] Ibid., 15.
[11] Ibid., 166.
[12] Gustavo Mesch and Ilan Talmud, "The Quality of Online and Offline Relationships: The Role of Multiplexity and Duration of Social Relationships," *The Information Society* 22:3 (2006): 147
[13] Simon Gottschalk, "The Presentation of Avatars in Second Life: Self and Interaction in Social Virtual Spaces," *Symbolic Interaction* 33:4 (2010): 507.

Literature

Clement, Matthew Thomas. 2010. "Urbanization and the Natural Environment: An Environmental Sociological Review and Synthesis." *Organization &Environment* 23, no. 3: 291-314, doi:10.1177/1086026610382621.

Ellul, Jacques. 1965. *The Technological Society*. Translated by John Wilkinson, New York: Vintage.

Gaspar, Christine Cerqueira. 2003. "From Alienated to Aleatory: Nature in the Post-Post-Modern World." *Thresholds* 26: 22–25. http://www.jstor.org/stable/43876154. Accessed 27 Apr. 2022.

Gottschalk, Simon. 2010. "The Presentation of Avatars in Second Life: Self and Interaction in Social Virtual Spaces." *Symbolic Interaction* 33, no. 4: 501-525.

Mesch, Gustavo and Ilan Talmud. 2006. "The Quality of Online and Offline Relationships: The Role of Multiplexity and Duration of Social Relationships." *The Information Society*, 22, no. 3: 137-148, DOI: 10.1080/01972240600677805 https://www.dhi.ac.uk/san/waysofbeing/data/communication-zangana-mesch-2006b.pdf. Accessed 27 April 2022.

Newell, Peter. 2012. *Globalization and the Environment: Capitalism, Ecology and Power*. Cambridge, UK: Polity. https://books.google.com/books?hl=en&lr=&id=c_ZOXqqI2KQC&oi=fnd&pg=PA1924&dq=capitalism+and+the+environment&ots=nyXZjhW3wQ&sig=nVWHRBz-6AADCzEh1EM75PlyUbI#v=onepage&q&f=false. Accessed 27 April 2022.

Parigi, Paolo, and Warner Henson. 2014. "Social Isolation in America." *Annual Review of Sociology*, 40: 153–71. http://www.jstor.org/stable/ 43049530. Accessed 27 Apr. 2022.

Quane, James M., and William Julius Wilson. 2012. "Critical Commentary. Making the Connection between the Socialisation and the Social Isolation of the Inner-City Poor." *Urban Studies* 49, no. 14: 2977–87. http://www.jstor.org/stable/26144125. Accessed 27 Apr. 2022.

Renken, Elena. 2020. "Most Americans Are Lonely, and Our Workplace Culture May Not Be Helping." *NPR*, NPR, 23 Jan. https://www.npr.org/sections/health-shots/2020/01/23/798676465/most-americans-are-lonely-and-our-workplace-culture-may-not-be-helping.

CHAPTER TWENTY-FOUR

ANIMAL EXPERIMENTATION AND THE ENVIRONMENT

COURTNEY FEREK

Animal experimentation is a controversial topic, both ethically and environmentally. People for the Ethical Treatment of Animals (PETA) defines animal testing as "the practice of performing unnatural and often painful experiments on animals held captive in stressful laboratory settings, often in the misguided belief that the results of the tests will be applicable to humans."[1] Animal experimentation often includes testing cosmetic, industrial, and medicinal products. These products are usually tested by being injected into their eyes or inhaling toxic chemicals. In this chapter, I am going to explain the negative effects that animal experimentation has on both the animals and the environment.

Peter Singer is a well-known philosopher, he is a Utilitarian, and he does not support the idea of animal experimentation. He does not focus specifically on the harm done to the environment, but on the suffering that is done to the animals during experimentation. His main goal is to extend moral standing to include animals. In Singer's article, "Animal Liberation", he begins by explaining Jeremy Bentham's argument. Bentham states that every being that has interests should be treated equally.[2] His overall question was can animals suffer, and if they can, there is no moral justification as to why we can experiment on them. Singer agrees with his perspective and further explains his personal views using three main arguments. His first argument is that animals do suffer, even though they do not have the cognitive ability for language.[3] The counterargument is that we cannot know if an animal is suffering because they cannot tell us. He then explains that there are other signs of suffering, such as forms of nonverbal communication. Also, since many humans such as infants and the disabled cannot speak, do we assume they also don't suffer? He concludes this first argument with the idea that it is speciesism to think it is ok to treat animals badly. Speciesism is the belief that humans are superior to animals. The next argument further explains the actual experimentation done on animals.[4] He states that there are many experiments done in universities for cosmetic and medical research. A lot of these experiments are senseless harm, and the idea that it is for medical research is deeply problematic. He then adds that we would never experiment on a 6-month-old human, just to save many other lives. This argument explains that animal experimentation is speciest discrimination and must cease. His last argument explains the process of farming and eating animals.[5] The idea that animals are our utilities pervades many peoples' thoughts. Many people believe that farming is just animals freely roaming around, but Singer says it is far from this. It is explained almost as a competition to see who can produce the most meat or animal products, causing more animals to be ruthlessly killed. Peter Singer's overall conclusion is that we are morally constrained to become vegetarians, for animals to receive the level of moral standing they deserve.

I would now like to focus on the negative impact animal experimentation has on the environment. First, it often uses the process of incineration to discard animal carcasses and laboratory supplies. The website "Animal Research: An Environmental Perspective" by Rebekah Corbett explains how incineration produces air pollution. These animal carcasses and laboratory supplies often contain many toxic chemicals.[6] Incineration is simply the destruction of any waste, often by burning. Air pollution is caused by the emission of gases produced by the process of incineration. Incinerators require a large amount of fuel to maintain the necessary temperatures. This has led environmental groups to conclude that incineration is not environmentally sound. Researchers were unable to calculate what percentage of pollution these incinerators produce, but it is important to realize that incineration contributes to these negative environmental effects. The process of incineration does not only cause harm to the environment but also to human health. Incineration has been proven to cause chronic illness and developmental delays. After the animal is experimented on the process of incineration alone has a negative impact on the environment.

Another negative impact animal experimentation has on the environment is the amount of environmental waste that is produced. The website "The Significant Impact of Animal Testing on the Environment" further explains how the amount of waste produced negatively impacts the environment. About 115-127 million animals are used each year for animal experimentation, these animals are then later discarded.[7] Animal decomposition and the process of transporting and disposing of this waste is the main responsibility for this large amount of environmental waste produced. These companies that conduct animal experiments often dispose of much more than just animal bodies. They are often getting rid of dangerous chemicals, food waste, and other supplies used such as needles, bedding, cages, and syringes. According to the National Institutes of Health (NIH), waste from their animal testing facilities totaled 1.5 million pounds from 2011 to 2013. Similarly, a pharmaceutical lab produced almost 15 tons of animal waste in just one year. Since animals are alive while being experimented on, they must still feed them. The amount of food necessary to feed these animals causes harm to the environment through agricultural processes. Unfortunately, many animals do not survive after experiencing these cruel conditions, and this results in animals being disposed of in large numbers. This introduces the environment to many toxic chemicals, diseases, and exposure to biohazards that could include radioactive materials. Animal experimentation produces many types of waste in the environment, such as animal waste, chemical waste, and food waste.

Since animal experimentation has been proven to have such negative effects, there has been recent research examining alternatives. PETA briefly describes these alternatives as they help fund and promote them. The "in vitro" method conducts experiments using human cells and tissues. Researchers replicate the structure and function of human organs; they are then placed on "chips." The chips are then used instead of animals for disease research and drug testing. This method has been proven more accurate than testing on live animals. Advanced computer-modeling techniques are also used; this is named the "in-silico" model. The in-silico method uses computer models to "simulate human biology and the progression of developing diseases."[8] This method is also proven to more accurately predict how drugs will react in humans. Both methods are not hindered by any differences between animals and humans. It also takes less time to complete compared to traditional animal testing. Although these methods are still relatively new, they seem to already be more beneficial and will cause less harm to the environment.

In conclusion, philosophers such as Peter Singer disagree with the idea of animal experimentation because he believes animals should not experience suffering. Animal experimentation also causes harm to the environment. The process of incineration produces pollution, and it produces a large amount of environmental waste. Animal experimentation produces many negative impacts on the environment and the animals themselves. These costs do not outweigh the small amount of benefit to humans. The overall conclusion is that animal experimentation is a cruel and senseless act that must cease.

Notes

[1] People for the Ethical Treatment of Animals (PETA), Accessed on 24 April, 2022, Retrieved from https://www.peta.org/

[2] Peter Singer, "Animal Liberation," in *Animal Rights: The Changing Debate* (London: Palgrave Macmillan UK, 1996), 136.

[3] Ibid., 137.

[4] Ibid., 138-139.

[5] Ibid., 140.

[6] Rebekah Corbett, "Animal Research: An Environmental Perspective," Accessed on 6 May, 2022, Retrieved from https://faunalytics.org/animal-research-an-environmental-perspective/

[7] Jai Shroff, "The Significant Impact of Animal Testing on the Environment," from Jai Shroff's Blog, Accessed on 6 May, 2022, Retrieved from https://jaishroff. Wordpress.com/2016/11/15/the-significant-impact-of-animal-testing-on-the-environment/

[8] People for the Ethical Treatment of Animals.

Literature

Corbett, Rebekah. 2019. "Animal Research: An Environmental Perspective." Faunalytics. Published September 16. Retrieved May 6, 2022, from https://faunalytics.org/animal-research-an-environmental perspective/

PETA. 2022. *People for the Ethical Treatment of Animals (PETA)*. Published April 24. Retrieved April 24, 2022, from https://www.peta.org/

Shroff, Jai. 2016. "The Significant Impact of Animal Testing on the Environment." Jai Shroff's Blog. Published November 11. Retrieved May 6, 2022, from https://jaishroff.wordpress.com/2016/11/15/the-significant-impact-of-animal-testing-on-the-environment/

Singer, Peter. 1996. "Animal Liberation." In *Animal Rights: The Changing Debate*, edited by Robert Garner. London, Palgrave Macmillan UK.: 135-142. https://doi.org/10.1007/978-1-349-25176-6_1

CHAPTER TWENTY-FIVE

AGRICULTURAL PESTICIDES AND MARINE WILDLIFE: A PHILOSOPHICAL PERSPECTIVE

KATELYNN PIDGORODETSKIY

Every year, about one billion pounds of pesticides are used in the United States to control weeds, insects, and other pests in agricultural farming. This makes the United States the second leading country in the usage of agricultural pesticides, making China the first.[1] In a report completed by USGS, they stated, "the use of pesticides has resulted in a range of benefits, including increased food production and reduction of insect-borne disease, but also raises questions about possible adverse effects on the environment including water quality" (USGS). They also revealed that ninety percent of water samples contain five or more different pesticides.[2] With this, it can be concluded that pesticides and nutrients that are used in agriculture run into bodies of water, depleting oxygen, and harming and killing off marine wildlife. Not only does this directly affect our oceans and other water sources, but also the wildlife that resides within them and the different forms of life that may use them as a resource. When it comes to looking at the environment, including water sources and their wildlife counterparts, we can look through a philosophical lens to help guide our beliefs about what is right and wrong. However, the most important question that we must first ask ourselves is, "How far are we willing to extend morality from an anthropocentric viewpoint?"

To begin with, Jeremy Bentham's Utilitarian ethics stance focuses on how much pain and pleasure are being created through a course of action.[3] With this in mind, Peter Singer, a utilitarian philosopher, believes that we must expand moral standing to include animals and that suffering is the issue not only when animals' lives are at risk, but all the time. In "Animal Liberation," he excludes Molluscs, which are invertebrates and are further away from being human-like than other species with vertebrates. Mollusks are part of marine wildlife, but there are many other animals that he does defend through his statements that play important roles in the marine ecosystem, such as fish, whales, sharks, and any sea animal that has a vertebrate or is not too far from being human-like.

Another part of his claim that helps support the wildlife that resides in the affected waters is that animals may not be able to speak for themselves and communicate their needs, but they are able to show signs that they are harmed or are suffering. Along with this claim, he compares these animals to babies, which also cannot speak for themselves, but give specific signals that they need something, so we know that we must provide something for them.[4]

However, Tom Regan, in “The Case for Animal Rights,” critiques utilitarianism and contractarianism. Even though he also concludes we should not harm animals, he thinks the issue is better thought about by focusing on animals as being subjects of their own lives, as having experiences and consciousness. This is what gives them inherent value. To argue against Utilitarianism, he states that it fails to explain why a good end does not justify an evil mean. For example, Regan argues that Utilitarians would believe that killing someone for money that would be used to benefit others would not be immoral. Concerning contractarianism, Regan states that because the environment and its animals are not able to sign a contract themselves, humans make the decisions and create the contracts on behalf of the environment, meaning they do not have a say in what policies and support are being given to them and their living spaces. This may not always allow for the best interest of the environment and wildlife.[5] Only a rights-based theory does the job, according to Regan.

The marine ecosystem not receiving benefits can also be seen through Stone’s “Should Trees Have Standing?” In his article, Stone talks about legal operational mechanisms: we must act on behalf of the environment within the legal system. However, often, the benefits from these actions do not go directly to what was harmed or damaged, but rather to the owner, so the thing that was damaged does not always get the help that it needs.[6] Therefore, we must make sure that the marine ecosystems are being acted on behalf of and receive the aid and care that it deserves. In addition, towards the end of Stone’s legal operational argument he states that once something is taken away, it costs a lot to bring back and is often not able to be brought back, especially to its original state.

In Paul Hawken’s, “Natural Capitalism,” he brings forward the argument that what nature does for humans is not taken into consideration. Our economy and lives rely heavily on many environmental functions that humans are sometimes not able to do for themselves.[7] For example, the bodies of water that are being polluted are affecting the amounts of drinking water and water for the resources that we use, that we have available to us. This can also be seen through the limitation of resources due to pollutants killing the animals that we use as a food source from these bodies of water.

Arne Naess, in his "Deep Ecology," presents the idea that if we harm the environment, we are harming ourselves. An example of this can be seen as we are causing harm to marine ecosystems, which will decrease the number of food resources that we can obtain from them. According to Naess, we must look at a total-field image, meaning we must look at the wildlife and its marine environment, and how both of them are being affected: without the water being up to living standards, the wildlife is no longer able to live in that space and it has less chance of survival.[8] As Hawken states in his article, failure to recognize what the environment, or the marine ecosystem, actually does for us, could lead us to possible extinction, so we must see ourselves in the environmental workplace and include nature as a part of the economy's market.

In William Baxter's, "People or Penguins," he takes a very anthropocentric viewpoint toward wildlife and the environment. However, towards the end of his argument, he lightly touches on Cost-Benefit-Analysis, which is important when talking about the marine ecosystem being negatively affected by human agricultural farming. Baxter states that when we spend more money to preserve the environment, we are essentially taking more away from other necessary life resources, such as a hospital.[9] However, if we decide to spend more money on finding alternative farming methods that will not harm marine life and its waters, it can benefit our health as we will have healthy drinking water and a continuation of food sources, so we will not need to spend as much money on healthcare facilities.

Ulrich Beck's, "From Industrial Society to Risk Society," also allows us to look at the risks that humans can endure through the use of pesticides in agricultural farming and its effects on marine wildlife. In his argument, he believes that the risk-to-society assessments that are completed are biased as those in the industry complete them, and it is hypothetical because they do not know the effects that a certain thing will have until it is created and tested. Concerning marine ecosystems and agricultural farming, we are not able to test how certain pesticides will affect the water and its wildlife until it is already being used, which may negatively affect it. Also, those creating the pesticides are the farmers that are using them, and they want their crops to grow so they can make a living, so they want the process of creating them to keep moving along, allowing for biased evaluations. In his conclusion, he mentions that we need more public input and collective decisions about what is being created and used in our society, and therefore we must all become educated on the topic itself, and the negative effects that it may allow for.[10] When we think about agricultural farming and its use of pesticides, we must think about the effects that it has on the environment as it directly affects humans and their ability to live a healthy lifestyle.

In conclusion, not only must we think about how the marine wildlife and its aquatic environment are being affected, but we must also think about how we must act on behalf of it through the legal system. We must also take into consideration what it does for us since it can do certain things that humans are not able to do for themselves. According to Aldo Leopold in his, "Land Ethic," we must see ourselves within the bio-spherical community: this will help us identify the issues. For example, personal experiences can teach us directly what harm pesticides may be causing for the ecosystem. He also states that the biosphere as an interconnected whole has moral standing,[11] which supports the claim that we rely on the functioning of the ecosystem, while it relies on us. With this, an approach that can be used to help decrease negative outcomes on marine wildlife includes the usage of organic pesticides. Some of the options that Jane Alice Liu presented in her blog entitled, "Organic Pesticides: Making Use of Nature," include the use of natural barriers that will allow for the containment of run-off into water sources and other bodies of water, such as mulch between plants and cloth crop covers, and Biocontrol, which presents pest's natural enemies, including other insects or bacteria strains, and lastly, Polyculture, which involves planting multiple types of crops and some that naturally repel specific pests of the crop. An example of Polyculture includes planting tomato plants, which repel larvae that eat the cabbage plants.[12] By doing this, farmers are able to produce healthier, more natural crops, while also protecting marine wildlife by reducing the amount of waste that is running into the ocean and other natural or important bodies of water and depleting the necessary oxygen.

Notes

[1] Lucía Fernández, "Global Pesticide Use by Country." *Statista*, 27 Sept. 2021, www.statista.com/statistics/1263069/global-pesticide-use-by-country/.

[2] "Pesticides Active," *Pesticides | U.S. Geological Survey*, www.usgs.gov/centers /ohio-kentucky-indiana-water-science-center/science/pesticides.

[3] J. Bentham, 1789, *An Introduction to the Principles of Morals and Legislation.*
[4] Peter Singer, "Animal Liberation," in Donald VanDeVeer and Christine Pierce, *The Environmental Ethics and Policy Book (*Wadsworth, Cengage Learning: 2003), 135-142.
[5] Tom Regan, "The Case for Animal Rights," in, Donald VanDeVeer and Christine Pierce, *The Environmental Ethics and Policy Book (*Wadsworth, Cengage Learning: 2003), 143-149.
[6] Christopher D.Stone, "Should Trees Have Standing? – Toward Legal Rights for Natural Objects," in, Donald VanDeVeer and Christine Pierce, *The Environmental Ethics and Policy Book (*Wadsworth, Cengage Learning: 2003), 189-200.
[7] Paul Hawken, Excerpts from *Natural Capitalism*, Boston: Little, Brown and Company, 1999, 227-237.
[8] Arne Naess, "The Shallow and the Deep, Long-Range Ecology Movement," *Inquiry 16,* University of Oslo, 1973: 95-100.
[9] William Baxter, *People or Penguins: The Case for Optimal Pollution*, New York: Columbia University Press, 1974: 381-384.
[10] Ulrich Beck, "From Industrial Society to Risk Society: Questions of Survival, Social Structure, and Ecological Enlightenment," *Theory, Culture, and Society* Vol. 9, No. 1 (1992): 97-123.
[11] Aldo Leopold, "The Land Ethic," in, Donald VanDeVeer and Christine Pierce, *The Environmental Ethics and Policy Book (*Wadsworth, Cengage Learning: 2003), 215-224.
[12] Jane Alice Liu,"Organic Pesticides: Making Use of Nature," *Organic Pesticides | Making Use of Nature*, 23 July 2019, www.foodunfolded.com/article/ pesticide-alternatives-organic-farming.

Literature

Baxter, William. 1974. *People or Penguins: The Case for Optimal Pollution*. New York: Columbia University Press, 381-384

Beck, Ulrich. 1992. "From Industrial Society to Risk Society: Questions of Survival, Social Structure, and Ecological Enlightenment." *Theory, Culture, and Society* Vol. 9, No. 1: 97-123.

Bentham, J. 1789. *An Introduction to the Principles of Morals and Legislation.*

Fernández, Lucía. 2021. "Global Pesticide Use by Country." *Statista*, 27 Sept. www.statista.com/statistics/1263069/global-pesticide-use-by-country/.

Hawken, Paul. 1999. *Natural Capitalism*. Boston: Little, Brown and Company: 227-237.

Leopold, Aldo. 2003. "The Land Ethic." In, Donald VanDeVeer and Christine Pierce, *The Environmental Ethics and Policy Book.* Wadsworth, Cengage Learning: 215-224.

Liu, Jane Alice. 2019. "Organic Pesticides: Making Use of Nature." *Organic Pesticides | Making Use of Nature*. 23 July. www.foodunfolded.com/article/pesticide-alternatives-organic-farming.

Naess, Arne. 1973. "The Shallow and the Deep, Long-Range Ecology Movement." *Inquiry 16.* University of Oslo: 95-100.

"Pesticides Active." N. d. *Pesticides | U.S. Geological Survey*, www.usgs.gov/centers/ohio-kentucky-indiana-water-science-center/science/pesticides.

Regan, Tom. 2003. "The Case for Animal Rights." In, Donald VanDeVeer and Christine Pierce, *The Environmental Ethics and Policy Book.* Wadsworth, Cengage Learning: 143-149.

Singer, Peter. 2003. "Animal Liberation." In, Donald VanDeVeer and Christine Pierce, *The Environmental Ethics and Policy Book.* Wadsworth, Cengage Learning: 135-142.

Stone, Christopher D. 2003. "Should Trees Have Standing? – Toward Legal Rights for Natural Objects." (1974). In, Donald VanDeVeer and Christine Pierce, *The Environmental Ethics and Policy Book.* Wadsworth, Cengage Learning: 189-200.

CHAPTER TWENTY-SIX

PRIMARY CAUSES AND IMPACTS OF CORAL BLEACHING & FUTURE CONSERVATION OUTLOOKS

HAILEY ADAMIK

Introduction

Over the past three decades, cases of coral bleaching have been described in nearly every coral reef-containing region.[1] Coral bleaching is defined as the process in which coral rapidly pale in color, resulting from the loss of their algae endosymbionts, called zooxanthellae.[2] These forms of algae produce carbon compounds that sustain coral reefs and provide almost 60% of their nutrients for some species.[3] Although this issue has recently gained mass attention, the problem of bleaching events remained unknown to many until now. The greatest observed bleaching events occurred in the years 1998, 2005, and 2010 with 48%, 32%, and 48% of all reefs impacted, respectively.[4] The number and magnitude of negative effects on other organisms, as well as entire marine ecosystems, have led these bleaching events to become an alarming concern within our oceans. For example, over 6,000 different species of fish are known to depend on coral reefs.[5] Should these fishes decline in population due to coral reef destruction, any organisms relying on those species as part of their diet will be influenced by a ripple effect. As shown in Figure 1, humans depend on coral reefs for a variety of goods and necessities, including food and resources, coastal protection, ocean recreation and tourism, and biogeochemical cycling.[6] In addition, healthy reefs are capable of producing 20,000 kg of fish and edible resources per square kilometer per year.[7]

This significant decline in reef health branches from a variety of factors, including growing urbanization, agriculture, and industrialization over the years of our nation's development.[8] As a result, mass-mortality coral bleaching events have increased in severity as well as frequency due to increased temperatures, primarily driven by climate change.[9] This review will indicate various other coral reef stressors today, highlight the great importance of reefs within a marine ecosystem, and provide a future outlook for conservation efforts as well as foreseeable recovery trends.

Major Stressors

Today, climate change is a major concern due to the alarming rate and failure to show signs of slowing.[10] Studies indicate such change as the primary factor in coral bleaching events. One of the major issues coupled with climate change is an increase in ocean acidification, which occurs when elevated levels of carbon dioxide cause a drop in the pH of ocean water. One study revealed that high levels of carbon dioxide had a greater effect on coral and their algal symbionts than temperature.[11] After 8 weeks, the research also showed that high carbon dioxide doses resulted in 40-50% bleaching for coralline algae and branching coral species, as well as a 50% plummet in calcification rates for coralline algae coral.[12] These studies demonstrate the importance of working to reduce greenhouse gas emissions and finding alternatives to burning fossil fuels.

A variety of recent research suggests that individual types of coral (and their symbiotic algae) are affected more intensely by specific combinations of different factors. For example, one study revealed that hard coral was primarily harmed by elevated temperature, overfishing, and salinity, while soft coral was most negatively impacted by a mixture of salinity, oxygen, and phosphate levels.[13] The same study also concluded that when incorporated with chlorophyll a and wind speed, the oxygen variable affected hard corals positively.[14] In another study performed on the Eastern Gulf of Thailand, it was found that *D. heliopora*, a type of hard coral, was the most resistant coral following the mass bleaching events of 2010.[15] It was also noted that coral species most affected by bleaching included *Montipora tuberculosa*, *Acropora millepora*, and *Favia pallida*—all variants of hard coral.[16] This research is important because it indicates not all coral types respond the same to certain factors, as well as the same types are affected differently. Therefore, additional research on how specific corals react is necessary for future conservation efforts.

One major contributor to coral bleaching events and marine damage, in general, is ocean pollution. A study conducted at the reefs of Christmas Island concluded that elevated phosphate levels resulted in a decrease in calcareous encrusting algae (an important form of algae that helps bind coral reefs together).[17] Although phosphorous is required by many organisms to survive, extensive levels in water lead to eutrophication, which includes the promotion of microalgae growth that has the potential to surpass and overtake slower-growing coral.[18] These increased levels of phosphate result from a variety of land-based activities including mining, fertilizer runoff, and sewage spills.[19] Another form of ocean pollution affecting coral reefs is microplastic. In a study conducted at the Nansha Islands, it was found that each of the tested water samples was contaminated with microplastic and that larger plastic (over 5 millimeters in size) polluted all but one site.[20] The results of these studies demonstrate the substantial amount of pollution within our oceans.

Recreational diving is one of the less immense stressors to negatively affect coral reefs. One study revealed that of 100 supervised recreational divers, an alarming 88% were witnessed making contact with the reef at least once per drive.[21] It was also noted that the majority of divers observed in the study had prior experienced – approximately half had dived at least 100 times.[22] Additionally, the research found that with increased group size, diving instructors were less likely to intervene or correct diving practice.[23] This particular study shows how humans directly impact reefs and marine life in general.

Reef Destruction Impacts

Due to the extensive reliance of thousands of species upon coral reefs for shelter, food, and spawning grounds, coral reefs are considered one of the most biologically productive ecosystems within our oceans (as shown in Figure 1). As a result of a reef's significance within a marine environment, various species have suffered and even become extinct following the widespread bleaching events and reef deterioration. For example, elevated temperatures have been shown to directly kill *Pocillopora* (a type of stony coral that protects other massive coral species).[24] As a result, various organisms have been affected by a chain reaction. Other coral species now face increased predation from starfish, which previously refrained from crossing *Pocillopera*.[25] Additionally, crustacean populations lessened as they relied on mucus produced by *Pocillopora* for food.[26] The population decline of butterflyfishes is another example of a species suffering directly from the loss of coral. A study concluded that all butterflyfish affected fed on stony or hard coral, and that mean densities of the presiding species of butterflyfish experienced a 97% drop in the same year.[27] Furthermore, reef fish populations not only suffer from loss of habitat and food in correlation with coral bleaching but from increases in toxic cyanobacteria, a factor associated with reef destruction.[28]

In addition to the extinctions and population declines of reef-relying organisms, feeding patterns and migrations of specific species have been observed as well. One study showed that following coral bleaching and destruction, larger anemone-fish interaction with smaller members increased.[29] As a result, the less-dominant fish were forced to migrate from their host anemones.[30] The research inferred these frequent attacks as consequences of drops in anemone population and size following coral damage. A different study revealed that various corallivorous (coral-feeding) fish species developed an increased preference for feeding on recently bleached corals, likely due to elevated mucous release.[31] These new feeding habits have the potential to accelerate coral destruction rates, which will continue to impact all other species relying on reefs for different reasons.[32]

Conservation Efforts

One of the primary ways to help restore and protect coral reefs worldwide is to continue new research. As data begins to grow outdated and new studies are conducted, unlikely discoveries with the potential to become solutions are uncovered. For example, recent research has shown that although changing climate and precipitation is one of the key factors leading to coral bleaching, these variations have proven beneficial in some cases. Storms and hurricanes help control bleaching because their resulting wave action leads to "vertical mixing", which then allows for excess heat to be transferred from shallow water.[33] Additionally, waves generate water motion that benefits coral reefs by expelling toxic oxygen radicals in the process.[34] This has been demonstrated by the observation of healthier corals found near tidal passes in lagoons in America Somoa, one of the warmest environments known to host coral reefs.[35] These results can be used to help shape plans for the placement of coral regrowth programs. Furthermore, elevated temperatures are arguably one of the most devastating contributors to coral deterioration. As a result, it is important to find adequate ways of monitoring ocean temperature levels. With newly improved satellite climatology technology, it is now possible to identify and describe areas of elevated water temperatures across oceans as a whole, allowing scientists to pinpoint coral reef locations susceptible to bleaching due to increased temperatures.[36] In addition, NESDIS (NOAA's National Environmental Satellite, Data, and Information Service) is working to create an SST (satellite-derived sea surface temperatures) chart, focusing on abnormalities in precipitation, that can be utilized to help predict likely bleaching patterns.[37] These new and upcoming innovations might not solve the bleaching crisis itself but will play an important role in assessing and finding solutions to climate change's effects on reefs.

Unlike more long-term factors such as rising ocean temperature, the rectification of more controllable issues, such as incorrect diving practices, can provide a quick way to help protect coral reefs and decrease destruction.[38] Those working towards coral reef and marine conservation are also calling for stricter laws. Despite current marine protection attempts, research now shows that to save remaining reefs and provide the chance for others to recover, nations must immediately adopt more aggressive reef management policies.[39]

Conclusion

In the final analysis, mass coral bleaching events stem from a variety of factors, but climate change and thermal stress remain the primary component of reef damage. Due to this decrease in reef health, marine ecosystems are at serious risk on account of many symbiotic relationships between reefs and other dependent organisms. In addition, humans depend on such reefs for a variety of resources and economic value. As a result, it is crucial to conduct new research to further our understanding of how specific factors affect different types of coral. Overall, research shows that reducing emissions, as well as social adaption and mitigation programs will be required to prevent further bleaching events and contribute to the protection of our reefs in the future.[40]

Notes

[1] Ronan C. Roche, et al., "Recreational Diving Impacts on Coral Reefs and the Adoption of Environmentally Responsible Practices within the SCUBA Diving Industry," *Environmental Management* 58, no. 1 (2016), https://doi.org/10.1007/s00267-016-0696-0.

Ibid.

[3] Barbara E. Brown and John C. Ogden, "Coral Bleaching," *Scientific American* 268, no. 1 (1993), https://doi.org/10.1038/scientificamerican0193-64.

[4] Scott F. Heron, et al., "Warming Trends and Bleaching Stress of the World's Coral Reefs 1985–2012," *Scientific Reports* 6 (2016), https://doi.org/10.1038/srep38402.

[5] Terry P. Hughes, et al., "Coral reefs in the Anthropocene," *Nature* 546, no. 7656 (2017), https://doi.org/10.1038/nature22901.

[6] Ibid.

[7] Alan T. White, Helge P. Vogt, and Tijen Arin, "Philippine Coral Reefs Under Threat: The Economic Losses Caused by Reef Destruction," *Marine Pollution*

Bulletin 40, no. 7 (2000/07/01/ 2000), https://doi.org/https://doi.org/10.1016/S0025-326X(00)00022-9.

[8] Ronan C. Roche, et al., "Recreational Diving Impacts on Coral Reefs and the Adoption of Environmentally Responsible Practices within the SCUBA Diving Industry," Environmental Management 58, no. 1 (2016), https://doi.org/10.1007/s00267-016-0696-0.

[9] Ibid.

[10] Terry L. Root, et al., "Fingerprints of global warming on wild animals and plants," *Nature* 421, no. 6918 (2003/01/01 2003), https://doi.org/10.1038/nature01333, https://doi.org/10.1038/nature01333.

[11] K. R. N. Anthony, et al., "Ocean acidification causes bleaching and productivity loss in coral reef builders," *Proceedings of the National Academy of Sciences* 105, no. 45 (2008), https://doi.org/10.1073/pnas.0804478105.

[12] Ibid.

[13] Joanne I. Ellis, et al., "Multiple stressor effects on coral reef ecosystems," Article, *Global Change Biology* 25, no. 12 (2019), https://doi.org/10.1111/gcb.14819.

[14] Ibid.

[15] Makamas Sutthacheep, et al., "Impacts of the 1998 and 2010 mass coral bleaching events on the Western Gulf of Thailand," *Deep Sea Research Part II: Topical Studies in Oceanography* 96 (2013/11/01/ 2013), https://doi.org/10.1016/j.dsr2.2013.04.018.

[16] Ibid.

[17] Daniel F. Martinez-Escobar and Jennie Mallela, "Assessing the impacts of phosphate mining on coral reef communities and reef development," *Science of The Total Environment* 692 (2019), https://doi.org/https://doi.org/10.1016/j.scitotenv.2019.07.139.

[18] Ibid.

[19] Ibid.

[20] Fei Tan, et al., "Microplastic pollution around remote uninhabited coral reefs of Nansha Islands, South China Sea," *Science of The Total Environment* 725 (2020), https://doi.org/https://doi.org/10.1016/j.scitotenv.2020.138383.

[21] Roche, et al.

[22] Ibid.

[23] Ibid.

[24] Brown and Ogden.

[25] Ibid.

[26] Ibid.

[27] M. S. Pratchett, S. K. Wilson, and A. H. Baird, "Declines in the abundance of Chaetodon butterflyfishes following extensive coral depletion," *Journal of Fish Biology* 69, no. 5 (2006), https://doi.org/10.1111/j.1095-8649.2006.01161.x.

[28] Sharon Wismer et al., "Spatial mismatch in fish and coral loss following 2016 mass coral bleaching," *Science of The Total Environment* 650 (2019/02/10/ 2019), https://doi.org/https://doi.org/10.1016/j.scitotenv.2018.09.114.

[29] Akihisa Hattori, "Small and large anemonefishes can coexist using the same patchy resources on a coral reef, before habitat destruction," *Journal of Animal Ecology* 71, no. 5 (2002), https://doi.org/10.1046/j.1365-2656.2002.00649.x.
[30] Ibid.
[31] A. J. Cole, M. S. Pratchett, and G. P. Jones, "Effects of coral bleaching on the feeding response of two species of coral-feeding fish," *Journal of Experimental Marine Biology and Ecology* 373, no. 1 (2009), https://doi.org/10.1016/j.jembe. 2009 .02.016.
[32] Ibid.
[33] Andrew C. Baker, Peter W. Glynn, and Bernhard Riegl, "Climate change and coral reef bleaching: An ecological assessment of long-term impacts, recovery trends and future outlook," *Estuarine, Coastal and Shelf Science* 80, no. 4 (2008), https://doi.org/https://doi.org/10.1016/j.ecss.2008.09.003.
[34] Ibid.
[35] Ibid.
[36] Alan Strong and C. Barrientos, "Improved Satellite Techniques for Monitoring Coral Reef Bleaching." (2020). https://coralreefwatch.noaa.gov/satellite/publications/crbpub_8icrs97txt.pdf.
[37] Ibid.
[38] Roche, et al.
[39] David J. Suggett and David J. Smith, "Coral bleaching patterns are the outcome of complex biological and environmental networking," Article, *Global Change Biology* 26, no. 1 (2020), https://doi.org/10.1111/gcb.14871.
[40] Roche, et al.

Literature

Anthony, K. R. N., D. I. Kline, G. Diaz-Pulido, S. Dove, and O. Hoegh-Guldberg. 2008. "Ocean Acidification Causes Bleaching and Productivity Loss in Coral Reef Builders." *Proceedings of the National Academy of Sciences* 105, no. 45: 17442. https://doi.org/10.1073/pnas. 0804478105.

Baker, Andrew C., Peter W. Glynn, and Bernhard Riegl. 2008. "Climate Change and Coral Reef Bleaching: An Ecological Assessment of Long-Term Impacts, Recovery Trends and Future Outlook." *Estuarine, Coastal and Shelf Science* 80, no. 4: 435-71. https://doi.org/10.1016/j.ecss. 2008.09.003.

Brown, Barbara E., and John C. Ogden. 1993. "Coral Bleaching." *Scientific American* 268, no. 1: 64-70. https://doi.org/10.1038/scientificamerican 0193-64.

Cole, A. J., M. S. Pratchett, and G. P. Jones. 2009. "Effects of Coral Bleaching on the Feeding Response of Two Species of Coral-Feeding Fish." *Journal of Experimental Marine Biology and Ecology* 373, no. 1: 11-15. https://doi.org/https://doi.org/10.1016/j.jembe.2009.02. 016.

Ellis, Joanne I., Tahira Jamil, Holger Anlauf, Darren J. Coker, Joao Curdia, Judi Hewitt, Burton H. Jones, et al. 2019. "Multiple Stressor Effects on Coral Reef Ecosystems." Article. *Global Change Biology* 25, no. 12: 4131-46. https://doi.org/10.1111/gcb.14819.

Hattori, Akihisa. 2002. "Small and Large Anemonefishes Can Coexist Using the Same Patchy Resources on a Coral Reef, before Habitat Destruction." *Journal of Animal Ecology* 71, no. 5: 824-31. https://doi.org/ 10.1046/j.1365-2656.2002.00649.x.

Heron, Scott F., Jeffrey A. Maynard, Ruben van Hooidonk, and C. Mark Eakin. 2016. "Warming Trends and Bleaching Stress of the World's Coral Reefs 1985–2012." *Scientific Reports* 6. https://doi.org/10.1038/srep38402.

Hughes, Terry P., Michele L. Barnes, David R. Bellwood, Joshua E. Cinner, Graeme S. Cumming, Jeremy B. C. Jackson, Joanie Kleypas, et al. "Coral Reefs in the Anthropocene." *Nature* 546, no. 7656 (2017): 82-90. https://doi.org/10.1038/nature22901.

Martinez-Escobar, Daniel F., and Jennie Mallela. 2019. "Assessing the Impacts of Phosphate Mining on Coral Reef Communities and Reef Development." *Science of The Total Environment* 692: 1257-66. https://doi.org/https://doi.org/10.1016/j.scitotenv.2019.07.139.

Pratchett, M. S., S. K. Wilson, and A. H. Baird. 2006. "Declines in the Abundance of Chaetodon Butterflyfishes Following Extensive Coral Depletion." *Journal of Fish Biology* 69, no. 5: 1269-80. https://doi.org/10.1111/j.1095-8649.2006.01161.x.

Roche, Ronan C., Chloe V. Harvey, James J. Harvey, Alan P. Kavanagh, Meaghan McDonald, Vivienne R. Stein-Rostaing, and John R. Turner. 2016. "Recreational Diving Impacts on Coral Reefs and the Adoption of Environmentally Responsible Practices within the Scuba Diving Industry." *Environmental Management* 58, no. 1: 107-16. https://doi.org/10.1007/s00267-016-0696-0.

Root, Terry L., Jeff T. Price, Kimberly R. Hall, Stephen H. Schneider, Cynthia Rosenzweig, and J. Alan Pounds. 2003. "Fingerprints of Global Warming on Wild Animals and Plants." *Nature* 421, no. 6918: 57-60. https://doi.org/10.1038/nature01333.

Strong, Alan, and C. Barrientos. 2020. "Improved Satellite Techniques for Monitoring Coral Reef Bleaching.". https://coralreefwatch.noaa.gov/satellite/publications/crbpub_8icrs97txt.pdf

Suggett, David J., and David J. Smith. 2020. "Coral Bleaching Patterns Are the Outcome of Complex Biological and Environmental Networking." Article. *Global Change Biology* 26, no. 1: 68-79.

https://doi. org/ 10.1111/gcb.14871.

Sutthacheep, Makamas, Mathinee Yucharoen, Wanlaya Klinthong, Sittiporn Pengsakun, Kanwara Sangmanee, and Thamasak Yeemin. 2013. "Impacts of the 1998 and 2010 Mass Coral Bleaching Events on the Western Gulf of Thailand." *Deep Sea Research Part II: Topical Studies in Oceanography* 96: 25-31. https://doi.org/10.1016/j.dsr2.2013.04.018.

Tan, Fei, Hongqiang Yang, Xiangrong Xu, Zhou Fang, Huilong Xu, Qi Shi, Xiyang Zhang, et al. 2020. "Microplastic Pollution around Remote Uninhabited Coral Reefs of Nansha Islands, South China Sea." *Science of The Total Environment* 725: 138383. https://doi.org/10.1016/j.scitotenv.2020.138383.

White, Alan T., Helge P. Vogt, and Tijen Arin. 2000. "Philippine Coral Reefs under Threat: The Economic Losses Caused by Reef Destruction." *Marine Pollution Bulletin* 40, no. 7: 598-605. https://doi.org/10.1016/S0025-326X(00)00022-9.

Wismer, Sharon, Sterling B. Tebbett, Robert P. Streit, and David R. Bellwood.2019. "Spatial Mismatch in Fish and Coral Loss Following 2016 Mass Coral Bleaching." *Science of The Total Environment* 650 : 1487-98. https://doi.org/10.1016/j.scitotenv.2018.09.114.

Chapter Twenty-Seven

California Wildfires of 2020

Hannah Cawley

Introduction

The climate in California has always been fire-prone, with low levels of rainfall and high heat in the summertime that dries out vegetation. By the time fall arrives the conditions are just right for the dry Santa Ana winds to spread any small fire that starts. Often, all that is needed is a spark to turn the dry brush on the ground into fuel.[1] While these conditions are not new to the area, the fires are worsening every season. In 2020, California experienced its worst wildfire season on record in both the number and size of the fires.[2] There was also significant destruction of property and loss of life that cannot be overlooked.

Multiple states in the western part of the country were significantly impacted by the fires in 2020. This chapter, however, will focus solely on California. The increasing severity of the fires can in part be attributed to the changing global climate, but also to the history of fire suppression in the area. Some unique features of the 2020 season, such as the new records it set and the COVID-19 pandemic, will be examined along with their effects on future seasons. The effectiveness of the emergency response will also be discussed.

Record Highs in 2020

In 2020, about 10,000 fires burned more than four percent of California's nearly 100 million acres of land, making it the largest wildfire season recorded in the state's history.[3] It was a typical season until mid-August, when a large storm hit Central and North California that delivered upwards of 15,000 lightning strikes that set forests ablaze, instead of sufficiently dousing them with needed rainfall. While thunderstorms are common for that region, that amount of lightning is not. The warm winds that succeeded the storm contributed to the rapid spread of those new fires.[4]

According to the National Interagency Coordination Center (NICC), while the national annual number of fires has been decreasing slightly (over the last 10 years), the number of acres burned has been increasing. In 2020 the number of acres burned nationally was over 10 million, which was well above the previous averages. Also unique to the year was the first reported fire incident over 1 million acres at the August Complex in Northern California. California alone also contributed to 38% of the national total acres burned.[5] All of the record-breaking statistics set this year apart from the past and may even be predictive for what is to come in future seasons.

In their review of the 2020 season, Cal Fire also reported that the five highest daily average air pollution readings that have ever been recorded in the state all occurred that year. This impacted both the state's carbon budget and the health of citizens.[6]

Environmental Challenges

The history of fire suppression in California has led to about a century's worth of fuel build-up from overgrown plant material and debris. This makes fires that would otherwise be easier to contain, catastrophic. Not all natural fires are so destructive and are often essential to a healthy ecosystem.[7] The Native Americans knew this, and historically used burns to control the environment by reducing the overgrown fuels on the forest floor.[8] In the areas where the Forest Service had done controlled burns and thinned the forest, the King Fire of 2017 burned less intensely and stayed mostly to the forest floor, not reaching the canopies and killing all the trees, as it had done in other areas.[9]

From an environmental standpoint, the impact of downing a few trees is much more favorable than losing an entire forest to a fire that could have been mitigated. The carbon emissions released during an uncontrolled burn can reach millions of metric tons, which doesn't even include the emissions that will be produced in the years to follow as the dead trees decay.[10]

Despite what fire ecologists and other researchers say, there is a disconnect between what has been proven to work historically and studied in academia, and what is effectively carried out on the ground. There have been increases in controlled burns over the past few years, but it is not enough. Part of this is because the Forest Service is hesitant to allow natural fires to burn, because of sporadic land ownership and more housing being built on private lands. More widespread public support and consistent funding would be needed to make this change.[11]

A peer-reviewed article titled "Climate change is increasing the likelihood of extreme autumn wildfire conditions across California", was able to quantify changes in the frequency and size of weather conditions that enable such extreme autumn wildfires in California. The researchers used climate model simulations to determine that there was a statistically significant link between worsening wildfire conditions and human-caused climate change. It was acknowledged that these factors are certainly not the only ones contributing to the fires, but state-wide increases in autumn temperature, and decreases in autumn precipitation over the past four decades have increased days with these fire weather conditions, which are projected to continue.[12] They also referenced a previous study titled "Attributing extreme fire risk in Western Canada to human emissions", which found that anthropogenic climate forcing has increased the probability of recent extreme fire seasons.[13]

While climate change is not the reason California experiences yearly wildfires, it cannot be ignored for the ways it has and will continue to make conditions worse. Multiple other concordant studies support the previously mentioned articles, such as "Global Emergence of Anthropogenic Climate Change in Fire Weather Indices", and "Intensified burn severity in California's northern coastal mountains by drier climactic condition."[14] The latter even provides suggestions such as fuel reduction to manage the landscapes prone to fires.

Impact of COVID-19

One of the most unique challenges in 2020 was trying to navigate an already novel fire season with the addition of coronavirus precautions. There had to be a balance between effectively fighting fires and evacuating people at risk and minimizing the spread of the virus as much as possible.

Firefighters and the personnel that support them during the peak season often work in close proximity and many travel from out of state to do so, which increases the likelihood of COVID-19 transmission and the spread of it to new locations.[15] To protect the health and wellness of first responders, the Fire Management Board established the Wildland Fire Medical and Public Health Advisory Team in April 2020. This team made modifications to standard operating procedures that included frequent hand washing/sanitizing, health screenings through self-reporting and temperature readings, use of PPE, and six feet distancing where possible.[16]

Instead of the usual buffet-style meals that could see around 100 firefighters during mealtimes at base camps, in California meals were staggered and pre-packed lunches were handed out instead. They also tried to limit the number of people on the frontlines by relying more heavily on aircraft for dousing flames, drones for observations, bulldozers for digging fire lines, and radio or online briefings.[17] And for any firefighters that did contract coronavirus while on the job, they were covered in Governor Newsom's executive order that provided worker's compensation during their time off.[18]

Every year, California employs thousands of inmates to support state and federal responses to wildland fires. The pandemic severely depleted this force, due to the early release granted to people serving time for non-violent crimes, and others being put into quarantine because of large outbreaks in conservation camps.[19] For those who were released, it was very difficult to continue working as a firefighter because of their criminal record, despite having the same training as the state's seasonal staff.[20] To make up for the shortages, Governor Newsom announced plans to spend $72.4 million from the emergency wildfire funds to hire over 850 more seasonal firefighters and six California Conservation Corps crews.[21]

In August when the wildfires began, California was still experiencing a high number of coronavirus cases that had many people staying inside their homes as much as possible.[22] The situation got worse when the air quality declined significantly due to the fire smoke, forcing people to keep windows closed in the middle of a heat wave.[23] These conditions were especially concerning for the elderly and sensitive groups who are at a higher risk for heat stroke and heat exhaustion.

Stopping the spread of the virus put a strain on evacuations as well. The traditional evacuation centers were not utilized like they were before, state officials instead opted for local hotels first. Those accommodations were funded by the Red Cross.[24] When those capacities were met and other open-room shelters were needed, they were de-densified, equipped with more handwashing stations, and had pre-packaged food. Before being allowed into any shelters, temperatures were checked and people with COVID-19 symptoms were isolated or sent to other treatment facilities.[25]

A study was published by Harvard Chan School, the John A. Paulson School of Engineering and Applied Sciences at Harvard University, and the Environmental Systems Research Institute in 2021 which quantified the degree to which PM2.5 pollution (fine particular matter caused by the fires), contributed to excess COVID-19 cases. The researchers examined the daily concentrations of PM2.5 in the air by county, and the number of cases, and built a statistical model to explain the data. They concluded that a daily increase of 10 ug/m3 in PM2.5 each day for 28 consecutive days resulted in an 11.7% increase in COVID-19 cases and an 8.4% increase in COVID-19 deaths. This data built upon the previous knowledge that linked PM2.5 to conditions such as asthma, COPD, and other respiratory illnesses.[26]

Even though many things changed about how California's emergency response had to be carried out, there was no evidence found that the firefighting was less effective because of virus precautions. There was also no evidence found that there were major COVID-19 outbreaks amongst evacuees, likely because of the hotel sheltering strategy and other distancing measures.

Recovery Efforts

In terms of environmental recovery, the federal government's spending on fire prevention has been fluctuating for years. While the overall Wildland Fire Management Budget has been increasing, the amount allocated for vegetation management has declined from $240 million in 2001 to $220 in 2021, with varying amounts in between.[27] This is not because the need for it is diminishing, but because as fires worsen, more resources are needed to fight them, and that takes away critical funding from efforts that would prevent this trend from continuing. Funding does not solely depend on the federal government, but as mentioned previously, more than half of the forests in California are federally owned.

The impact of the 2020 season led to an increase in California's state budget for prevention in 2021. Nearly $1 billion was allocated for the year, with guarantees of at least $200 million for the following six years, which will come from the Greenhouse Gas Reduction Fund.[28] While this is a step in the right direction, this type of funding needs to be kept up with for years to come, and costs will likely go up annually.[29]

This news in budget increase came after it was revealed that Governor Newsom had overstated fire prevention efforts in 2020 and that Cal Fire's fuel reduction efforts halved. In early 2020 they both stated that 90,000 acres of forest had been treated, but investigations found roughly 12,000 acres had been worked on. There was no official comment made by the Governor addressing these claims. Whether or not the claims hold true does not change the fact that more acreage needs to be worked on annually to see the types of changes needed to more effectively mitigate fires.[30]

Environmental recovery is another important consideration after such a disaster. California relies on these forests for sustainable wood products, clean air, clean water, and a diversity of ecosystems. There are conflicting opinions on how to best deal with forests after a large fire. Some say when the soil has crystallized after a high heat burn, there is less of a chance that rainwater can effectively be absorbed, and the lack of living trees increases the risk of landslides. Efforts like tilling the landscape and clearing away dead trees can help to mitigate more negative effects.[31] Others argue that allowing forests to regenerate on their own is the best way to maintain diversity in the ecosystem, and post-fire logging should not be allowed.[32]

Although there are conflicting opinions, any restoration effort in the long-term would cost a lot of money and take a lot of time, and there is already a backlog in the Forest Service's restoration efforts.[33] The most significant effort made in 2021 were the over one million trees planted, which hopefully do not fall victim to fires in this year's season.[34]

Notes

[1] Kendra Pierre-Louis and John Schwartz, "Why Does California Have So Many Wildfires?" New York Times (Online) (New York, United States: New York Times Company, August 21, 2020),
https://www.proquest.com/docview/2435775530/citation/876FF3A7B2C14D31PQ/1.
[2] Dani Anguiano, "California's Wildfire Hell: How 2020 Became the State's Worst Ever Fire Season," *The Guardian*, December 30, 2020, sec. US news, https://www.theguardian.com/us-news/2020/dec/30/california-wildfires-north-complex-record.
[3] "2020 Fire Season," Cal Fire, n.d., accessed February 6, 2022.
[4] Julie Cart, "California's 2020 Fire Siege: Wildfires by the Numbers," *CalMatters*, July 29, 2021, sec. Environment,
http://calmatters.org/environment/2021/07/california-fires-2020/.
[5] NICC, Annual Report: "Wildland Fire Summary and Statistics" accessed April 21, 2022,
https://www.predictiveservices.nifc.gov/intelligence/2020_statssumm/intro_summary20.pdf.
[6] Cal Fire, "Cal-Fire-2020-Fire-Siege.Pdf," accessed April 24, 2022,
https://www.fire.ca.gov/media/hsviuuv3/cal-fire-2020-fire-siege.pdf.
[7] Elizabeth Shogren, "What Fire Researchers Learned from California's Blazes," December 11, 2017,
https://www.hcn.org/issues/49.21/wildfire-what-fire-researchers-learned-from-northern-california-blazes.
[8] Anguiano.s
[9] Shogren.
[10] Ibid.
[11] Ibid.
[12] Michael Goss et al., "Climate Change Is Increasing the Likelihood of Extreme Autumn Wildfire Conditions across California," *Environmental Research Letters* 15, no. 9 (August 2020): 094016, https://doi.org/10.1088/1748-9326/ab83a7.

[13] Megan C. Kirchmeier-Young et al., "Attributing Extreme Fire Risk in Western Canada to Human Emissions," *Climatic Change* 144, no. 2 (September 1, 2017): 365–79, https://doi.org/10.1007/s10584-017-2030-0.
[14] John T. Abatzoglou, A. Park Williams, and Renaud Barbero., "Global Emergence of Anthropogenic Climate Change in Fire Weather Indices," *Geophysical Research Letters* 46, no. 1 (2019), https://doi.org/10.1029/2018GL080959.; Yuhan Huang et al., "Intensified Burn Severity in California's Northern Coastal Mountains by Drier Climatic Condition," *Environmental Research Letters* 15, no. 10 (September 2020): 104033, https://doi.org/10.1088/ 1748-9326/aba6af.
[15] "Federal Assistance for Wildfire Response and Recovery," accessed February 20, 2022, https://crsreports.congress.gov/product/pdf/IF/IF10732.
[16] "Fire Management Board | NWCG," accessed February 20, 2022, https://www.nwcg.gov/partners/fmb/.
[17] Paul Vercammen CNN, "California Firefighters Brace for Raging Flames and a Burgeoning Pandemic," CNN, accessed April 23, 2022, https://www.cnn.com/2020/08/02/us/california-firefighters-covid-pandemic/index.html.
[18] "New California Law SB 1159 Creates Workers' Compensation Presumption for Certain Eligible Employees," The National Law Review, accessed April 23, 2022, https://www.natlawreview.com/article/new-california-law-sb-1159-creates-workers-compensation-presumption-certain-eligible.
[19] CNN, "California Firefighters Brace for Raging Flames and a Burgeoning Pandemic."
[20] Isabelle Chapman, "Prison Inmates Are Fighting California's Fires, but Are Often Denied Firefighting Jobs after Their Release," CNN, October 31, 2019, https://www.cnn.com/2019/10/31/us/prison-inmates-fight-california-fires-trnd/index.html.
[21] Taryn Luna, "Temperature Checks and Hotels: Newsom Details New Precautions for Wildfire Evacuations," Los Angeles Times, July 9, 2020, https://www.latimes.com/california/story/2020-07-09/newsom-coronavirus-california-fire-prevention-evacuations-covid-19.
[22] Matt Simon, "California's Wildfire and Covid-19 Disasters Just Collided," *Wired*, accessed February 14, 2022, https://www.wired.com/story/californias-wildfire-and-covid-19-disasters-just-collided/.
[23] Ibid.
[24] "California Wildfire Victims Sent to Motels Instead of Shelters, to Lessen COVID-19 Spread," *The Sacramento Bee*, accessed April 25, 2022, https://www.sacbee.com/news/california/fires/article245092745.html.
[25] Louis Sahagun and Joseph Serna, "Coronavirus Fears Complicate Apple Fire Evacuation Plans," Los Angeles Times, August 5, 2020, https://www.latimes.com/california/story/2020-08-05/la-me-apple-fire-evacuations-coronavirus.
[26] Harvard T.H. Chan School of Public Health, "Wildfire Smoke May Have Contributed to Thousands of Extra COVID-19 Cases and Deaths in Western U.S. in 2020," News, August 13, 2021,

https://www.hsph.harvard.edu/news/press-releases/wildfire-smoke-may-have-contributed-to-thousands-of-extra-covid-19-cases-and-deaths-in-western-u-s-in-2020/.

[27] U.S. Department of the Interior, "Office of Wildland Fire: Budget," July 17, 2015, https://www.doi.gov/wildlandfire/budget; Jim Lyons, "Trump Blames California for Fires. He Should Check to See Whose Land They're On.," POLITICO, accessed April 21, 2022, https://www.politico.com/news/agenda/2020/09/15/trump-fires-california-federal-land-415431; "Fuels Management," August 20, 2015, https://www.doi.gov/wildlandfire/fuels.

[28] Scott Rodd, "Budget Package Guarantees $1B In Fire Prevention Funding This Year, At Least $200M Annually Moving Forward," accessed April 24, 2022, https://www.capradio.org/169616.

[29] Gabrielle Canon. "Gavin Newsom Oversold California's Fire Prevention Efforts. A Journalist Uncovered the Truth," *The Guardian*, July 2, 2021, sec. Environment, https://www.theguardian.com/us-news/2021/jul/02/gavin-newsom-california-fire-prevention-investigation.

[30] Ibid.

[31] "WILDFIRE," The Forest Foundation, accessed April 25, 2022, http://www.calforestfoundation.org/wildfire.

[32] "Does a Fire-Ravaged Forest Need Human Help to Recover?," Yale E360, accessed April 25, 2022, https://e360.yale.edu/features/does-a-fire-ravaged-forest-need-human-help-to-recover.

[33] Martha Conklin and Roger Bales, "Restoring California's Forests to Reduce Wildfire Risks Will Take Time, Billions of Dollars and a Broad Commitment," The Conversation, accessed April 25, 2022, http://theconversation.com/restoring-californias-forests-to-reduce-wildfire-risks-will-take-time-billions-of-dollars-and-a-broad-commitment-146532.

[34] One Tree Planted, "1.2 Million Trees Planted in California for Forest Fire Restoration in 2021," One Tree Planted, accessed April 25, 2022, https://onetreeplanted.org/blogs/stories/california-forest-fire-restoration-2021.

Literature

Abatzoglou, John T., A. Park Williams, and Renaud Barbero. 2019. "Global Emergence of Anthropogenic Climate Change in Fire Weather Indices." *Geophysical Research Letters* 46, no. 1: 326–36. https://doi.org/10.1029/2018GL080959.

Anguiano, Dani. 2020. "California's Wildfire Hell: How 2020 Became the State's Worst Ever Fire Season." *The Guardian*, December 30. sec. US news. https://www.theguardian.com/us-news/2020/dec/30/california-wildfires-north-complex-record.

"Cal-Fire-2020-Fire-Siege.Pdf." 2020. Accessed April 24, 2022. https://www. fire.ca.gov/media/hsviuuv3/cal-fire-2020-fire-siege.pdf.

The Sacramento Bee. N. d. "California Wildfire Victims Sent to Motels Instead of Shelters, to Lessen COVID-19 Spread." Accessed April 25, 2022. https://www.sacbee.com/news/california/fires/article245092745.html.

Canon, Gabrielle. 2021. "Gavin Newsom Oversold California's Fire Prevention Efforts. A Journalist Uncovered the Truth." *The Guardian*, July 2., sec. Environment. https://www.theguardian.com/us-news/2021 /jul/02/gavin-newsom-california-fire-prevention-investigation.

Cart, Julie. 2021. "California's 2020 Fire Siege: Wildfires by the Numbers." *CalMatters*, July 29. sec. Environment. http://calmatters.org/ environment/2021/07/california-fires-2020/.

Chapman, Isabelle. 2019. "Prison Inmates Are Fighting California's Fires, but Are Often Denied Firefighting Jobs after Their Release." CNN, October 31. https://www.cnn.com/2019/10/31/us/prison-inmates -fight-california-fires-trnd/index.html.

CNN, Paul Vercammen. 2020. "California Firefighters Brace for Raging Flames and a Burgeoning Pandemic." CNN. Accessed April 23, 2022. https://www.cnn.com/2020/08/02/us/california-firefighters-covid-pandemic/index.html.

Conklin, Martha, and Roger Bales. N. d. "Restoring California's Forests to Reduce Wildfire Risks Will Take Time, Billions of Dollars and a Broad Commitment." The Conversation. Accessed April 25, 2022. http://theconversation.com/restoring-californias-forests-to-reduce-wildfire-risks-will-take-time-billions-of-dollars-and-a-broad-commitment-146532.

Yale E360. N. d. "Does a Fire-Ravaged Forest Need Human Help to Recover?" Accessed April 25, 2022. https://e360.yale.edu/features/does-a-fire-ravaged-forest-need-human-help-to-recover.

Elizabeth Shogren. 2017. "What Fire Researchers Learned from California's Blazes," December 11. https://www.hcn.org/issues/49.21/ wildfire-what-fire-researchers-learned-from-northern-california-blazes.

"Federal Assistance for Wildfire Response and Recovery." N. d. Accessed February 20, 2022. https://crsreports.congress.gov/product/pdf/IF/I F10732.

"Fire Management Board | NWCG." N. d. Accessed February 20, 2022. https:// www.nwcg.gov/partners/fmb/.

"Fuels Management," 2015. U.S. Department of the Interior. August 20. https://www.doi.gov/wildlandfire/fuels.

Goss, Michael, Daniel L. Swain, John T. Abatzoglou, Ali Sarhadi, Crystal A. Kolden, A. Park Williams, and Noah S. Diffenbaugh. 2020. "Climate

Change Is Increasing the Likelihood of Extreme Autumn Wildfire Conditions across California." *Environmental Research Letters* 15, no. 9: 094016. https://doi.org/10.1088/1748-9326/ab83a7.

Harvard T.H. Chan School of Public Health. 2021. "Wildfire Smoke May Have Contributed to Thousands of Extra COVID-19 Cases and Deaths in Western U.S. in 2020." News, August 13. https://www.hsph.harvard.edu/news/press-releases/wildfire-smoke-may-have-contributed-to-thousands-of-extra-covid-19-cases-and-deaths-in-western-u-s-in-2020/.

Huang, Yuhan, Yufang Jin, Mark W. Schwartz, and James H. Thorne. 2020. "Intensified Burn Severity in California's Northern Coastal Mountains by Drier Climatic Condition." *Environmental Research Letters* 15, no. 10: 104033. https://doi.org/10.1088/1748-9326/ aba6af.

"Intro summary20.Pdf." N. d. Accessed April 21, 2022. https://www.predictiveservices.nifc.gov/intelligence/2020_statssumm/intro_summary20.pdf.

Kirchmeier-Young, Megan C., Francis W. Zwiers, Nathan P. Gillett, and Alex J. Cannon. 2017. "Attributing Extreme Fire Risk in Western Canada to Human Emissions." *Climatic Change* 144, no. 2: 365–79. https://doi.org/10.1007/s10584-017-2030-0.

Lyons, Jim. 2020. "Trump Blames California for Fires. He Should Check to See Whose Land They're On." POLITICO. Accessed April 21, 2022. https://www.politico.com/news/agenda/2020/09/15/trump-fires-california-federal-land-415431.

The National Law Review. N. d. "New California Law SB 1159 Creates Workers' Compensation Presumption for Certain Eligible Employees." Accessed April 23, 2022. https://www.natlawreview.com/article/new-california-law-sb-1159-creates-workers-compensation-presumption-certain-eligible.

Pierre-Louis, Kendra, and John Schwartz. 2020. "Why Does California Have So Many Wildfires?" New York Times (Online). New York, United States: New York Times Company, August 21. https://www.proquest.com/docview/2435775530/citation/876FF3A7B2C14D31PQ/1.

Planted, One Tree. 2022. "1.2 Million Trees Planted in California for Forest Fire Restoration in 2021." One Tree Planted. Accessed April 25, 2022. https://onetreeplanted.org/blogs/stories/california-forest-fire-restoration-2021.

Rodd, Scott. N. d. "Budget Package Guarantees $1B In Fire Prevention Funding This Year, At Least $200M Annually Moving Forward." Accessed April 24, 2022. https://www.capradio.org/169616.

Sahagun, Louis and Joseph Serna. 2020. "Coronavirus Fears Complicate Apple Fire Evacuation Plans." Los Angeles Times, August 5. https://www.latimes.com/california/story/2020-08-05/la-me-apple-fire-evacuations-coronavirus.

Simon, Matt. N. d. "California's Wildfire and Covid-19 Disasters Just Collided." *Wired*. Accessed February 14, 2022. https://www.wired.com/story/californias-wildfire-and-covid-19-disasters-just-collided/.

Taryn, Luna. 2020. "Temperature Checks and Hotels: Newsom Details New Precautions for Wildfire Evacuations." *Los Angeles Times*, July 9. https://www.latimes.com/california/story/2020-07-09/newsom-coronavirus-california-fire-prevention-evacuations-covid-19.

U.S. Department of the Interior. 2015. "Office of Wildland Fire: Budget," July 17. https://www.doi.gov/wildlandfire/budget.

The Forest Foundation. N. d. "WILDFIRE." Accessed April 25, 2022. http:// www.calforestfoundation.org/wildfire.

V. CONCLUSION

CHAPTER TWENTY-EIGHT

THE ETHICAL IMPERATIVE TO FOSTER RESILIENCE IN ECOLOGY AND HEALTH

GERARD MAGILL

To Conclude

This book, *Resilience in Ecology and Health*, contains presentations from the 7th conference in 2022 of an annual endowed series at Duquesne University in Pittsburgh, Pennsylvania, in the United States. The preceding chapters have explored how discourse on resilience about climate can inspire a focus on resilience in other fields. This conclusion emphasizes the ongoing ethical imperative to foster resilience across fields and disciplines to build a more resilient world. The discussion adopts a similar approach in this final chapter on ethics as in the concluding chapters of the previous books in this series: the urgency of climate change, integral ecology, the global water crisis, the global sustainability challenge, toward a healthy planet, and strands of sustainability. The ecology encyclical of Pope Francis continues to inspire the ethical analysis in this chapter.

The Inspiration of Pope Francis

As mentioned in the proceedings of the 1st conference on the topic of *Climate Change*,[1] Pope Francis enunciated a bold and dramatic vision for the environment in his encyclical *Laudato Si'* published in 2015. It is impressive to note how he is unwilling to be evasive about the ecological threat facing our planet. He writes provocatively: "doomsday predictions can no longer be met with irony or disdain … our contemporary lifestyle, unsustainable as it is, can only precipitate catastrophes." In the face of this bleak outlook he courageously challenges the "ethical and cultural decline which has accompanied the deterioration of the environment" and is emphatic that "halfway measures simply delay the inevitable disaster."[2] The subtitle of this Papal Encyclical presents a clarion call for advocacy "On

Care for Our Common Home."[3] His inspiring vision has inspired this book's discussion.

Pope Francis emphasized that we must engage the "relationship existing between nature and the society in which it lives;" he did so to seek "comprehensive solutions which consider the interactions within natural systems themselves and with social systems;" his point here is breathtaking insofar as he integrates the environmental and social components: "we are faced not with two separate crises, one environmental and the other social, but rather with one complex crisis which is both social and environmental."[4] He explains the respect for "the principle of the common good" not only applies "the principle of subsidiarity" (to "develop the capabilities at every level of society") but also has a "particular concern for distributive justice" as "a summons to solidarity and a preferential option for the poorest."[5] This approach includes a commitment to "justice between generations" that promotes "intergenerational solidarity" and "intragenerational solidarity."[6] To implement this vision requires "major paths of dialogue" to address the "great cultural, spiritual and educational challenge" that we encounter.[7] As this book goes to press, Pope Francis is about to promulgate a new Apostolic Exhortation called *Laudate Deum*. Subsequent books in this series will discuss its significance for global climate, sustainability and resilience. The insights of *Laudato Si'* continue to enlighten the discussion in this chapter on the ethical imperative to foster resilience across fields and disciplines in order to build a more resilient world.

Resilience

Discourse on resilience about climate can inspire a focus on resilience in other fields—hence the title of the book, *Resilience in Ecology and Health*. The organization of chapters adopts the following layout, as mentioned in the introduction to the book: resilience and ecology, resilience and health, resilience and biotechnology, and generational challenges on ecology and climate and on environment and conservation. In this concluding chapter, the discussion identifies different perspectives on resilience that have emerged with prominence in recent literature to delineate the ongoing ethical imperative to foster a more resilient world.

Pursuing resilience here occurs within the context of our unprecedented global environmental and climate change that intertwines systems of nature and human populations.[8] Resilience is a key concept for understanding the world with multiple theoretical and practical meanings.[9] Here, the concept of resilience is adopted to entail a new way of thinking about human-environmental relations and their ethical governance. This approach means emphasizing an appreciation of design that encourages collaborative and cross-boundary solutions to complex problems. Specifically, resilience approaches seek to produce more sustainable futures for complex social and ecological systems.[10] The focus on resilience is a workable community-oriented approach that is pivotal for effective environment governance; this is especially the case in our world that is so characterized by extreme complexity, radical uncertainty and unprecedented change.[11] One of the constructive outcomes of the dreadful Covid-19 global pandemic has been to recognize this new emphasis on resilience.[12]

The focus on the ethical imperative to foster the reciprocity between resilience, ecology and health honors the Sustainable Development Goals adopted by the United Nations Development Programme in 2015.[13]

Resilience and Ecology

The reciprocity between resilience and ecology is expansive. Naturally, the circumstances of the climate conundrum that we face is of paramount significance. Climate damage envelops the entire geobiological system within its interrelated forms of life, creating catastrophic fault lines that threaten all species, including humanity.[14] The urgent need for climate resilience requires a comprehensive understanding of and commitment to climate care.[15] To counter arguments and strategies used to obstruct and avert climate action, there needs to be systemic cultural changes to mitigate the climate crisis, addressing both climate resistance and renewal.[16]

Not surprisingly, a resilience-focused approach engages the major risks to climate in as prominent a way as possible. Combining risk and resilience enables us to address the causes of climate catastrophes (such as carbon emissions) rather than the symptoms, as well as to focus on prevention as being critical for resilience building.[17] This proactive stance recognizes the significance of ecosystem restoration as being fundamental to resilient development, including sustainable policy making with regard to disaster management.[18] For example, ecosystem resilience is fundamental for agriculture in the water-ecosystem-agriculture nexus; resilience occurs by addressing multiple issues simultaneously, including soil degradation, crop productivity, and economic stagnation.[19]A critical aspect of resilient development must be resilient energy systems and resilient landscapes.

For energy systems, it is indispensable to nurture stakeholder cooperation in order to have an influence on effective policymaking for innovation within the food-energy-climate nexus.[20] In turn, resilient landscapes must include agricultural, island, forest, coastal, and urban scenarios. Solution oriented approaches can generate ecological hope, connecting sustainable food systems, biodiversity and climate change adaptation with green infrastructure, nature-based architecture, green-tech, and ecosystem services.[21] It is by creating and sharing ecosystem services that we build a resilient society,[22] as illustrated for example by hybrid ecosystems in the field of urban ecology.[23]

Critically, landscape resilience seeks to help landscapes emerge from stresses and crises, being strengthened rather than weakened.[24] Landscape ecology enables us to address challenges over large areas, across time and space, to facilitate practical decision-making on a global scale.[25] For example, constructing resilient cities entails being attentive to metropolitan landscapes that respect the interface space among a city's infrastructure, agriculture, and nature. The massive growth of urban areas highlights the need for sustainable cities that minimize negative climate-change impacts, such as by balancing production and energy consumption.[26] Urban environmental sustainability requires a close link between climate change, urban development, and ecology from complex economic and social perspectives.[27] That is, a steadfast focus on resilience can lead help to create living urban organisms.[28]

Also, community-based urban development, with astute urban planning and design support, is indispensable for social sustainability regarding climate resilience.[29] Hence, resilience connects with environmental justice when considering cities to embed social vulnerabilities and resilience in urban planning.[30] As a result, community renewal and climate justice are inextricably linked in efforts to advance urban policy and climate-related energy democracy.[31] In this sense, community resilience can encourage innovation and creativity for grassroots engagement to address climate change globally.[32] This community-focused approach echoes the landmark clarion call of Mary Robinson, the former President of Ireland, for justice, hope and resilience to connect climate change with the global struggle for human rights,[33] such as constructively engaging migrants and minorities. Undoubtedly, the escalation of climate-induced migration and cross-border interaction means that it is increasingly critical to prepare comprehensively for displaced persons resiliently.[34]

Naturally, resilient cities require complex urban food systems supported by policies that increase food security and decrease ecological negative impacts. This necessarily entails addressing the twin problems of economic and nutritional distress to be able to sustainably develop resilient urban food systems.[35] Hence, agricultural resilience requires a necessary link between ecology and economics. It is only by viewing agriculture through the lens of resilience that the economic, environmental, and social uncertainties are appreciated,[36] especially regarding the ongoing inter-related crises of food, fossil fuels, and finance.[37]

Resilience and Health

Just as resilience is critical for global ecology, resilience is similarly crucial for human health. As we recover from the effects of the Covid-19 pandemic, the constructive impact of resilience in relationships, especially long-term relationships, has been confirmed.[38] The capacity for building resilience to negative experiences, such as trauma, is becoming increasingly appreciated.[39] Most especially, community resilience is being recognized as being of paramount importance for recovering when adversity occurs, such as in promoting health equity.[40] Community resilience can engender coping skills in families and organizations alike.[41] The point here is to shift from a more historical view of resilience as driven by individuals; the new focus is to hinge resilience on the community context that is the mainstay of public health,[42] thereby fostering more resilient public health systems post-pandemic.[43]

The professional burnout from the Covid-19 pandemic has highlighted the importance of resilience not only for health professionals but also for healthcare organizations,[44] and health systems globally.[45] Closely connected is the need for resilience in healthcare leadership to foster resilience throughout their organizations in order to adapt and flourish.[46] This recognition of resilience among healthcare professionals is celebrated in the growing understanding of the need for moral resilience in the face of distress or suffering scenarios when professionals know the right course of action but are prevented from pursuing it.[47] The widespread focus on developing resilient healthcare promotes efforts for interdisciplinary care and health systems improvements,[48] most especially through resilience engineering to health care and patient safety.[49]

Resilience is notably significant in mental health,[50] where, for example, resilience related to cognitive behavior therapy is widely acknowledged.[51] Importantly, cultivating resilience in early childhood is becoming much more widespread where identifying mental health and wellbeing is critical, especially with vulnerable children.[52] Fostering resilience is particularly helpful with children who have experienced trauma; here trauma-informed care is a critical component of pediatric best practices.[53] In this context, building community resilience is all the more helpful to support families in crisis.[54]

Resilience and Biotechnology

Insofar as resilience is crucial for ecology and health, resilience in biotechnology is emerging as critically significant for both health and ecology.

Regarding health, resilience requires new standards of excellence that focus on a deep understanding of suffering and build the trust of patients, the healthcare workforce, and a culture of safety that is patient-centered.[55] The Covid-19 pandemic provided an opportunity to showcase the capacity of biotechnology to create a vaccine that would respond effectively to the virus globally.[56] Another global effort for health can be found in the World Health Organization's global strategy for digital health; this seeks to foster computable clinical data and knowledge to improve patient, organizational and global health outcomes.[57] Of course, digital health technology, which is one of the fastest growing sectors of healthcare, has both benefits and risks, such as abuses pervasive to cyber space.[58] Nonetheless, the emerging and fast-developing efforts between medical biotechnology and human health sciences continue to leverage multidisciplinary research to contribute to patient diagnosis, treatment and well-being.[59] This interface between medical biotechnology and human health is particularly evident in pharmaceutical biotechnology, such as with the large number of monoclonal antibodies and therapeutic proteins that have been approved.[60] Hence, international regulation becomes all the more urgent to protect human rights in the face of landmark biotechnology breakthroughs, such as in stem cell research, human cloning, or gene therapy,[61] as well as nanotech and enzyme replacement therapies.[62]

Regarding ecology and the environment, agriculture and food is highly prominent in biotechnology development. This is especially the case in the climate crisis where climate-smart agriculture requires agroecosystems and cropping systems to improve cultivated crops.[63] In this emergent time for climate resilient agriculture, it is clear that agricultural systems will be transformed in coming decades, especially by using biotechnologically and environmentally relevant microbes.[64] The theme of climate resilient agriculture is indeed dominant in such discussions. Today's harsh climate environments affect crop productivity in many complex ways. Specifically, under continuing climate change, extensive disturbance occurs with both natural and cultivated environments of major food crops.[65] For example, crucial breakthroughs are occurring for next generation breeding of resistant legume crops, such as chickpea lentil, and soybean. These versatile high-nutrition crops are indispensable for global food security.[66] Generally, plant breeding is crucial for stress resilience in the context of climate change that brings pests, diseases, and multiple types of stresses. Mutation breeding, that has released thousands of mutant varieties in most of the crop plants with great economic impact, has become a major characteristic of sustainable food production with climate resilience.[67] Hence, for example, given high cereal losses in the era of climate change, there is an urgent need to develop stress-resilient cereal cultivars,[68] as well as climate resilient grain to ensure plant-based dietary protein security globally.[69] This emerging process of cultivation for climate change resilience pertains to many other areas such as tropical fruit tree species management.[70] Such examples could be readily multiplied.

Conclusion

In sum, ethics discourse on this book's topic, *Resilience in Ecology and Health*, contributes to the ethical discussion in the previous books related to the annual conference series that inspires these proceedings: the urgency of climate change, integral ecology, the global water crisis, the global sustainability challenge, toward a healthy planet, and most recently, strands of sustainability. Hopefully, these perspectives will enable the reader to grasp and engage this global ethical imperative to foster resilience across fields and disciplines in order to build a more resilient world.

Notes

[1] Gerard Magill, "The Urgency of Climate Change," in Gerard Magill, Kia Aramesh, eds., *The Urgency of Climate Change* (Newcastle Upon Tyne, UK: Cambridge Scholars Publishing), 2016, 450-459.

[2] Pope Francis, *Laudato Si': Praise Be To You. On Care for Our Common Home*. Vatican City: Libreria Editrice Vatican, 2015), §161, 194.

[3] Pope Francis, *Laudato Si'*, also see, §13. This call for advocacy is related to the Pope's view of integral ecology, as explained in the Conference Proceedings of the 2nd annual conference. See, Gerard Magill, "Ethics and Integral Ecology," in Gerard Magill, Jordan Potter, *Integral Ecology: Protecting Our Common Home* (Cambridge Scholars Publishing, 2018), Chapter 10, 244-251.

[4] Pope Francis, *Laudato Si'*, §139.

[5] Pope Francis, *Laudato Si'*, §157-158, 196.

[6] Pope Francis, *Laudato Si'*, §159, 162.

[7] Pope Francis, *Laudato Si'*, §163, 202.

[8] Figret Berkes, *Advanced Introduction to Resilience*, Edward Elgar Publishing, 2023.

[9] Kelly A. Parker, Heather E. Keith, *Pragmatist and American Philosophical Perspectives on Resilience*, Lexington Books, 2019.

[10] Kevin Grove, *Resilience*, Routledge, 2018.

[11] Melinda Harm Benson, Robin Kundis Craig, *The End of Sustainability: Resilience and the Future of Environmental Governance in the Anthropocene*, University Press of Kansas, 2017.
[12] Indian Institute of Technology, *Integrated Risk of Pandemic: Covid-19 Impacts, Resilience and Recommendations*, Springer, 2020.
[13] See, United Nations Development Programme, *Sustainable Development Goals* (2015), at, https://www.undp.org/content/undp/en/home/sustainable-development-goals.html.
[14] Dipesh Chakrabarty, *One Planet, Many Worlds: The Climate Parallax* (The Mandel Lectures in the Humanities at Brandeis University), Brandeis University Press, 2023.
[15] Kylie Flanagan, *Climate Resilience: How We Keep Each Other Safe, Care for Our Communities, and Fight Back Against Climate Change*, North Atlantic Books, 2023.
[16] Kristoffer Ekberg, et al., *Climate Obstruction*, Routledge, 2022.
[17] Vinod Thomas, *Risk and Resilience in the Era of Climate Change*, Palgrave MacMillen, 2023.
[18] Anil Kumar Gupta, et al, eds., Ecosystem Restoration: Towards Sustainability and Resilient Development, Springer, 2023.
[19] Yazidhi Bamutaze, et al, eds., *Agriculture and Ecosystem Resilience in Sub Saharan Africa: Livelihood Pathways Under Changing Climate*, Springer, 2019.
[20] Jennifer F. Sklarew, *Building Resilient Energy Systems*, Routledge, 2022.
[21] Amin Rastandeh, Meghann Jarchow, eds., *Creating Resilient Landscapes in an Era of Climate Change: Global Case Studies and Real-World Solutions*, Routledge, 2022.
[22] Osamu Saito, ed., *Sharing Ecosystem Services: Building More Sustainable and Resilient Society*, Springer, 2020.
[23] Marina Alberti, *Cities That Think like Planets: Complexity, Resilience, and Innovation in Hybrid Ecosystems*, University of Washington Press, 2018.
[24] Catrin Schmidt, *Landscape Resilience: Basics, Case Studies, Practical Recommendations*. Springer, 2021.
[25] Sarah E. Gergel, Monica G. Turner, eds., *Learning Landscape Ecology: A Practical Guide to Concepts and Techniques*, Springer, 2017.
[26] Antonella Contin, ed., *Metropolitan Landscapes: Towards a Shared Construction of the Resilient City of the Future*, Springer, 2021.
[27] Pathak Bhawana, Rama Shanker Dubey, eds., *Climate Change and Urban Environment Sustainability*, Springer, 2023.
[28] Zaheer Allam, *Biotechnology and Future Cities: Towards Sustainability, Resilience and Living Urban Organisms*, Routledge 2020.
[29] Cathy Baldwin, Robert King, *Social Sustainability, Climate Resilience and Community-Based Urban Development*, Routledge, 2018
[30] Beth Caniglia, *Resilience, Environmental Justice and the City*, Routledge, 2018.
[31] Brian Tokar, *Climate Justice and Community Renewal*, Routledge, 2020.
[32] A. K. Enamul Haque, et al, eds., *Climate Change and Community Resilience: Insights from South Asia*, Springer, 2021.

[33] Mary Robinson, *Climate Justice: Hope, Resilience, and the Fight for a Sustainable Future*, Bloomsbury Publishing, 2019.
[34] Peter H. Koehn, et al., *Migrant Health and Resilience*, Routledge, 2023.
[35] Christopher Bosso, *Feeding Cities: Improving Local Food Access, Security, and Resilience*, Routledge, 2019.
[36] Sarah M. Gardner, Stephen J. Ramsden, eds, *Agricultural Resilience: Perspectives from Ecology and Economics*, Cambridge University Press, 2019.
[37] Mark Tilzey, *Political Ecology, Food Regimes, and Food Sovereignty: Crisis, Resistance, and Resilience*, Palgrave MacMillen, 2017.
[38] Christian Heim, Caroline Heim, *Resilient Relationships*, Routledge, 2023.
[39] Elaine Miller-Karas, *Building Resilience to Trauma*, Routledge, 2023; Stephen M. Southwick, *Resilience*, Cambridge University Press, 2023.
[40] Alonzo L. Plough, *Community Resilience: Equitable Practices for an Uncertain Future*, Oxford University Press, 2021.
[41] Jack Saul, *Collective Trauma, Collective Healing: Promoting Community Resilience in the Aftermath of Disaster*, Routledge, 2022.
[42] Kelli E. Canada, Clark M. Peters, *Community-Engaged Research for Resilience and Health*, University of Cincinnati Press, 2023.
[43] Jennifer A. Horney, *Covid-19, Frontline Responders and Mental Health: A Playbook for Delivering Resilient Public Health Systems Post-pandemic*, Emerald Group Publishing, 2023.
[44] George Mayzell, *The Resilient Healthcare Organization*, Productivity Press, 2020.
[45] Organization for Economic Co-operation and Development (OECD), *Primary Health Care for Resilient Health Systems in Latin America*, OECD Health Policy Studies, 2023.
[46] Alan Belasen, *Resilience in Healthcare Leadership*, Routledge, 2021.
[47] Cynda Rushton, *Moral Resilience: Transforming Moral Suffering in Healthcare*, Oxford University Press, 2018.
[48] Jeffrey Braithwaite, et al, eds., *Working Across Boundaries: Resilient Health Care*, CRC Press, 2019.
[49] Erik Hollnagel, Jeffrey Braithwaite, et al, eds, *Delivering Resilient Health Care*, Routledge, 2018.
[50] Jennifer M. Ossege, Richard W. Sears, *The Resilient Mental Health Practice*, Routledge, 2017.
[51] Michael Neenan, *Developing Resilience: A Cognitive-Behavioural Approach*, Routledge, 2017.
[52] Louise Jackson, *Cultivating Resilience in Early Childhood: A Practical Guide to Support the Mental Health and Wellbeing of Young Children*, Routledge, 2021.
[53] Heather R. Forkey, et al., *Childhood Trauma and Resilience: A Practical Guide*, American Academy of Pediatrics, 2021.
[54] Mike Stout, *Building Community and Family Resilience: Research, Policy, and Programs*, Springer, 2021.
[55] Thomas Lee, *Healthcare's Path Forward: How Ongoing Crises Are Creating New Standards for Excellence*, McGraw Hill, 2023.

[56] Peter Loftus, *The Messenger: Moderna, the Vaccine, and the Business Gamble That Changed the World*, Harvard Business Review Press, 2022.
[57] Evelyn Hovenga, Heather Grain, eds., *Roadmap to Successful Digital Health Ecosystems: A Global Perspective*, Academic Press, 2022.
[58] Erik D. Perakslis, et al., *Digital Health: Understanding the Benefit-Risk Patient-Provider Framework*, Oxford University Press, 2021.
[59] Sanket Joshi, et al, eds., *Contemporary Medical Biotechnology Research for Human Health*, Academic Press, 2022.
[60] Jayanta Kumar Patra, et al, eds., *Advances in Pharmaceutical Biotechnology: Recent Progress and Future Applications*, Springer, 2020.
[61] Matthias Herdegen, *The International Law of Biotechnology: Human Rights, Trade, Patents, Health and the Environment*, Edward Elgar Publishing, 2020.
[62] Daan Crommelin, et al, eds., *Pharmaceutical Biotechnology: Fundamentals and Applications*, Springer, 2019.
[63] Noureddine Benkeblia, ed., *Climate Change and Agriculture: Perspectives, Sustainability and Resilience*, Wiley, 2022.
[64] Prem Lal Kashyap, et al, eds., *Microbes for Climate Resilient Agriculture*, Wiley-Blackwell, 2018.
[65] Mirza Hasanuzzaman, ed., *Climate Resilient Agriculture*, Vol 1: *Crop Responses and Agroecological Perspectives*, 2023; Vol 2: *Agro-Biotechnological Advancement for Crop Production*, Springer, 2023.
[66] Uday C. Jha, et al, eds., *Diseases in Legume Crops: Next Generation Breeding Approaches for Resistant Legumes Crops*, Springer, 2023
[67] Suprasanna Penna, et al, eds., *Mutation Breeding for Sustainable Food Production and Climate Resilience*, Springer, 2023.
[68] Mallana Gowdra Mallikarjuna, et al, ed., *Next-Generation Plant Breeding Approaches for Stress Resilience in Cereal Crops*, Springer, 2022.
[69] Uday C. Jha, et al, eds., *Developing Climate Resilient Grain and Forage Legumes*, Springer, 2022.
[70] Adel A. Abdul-Soad, Jameel M. Al-Khayri, eds., *Cultivation for Climate Change Resilience*, CRC Press, Vol. 2, 2023, Vol.1, 2021.

Literature

Abdul-Soad, Adel A., Jameel M. Al-Khayri, eds. 2023. *Cultivation for Climate Change Resilience*. CRC Press, Vol. 2; Vol.1, 2021.

Alberti, Marina. 2018. *Cities That Think like Planets: Complexity, Resilience, and Innovation in Hybrid Ecosystems*. University of Washington Press.

Baldwin, Cathy, Robert King. 2018. *Social Sustainability, Climate Resilience and Community-Based Urban Development*. Routledge.

Belasen, Alan. 2021. *Resilience in Healthcare Leadership*. Routledge.

Benkeblia, Noureddine, ed. 2022. *Climate Change and Agriculture: Perspectives, Sustainability and Resilience*. Wiley.

Benson, Melinda Harm, Robin Kundis Craig. 2017. *The End of Sustainability: Resilience and the Future of Environmental Governance in the Anthropocene*. University Press of Kansas.

Berkes, Figret. 2023. *Advanced Introduction to Resilience*. Edward Elgar Publishing.

Bosso, Christopher. 2019. *Feeding Cities: Improving Local Food Access, Security, and Resilience*. Routledge.

Braithwaite, Jeffrey, et al, eds. 2019. *Working Across Boundaries: Resilient Health Care*. CRC Press.

Canada, Kelli E., Clark M. Peters. 2023. *Community-Engaged Research for Resilience and Health*. University of Cincinnati Press.

Caniglia, Beth. 2018 *Resilience, Environmental Justice and the City*. Routledge.

Chakrabarty, Dipesh. 2023. *One Planet, Many Worlds: The Climate Parallax*. (The Mandel Lectures in the Humanities at Brandeis University). Brandeis University Press.

Contin, Antonella, ed. 2021. *Metropolitan Landscapes: Towards a Shared Construction of the Resilient City of the Future*. Springer.

Crommelin, Daan, et al, eds. 2019. *Pharmaceutical Biotechnology: Fundamentals and Applications*. Springer.

Ekberg, Kristoffer, et al. 2022. *Climate Obstruction*, Routledge.

Flanagan, Kylie. 2023. *Climate Resilience: How We Keep Each Other Safe, Care for Our Communities, and Fight Back Against Climate Change*. North Atlantic Books.

Forkey, Heather R., et al. 2021. *Childhood Trauma and Resilience: A Practical Guide*. American Academy of Pediatrics.

Gardner, Sarah M, Stephen J. Ramsden, eds. 2019. *Agricultural Resilience: Perspectives from Ecology and Economics*. Cambridge University Press.

Gergel, Sarah E., Monica G. Turner, eds. 2017. *Learning Landscape Ecology: A Practical Guide to Concepts and Techniques*. Springer.

Grove, Kevin. 2018. *Resilience*. Routledge.

Gupta, Anil Kumar Gupta, et al, eds. 2023. *Ecosystem Restoration: Towards Sustainability and Resilient Development*. Springer.

Haque, A. K. Enamul, et al, eds. 2021. *Climate Change and Community Resilience: Insights from South Asia*. Springer.

Hasanuzzaman, Mizra, ed. 2023. *Climate Resilient Agriculture*, Vol 1: *Crop Responses and Agroecological Perspectives*; Vol 2: *Agro-Biotechnological Advancement for Crop Production*. Springer.

Heim, Christian, Caroline. 2023. *Resilient Relationships*. Routledge.

Herdegen, Matthias. 2020. *The International Law of Biotechnology: Human Rights, Trade, Patents, Health and the Environment*. Edward Elgar Publishing.

Hollnagel, Erik, Jeffrey Braithwaite, et al, eds. 2018. *Delivering Resilient Health Care*. Routledge.

Horney, Jennifer A. 2023. *Covid-19, Frontline Responders and Mental Health: A Playbook for Delivering Resilient Public Health Systems Post-pandemic*. Emerald Group Publishing.

Hovenga, Evelyn. 2022. Heather Grain, eds., *Roadmap to Successful Digital Health Ecosystems: A Global Perspective*. Academic Press.

Indian Institute of Technology. 2020. *Integrated Risk of Pandemic: Covid-19 Impacts, Resilience and Recommendations*. Springer.

Jackson, Louise. 2021. *Cultivating Resilience in Early Childhood: A Practical Guide to Support the Mental Health and Wellbeing of Young Children*. Routledge.

Jha, Uday C., et al, eds. 2023. *Diseases in Legume Crops: Next Generation Breeding Approaches for Resistant Legumes Crops*. Springer.

Jha, Uday C., et al, eds. 2022. *Developing Climate Resilient Grain and Forage Legumes*. Springer.

Joshi, Sanket, et al, eds. 2022. *Contemporary Medical Biotechnology Research for Human Health*. Academic Press.

Kayshap, Prem Lal, et al, eds. 2018. *Microbes for Climate Resilient Agriculture*. Wiley-Blackwell.

Koehn, Peter, et al. 2023. *Migrant Health and Resilience*. Routledge.

Lee, Thomas. 2023. *Healthcare's Path Forward: How Ongoing Crises Are Creating New Standards for Excellence*. McGraw Hill.

Loftus, Peter. 2022. *The Messenger: Moderna, the Vaccine, and the Business Gamble That Changed the World*. Harvard Business Review Press.

Magill, Gerard, Kia Aramesh, eds. 2017. *The Urgency of Climate Change*. Newcastle Upon Tyne, UK: Cambridge Scholars Publishing

Mallikarjuna, Mallana Gowdra, et al,, ed. *2022. Next-Generation Plant Breeding Approaches for Stress Resilience in Cereal Crops*. Springer.

Mayzell, George. 2020. *The Resilient Healthcare Organization*. Productivity Press.

Miller-Karas, Elaine. 2023. *Building Resilience to Trauma*. Routledge.

Neenan, Michael. 2017. *Developing Resilience: A Cognitive-Behavioural Approach*. Routledge.

Organization for Economic Co-operation and Development (OECD). 2023. *Primary Health Care for Resilient Health Systems in Latin America*. OECD Health Policy Studies.

Ossege, Jennifer M., Richard W. Sears. 2017. *The Resilient Mental Health Practice*. Routledge.

Parker, Kelly A, Heather E. Keith. 2019. *Pragmatist and American Philosophical Perspectives on Resilience*. Lexington Books.

Pathak, Bhawana, Rama Shanker Dubey, eds. 2023. *Climate Change and Urban Environment Sustainability*. Springer.

Patra, Jayanta Kumar, et al, eds. 2020. *Advances in Pharmaceutical Biotechnology: Recent Progress and Future Applications*. Springer.

Penna, Suprasanna, et al, eds. 2023. *Mutation Breeding for Sustainable Food Production and Climate Resilience*. Springer.

Perakslis, Erik D., et al. 2021. *Digital Health: Understanding the Benefit-Risk Patient-Provider Framework*. Oxford University Press.

Plough, Alonzo L. 2021. *Community Resilience: Equitable Practices for an Uncertain Future*. Oxford University Press.

Pope Francis. 2015. *Laudato Si'*: *Praise Be To You. On Care for Our Common Home*. Vatican City: Libreria Editrice Vatican.

Rastandeh, Amin, Meghann Jarchow, eds. 2022. *Creating Resilient Landscapes in an Era of Climate Change: Global Case Studies and Real-World Solutions*. Routledge.

Robinson, Mary. 2019. *Climate Justice: Hope, Resilience, and the Fight for a Sustainable Future*. Bloomsbury Publishing.

Rushton, Cynda. 2018. *Moral Resilience: Transforming Moral Suffering in Healthcare*. Oxford University Press.

Saito, Osamu, ed. 2020. *Sharing Ecosystem Services: Building More Sustainable and Resilient Society*. Springer.

Saul, Jack. 2022. *Collective Trauma, Collective Healing: Promoting Community Resilience in the Aftermath of Disaster*. Routledge.

Schmidt, Catrin. 2021. *Landscape Resilience: Basics, Case Studies, Practical Recommendations*. Springer.

Sklarew, Jennifer F. 2022. *Building Resilient Energy Systems*. Routledge.

Southwick, Stephen M. 2023. *Resilience*. Cambridge University Press.

Stout, Mike. 2021. *Building Community and Family Resilience: Research, Policy, and Programs*. Springer.

Thomas, Vinod. 2023. *Risk and Resilience in the Era of Climate Change*. Palgrave MacMillen.

Tilzey, Mark. 2017. *Political Ecology, Food Regimes, and Food Sovereignty: Crisis, Resistance, and Resilience*. Palgrave MacMillen.

Tokar, Brian. 2020. *Climate Justice and Community Renewal*. Routledge.

United Nations Development Programme. 2015. *Sustainable Development Goals* (2015). https://www.undp.org/content/undp/en/home/sustainable-development-goals.html

BIBLIOGRAPHY

"2 out of 3 Catholic Universities in 3rd World." 2003. Collegium Newsletter, May, 1. www.collegium.org/sites/default/files/ resources/fall04.pdf.

23andMe. 2022. "DNA Genetic Testing For Health, Ancestry And More - 23andMe." Accessed March 13. https://www.23andme.com.

"A Practical Guide to Newborn Bloodspot Screening in Ireland." 2022. 68. https://www.hse.ie/eng/health/child/newbornscreening/newbornbloodspotscreening/information-for-professionals/a-practical-guide-to-newborn-bloodspot-screening-in-ireland.pdf

Abatzoglou, John T., A. Park Williams, and Renaud Barbero. 2019. "Global Emergence of Anthropogenic Climate Change in Fire Weather Indices." *Geophysical Research Letters* 46, no. 1: 326–36. https://doi. org/10.1029/2018GL080959.

Abbing, Annemarie, Leo de Sonneville, Erik Baars, Daniëlle Bourne, and Hanna Swaab. 2019. "Anxiety Reduction through Art Therapy in Women. Exploring Stress Regulation and Executive Functioning as Underlying Neurocognitive Mechanisms." *PLOS ONE* 14, no. 12 (December 3): e0225200. https://doi.org/10.1371/journal.pone.0225200.

Abdul-Soad, Adel A., Jameel M. Al-Khayri, eds. 2023. *Cultivation for Climate Change Resilience*. CRC Press, Vol. 2; Vol.1, 2021.

Acton, H. B. 1975. "Introduction." In, G. W. F. Hegel, *Natural Law*. Translated by T. M. Knox, Introduction by H. B. Acton, Pennsylvania: University of Pennsylvania Press, 9-47.

Adamiak, Czeslaw, and Barbara Szyda. 2022. "Combining Conventional Statistics and Bid Data to Map Global Tourism Destinations Before COVID-19." *Journal of Travel Research* 61, 8:1848-71. doi: 10.1177/00472875211051418

Adekoya, Olatunji David, Toyin Ajibade Adisa, and Opeoluwa Aiyenitaju. 2022. "Going Forward: Remote Working in the Post-COVID-19 Era." Employee Relations 44, 6: 1410-27. doi: 10.1108/ER-04-2021-0161.

Agency for Healthcare Research and Quality. 2015. *National Healthcare Quality and Disparities Report*. AHRQ pub no. 15-0007.

Aggarwal, Ajay, Davies, Joanna, Sullivan, Richard. 2014. "'Nudge' in the clinical consultation-an acceptable form of medical paternalism?" *BMC Medication Ethics.* 15, no. 31: 1-6.

Aiuti A, Roncarolo MG, Naldini L. 2017. "Gene therapy for ADA-SCID, the first marketing approval of an ex vivo gene therapy in Europe: paving the road for the next generation of advanced therapy medicinal products." *EMBO Mol Med.* 9(6):737-740. doi:10.15252/emmm. 201707573

Alberti, Marina. 2018. *Cities That Think like Planets: Complexity, Resilience, and Innovation in Hybrid Ecosystems*. University of Washington Press.

Ali, Hossamaldein Gaber, Khalid Ibrahim, Mahmoud Fawzi Elsaid, Reem Babiker Mohamed, Mahmoud I. A. Abeidah, Azhar Othman Al Rawwas, Khaled Elshafey, et al. 2021. "Gene Therapy for Spinal Muscular Atrophy: The Qatari Experience." *Gene Therapy* 28, no. 10 (November): 676–80. https://doi.org/10.1038/s41434-021-00273-7.

Allen, Mary J., Sarah Sabir, and Sandeep Sharma. 2022. "GABA Receptor." In *StatPearls*. Treasure Island (FL): StatPearls Publishing. http://www.ncbi.nlm.nih.gov/books/NBK526124/.

Alliance, Genetic, and District of Columbia Department of Health. 2010. Single-Gene Disorders. Understanding Genetics: A District of Columbia Guide for Patients and Health Professionals. Genetic Alliance. http://www.ncbi.nlm.nih.gov/books/NBK132154/.

Ambagtsheer, Frederike. 2021. "Understanding the Challenges to Investigating and Prosecuting Organ Trafficking: A Comparative Analysis of Two Cases." *Trends in Organized Crime*, 1–28.

"AncestryDNA® | DNA Tests for Ethnicity & Genealogy DNA Test." https://www.ancestry.com/dna/.

Andersen, Kristian G., Andrew Rambaut, W. Ian Lipkin, Edward C. Holmes, and Robert F. Garry. 2020. "The Proximal Origin of SARS-CoV-2." *Nature Medicine* 26, no. 4: 450–52.

Anderson, Cindy, Zaharchuk, Dave, Cheung, Jane, Mertens, Janet, and Peterson, Steven. 2020. Home for the holidays. Armonk, NY: IBM Corporation. www.ibm.com/thought-leadership/institute-businessvalue/report/2020-holiday-consumer-trends#.

Anderson, Laurie M., Susan C. Scrimshaw, Mindy T. Fullilove, Jonathan E. Fielding, Jacques Normand, and Task Force on Community Preventive Services. 2003. "Culturally Competent Healthcare Systems: A Systematic Review." *American Journal of Preventive Medicine* 24, no. 3: 68-79.

Anderson, Meg & McMinn, Sean. 2019. "As Rising Heat Bakes U.S. Cities, The Poor Often Feel It Most." *NPR*.

Anderson, W. French. 1989. "Human Gene Therapy: Why Draw a Line?" *The Journal of Medicine and Philosophy: A Forum for Bioethics and Philosophy of Medicine* 14, no. 6 (December 1): 681–93. https://doi. org/10.1093/jmp/14.6.681.

Anguiano, Dani. 2020. "California's Wildfire Hell: How 2020 Became the State's Worst Ever Fire Season." *The Guardian*, December 30. sec. US news. https://www.theguardian.com/us-news/2020/dec/30/california-wildfires-north-complex-record.

Anthony, K. R. N., D. I. Kline, G. Diaz-Pulido, S. Dove, and O. Hoegh-Guldberg. 2008. "Ocean Acidification Causes Bleaching and Productivity Loss in Coral Reef Builders." *Proceedings of the National Academy of Sciences* 105, no. 45: 17442. https://doi.org/10.1073/pnas. 0804478105.

Anton-Paduraru, Dana-Teodora. 2017. "OC-24 Newborn Screening for Cystic Fibrosis: To Screen or Not to Screen (Pros and Cons)." Archives of Disease in Childhood 102, no. Suppl 2 (June 1): A9–A9.

Antoniou, Stavros A., George A. Antoniou, Frank A. Granderath, Anna Mavroforou, Athanasios D. Giannoukas and Athanasios I. Antoniou. 2010. "Reflections of the Hippocratic Oath in Modern Medicine." *World Journal of Surgery* 34: 3075–3079. doi: 10.1007/s00268-010-0604-3.

"APA Dictionary of Psychology." 2022. Accessed July 10. https:// dictionary.apa.org/.

Aristotle. 1988. *The Politics.* Cambridge Texts in the History of Political Thought. Cambridge: Cambridge University Press.

Asbury Theological Seminary. 2022. "Master of Divinity (M.Div.)." Asbury Theological Seminary. Accessed October 19.

Ashcroft, Richard. 2012. "Smoking, health and ethics," in Angus Dawson *Public Health Ethics: Key Concepts and Issues in Policy and Practice.* New York: Cambridge University Press.

Association for Clinical Pastoral Education (ACPE). 2020. "Objectives and Outcomes for Level I/Level II CPE," in "2020 Accreditation Manual," ACPE, accessed October 24, 2022. https://www.manula.com/manuals/acpe/acpe-manuals/2016/en/topic/ objectives-and-outcomes-for-level-i-level-ii-cpe.

Association for Clinical Pastoral Education (ACPE). 2022. "CPE Center Accreditation." ACPE: The Standard Spiritual Care and Education. Accessed October 19. https://acpe.edu/programs/accreditation.

Association of Professional Chaplains (APC). 2000. "Code of Ethics," Association of Professional Chaplains, revised September 24, https://www.professionalchaplains.org/Files/professional_standards/professional_ethics/apc_code_of_ethics.pdf.

Association of Professional Chaplains (APC). 2000. "Guidelines for the Chaplain's Role in Health Care Ethics." Approved by the APC Board of Directors 10. https://033012b.membershipsoftware.org/files/application_materials/guidelines_for_chaplains_role_in_healthcare_ethics.pdf.

Austriaco, Nicanor Pier Giorgio (O.P.). 2011. *Abortion in a Case of Pulmonary Hypertension*. Philadelphia, PA: The National Catholic Bioethics Quarterly, 505.

Avci, Ercan. 2017. "A Normative Analysis to Determine the Goals of Ethics Education through Utilizing Three Approaches: Rational Moral Education, Ethical Acculturation, and Learning Throughout Life." *International Journal of Ethics Education*: 125-145.

Axelrod, David A., Mark A. Schnitzler, Huiling Xiao, William Irish, Elizabeth Tuttle-Newhall, Su-Hsin Chang, Bertram L. Kasiske, Tarek Alhamad, and Krista L. Lentine. 2018. "An Economic Assessment of Contemporary Kidney Transplant Practice." *American Journal of Transplantation* 18, no. 5: 1168–76.

Ayton-Shenker, Diana. 1995. "The Challenge of Human Rights and Cultural Diversity." United Nations Background Note DPI/1627/HR, March 1995. New York: United Nations Department of Public Information. https://digitallibrary.un.org/record/205090.

Bahta, Merhawi, Azieb Ogbaghebriel, Mulugeta Russom, Eyasu H. Tesfamariam, and Tzeggai Berhe. 2021. "Impact of Adverse Reactions to First-Generation Antipsychotics on Treatment Adherence in Outpatients with Schizophrenia: A Cross-Sectional Study." *Annals of General Psychiatry* 20, no. 1 (December): 27. https://doi.org/ 10.1186/s12991-021-00348-0.

Bailey, Alexander M., Judith Arcidiacono, Kimberly A. Benton, Zenobia Taraporewala, and Steve Winitsky. 2015. "United States Food and Drug Administration Regulation of Gene and Cell Therapies." In *Regulatory Aspects of Gene Therapy and Cell Therapy Products: A Global Perspective*, edited by Maria Cristina Galli and Mercedes Serabian, 1–29. Advances in Experimental Medicine and Biology. Cham: Springer International Publishing,

Bailey, R. 2020. *Gram positive vs. Gram negative bacteria*. ThoughtCo. Retrieved December 5, 2022.

https://www.thoughtco.com/gram-positive-gram-negativebacteria 4174239#:~:text=The%20differences%20between%20Gram%20positive%20and%20Gram%20negative,murein.%20These%20bacteria%20stain%20purple%20after%20Gram%20staining.

Baker, Amanda, Naomi Simon, Aparna Keshaviah, Amy Farabaugh, Thilo Deckersbach, John J Worthington, Elizabeth Hoge, Maurizio Fava, and Mark P Pollack. 2019. "Anxiety Symptoms Questionnaire (ASQ): Development and Validation." *General Psychiatry* 32, no. 6 (December 18): e100144. https://doi.org/10.1136/gpsych-2019-100144.

Baker, Andrew C., Peter W. Glynn, and Bernhard Riegl. 2008. "Climate Change and Coral Reef Bleaching: An Ecological Assessment of Long-Term Impacts, Recovery Trends and Future Outlook." *Estuarine, Coastal and Shelf Science* 80, no. 4: 435-71. https://doi.org/10.1016/ j.ecss. 2008.09.003.

Baldwin, Cathy, Robert King. 2018. *Social Sustainability, Climate Resilience and Community-Based Urban Development*. Routledge.

Barker, Clyde F., and James F. Markmann. 2013. "Historical Overview of Transplantation." *Cold Spring Harbor Perspectives in Medicine* 3, no. 4.

Barnhill, Anne. 2014. "What is Manipulation?" *Manipulation: Theory and Practice.* Eds: Christian Coons and Michael Weber. Oxford University Press, 51-72.

Baruch, Susannah, David Kaufman, and Kathy L. Hudson. 2008. "Genetic Testing of Embryos: Practices and Perspectives of US in Vitro Fertilization Clinics." Fertility and Sterility 89, no. 5 (May 1): 1053–58. https://doi.org/10.1016/j.fertnstert.2007.05.048.

Bates, Jennifer. 2003. "An Inquiry Into The Nature of Environmentally Sound Thinking." *Environmental Ethics* 25 (2):193-197.

Baxter, William. 1974. *People or Penguins: The Case for Optimal Pollution*. New York: Columbia University Press, 381-384

Bayefsky, Michelle J. 2016. "Comparative Preimplantation Genetic Diagnosis Policy in Europe and the USA and Its Implications for Reproductive Tourism." Reproductive Biomedicine & Society Online 3 (December): 41–47. https://doi.org/10.1016/j.rbms.2017.01.001.

Beauchamp, Tom L. and James F. Childress. *Principles of Biomedical Ethics*. 2013. Seventh edition. New York: Oxford University Press.

Beauchamp, Tom L., and James F. Childress. 2019. *Principles of Biomedical Ethics*. 8th edition. New York: Oxford University Press.

Beck, U. 1992. "The Role of Technology and the Natural Sciences in the Risk Society." *Theory, Culture, and Society*: 249-253.

Beck, Ulrich. 1992. "From Industrial Society to Risk Society: Questions of Survival, Social Structure, and Ecological Enlightenment." *Theory, Culture, and Society* Vol. 9, No. 1: 97-123.

Belasen, Alan. 2021. *Resilience in Healthcare Leadership*. Routledge.

Bender, Eric. 2018. "Regulating the Gene-Therapy Revolution." *Nature* 564, no. 7735 (December 12): S20–22. https://doi.org/10.1038/d41586 -018-07641-1.

Benkeblia, Noureddine, ed. 2022. *Climate Change and Agriculture: Perspectives, Sustainability and Resilience*. Wiley.

Benson, Melinda Harm, Robin Kundis Craig. 2017. *The End of Sustainability: Resilience and the Future of Environmental Governance in the Anthropocene*. University Press of Kansas.

Bentham, J. 1789. *An Introduction to the Principles of Morals and Legislation*.

Berkes, Figret. 2023. *Advanced Introduction to Resilience*. Edward Elgar Publishing.

Berman, Brian D. "Neuroleptic Malignant Syndrome." 2011. *The Neurohospitalist* 1, no. 1 (January): 41–47. https://doi.org/10.11 77/1941875210386491.

Bhargava, Anushka, and Praveen Dahiya. 2022. "COVID-19 Pandemic: Assessment of Current Strategies and Socio-Economic Impact." Journal of Health Management 24, 3: 466-77. doi: 10.1177/ 09720634221109295.

Bhatia, Saurabh. 2018. "History, Scope and Development of Biotechnology." *Introduction to Pharmaceutical Biotechnology* 1: 1–61.

Biesecker, Leslie G. 1996. "Orphan Tests." Cambridge Quarterly of Healthcare Ethics 5, no. 2: 300–306. https://doi.org/10.1017/ S0963180100007076.

Birnbacher, Dieter. 2007. "Prenatal Diagnosis Yes, Preimplantation Genetic Diagnosis No: A Contradictory Stance?" Reproductive BioMedicine Online 14 (January 1): 109–13. https://doi.org/10.1016/S 1472-6483 (10)60741-9.

Black, Cara K., Kareem M. Termanini, Oswaldo Aguirre, Jason S. Hawksworth, and Michael Sosin. 2018. "Solid Organ Transplantation in the 21st Century." *Annals of Translational Medicine* 6, no. 20.

Blankenship, Crystal S. 2008. "To Manage Costs of Hemophilia, Patients Need More Than Clotting Factor." Biotechnology Healthcare 5, no. 4: 37–40.

Board of Chaplaincy Certification, Inc. (BCCI). 2022. "2021, January28 Standards for BCCACC," Association of Professional Chaplains, accessed October 20.

https://view.officeapps.live.com/op/view.aspx?src=https%3A%2F%2F033012b.membershipsoftware.org%2Ffiles%2F2021January28%2520Standards%2520for%2520BCCACC.docx&wdOrigin=BROWSELINK.

Board of Chaplaincy Certification, Inc. (BCCI). 2020. "competencies _writing_ guide," Association of Professional Chaplains, last updated May 2020. https://033012b.membershipsoftware.org/files/application_materials/competencies_writing_guide.pdf.

Board of Chaplaincy Certification, Inc. (BCCI). 2022. "BCCI Application Materials," Association of Professional Chaplains, accessed October 19. https://bcci.professionalchaplains.org/content.asp?admin=Y&pl=16&sl =16&contentid=30.

Board of Chaplaincy Certification, Inc. (BCCI). 2022. "Equivalency," Association of Professional Chaplains, accessed October 1. https://bcci.professionalchaplains.org/content.asp?pl=19&contented=19.

Borowy, Iris. 2013. *Defining Sustainable Development for Our Common Future: A History of the World Commission on Environment and Development (Brundtland Commission)*. London: Routledge.

Bosso, Christopher. 2019. *Feeding Cities: Improving Local Food Access, Security, and Resilience*. Routledge.

Bostrom, Nick 2003. "Transhumanist FAQ." 2022. Accessed 16 September. https://nickbostrom.com/views/transhumanist.pdf.

Bostrom, Nick and Sandberg, Anders. 2009. "Cognitive Enhancements: Methods, Ethics, Regulatory Challenges" *Sci Eng Ethics.* 15: 311-341, 323.

Bostrom, Nick. 2005. "A History of Transhumanist Thought." *Journal of Evolution and Technology.* 14, no. 1. April.

Bostrom, Nick. 2005. "Transhumanist Values", *Journal of Philosophical Research*. 30 (Supplement): 3-14.

Brach, Cindy and Irene Fraser. 2000. "Can Cultural Competency Reduce Racial and Ethnic Health Disparities? A Review and Conceptual Model." *Med Care Res Rev* 57 Suppl : 181-217.

Braithwaite, Jeffrey, et al, eds. 2019. *Working Across Boundaries: Resilient Health Care*. CRC Press.

Braveman P, Gottlieb L. 2014. "The Social Determinants of Health: It's Time to Consider the Causes of the Causes." *Public Health Reports*. 129 (1, suppl. 2):19-31.

Brennan Center for Justice. https://www.brennancenter.org/our-work/analysis-opinion/why-climate-crisis-demands-democracy- reform

Brezina, P. R., D. S. Brezina, and W. G. Kearns. 2012. "Preimplantation Genetic Testing." *BMJ* 345, no. sep18 2 (September 18): e5908–e5908. https://doi.org/10.1136/bmj.e5908.

Brighouse, Harry, and Ingrid Robeyns. 2010. *Measuring Justice: Primary Goods and Capabilities*. Cambridge University Press.

Broad Institute. 2014. "Questions and Answers about CRISPR," December 17. https://www.broadinstitute.org/what-broad/areas-focus/project-spotlight/questions-and-answers-about-crispr.

Brown, Adrienne Maree. 2012. "Gifting my attention." WordPress, 9 August. https://adriennemareebrown.net/2012/08/09/giftingmyattention/.

Brown, Barbara E., and John C. Ogden. 1993. "Coral Bleaching." *Scientific American* 268, no. 1: 64-70. https://doi.org/10.1038/scientificamerican 0193-64.

Brown, Brené. 2013. Daring Greatly: How the Courage to Be Vulnerable Transforms the Way We Live, Love, Parent and Lead. London: Portfolio Penguin.

Brunner, H G, et al. 1993. "X-linked borderline mental retardation with prominent behavioral disturbance: phenotype, genetic localization, and evidence for disturbed monoamine metabolism." *American Journal of Human Genetics* vol. 52, no. 6: 1032-9.

Buckner, Tyler W., Iryna Bocharova, Kaitlin Hagan, Arielle G. Bensimon, Hongbo Yang, Eric Q. Wu, Eileen K. Sawyer, and Nanxin Li. 2021. "Health Care Resource Utilization and Cost Burden of Hemophilia B in the United States." Blood Advances 5, no. 7 (April 8): 1954–62. https://doi.org/ 10.1182/bloodadvances.2020003424.

Bulaklak, Karen, and Charles A. Gersbach. 2020. "The Once and Future Gene Therapy." *Nature Communications* 11, no. 1 (November 16): 5820. https://doi.org/10.1038/s41467-020-19505-2.

Burgess, Diana Jill, Megan Crowley-Matoka, Sean Phelan, John F. Dovidio, Robert Kerns, Craig Roth, Somnath Saha, and Michelle van Ryn. 2008. "Patient Race and Physicians' Decisions to Prescribe Opioids for Chronic Low Back Pain." *Social Science & Medicine* 67, no. 11: 1852-1860.

Caggana, Michele, Elizabeth A. Jones, S.I. Shahied, Susan Tanksley, Cheryl A. Hermerath, and Ira M. Lubin. 2013. "Newborn Screening: From Guthrie to Whole Genome Sequencing." Public Health Reports 128, no. Suppl 2: 14–19.

"Cal-Fire-2020-Fire-Siege.Pdf." 2020. Accessed April 24, 2022. https:// www.fire.ca.gov/media/hsviuuv3/cal-fire-2020-fire-siege.pdf.

Callahan, Daniel. 2003. "Individual Good and Common Good: A Communitarian Approach to Bioethics." *Perspectives in Biology and Medicine* 46, no. 4: 496.

Callicott, J. B. 1980. "Animal Liberation: A Triangular Affair." *Environmental Ethics*: 115-132.

Callicott, J. B. 1987. "Second Thoughts on 'A Triangular Affair.'" *The Conceptual Foundations of the Land Ethic*: 133-135.

Canada, Kelli E., Clark M. Peters. 2023. *Community-Engaged Research for Resilience and Health*. University of Cincinnati Press.

Caniglia, Beth. 2018 *Resilience, Environmental Justice and the City*. Routledge.

Canon, Gabrielle. 2021. "Gavin Newsom Oversold California's Fire Prevention Efforts. A Journalist Uncovered the Truth." *The Guardian*, July 2., sec. Environment. https://www.theguardian.com/us-news/2021/jul/02/gavin-newsom-california-fire-prevention-investigation.

Caplan, Arthur. 2011. "The Use of Prisoners as Sources of Organs–an Ethically Dubious Practice." *The American Journal of Bioethics* 11, no. 10: 1–5.

Caravaca-Sanchez, F., Garcia-Jarillo, M. 2020. "Perceived Social Support, Resilience, and Consumption of Psychoactive Substances Amongst Inmates in Prisons." *Esp. Sanid Penit*. 22, 2: 75-79.

Carr, Paul H. 2018. "What is Climate Change Doing to Us and for Us?" *Zygon* 53, no. 2 (June): 443-461. https://search-ebscohost-com.authenticate.library.duq.edu/login.aspx?direct=true&db=rfh&AN=ATLAiE58180528001102&site=eds-live&scope=site.

Carruthers, Peter. 1992. *The Animals Issue: Moral Theory in Practice*. Cambridge University Press.

Cart, Julie. 2021. "California's 2020 Fire Siege: Wildfires by the Numbers." *CalMatters*, July 29. sec. Environment. http://calmatters.org/ environment/2021/07/california-fires-2020/.

Catholic Church (promulgated by Pope John Paul II). 2016. *Catechism of the Catholic Church*, Second Edition. Washington, DC: United States Conference of Catholic Bishops, 1755.

Catholic Health Association. 2022. *Ethical and Religious Directives.* Accessed December 2. https://www.chausa.org/ethics/ethical-andreligiousdirectives#:~:text=The%20Ethical%20and%20Religious%20Directives,aspects%20of%20health%20care%20delivery.

Cavanaugh, Thomas A. 2006. "The History of DER," *Double Effect Reasoning.* New York, NY: Oxford University Press,.

Cavanaugh, Thomas A. 2011. *Double-effect Reasoning, Craniotomy, and Vital* Conflicts. Philadelphia, PA: The National Catholic Bioethics Quarterly, 455.

Cavazzana-Calvo, Marina, and Alain Fischer. 2007. "Gene Therapy for Severe Combined Immunodeficiency: Are We There Yet?" *Journal of Clinical Investigation* 117, no. 6 (June 1): 1456–65. https://doi.org/10.1172/JCI30953.

Cengiz, N., and C. S. Wareham. 2019. "Pig-to-Human Xenotransplantation: Overcoming Ethical Obstacles." *South African Journal of Bioethics and Law* 12, no. 2: 66–70.

Cengiz, Nezerith, and Christopher Simon Wareham. 2020. "Ethical Considerations in Xenotransplantation: A Review." *Current Opinion in Organ Transplantation* 25, no. 5: 483–88.

Centers for Disease Control and Prevention (CDC). 2019. *Pseudomonas aeruginosa infection*. Centers for Disease Control and Prevention. Retrieved December 5, 2022. https://www.cdc.gov /hai/organisms/ pseudomonas.html

Centers for Disease Control and Prevention (CDC). 2021. "2019 Antibiotic Resistance Threats Report." *Centers for Disease Control and Prevention*. Centers for Disease Control and Prevention 23 Nov. https://www. cdc.gov/drugresistance/biggest-threats.html.

Chakrabarty, Dipesh. 2023. *One Planet, Many Worlds: The Climate Parallax*. (The Mandel Lectures in the Humanities at Brandeis University). Brandeis University Press.

Challe, Tiffany. 2021. News.climate.columbia.edu/2021/004/22/rights-of-nature-lawsuits. April 22.

Chandler PhD, Redonna K., Fletcher PhD, Bennett W., & Volkow MD, Nora D. 2009. "Treating Drug Abuse and Addiction in the Criminal Justice System: Improving Public Health and Safety." *JAMA: Journal of the American Medical Association* 301, no. 2: 183-90.

Chapman, Isabelle. 2019. "Prison Inmates Are Fighting California's Fires, but Are Often Denied Firefighting Jobs after Their Release." CNN, October 31. https://www.cnn.com/2019/10/31/us/prison-inmates-fight-california-fires-trnd/index.html.

Charlesworth, Max. 2005. "Don't Blame the 'Bio'—Blame the 'Ethics': Varieties of (Bio) Ethics and the Challenge of Pluralism." *Journal of Bioethical Inquiry* 2: 10-17.

Chiang, Mathew, William Bernard Reid-Varley, and Xiaoduo Fan. 2019. "Creative Art Therapy for Mental Illness." *Psychiatry Research* 275 (May): 129–36. https://doi.org/10.1016/j.psychres.2019.03.025.

Childress, James F., and J. Steinbock. 1997. "Fairness in the Allocation and Delivery of Health Care: A Case Study in Organ Transplantation." *Practical Reasoning in Bioethics*, 214–36.

Chin, Jean Lau. 2000. "Culturally Competent Health Care." *Public Health Report* 115, no:1: 25-26.

"Choose Your Baby's Eye Color | The Fertility Institutes Offers Eye Color Screening and Selection." 2022. Accessed April 25. http://www.fertility-docs.com/programs-and-services/pgd-screening/choose-your-babys-eye-color.php.

Chu, Andrew, and Roopma Wadhwa. 2022. "Selective Serotonin Reuptake Inhibitors." In *StatPearls*. Treasure Island (FL): StatPearls Publishing. http://www.ncbi.nlm.nih.gov/books/NBK554406/.

Cifuentes-Faura, Javier, 2022. "Is Environmental Pollution Associated with an Increased Number of COVID-19 Cases in Europe?" *International Journal of Environmental Research and Public Health* 19, no. 2: 703. https://doi.org/10.3390/ijerph19020703

Clayton, S. 2020. "Climate Anxiety: Psychological Responses to Climate Change." *Journal of Anxiety Disorders* 74. https://doi.org/10.1016/j. janxdis.2020.102263

Clement, Matthew Thomas. 2010. "Urbanization and the Natural Environment: An Environmental Sociological Review and Synthesis." *Organization & Environment* 23, no. 3: 291-314,

Cleveland Clinic. 2022. *Gram stain: What it is, purpose, procedure & results*. Cleveland Clinic. Retrieved December 8. https://my.clevelandclinic.org/health/diagnostics/22612-gram-stain.

Cleveland Clinic. 2020. "Why Do They Prick Your Newborn Baby's Heel?," October 6. https://health.clevelandclinic.org/why-do-they-prick-your-newborn-babys-heel/.

Cleveland Clinic. 2022. "Spinal Muscular Atrophy (SMA): Types, Symptoms & Treatment." Accessed April 25. https://my.clevelandclinic.org/health/diseases/14505-spinal-muscular-atrophy-sma.

Climate Change, Water and Agriculture: Towards Resilient Systems. 2014. OECD Studies on Water. OECD, 17. https://search-ebscohost-com.authenticate.library.duq.edu/login.aspx?direct=true&db=cat05821a&AN=ecp.EBC3121257&site=eds-live&scope=site.

CNN, Paul Vercammen. 2020. "California Firefighters Brace for Raging Flames and a Burgeoning Pandemic." CNN. Accessed April 23, 2022. https://www.cnn.com/2020/08/02/us/california-firefighters-covid-pandemic/index.html.

Coe, Taylor M., Danielle Detelich, Charles G. Rickert, Nikolaos Serifis, Rudy Matheson, Ivy Rosales, Heidi Yeh, Wenning Qin, Yinan Kan, and Jacob Layer. 2020. "Survival Benefit of Pig-to-Baboon Liver Xenotransplantation Utilizing Genetically Modified Swine." *Transplantation* 104, no. S3: S639–40.

Cognitive Behavioral Therapy. 2016. *InformedHealth.Org*. Institute for Quality and Efficiency in Health Care (IQWiG). http://www.ncbi.nlm.nih.gov/books/NBK279297/.

Cole, A. J., M. S. Pratchett, and G. P. Jones. 2009. "Effects of Coral Bleaching on the Feeding Response of Two Species of Coral-Feeding Fish." *Journal of Experimental Marine Biology and Ecology* 373, no. 1: 11-15. https://doi.org/https://doi.org/10.1016/j.jembe.2009.02.016.

Collins, Karen Scott, Dora L. Hughes, Michelle M. Doty, Brett L. Ives, Jennifer N. Edwards, and Katie Tenney. 2002. *Diverse Communities, Common Concerns: Assessing Health Care Quality for Minority Americans*. New York: Commonwealth Fund.

Contin, Antonella, ed. 2021. *Metropolitan Landscapes: Towards a Shared Construction of the Resilient City of the Future*. Springer.

Comer, R. J., & Comer, J. S. 2018. *Abnormal Psychology.* New York: Worth Publishers.

Committee on Human Gene Editing. 2017. Scientific, Medical, and Ethical Considerations, National Academy of Sciences, National Academy of Medicine, and National Academies of Sciences, Engineering, and Medicine. *Human Genome Editing: Science, Ethics, and Governance*. Washington, D.C.: National Academies Press. https://doi. org/ 10.17226/24623.

Conell, F. J. 2022. "Principle of Double Effect." Encyclopedia.com. Accessed December 7. https://www.encyclopedia.com/religion/encyclopedias-almanacs-transcripts-and-maps/double-effect-principle

Conklin, Martha, and Roger Bales. 2022. "Restoring California's Forests to Reduce Wildfire Risks Will Take Time, Billions of Dollars and a Broad Commitment." The Conversation. Accessed April 25. http://theconversation.com/restoring-californias-forests-to-reduce-wildfire-risks-will-take-time-billions-of-dollars-and-a-broad-commitment-146532.

"Consumer Genetic Testing Is Booming: But What Are the Benefits and Harms to Individuals and Populations? | Blogs | CDC." 2022. Accessed March 13. https://blogs.cdc.gov/genomics/2018/06/12/consumer-genetic-testing/.

Cook, Lola, Jeanine Schulze, Jennifer Verbrugge, James C Beck, Karen S Marder, Rachel Saunders-Pullman, Christine Klein, Anna Naito, and

Roy N Alcalay. 2021. "The Commercial Genetic Testing Landscape for Parkinson's Disease." Parkinsonism & Related Disorders 92 (November): 107–11.https://doi.org/10.1016/j.parkreldis.2021. 10.001.

Cook, Nathan J., Grillos, Tara and Andersson, Krister P. 2019. "Gender Quotas Increase the Equality and Effectiveness of Climate Policy Interventions." Nature Climate Change 9: 330–34.

Cooper, D. K. C., R. Gaston, D. Eckhoff, J. Ladowski, T. Yamamoto, L. Wang, H. Iwase, H. Hara, M. Tector, and A. J. Tector. 2018. "Xenotransplantation—the Current Status and Prospects." *British Medical Bulletin* 125, no. 1 (March): 5–14.

Cooper, David K.C., and Hidetaka Hara. 2021. "'You Cannot Stay in the Laboratory Forever': Taking Pig Kidney Xenotransplantation from the Laboratory to the Clinic." *EBioMedicine* 71: 103562.

Cooper, David K.C., Bernd Gollackner, and David H. Sachs. 2002. "Will the Pig Solve the Transplantation Backlog?" *Annual Review of Medicine* 53, no. 1: 133–47.

Cooper, David K.C., Mohamed B. Ezzelarab, Hidetaka Hara, Hayato Iwase, Whayoung Lee, Martin Wijkstrom, and Rita Bottino. 2016. "The Pathobiology of Pig-to-Primate Xenotransplantation: A Historical Review." *Xenotransplantation* 23, no. 2: 83–105.

Cooper, David KC. 2020. "What Will Be the Cost of a Genetically-Engineered Pig Organ for Clinical Xenotransplantation?" *Xenotransplantation* 27, no. 5: e12606.

Corbett, Rebekah. 2019. "Animal Research: An Environmental Perspective." Faunalytics. Published September 16. Retrieved May 6, 2022, https://faunalytics.org/animal-research-an-environmental perspec tive/

Cosci, Fiammetta. 2017. "Withdrawal Symptoms after Discontinuation of a Noradrenergic and Specific Serotonergic Antidepressant: A Case Report and Review of the Literature." *Personalized Medicine in Psychiatry* 1–2 (March): 81–84.
https://doi.org/ 10.1016/j.pmip .2016.11.001.

Cossu, Fausto. 2010. "Genetics of SCID." *Italian Journal of Pediatrics* 36, no. 1 (November 15): 76. https://doi.org/10.1186/1824-7288-36-76.

Courtright-Lim, Amanda, and Mary Drago. 2020. "Ethics of Genetic Testing." Medicine 48, no. 10 (October 1): 675–79.
https://doi.org/10. 1016/j.mpmed.2020.07.014.

Crockett, Molly J., Clark, Luke, Hauser, Marc D., et al. 2010. "Serotonin selectively influences moral judgment and behavior through effects on harm aversion." *Proceedings of the National Academy of Sciences of the United States of America* 107, no. 40: 17433-8.

Crommelin, Daan, et al, eds. 2019. *Pharmaceutical Biotechnology: Fundamentals and Applications*. Springer.

Crowe, John. 2022. "Plan to Destroy Heelprick Cards Is a Serious Mistake." The Irish Times. Accessed February 6. https://www.irishtimes.com/news/health/plan-to-destroy-heelprick-cards-is-a-serious-mistake-1.1330334.

Cuijpers, Pim, Mirjam Reijnders, and Marcus J. H. Huibers. 2019. "The Role of Common Factors in Psychotherapy Outcomes." *Annual Review of Clinical Psychology* 15, no. 1 (May 7): 207–31. https://doi.org/ 10.1146/annurev-clinpsy-050718-095424.

Cullum, Lauren. 2019. "An Analysis of the International Climate Change Adaptation Regime and Its Response to Global Public Health Concerns." *UCLA Journal of Environmental Law and Policy,* vol. 37, no. 2, (Jan.): 191-228. https://search-ebscohost-com.authenticate.library.duq.edu/login.aspx?direct=true&db=edshol&AN=edshol.hein.journals.uclalp37.11&site=eds-live&scope=site.

Dalai Lama. 2011. *My Spiritual Journey: Personal Reflections, Teachings, and Talks*. Collected by Sofia Stril-Rever and translated by Charlotte Mandell. HarperCollins ebooks.

Dalai Lama. 2022. His Holiness the 14th Dalai Lama of Tibet, "The Dalai Lama," The Office of His Holiness The Dalai Lama, accessed October 25. https://www.dalailama.com/the-dalai-lama.

Dallas Theological Seminary. 2022. "Master of Arts in Chaplaincy and Ministry Care-Dallas Theological Seminary." Dallas Theological Seminary. Accessed October 20. https://www.dts.edu/academics/degrees-programs/master-of-arts-in-chaplaincy-and-ministry-care/.

Dangouloff, Tamara, and Laurent Servais. 2019. "Clinical Evidence Supporting Early Treatment Of Patients With Spinal Muscular Atrophy: Current Perspectives." *Therapeutics and Clinical Risk Management* 15 (October 2): 1153–61. https://doi.org/10.2147/TCRM.S172291.

Dangouloff, Tamara, Camille Botty, Charlotte Beaudart, Laurent Servais, and Mickaël Hiligsmann. 2021. "Systematic Literature Review of the Economic Burden of Spinal Muscular Atrophy and Economic Evaluations of Treatments." Orphanet Journal of Rare Diseases 16, no. 1 (January 23): 47. https://doi.org/10.1186/s13023-021-01695-7.

Davenport, Coral & Campbell, Robertson. 2016. "Resettling the First American 'Climate Refugees.'" *The New York Times*.

Davis, Brian R., and Nicole L. Prokopishyn. 2009. "Chapter 65 - Stem Cell Gene Therapy." In *Essentials of Stem Cell Biology (Second Edition)*, edited by Robert Lanza, John Gearhart, Brigid Hogan, Douglas Melton, Roger Pedersen, E. Donnall Thomas, James Thomson, and Ian Wilmut, 591–97. San Diego: Academic Press.

Davis, John K. 2018. *New Methuselahs: The Ethics of Life Extension*. Massachusetts Institute of Technology.

Davydov, Dmitry M., Stewart, Robert, Ritchie, Karen, & Chaudieu, Isabelle. 2010. "Resilience and Mental Health." *Clinical Psychoology Review* 30, no. 5: 479-95.

Dawson, Angus. 2012. "Resetting the Parameters: Public health as the foundation for public health ethics," in *Public Health Ethics: Key Concepts and Issues in Policy and Practice.* New York: Cambridge University Press.

"Default - Stanford Children's Health." 2022. Accessed February 7. https://www.stanfordchildrens.org/en/topic/default?id=newborn-screening-tests-90-P01967.

"Definition of ALGORITHM." 2022. Accessed March 13. https:// www.merriam-webster.com/dictionary/algorithm.

"Degradation." *International Studies Quarterly*, vol. 50, no. 4: 935–56. http://www.jstor.org/stable/4092786.

Degrazia, David and Tom L. Beauchamp. 2010. "Philosophy: Ethical Principles and Common Morality." In *Methods in Medical Ethics*, 2nd ed. Edited by Jeremy Sugarman and Daniel P. Sulmasy, 37-54. Washington, DC: Georgetown University Press.

Dehlendorf, Christine, Allison S. Bryant, Heather G. Huddleston, Vanessa L. Jacoby, and Victor Y. Fujimoto. 2010. "Health Disparities: Definitions and Measurements." *American Journal of Obstetrics & Gynecology* 202, no. 3: 212-213.

Delfin, Carl, Hedvig Krona, Peter Andiné, Erik Ryding, Märta Wallinius, and Björn Hofvander. 2019. "Prediction of Recidivism in a Long-Term Follow-up of Forensic Psychiatric Patients: Incremental Effects of Neuroimaging Data." Edited by Stephan Doering. *PLOS ONE* 14, no. 5 (May 16): e0217127. https://doi.org/10.1371/journal.pone. 0217127.

Demangeot, Catherine, and Kizhekepat Sankaran. 2012. "Cultural pluralism: Uncovering Consumption Patterns in a Multicultural Environment." *Journal of Marketing Management* 28, no. 7-8): 760-783.

Denner, Joachim, and Nicolas J. Mueller. 2015. "Preventing Transfer of Infectious Agents." *International Journal of Surgery* 23: 306–11.

Denner, Joachim. 2016. “How Active Are Porcine Endogenous Retroviruses (PERVs)?” *Viruses* 8, no. 8: 215.

Denner, Joachim. 2016. “Recent Progress in Xenotransplantation, with Emphasis on Virological Safety,” no. 21: 717–27.

DeRossett, Tommy, Donna J. LaVoie, and Destiny Brooks. 2021. “Religious Coping amidst a Pandemic: Impact on COVID-19-Related Anxiety. In *Journal of Religion and Health* 60, 5: 3161-76.

Díaz, S., J. et al, eds. 2019. Summary for policymakers of the global assessment report on biodiversity and ecosystem services of the Intergovernmental Science-Policy Platform on Biodiversity and Ecosystem Services. Bonn: IPBES Secretariat.

Dinc, Leyla, and Fusun Terzioglu. 2006. “The Psychological Impact of Genetic Testing on Parents.” Journal of Clinical Nursing 15, no. 1 (January): 45–51. https://doi.org/10.1111/j.1365-2702.2005.01228.x.

Ditlefsen, Ingrid Tande, Helene Amundsen Nissen-Lie, Agnes Andenæs, Eivind Normann-Eide, Merete Selsbak Johansen, and Elfrida Hartveit Kvarstein. 2021. “’Yes, There Is Actually Hope!’—A Qualitative Investigation of How Patients Experience Mentalization-Based Psychoeducation Tailored for Borderline Personality Disorder.” *Journal of Psychotherapy Integration* 31, no. 3 (September): 257–76. https://doi.org/10.1037/int0000243.

Douglas, Thomas, and Julian Savulescu. 2009. “Destroying Unwanted Embryos in Research. Talking Point on Morality and Human Embryo Research.” EMBO Reports 10, no. 4 (April): 307–12. https://doi.org/ 10.1038 /embor.2009.54.

Drago, Daniela, Betsy Foss-Campbell, Keith Wonnacott, David Barrett, and Adora Ndu. 2021. “Global Regulatory Progress in Delivering on the Promise of Gene Therapies for Unmet Medical Needs.” *Molecular Therapy. Methods & Clinical Development* 21 (April 5): 524–29.

Duda, Jeremy 2020. "Sentencing Reform Debate Shines Light on Lack of Substance Abuse Treatment in Prisons " *AZ Mirror*, 1.

Dupré, Louis. 1993. "The Common Good and the Open Society." *The Review of Politics* 55, no. 4: 687-712.

Durante, Chris. 2009. "Bioethics in a pluralistic society: Bioethical Methodology in lieu of Moral Diversity." *Medicine, Health Care and Philosophy* 12: 35-47.

Durns, Tyler A., Patrick H. O’Connell, Anna Shvartsur, Jessica S. Grey, and Brent M. Kious. 2021. “Effects of Temporary Psychiatric Holds on Length of Stay and Readmission Risk among Persons Admitted for Psychotic Disorders.” *International Journal of Law and Psychiatry* 76 (May): 101695. https://doi.org/10.1016/j.ijlp.2021.101695.

Dyer, Allen R. 1997. "The Ethics of Human Genetic Intervention: A Postmodern Perspective." *Experimental Neurology* 144, no. 1 (March): 168–72. https://doi.org/10.1006/exnr.1996.6404.

Earth Charter. 2000. The Hague. At, https://earthcharter.org/discover/what-is-the-earth-charter.

Eckelman, Matthew J., Kaixin Huang, Robert Lagasse, Emily Senay, Robert Dubrow, and Jodi D. Sherman. 2020. "Health Care Pollution And Public Health Damage In The United States: An Update." Health Affairs 39, no. 12 (December): 2071–79. https://doi.org/10. 1377/ hlthaff.2020.01247.

Edgar, R. H., Cook, J., Noel, C., Minard, A., Sajewski, A., Fitzpatrick, M., Fernandez, R., Hempel, J. D., Kellum, J. A., and Viator, J. A. 2019. "Bacteriophage-mediated identification of bacteria using photoacoustic flow cytometry." *Journal of biomedical optics* 24, no. 11: 1–7. https://doi.org/10.1117/1.JBO.24.11.115003

Edgar, R., Samson, A.-P., Cook, J., Dougals, M., Urish, K., Kellum, J., Hempel, J., & Viator, J. 2021. *Photoacoustic discrimination of antibiotic-resistant and sensitive staphylococcus aureus isolates*. Lasers in surgery and medicine. Retrieved November 8, 2022. https:// pubmed.ncbi.nlm.nih.gov/34940986/

Edinoff, Amber N., Catherine A. Nix, Janice Hollier, Caroline E. Sagrera, Blake M. Delacroix, Tunde Abubakar, Elyse M. Cornett, Adam M. Kaye, and Alan D. Kaye. 2021. "Benzodiazepines: Uses, Dangers, and Clinical Considerations." *Neurology International* 13, no. 4 (November 10): 594–607. https://doi.org/10.3390/neurolint 13040059.

Edinoff, Amber N., Haseeb A. Akuly, Tony A. Hanna, Carolina O. Ochoa, Shelby J. Patti, Yahya A. Ghaffar, Alan D. Kaye, et al. 2021. "Selective Serotonin Reuptake Inhibitors and Adverse Effects: A Narrative Review." *Neurology International* 13, no. 3 (September): 387–401. https://doi.org/10.3390/neurolint13030038.

Ekberg, Kristoffer, et al. 2022. *Climate Obstruction*, Routledge.

Ekser, Burcin, David K.C. Cooper, and A. Joseph Tector. 2015. "The Need for Xenotransplantation as a Source of Organs and Cells for Clinical Transplantation." *International Journal of Surgery* 23: 199–204.

Elizabeth Shogren. 2017. "What Fire Researchers Learned from California's Blazes," December 11. https://www.hcn.org/issues/49.21/wildfire-what-fire-researchers-learned-from-northern-california-blazes.

Ellis, Joanne I., Tahira Jamil, Holger Anlauf, Darren J. Coker, Joao Curdia, Judi Hewitt, Burton H. Jones, et al. 2019. "Multiple Stressor

Effects on Coral Reef Ecosystems." Article. *Global Change Biology* 25, no. 12: 4131-46. https://doi.org/10.1111/gcb.14819.

Ellul, Jacques. 1965. *The Technological Society*. Translated by John Wilkinson, New York: Vintage.

Engelen, Bart. 2019. "Ethical Criteria for Health-Promoting Nudges: A Case-by-Case Analysis." *The American Journal of Bioethics.* 19(5), 48-59.

Erie County, PA, USA. 2022. "Voluntary and Involuntary Commitment." Accessed July 29. https://eriecountypa.gov/departments/human-services/mental-health/voluntary-and-involuntary-commitment/.

Esang, Michael, Sabina Goldstein, and Ravina Dhami. 2019. "The Role of Physical Examinations in Psychiatry as Illustrated in a Case of Neuroleptic Malignant Syndrome Versus Viral Encephalitis: A Case Report and Literature Review." *Cureus* 11, no. 6: e4840. https://doi.org/10.7759/cureus.4840.

Eternal Word Television Network. 2022. "The Principle of Double Effect." Accessed November 1. https://www.ewtn.com/catholicism/answers/principle-of-double-effect-23224

Etzioni, Amitai. 2021. "For Faster Science: Accelerated Genetic Engineering." *The Journal of Clinical Ethics* 32, no. 4: 349–57.

Evans, John H. 2021. "Setting Ethical Limits on Human Gene Editing after the Fall of the Somatic/Germline Barrier." *Proceedings of the National Academy of Sciences* 118, no. 22 (June). e2004837117. https: //doi.org/10.1073/pnas.2004837117.

Evans, Simon. 2021. "Analysis: Which countries are historically responsible for climate change?" *Carbon Brief.*

Ezzelarab, Mohamed B., Burcin Ekser, Agnes Azimzadeh, Chih Che Lin, Yuming Zhao, Rachael Rodriguez, Gabriel J. Echeverri, Hayato Iwase, Cassandra Long, and Hidetaka Hara. 2015. "Systemic Inflammation in Xenograft Recipients Precedes Activation of Coagulation." *Xenotransplantation* 22, no. 1: 32–47.

Facchini, Maria, and Chiara Ruini. 2021. "The Role of Music Therapy in the Treatment of Children with Cancer: A Systematic Review of Literature." *Complementary Therapies in Clinical Practice* 42 (February): 101289. https://doi.org/10.1016/j.ctcp.2020.101289.

Fagan, Andrew. 2004. "Challenging the Bioethical Application of the Autonomy Principle within Multicultural Societies." *Journal of Applied Philosophy* 21, no. 1: 15-31.

Fairchild, Amy L., Ronald Bayer, James Keith Colgrove, and Daniel Wolfe. 2007. *Searching Eyes : Privacy, the State, and Disease Surveillance in America.* Berkeley: University of California Press.

Fallon, L., G. L. Harton, M. E. Sisson, E. Rodriguez, L. K. Field, E. F. Fugger, M. Geltinger, et al. 1999. "Preimplantation Genetic Diagnosis for Spinal Muscular Atrophy Type I." Neurology 53, no. 5 (September 22): 1087–90. https://doi.org/10.1212/wnl.53.5.1087.

Farhud, D. Darius. 2015. Impact of Lifestyle on Health. *Iranian Journal of Public Health*, *44* (11), 1442-1444. https://doi.org/https:// www.ncbi.nlm.nih.gov/pmc/articles/PMC4703222/

Farris, Joshua R. 2017. "An Immaterial Substance View: Imago Dei in Creation and Redemption." *The Heythrop Journal* 58, no. 1: 108–23.

Fava, Giovanni A., Giada Benasi, Marcella Lucente, Emanuela Offidani, Fiammetta Cosci, and Jenny Guidi. 2018. "Withdrawal Symptoms after Serotonin-Noradrenaline Reuptake Inhibitor Discontinuation: Systematic Review." *Psychotherapy and Psychosomatics* 87, no. 4: 195–203. https://doi.org/10.1159/000491524.

"Federal Assistance for Wildfire Response and Recovery." 2022. Accessed February 20. https://crsreports.congress.gov/product/pdf/IF/I F10732.

Feero, W. Gregory, Alan E. Guttmacher, and Francis S. Collins. 2010. "Genomic Medicine--an Updated Primer." The New England Journal of Medicine 362, no. 21 (May 27): 2001–11. https://doi.org/10. 1056/NEJMra 0907175.

Fernández, Lucía. 2021. "Global Pesticide Use by Country." *Statista*, 27 Sept. www.statista.com/statistics/1263069/global-pesticide-use-by-country/.

Ferrando, Francesca. 2019 *Philosophical Posthumanism*. 2. Bloomsbury.

"Fire Management Board | NWCG." 2022. Accessed February 20. https://www.nwcg.gov/partners/fmb/.

Fiscella, Kevin, Peter Franks, Mark P. Doescher, and Barry G. Saver. 2022. "Disparities in Health Care by Race, Ethnicity, and Language Among the Insured: Findings from a National Sample." *Medical Care*: 52-59.

Flaherty, Mary Grace. 2018. *Promoting Individual and Community Health at the Library*, Chicago: American Library Association.

Flanagan, Kylie. 2023. *Climate Resilience: How We Keep Each Other Safe, Care for Our Communities, and Fight Back Against Climate Change*. North Atlantic Books.

Flinn, Aisling M., and Andrew R. Gennery. 2018. "Adenosine Deaminase Deficiency: A Review." *Orphanet Journal of Rare Diseases* 13 (April 24): 65. https://doi.org/10.1186/s13023-018-0807-5.

Foege, William H., Robert C. Hogan, and Ladene H. Newton. 1976. "Surveillance Projects for Selected Diseases." *International Journal of Epidemiology* 5, no. 1 (01): 29-37.

Folke, Carl, Stephen Polasky, Johan Rockstrom, Victor Galaz, Frances Westley, Michele Lamont, Marten Schefer, et. al. 2021. "Our Future in the Anthropocene Biosphere." *Ambio: A Journal of Environment and Society* 50, no. 4 (April 1): 834-869. https://doi.org/10.1007/s 13280-021-01544-8.

Forkey, Heather R., et al. 2021. *Childhood Trauma and Resilience: A Practical Guide*. American Academy of Pediatrics.

Foster, John Bellamy, Brett Clark, and Richard York. 2011. *The Ecological Rift: Capitalism's War on the Earth*. NYU Press.

Foucault, Michel. 2001. *Power.* Edited by James D. Faubion. The New Press.

Fox, Renée C. 1989. *The Sociology of Medicine : A Participant Observer's View.* Prentice Hall Foundations of Modern Sociology Series. Englewood Cliffs, N.J.: Prentice Hall.

Francis, Richard C. 2012. *Epigenetics: How Environment Shapes Our Genes*. New York: W.W Norton & Company, xi.

Frias, Lopez, Francisco Javier, and Donald B. Thompson. 2022. "Solidarity and Public Health." *Medicine, Health Care, and Philosophy: A European Journal*, 6, 9, 1-12. doi: 10.1007/s11019-022-10084-1

Fu, Y., Fischer, J., Pan, K., Yeoh, G. H., & Peng, Z. 2020. "Underwater sound absorption properties of polydimethylsiloxane/ carbon nanotube composites with steel plate backing." *Applied Acoustics*. Retrieved November 7, 2022. https://www.sciencedirect.com/science/article/pii/S0003682X20307726#:~:text=Also%2C%20PDMS%20is%20acoustically%20transparent%2C%20which%20means%20that,studies%20have%20chosen%20PDMS%20as%20the%20matrix%20material.

"Fuels Management." 2015. U.S. Department of the Interior. August 20. https://www.doi.gov/wildlandfire/fuels.

Fuller Theological Seminary. 2022. "MDiv Degree Outline, Fuller Seminary." Fuller Theological Seminary. Accessed October 20. https://www.fuller.edu/master-of-divinity-degree-outline/.

Galea, Sandro. 2019. *Well: What We Need to Talk About When We Talk About Health*. Oxford: Oxford University Press.

Gallagher, Ryan. 2013. "Software That Tracks People on Social Media Created by Defense Firm." *The Guardian*.

Galloway McLean, Kirsty. 2012. "Land Use, Climate Change Adaptation and Indigenous Peoples." *United Nations University*.

Garbutt, Gerard, and Peter Davies. 2011. "Should the Practice of Medicine Be a Deontological or Utilitarian Enterprise?" Journal of Medical Ethics 37, no. 5 (May): 267–70. https://doi.org/10.1136/jme.2010. 036111.

Garden, Gill. 2005. "Physical Examination in Psychiatric Practice." *Advances in Psychiatric Treatment* 11, no. 2 (March): 142–49. https://doi.org/ 10.1192/apt.11.2.142.

Gardner, Sarah M, Stephen J. Ramsden, eds. 2019. *Agricultural Resilience: Perspectives from Ecology and Economics*. Cambridge University Press.

Gardner, William, Charles W. Lidz, Steven K. Hoge, John Monahan, Marlene M. Eisenberg, Nancy S. Bennett, Edward P. Mulvey, and Loren H. Roth. 1999. "Patients' Revisions of Their Beliefs About the Need for Hospitalization." *American Journal of Psychiatry* 156, no. 9 (September): 1385–91. https://doi.org/10.1176/ajp.156.9.1385.

Gaspar, Christine Cerqueira. 2003. "From Alienated to Aleatory: Nature in the Post-Post-Modern World." *Thresholds* 26: 22–25. http://www.jstor .org/stable/43876154. Accessed 27 Apr. 2022.

"Gene Therapy Regulation: A Proactive Approach." 2022. Accessed March 9. https://www.ndsu.edu/pubweb/~mcclean/plsc431/students/ross.htm.

"Gene Therapy." 2022. Accessed January 24. https://eds-p-ebscohost-com.authenticate.library.duq.edu/eds/ebook viewer/ebook/bmxlYmtfXzYwOTE4OF9fQU41?sid=f22a596b-ff96-41d6-9907ec092311f64f@ redis&vid=4&format=EB&rid=1.

Genome.gov. 2022. "The Human Genome Project." Accessed March 12. https://www.genome.gov/human-genome-project.

Gergel, Sarah E., Monica G. Turner, eds. 2017. *Learning Landscape Ecology: A Practical Guide to Concepts and Techniques*. Springer.

Gert, Bernard, Charles M. Culver, and K. Danner Clouser. 2006. *Bioethics: A Systematic Approach*, second edition. New York: Oxford University Pres.

Ghaheri, Ali, Saeed Shoar, Mohammad Naderan, and Sayed Shahabuddin Hoseini. 2015. "The Applications of Genetic Algorithms in Medicine." Oman Medical Journal 30, no. 6 (November): 406–16. https://doi .org/ 10. 5001/omj.2015.82.

Gibbs, E. Paul J. 2014. "The Evolution of One Health: A Decade of Progress and Challenges for the Future." *Veterinary Record* 174, no. 4: 85–91.

Gilroy, Andrea, and Colin Lee. 2019. *Art and Music: Therapy and Research*. Routledg.

"Global Climate Change Policies," 2020. *European Environment Agency*, 23 Nov. https://www.eea.europa.eu/themes/climate-change-adaptation/global-climate-change-policies.

Goldsmith, Connie. 2018. *Pandemic: How Climate, the Environment, and Superbugs Increase the Risk*. Minneapolis, MN: Twenty-First Century Books.

Gonçalves, Giulliana Augusta Rangel, and Raquel de Melo Alves Paiva. 2017. "Gene Therapy: Advances, Challenges and Perspectives." *Einstein* 15, no. 3: 369–75. https://doi.org/10.1590/S167945082017RB4024.

Gordon, Diane. 2020. "Sustainability in the Operating Room: Reducing Our Impact on the Planet." Anesthesiology Clinics 38, no. 3 (September 1): 679–92. https://doi.org/10.1016/j.anclin.2020.06.006.

Gordon, John-Stewart. 2011. "Global Ethics and Principlism." *Kennedy Institute of Ethics Journal* 21, no. 3: 251-276.

Goss, Michael, Daniel L. Swain, John T. Abatzoglou, Ali Sarhadi, Crystal A. Kolden, A. Park Williams, and Noah S. Diffenbaugh. 2020. "Climate Change Is Increasing the Likelihood of Extreme Autumn Wildfire Conditions across California." *Environmental Research Letters* 15, no. 9: 094016. https://doi.org/10.1088/1748-9326/ab83a7.

Goto, Tatenobu, Takeshi Inoue, Chinami Kamiya, Hiroyuki Kawabe, Machiko Higuchi, Megumi Suyama, Tomoki Goto, et al. 2020. "Neonatal Pain Response to Automatic Lancet versus Needle Heel-prick Blood Sampling: A Prospective Randomized Controlled Clinical Trial." Pediatrics International 62, no. 3 (March): 357–62.

Gottesman, M. M., and F. S. Collins. 1994. "The Role of the Human Genome Project in Disease Prevention." Preventive Medicine 23, no. 5 (September): 591–94. https://doi.org/10.1006/pmed.1994.1094.

Gottschalk, Simon. 2010. "The Presentation of Avatars in Second Life: Self and Interaction in Social Virtual Spaces." *Symbolic Interaction* 33, no. 4: 501-525.

Gozum, Ivan Efreaim A., Jr., Abelardo E. Garcial, and John Lu Allan M. Nucum. 2022. "Laudato Si and the Role of the Church in Promoting Environmental Awareness Toward a Better Health-Related Quality of Life." *Journal of Religion and Health*, 1–15. doi. 10.1007/s10943-01631-4.

Green, Graeme. 2022. "Numbers Game: Can AI Save the Elephant?" *The Guardian Weekly* March 4, 24-25.

Gregory-Evans, Kevin, A. M. A. Emran Bashar, and Malcolm Tan. 2012. "Ex Vivo Gene Therapy and Vision." *Current Gene Therapy* 12, no. 2 (April 1): 103–15. https://doi.org/10.2174/156652312800099 607.

Grisch-Chan, Hiu Man, Gerald Schwank, Cary O. Harding, and Beat Thöny. 2019. "State-of-the-Art 2019 on Gene Therapy for Phenylketonuria." *Human Gene Therapy* 30, no. 10: 1274–83.

Griswold, Hali, Debbie Green, Keith Cruise, Brian Belfi, Jonathan Lam, Laura Grossi, Elena Cucco, and Elisabeth Iskander. 2018. "Assessing Risk for Victimization in a Forensic Psychiatric Setting Using the Short-Term Assessment of Risk and Treatability." *Violence and Victims* 33, no. 6: 1012–35.
https://doi.org/10.1891/0886-6708.33.6.1012.

Groseclose, S.L., R.R. German, and P. Nsubuga. "Evaluating Public Health Surveillance." 2010. In *Principles and Practice of Public Health Surveillance*, edited by Lisa M. Lee, S.M. Teutsch, Stephen B. Thacker and Michael E. St. Louis, New York: Oxford University Press, 166-197.

Grove, Kevin. 2018. *Resilience*. Routledge.

Groves, Anna. 2019. "Repairing the Future. (Cover Story)." Discover 40, no. 4 (May): 28–33.

Gupta, Anil Kumar Gupta, et al, eds. 2023. *Ecosystem Restoration: Towards Sustainability and Resilient Development*. Springer.

Gustafson, J. Perry Peter H. Raven, and Paul R. Ehrlich. 2020. *Population, Agriculture, and Biodiversity: Problems and* Prospects. Columbia: University of Missouri.
https://search-ebscohost-com.authenticate.library.duq.edu/login.aspx?direct=true&db=nlebk&AN=2462650&site=eds-live&scope=site.

Guttmacher Institute, *Medication Abortion Now Accounts for More Than Half of All US Abortions.* 2022. Accessed December 7.
https://www.guttmacher.org/article/2022/02/medication-abortion-now-accounts-more-half-all-us-abortions

H.R. 2019. "Recognizing the duty of the Federal Government to create a Green New Deal." 116th Cong., H. Res 109.

Habermas, Jurgen. 2003. *The Future of Human Nature. Polity Press.*

Haeyen, Suzanne, Susan van Hooren, William van der Veld, and Giel Hutschemaekers. 2018. "Efficacy of Art Therapy in Individuals With Personality Disorders Cluster B/C: A Randomized Controlled Trial." *Journal of Personality Disorders* 32, no. 4 (August): 527–42.
https://doi.org/10.1521/pedi_2017_31_312.

Half-earthproject.org/discover-half-earth/#why-half. https://map.half-earthproject.org/.

Hall, Andrea, Talha Bin Farooq, and Renato Alcaraz Jr. 2022. "Hypocalcemia-Induced Reversible Psychosis." *Cureus* 14, no. 1 (January 2). https://doi.org/10.7759/cureus.20874.

Hall, Mark A., and Stephen S. Rich. 2000. "Laws Restricting Health Insurers' Use of Genetic Information: Impact on Genetic Discrimination." The American Journal of Human Genetics 66, no. 1 (January 1): 293–307. https:// doi.org/10.1086/302714.

Hall, Melinda C. 2017. *The Bioethics of Enhancement: Transhumanism, Disability, and Biopolitics*. Lexington Books.

Hall, Melinda. 2013. "Vile Sovereigns in Bioethical Debate." *Disability Studies Quarterly*, 33, no. 4, 1-23.

Hamid, Najmeh, and Ali Biat. 2019. "The Effectiveness of Music Therapy on Depression and Happiness of Depressed Women." *NeuroQuantology* 17, no. 12: 19–26. https://doi.org/10.14704/nq.2019.17. 12.NQ 19110.

Hand, Carol. 2015. *Climate Change: Our Warming Earth. History of* Science. Minneapolis: Essential Library. https://search-ebscohost-com.authenticate.library.duq.edu/login.aspx?direct=true&db=nlebk&AN=878389&site=eds-live&scope=site.

Handyside, A. H., J. G. Lesko, J. J. Tarín, R. M. Winston, and M. R. Hughes. 1992. "Birth of a Normal Girl after in Vitro Fertilization and Preimplantation Diagnostic Testing for Cystic Fibrosis." The New England Journal of Medicine 327, no. 13 (September 24): 905–9. https://doi.org/10. 1056/NEJM199209243271301.

Hanna, Eve, Cécile Rémuzat, Pascal Auquier, and Mondher Toumi. 2017. "Gene Therapies Development: Slow Progress and Promising Prospect." *Journal of Market Access & Health Policy* 5, no. 1 (January 3): 1265293. https://doi.org/10.1080/20016689.2017.1265293.

Haque, A. K. Enamul, et al, eds. 2021. *Climate Change and Community Resilience: Insights from South Asia*. Springer.

Harden, Kathryn Paige. 2021. *The Genetic Lottery: Why DNA Matters for Social Equality*. 1st edition. Princeton Oxford: Princeton University Press.

Hardy, Philip John, and Kathy Hardy. 2018. "Chromosomal Instability in First Trimester Miscarriage: A Common Cause of Pregnancy Loss?" Translational Pediatrics 7, no. 3 (July): 211–18. https://doi.org/10. 21037/tp.2018.03.02.

Harris, H. S. 1997. *Hegel's Ladder*. Volumes 1& 2. Indianapolis, Cambridge: Hackett Publishing Company.

Harrison, Joseph D. and Patel, Mitesh S. 2020. "Designing Nudges for Success in Health Care" *AMA Journal of Ethics.* 22, no. 9: 796-801.

Hartnack, Sonja. 2022. "Animal Health Is Often Ignored, but Indispensable to the Human Right to Health." *International Journal for Equity in Health* 21, no. 1: 1–3.

Harvard T.H. Chan School of Public Health. 2021. "Wildfire Smoke May Have Contributed to Thousands of Extra COVID-19 Cases and Deaths in Western U.S. in 2020." News, August 13. https://www.hsph.harvard .edu/news/press-releases/wildfire-smoke-may-have-contributed-to-thousands-of-extra-covid-19-cases-and-deaths-in-western-u-s-in-2020/.

Hassan, Shermarke, Rory C. Monahan, Evelien P. Mauser-Bunschoten, Lize F. D. van Vulpen, Jeroen Eikenboom, Erik A. M. Beckers, Louise Hooimeijer, et al. 2021. "Mortality, Life Expectancy, and Causes of Death of Persons with Hemophilia in the Netherlands 2001–2018." Journal of Thrombosis and Haemostasis 19, no. 3 (March): 645–53.

Hasanuzzaman, Mizra, ed. 2023. *Climate Resilient Agriculture*, Vol 1: *Crop Responses and Agroecological Perspectives*; Vol 2: *Agro-Biotechnological Advancement for Crop Production*. Springer.

Hattori, Akihisa. 2002. "Small and Large Anemonefishes Can Coexist Using the Same Patchy Resources on a Coral Reef, before Habitat Destruction." *Journal of Animal Ecology* 71, no. 5: 824-31. https://doi. org/10.1046/j.1365-2656.2002.00649.x.

Haupt, Meighan B. 2014. "The Vital Role of State Psychiatric Hospitals." National Association of State Mental Health Program Directors, July.

Have, Henk ten. 2016. *Global Bioethics: An Introduction.* London; New York, NY: Routledge, Taylor & Francis Group.

Hawken, Paul. 1997. "excerpts from Natural Capitalism." In, *Mother Jones* March/April. https://www.motherjones.com/politics/1997/03/ natural-capitalism/.

Hawken, Paul. 1999. *Natural Capitalism*. Boston: Little, Brown and Company: 227-237.

"Healthcare's Climate Footprint." 2022. Accessed March 29. https://www.arup.com/perspectives/publications/research/section/healthcares-climate-footprint.

Heape, Walter, and Michael Foster. 1891. "III. Preliminary Note on the Transplantation and Growth of Mammalian Ova within a Uterine Foster-Mother." Proceedings of the Royal Society of London 48, no. 292–295 (January 1): 457–58. https://doi.org/10.1098/rspl.1890 .0053.

Heard, Clark Patrick, Jared Scott, Allan Tetzlaff, and Heather Lumley. 2019. "Transitional Housing in Forensic Mental Health: Considering Consumer Lived Experience." *Health & Justice* 7, no. 1 (May 20): 8.

https://doi.org/10.1186/s40352-019-0091-z.

Hedman, Leslie C., John Petrila, William H. Fisher, Jeffrey W. Swanson, Deirdre A. Dingman, and Scott Burris. 2016. "State Laws on Emergency Holds for Mental Health Stabilization." *Psychiatric Services* 67, no. 5 (May): 529–35. https://doi.org/10.1176/appi.ps.201500205.

Hegel, G. W. F. 1952. *Hegel's Philosophy of Right*. Translated by T. M. Knox. London, Oxford, New York: Oxford University Press.

Hegel, G. W. F. 1977. *Phenomenology of Spirit*. Translated by A. V. Miller. Oxford: Oxford University Press.

Hegel, G. W. F. 1969. *Hegel's Science of Logic*. Editor Lewis, H. D. Translator Miller, A. V.. Humanities Paperback Library. Atlantic Highlands, NJ: Humanities Press International, Inc.

Heim, Christian, Caroline. 2023. *Resilient Relationships*. Routledge.

Hellsten, Sirkku K. 2008. "Global Bioethics: Utopia or Reality?." *Developing World Bioethics* 8, no. 2: 70-81.

Hendricks, Marcus D. 2002. "Leveraging Critical Infrastructure Within an Environmental Justice Framework for Public Health Prevention," *American Journal of Public Health*, 112 (7), 972-74.

Hens, Kristien, Wybo Dondorp, Alan H. Handyside, Joyce Harper, Ainsley J. Newson, Guido Pennings, Christoph Rehmann-Sutter, and Guido de Wert. 2013. "Dynamics and Ethics of Comprehensive Preimplantation Genetic Testing: A Review of the Challenges." *Human Reproduction Update* 19, no. 4 (July 1): 366–75. https://doi. org/10.1093/humupd/dmt009.

Herdegen, Matthias. 2020. *The International Law of Biotechnology: Human Rights, Trade, Patents, Health and the Environment*. Edward Elgar Publishing.

Heron, Scott F., Jeffrey A. Maynard, Ruben van Hooidonk, and C. Mark Eakin. 2016. "Warming Trends and Bleaching Stress of the World's Coral Reefs 1985–2012." *Scientific Reports* 6. https://doi.org/10. 1038/srep38402.

"History of the Monash IVF Research Program from 1971 to 1989 | The Embryo Project Encyclopedia." 2022. Accessed February 6. https://embryo.asu.edu/pages/history-monash-ivf-research-program-1971-1989.

Hobijn, Bart, Russell A. Miles, James Royal, and Jing Zhang. 2022. "What is Driving U.S. Inflation amid a Global Inflation Surge?" *Chicago Fed Letter* 470: 1-6. doi:10.21033/cfl-2022-470.

Holger, Dieter. 2019. "Dna Testing for Ancestry Is More Detailed for White People." PCWorld 37, no. 1 (January): 96–103.

Hollnagel, Erik, Jeffrey Braithwaite, et al, eds. 2018. *Delivering Resilient Health Care*. Routledge.

Horney, Jennifer A. 2023. *Covid-19, Frontline Responders and Mental Health: A Playbook for Delivering Resilient Public Health Systems Post-pandemic*. Emerald Group Publishing.

Hospital Information Services for Jehovah's Witnesses International Office. 2012. "Religious and Ethical Position on Medical Therapy and Related Matters." Wallkill, NY: Watch Tower Bible and Tract Society of Pennsylvania. Accessed November 2, 2022. https://www.jw.org/en/medical-library/strategies-downloads/religious-and-ethical-position-medical-therapy/.

Hovenga, Evelyn. 2022. Heather Grain, eds., *Roadmap to Successful Digital Health Ecosystems: A Global Perspective*. Academic Press.

Howard, Heidi C., Carla G. van El, Francesca Forzano, Dragica Radojkovic, Emmanuelle Rial-Sebbag, Guido de Wert, Pascal Borry, and Martina C. Cornel. 2018. "One Small Edit for Humans, One Giant Edit for Humankind? Points and Questions to Consider for a Responsible Way Forward for Gene Editing in Humans." *European Journal of Human Genetics* 26, no. 1 (January): 1–11. https://doi.org/ 10.1038/s41431-017-0024-z.

Hryhorowicz, Magdalena, Joanna Zeyland, Ryszard S\lomski, and Daniel Lipiński. 2017. "Genetically Modified Pigs as Organ Donors for Xenotransplantation." *Molecular Biotechnology* 59, no. 9: 435–44.

Huang, Yuhan, Yufang Jin, Mark W. Schwartz, and James H. Thorne. 2020. "Intensified Burn Severity in California's Northern Coastal Mountains by Drier Climatic Condition." *Environmental Research Letters* 15, no. 10: 104033. https://doi.org/10.1088/1748-9326/ aba6af.

Huberfeld, Nicole, Sarah H. Gordon, and David K. Jones. 2020. "Federalism Complicates the Response to the COVID-19 Health and Economic Crisis: What Can Be Done?" *Journal of Health Politics, Policy & Law* 45, 6: 951-65. doi: 10.1215/036168-8641493.

Hughes, Terry P., Michele L. Barnes, David R. Bellwood, Joshua E. Cinner, Graeme S. Cumming, Jeremy B. C. Jackson, Joanie Kleypas, et al. 2017. "Coral Reefs in the Anthropocene." *Nature* 546, no. 7656: 82-90. https://doi.org/10.1038/nature22901.

Human Genome Editing: Science, Ethics, and Governance. 2017. The National Academies Press.

"Human Genome Project: Sequencing the Human Genome | Learn Science at Scitable." 2022. Accessed March 13.. http://www.nature.com/scitable/topicpage/dna-sequencing-technologies-key-to-the-human-828.

Huq, Aamira J, Adrienne Sexton, Paul Lacaze, Colin L Masters, Elsdon Storey, Dennis Velakoulis, Paul A James, and Ingrid M Winship. 2021. "Genetic Testing in Dementia-A Medical Genetics Perspective." International Journal of Geriatric Psychiatry 36, no. 8 (August): 1158–70. https://doi.org/10.1002/gps.5535.

Hyman, Steven E. 2000. "The Genetics of Mental Illness: Impliccatioons for Practice ". *Bulletin of the World Health Organization* 78, no. 4: 455-63.

Hyun, D. 2017. "Why doctors prescribe antibiotics-even when they shouldn't." The Pew Charitable Trusts. Retrieved October 26, 2022. https://www.pewtrusts.org/en/research-andanalysis/articles/2017/06/30/why-doctors-prescribe-antibiotics-even-when-they-shouldnt.

Iglewski, B. H. 1996. Chapter *27 Pseudomonas*. National Library of Medicine. Retrieved October 7, 2022. https://www.ncbi.nlm.nih.gov/ books/NBK8326/

Igra, Libby, David Roe, Adi Lavi-Rotenberg, Paul H. Lysaker, and Ilanit Hasson-Ohayon. 2021. "'Making Sense of My Diagnosis': Assimilating Psychoeducation into Metacognitive Psychotherapy for Individuals with Schizophrenia." *Journal of Psychotherapy Integration* 31, no. 3 (September): 277–90. https://doi.org/10.1037/int0000207.

Indian Institute of Technology. 2020. *Integrated Risk of Pandemic: Covid-19 Impacts, Resilience and Recommendations*. Springer.

"Indigenous Philosophy and Creating a More Sustainable Future for Georgetown." 2022. 25 April. Georgetown College. https://college.georgetown.edu/news-story/indigenous-philosophy-sustainability/.

Institute of Medicine. 2007. *The Learning Healthcare System: Workshop Summary*. Washington, DC: The National Academies Press. https://doi.org/10.17226/11903.

International Energy Agency, *Net Zero by 2050*. https://www.iea.org/reports/net-zero-by-2050.

"Intro summary20.Pdf." 2022. Accessed April 21. https://www.predictiveservices.nifc.gov/intelligence/2020_statssumm/intro_summary20.pdf.

Ioannidis, John PA. 2005. "Why most published research findings are false." *PLoS Med* 2, no. 8: e124.

Jackson, Louise. 2021. *Cultivating Resilience in Early Childhood: A Practical Guide to Support the Mental Health and Wellbeing of Young Children*. Routledge.

Jackson-Koku, Gordon. 2016. "Beck Depression Inventory." *Occupational Medicine* 66, no. 2 (March): 174–75. https://doi.org/10.1093/ occmed/kqv087.

Jacobsen, Kathryn H. 2019. *Introduction to Global Health*, Vol. Third edition. Burlington, MA: Jones & Bartlett Learning.

Jafarlou, M., B. Baradaran, T. A. Saedi, V. Jafarlou, D. Shanehbandi, M. Maralani, and F. Othman. 2016. "An overview of the history, applications, advantages, disadvantages and prospects of gene therapy." Journal of Biological Regulators and Homeostatic Agents 30, no. 2: 315-321.

Janice Berliner. 2015. Ethical Dilemmas in Genetics and Genetic Counseling : Principles Through Case Scenarios. Oxford: Oxford University Press. https://authenticate.library.duq.edu/login?url=https://search.ebscohost.com/login.aspx?direct=true&db=nlebk&AN=833180&site=eds-live&scope=site.

Jedan, Christoph, 2021. "Proportionate Universalism and Public Health: An analysis of WRR Policy Brief From Disparity to Potential." *NTT* 75 (1), 73-93.

Jensen, Derrick, and Aric McBay. 2011. *What We Leave Behind*. Seven Stories Press.

Jha, Uday C., et al, eds. 2023. *Diseases in Legume Crops: Next Generation Breeding Approaches for Resistant Legumes Crops*. Springer.

Jha, Uday C., et al, eds. 2022. *Developing Climate Resilient Grain and Forage Legumes*. Springer.

Jia, Fangzhi, Avi Fellner, and Kishore Raj Kumar. 2022. "Monogenic Parkinson's Disease: Genotype, Phenotype, Pathophysiology, and Genetic Testing." Genes 13, no. 3 (March 2022): 471-N.PAG. https:// doi.org/10.3390/genes 13030471.

John Paul II. 2022. *Veritatis Splendor*, 75. Accessed November 1. https://www.vatican.va/content/john-paulii/en/encyclicals/documents/hf_jp-ii_enc_06081993_veritatis-splendor.html

John, Stephen. 2020. "Artificial Ignorance, Epistemic Paternalism, and Epistemic Obligations." *Epistemic Paternalism: Conceptions, Justifications and Implications*. Amiel Bernal and Guy Axtell, eds. 17-27.

Jonas, Hans. 1984. *The Imperative of Responsibility*. University of Chicago Press.

Jonsen, Albert R. 1998. *The Birth of Bioethics*. New York: Oxford University Press.

Jonsen, Albert R. 2008. *A Short History of Medical Ethics.* New York: Oxford University Press.

Jonsen, Albert R. 1998. *The Birth of Bioethics*. New York: Oxford University Press, 83-84, cited in Daniel Callahan, personal communication, December 11, 1996.

Jorqui-Azofra, María. 2020. "Regulation of Clinical Xenotransplantation: A Reappraisal of the Legal, Ethical, and Social Aspects Involved." In *Xenotransplantation*: 315–58.

Joshi, Sanket, et al, eds. 2022. *Contemporary Medical Biotechnology Research for Human Health*. Academic Press.

Jossey-Bass, 2015. https://search-ebscohost-com.authenticate.library.duq.edu/login.aspx?direct=true&db=cat03556a&AN=duebr.EBC4187157&site=eds-live&scope=site.

Juckett, Gregory. 2005 "Cross-Cultural Medicine." *American Family Physician* 72, no. 11: 2267-2274.

Justia Dockets & Filings. 2022. "Beleno et al v. Texas Department of State Health Services et Al." Accessed February 7.
https://dockets.justia.com/docket/texas/txwdce/5:2009cv00188/353950.

Justia Law. 2022. "Bearder v. State." N. d. Accessed February 7.
https://law.justia.com/cases/minnesota/supreme-court/2011/a10-101.html.

Kaiser, G. 2022. "The lytic life cycle of bacteriophages." *Biology LibreTexts*. November 2.
https://bio.libretexts.org/Bookshelves/Microbiology/Book%3A_Microbiology_(Kaiser)/Unit_4%3A_Eukaryotic_Microorganisms_and_Viruses/10%3A_Viruses/10.07%3A_Bacteriophage_Life_Cycles%3A_An_Overview/10.7A%3A_The_Lytic_Life_Cycle_of_Bacteriophages

Kanavakis, E., and J. Traeger-Synodinos. 2002. "Preimplantation Genetic Diagnosis in Clinical Practice." *Journal of Medical Genetics* 39, no. 1 (January 1): 6–11. https://doi.org/10.1136/jmg.39.1.6.

Kant, Immanuel. 1981. *Grounding for the Metaphysics of Morals*. Third Edition. Translator W. Ellington. James. Indianapolis/Cambridge: Hackett Publishing Company, Inc.

Kaplowitz, Eliana, Truong, Ashley Q., Macmadu, Alexndria, Peterson, Meghan, Brinkley-Rubinstein, Lauren, Potter, Nathan, Green, Traci C., Clarke, Jennifer G., and Rich, Josiah D. 2021. "Fentanyl-Related Overdose During Incarceration: A Comprehensive Review." *Health and Justice* 9, no. 13: 1-11.

Karpati, G., and H. Lochmüller. 1997. "The Scope of Gene Therapy in Humans: Scientific, Safety and Ethical Considerations." *Neuromuscular Disorders: NMD* 7, no. 5 (July): 273–76.

https://doi. org/10. 1016/s0960-8966(97)00051-5.

Kayshap, Prem Lal, et al, eds. 2018. *Microbes for Climate Resilient Agriculture*. Wiley-Blackwell.

Kellert, Stephen R. 1980. *Knowledge, Affection, and Basic Attitudes toward Animals in American Society: Phase III*. US Department of the Interior, Fish and Wildlife Service.

Kelly, David F., Gerard Magill, and Henk Ten Have. 2013. *Contemporary Catholic Health Care Ethics*, 2nd edition. Washington, D.C: Georgetown University Press.

Kemper, James M, Christopher Gyngell, and Julian Savulescu. 2019. "Subsidizing PGD: The Moral Case for Funding Genetic Selection." Journal of Bioethical Inquiry 16, no. 3 (September): 405–14. https:// doi.org/10 .1007/s11673-019-09932-2.

Kendler, Kenneth S., Kathryn Tabb, and John Wright. 2022. "The Emergence of Psychiatry: 1650–1850." *American Journal of Psychiatry* 179, no. 5 (May): 329–35. https://doi.org/10.1176/appi.ajp.21060614.

Kendrick, Tony. 2021. "Strategies to Reduce Use of Antidepressants." *British Journal of Clinical Pharmacology* 87, no. 1: 23–33. https://doi.org/10.1111/bcp.14475.

Kenter, Marcel J. H. 2019. "Is It Time to Reform Oversight of Clinical Gene Therapy in the EU?" *British Journal of Clinical Pharmacology* 85, no. 1 (January): 8–10. https://doi.org/10.1111/bcp.13795.

Kessel, Anthony, and Stephens, Carolyn, "Environment, Ethics and Public Health." 2012. In Angus Dawson *Public Health Ethics: Key Concepts and Issues in Policy and Practice*. New York: Cambridge University Press.

Kimmerer, Robin Wall. 2003. *Gathering Moss: A Natural and Cultural History of Mosses*. Corvallis: Oregon State University Press.

Kirchmeier-Young, Megan C., Francis W. Zwiers, Nathan P. Gillett, and Alex J. Cannon. 2017. "Attributing Extreme Fire Risk in Western Canada to Human Emissions." *Climatic Change* 144, no. 2: 365–79.

Knapp, Samuel, Michael Gottlieb, and Mitchell M. Handelsman. 2021. "Talking to Patients Who Fail to Follow Public Health Measures during the COVID-19 Pandemic: A Principle-Based Approach." *Practice Innovations* 6, 3: 139-48. doi: 10.1037/pin0000144.

Knight, Dacre, Daniel V Dudenkov, and William P Cheshire Jr. 2021. "Religion in the US during the Time of a Pandemic: A Medical Perspective." *Journal of Religion and Health* 60, 5: 3177-92. doi:10.1007/s10943-021-01366-8.

Knoll, Michael F., David KC Cooper, and Rita Bottino. 2020. "How the COVID-19 Pandemic May Impact Public Support for Clinical Xenotransplantation in the United States?" *Xenotransplantation* 27, no. 5: e12623.

Knoppers, Bartha Maria, and Ruth Chadwick. 2005. "Human Genetic Research: Emerging Trends in Ethics." *Nature Reviews Genetics* 6, no. 1: 75-79.

Knutson, Brian, Wolkowitz, Owen M., Cole, Steve W. 1998. "Selective Alteration of Personality and Social Behavior by Serotonergic Intervention" *The American journal of psychiatry* vol. 155, no. 3: 373-9.

Kocsis, T. 2022. "Labeling Melanoma Cells With Black Microspheres for Improved Sensitivity in Detection via Photoacoustic Flow Cytometry." MA Thesis., Duquesne University.

Koehn, Peter, et al. 2023. *Migrant Health and Resilience*. Routledge.

Koenig, Harold G. 2002. *Spirituality in Patient Care: Why, How, When, and What.* Philadelphia: Templeton Foundation Press.

Kögel, Johannes, and Georg Marckmann. 2020. "'Xenotransplantation Challenges Us as a Society' What Well-Informed Citizens Think about Xenotransplantation." *EMBO Reports* 21, no. 9: e50274.

Kohn, Robert, Ali Ahsan Ali, Victor Puac-Polanco, Chantal Figueroa, Victor López-Soto, Kristen Morgan, Sandra Saldivia, and Benjamín Vicente. 2018. "Mental Health in the Americas: An Overview of the Treatment Gap." *Revista Panamericana de Salud Pública* 42 (October 18): e165. https://doi.org/10.26633/rpsp.2018.165.

Kolind, Torsten &. Duke, Karen 2016. "Drugs in Prison: Exploring Use, Control, Treatment, and Policy ". Drugs: *Educattion, Prevention, and Policy* 23, no. 2: 89-92.

Kolmes, Sara, Christine Ha, and Jordan Potter. 2022. "Responding to Cultural Limitations on Patient Autonomy: A Clinical Ethics Case Study." *HEC Forum* (August 9). doi: 10.1007/s10730-022-09490-y.

Kosicki, Michael, Kärt Tomberg, and Allan Bradley. 2018. "Repair of Double-Strand Breaks Induced by CRISPR-Cas9 Leads to Large Deletions and Complex Rearrangements." Nature Biotechnology 36, no. 8 (September): 765–71. https://doi.org/10.1038/nbt.4192.

Kotulska, Katarzyna, Aviva Fattal-Valevski, and Jana Haberlova. 2021. "Recombinant Adeno-Associated Virus Serotype 9 Gene Therapy in Spinal Muscular Atrophy." *Frontiers in Neurology* 12 (October 1).

Kratzer, Kateryna, Landon J. Getz, Thibaut Peterlini, Jean-Yves Masson, and Graham Dellaire. 2021. "Addressing the Dark Matter of Gene

Therapy: Technical and Ethical Barriers to Clinical Application." *Human Genetics*, April 8.https://doi.org/10.1007/s00439-021-02272-5.

Kruse, Judith, Regina Mueller, Ali A. Aghdassi, Markus M. Lerch, and Sabine Salloch. 2022. "Genetic Testing for Rare Diseases: A Systematic Review of Ethical Aspects." Frontiers in Genetics 13 (January 26): 1–14.

Kukla, Quill. 2014. "Living with Pirates: Common Morality and Embodied Practice" *Cambridge quarterly of healthcare ethics : CQ : the international journal of healthcare ethics committees* 23, no. 1: 75-85.

Kulka, Rebecca. 2005. "Conscientious Autonomy: Displacing Decisions in Health Care." *Hastings Center Report* (March-April): 37-41. doi: 10.1353/hcr.2005.0025.

Kwok, Sylvia Y. C. L. 2019. "Integrating Positive Psychology and Elements of Music Therapy to Alleviate Adolescent Anxiety." *Research on Social Work Practice* 29, no. 6 (September 1): 663–76. https://doi.org /10.1177/1049731518773423.

Lachner, Christian, Michael J. Maniaci, Tyler F. Vadeboncoeur, Nancy L. Dawson, Teresa A. Rummans, Archana Roy, Lorrina L. Hall, and M. Caroline Burton. 2020. "Are Pre-Existing Psychiatric Disorders the Only Reason for Involuntary Holds in the Emergency Department?" *International Journal of Emergency Medicine* 13, no. 1 (February 3). https://doi.org/10.1186/s12245-020-0265-4.

LaFollette, Hugh. 2014. *Ethics in Practice: An Anthology*. Vol. Fourth Edition. Hoboken: Wiley-Blackwell.

Lagay, Faith. 2001. "Preimplantation Genetic Diagnosis." *AMA Journal of Ethics* 3, no. 8 (August 1): 265–68. https://doi.org/10.1001/ virtualmentor.2001.3.8.gnth1-0108.

Lander, Eric, Baylis, Francoise, Zhang, Feng, et. al, 2019. "Adopt a moratorium on heritable genome editing." *Nature.*

Landhuis, Esther. 2021. "The Gene Fix." *Scientific American* 325, no. 5 (November): S3–5.

Lane, Haylee, Mitchell Sarkies, Jennifer Martin, and Terry Haines. 2017. "Equity in Healthcare Resource Allocation Decision Making: A Systematic Review." *Social Science & Medicine* 175: 11–27.

Lang, Graeme. 2022. "Religious Responses to a Pandemic: Explanation, Compliance, and Defiance." Journal for the Study of Religion, Nature & Culture 16, 1: 23-49. doi:10.1558/jsrnc.19456.

Lappé, M. 1991. "Ethical Issues in Manipulating the Human Germ Line." *The Journal of Medicine and Philosophy* 16, no. 6 (December): 621–39. https://doi.org/10.1093/jmp/16.6.621.

Larmuseau, Maarten H. D. 2019. "Growth of Ancestry DNA Testing Risks Huge Increase in Paternity Issues." Nature Human Behaviour 3, no. 1 (January): 5–5. https://doi.org/10.1038/s41562-018-0499-9.

Lau, Tim. 2021. "Why the Climate Crisis Demands Democracy Reform." https://www.brennancenter.org/our-work/analysis-opinion/why-climate-crisis-demands-democracy-reform.

LaVeist, Thomas A., Darrell J. Gaskin, and Patrick Richard. 2009. "The Economic Burden of Health Inequalities in the United States."

Le Page, Michael. 2018. "What Is Crispr?" New Scientist 240, no. 3207 (December 8): 7–7.

Leal Filho, Walter, Linda Ternova, Muhammad Muddassir, et al. 2022. "An Analysis of Climate Change and Health Hazards: Results from an International Study," *International Journal of Climate Change Strategies and Management* 14, no.4: 375-98. doi:10.1108/IJCCSM-08-2021-0090.

Lebacqz, Karen and Joseph D. Driskill. 2000. *Ethics and Spiritual Care: A Guide for Pastors, Chaplains, and Spiritual Directors.* Nashville: Abingdon Press.

Lee, Joseph. 2022. Grist.org/indigenous/do-salmon-have-rights/. March 1.

Lee, Lisa M., and Lawrence O. Gostin. 2009. "Ethical Collection, Storage, and Use of Public Health Data: A Proposal for a National Privacy Protection." *JAMA: The Journal of the American Medical Association* 302, no. 1 (07/01): 82-84.

Lee, Lisa M., and Stephen B. Thacker. 2011. "Public Health Surveillance and Knowing About Health in the Context of Growing Sources of Health Data." *American Journal of Preventive Medicine* 41, no. 6: 636-640.

Lee, Lisa M., Charles M. Heilig, and Angela White. 2012. "Ethical Justification for Conducting Public Health Surveillance without Patient Consent." *American Journal of Public Health* 102, no. 1 (Jan): 38-44.

Lee, Thomas. 2023. *Healthcare's Path Forward: How Ongoing Crises Are Creating New Standards for Excellence*. McGraw Hill.

Lehmann, Lisa Soleymani, Willard S. Kasoff, Phoebe Koch, and Daniel D. Federman. 2004. "A Survey of Medical Ethics Education at US And Canadian Medical Schools." *Academic Medicine* 79, no. 7: 682-689.

Leiman, P. G., Arisaka, F., Van Raaij, M. J., Kostyuchenko, V. A., Aksyuk, A. A., Kanamaru, S., & Rossmann, M. G. 2010. "Morphogenesis of the T4 tail and tail fibers." *Virology Journal* 7, 355. https://doi.org/10.1186/1743-422X-7-355

Leist, Anton, and Peter Singer. 2010. *JM Coetzee and Ethics: Philosophical Perspectives on Literature*. Columbia University Press.

Lenzen, Manfred, Mengyu Li, Arunima Malik, Francesco Pomponi, et al. 2020. "Global Socio-Economic Losses and Environmental Gains from the Coronavirus Pandemic." *PloS One* 15, 7: e025654. doi: 10. 1371/journal.pone.0235654.

Leopold, Aldo. 2003. "The Land Ethic." In, Donald VanDeVeer and Christine Pierce, *The Environmental Ethics and Policy Book.* Wadsworth, Cengage Learning: 215-224.

Leucht, Stefan, Caroline Corves, Dieter Arbter, Rolf R Engel, Chunbo Li, and John M Davis. 2009. "Second-Generation versus First-Generation Antipsychotic Drugs for Schizophrenia: A Meta-Analysis." *The Lancet* 373, no. 9657 (January): 31–41. https://doi.org/10.1016/S0140-6736(08)61764-X.

Leucht, Stefan, Maximilian Huhn, and John M. Davis. 2021. "Should 'Typical', First-Generation Antipsychotics No Longer Be Generally Used in the Treatment of Schizophrenia?" *European Archives of Psychiatry and Clinical Neuroscience* 271, no. 8 (December 1): 1411–13. https://doi.org/10.1007/s00406-021-01335-y.

Levy, Carly R., Philips, Lynelle M., Murray, Carolyn J., et al. 2022. "Addressing Gaps in Public. Health Education to Advance Environmental Justice: Time for Action," *American Journal of Public Health*, 112 (1): 69-74.

Levy, Neil, Douglas, Thomas, Kahane, Guy, et. al. 2014. "Are you Morally Modified? The Moral Effects of Widely Used Pharmaceuticals." *Philos Psychiatr Psychol.* 21, no. 2: 111-125.

Lewis, Michelle H., Aaron Goldenberg, Rebecca Anderson, Erin Rothwell, and Jeffrey Botkin. 2011. "State Laws Regarding the Retention and Use of Residual Newborn Screening Blood Samples." Pediatrics 127, no. 4 (April): 703–12. https://doi.org/10.1542/peds. 2010-1468.

Lexchin, Joel, et al., 2003. "Pharmaceutical industry sponsorship and research outcome and quality: systematic review." *BMJ (Clinical research ed.)* 326, no. 7400: 1167-70.

Li, Quan, and Rafael Reuveny. 2006. "Democracy and Environmental Degradation." *International studies quarterly* 50, no. 4: 935-956.

Liao, Chun-Hua, Ming-Yuh Chang, Gwo-Chin Ma, Shun-Ping Chang, Chi-Fang Lin, Wen-Hsiang Lin, Hsin-Fu Chen, et al. 2019. "Preimplantation Genetic Diagnosis of Neurodegenerative Diseases: Review of Methodologies and Report of Our Experience as a Regional Reference Laboratory." Diagnostics 9, no. 2 (April 23): 44. https://doi. org /10.3390/ diagnostics9020044.

Liberty University John W. Rawlings School of Divinity. 2022. "Master of Divinity in Professional Chaplaincy: Healthcare | Rawlings School of Divinity." Liberty University. Accessed October 19. https://www.liberty.edu/divinity/masters/mdiv/healthcare-chaplaincy/.

Lieberman, Alexis, and Joav Merrick. 2009. *Poverty and Children: A Public Health Concern*, Nova Science Publishers, Incorporated.

Lim, Michelle ML, Peter Søgaard Jørgensen, and Carina A. Wyborn. 2018. "Reframing the Sustainable Development Goals to Achieve Sustainable Development in the Anthropocene—a Systems Approach." *Ecology and Society* 23, no. 3.

Linda Farber Post, and Jeffrey Blustein. 2021. *Handbook for Health Care Ethics Committees*. Third Edition, Baltimore: John Hopkins University Press.

Lindsey, Rebecca. 2020. "Climate Change: Global Sea Level." *Climate Change: Global Sea Level | NOAA Climate.gov*.

Liu, Jane Alice. 2019. "Organic Pesticides: Making Use of Nature." *Organic Pesticides | Making Use of Nature*. 23 July. www.foodunfolded.com/article/pesticide-alternatives-organic-farming.

Locher, Cosima, Sibylle Meier, and Jens Gaab. 2019. "Psychotherapy: A World of Meanings." *Frontiers in Psychology* 10 (March 22): 460. https://doi.org/10.3389/fpsyg.2019.00460.

Loftus, Peter. 2022. *The Messenger: Moderna, the Vaccine, and the Business Gamble That Changed the World*. Harvard Business Review Press.

Loike, John D., and Alan Kadish. 2018. "Ethical Rejections of Xenotransplantation? The Potential and Challenges of Using Human-Pig Chimeras to Create Organs for Transplantation." *EMBO Reports* 19, no. 8: e46337.

Lovin, Robin W. 2000. *Christian Ethics: An Essential Guide.* Nashville: Abingdon Press.

Luber, George, and Jay Lemery. 2021. *Global Climate and Human Health: From Science to Practice.* Jossey-Bass.

Lunshof, Jeantine E., Ruth Chadwick, Daniel B. Vorhaus, and George M. Church. 2008. "From Genetic Privacy to Open Consent." *Nature Reviews Genetics* 9, no. 5 (05/01): 406.

Lyons, Jim. 2020. "Trump Blames California for Fires. He Should Check to See Whose Land They're On." POLITICO. Accessed April 21, 2022. https://www.politico.com/news/agenda/2020/09/15/trump-fires-california-federal-land-415431.

Lysaught, Therese. 2011. "Moral Analysis of Procedure at Phoenix Hospital." Loyola cCommons, Loyola University Chicago), 537.

M'hamdi, Hafez Ismaili, Hilhorst, Medard, Steegers, Eric A P. et. al., 2017. "Nudge me, help my baby: on other-regarding nudges" *Journal of Medical Ethics.* 43, 702-706.

MacIntyre, Alasdair. 1999. *Dependent Rational Animals.* Carus Publishing Company.

Mackenzie, John S., Martyn Jeggo, Peter Daszak, and Juergen A. Richt. 2013. *One Health: The Human-Animal-Environment Interfaces in Emerging Infectious Diseases*. Vol. 366. Springer.

Mackler, Aaron L. 2003. *Introduction to Jewish and Catholic Bioethics: A Comparative Analysis*. Washington, D.C: Georgetown University Press.

Magill, Gerard, James Benedict, eds., forthcoming. 2023. *Resilience in Ecology and Health*. Newcastle Upon Tyne, UK: Cambridge Scholars Publishing.

Magill, Gerard, James Benedict, eds. 2023. *Strands of Sustainability*. Newcastle Upon Tyne, UK: Cambridge Scholars Publishing.

Magill, Gerard, James Benedict, eds. 2021. *Toward a Healthy Planet*. Newcastle Upon Tyne, UK: Cambridge Scholars Publishing.

Magill, Gerard, James Benedict, eds. 2020. *The Global Sustainability Challenge*. Newcastle Upon Tyne, UK: Cambridge Scholars Publishing.

Magill, Gerard, James Benedict, eds. 2019. *Cascading Challenges in the Global Water Crisis*. Newcastle Upon Tyne, UK: Cambridge Scholars Publishing.

Magill, Gerard, Jordan Potter, eds. 2018. *Integral Ecology: Protecting Our Common Home*. Newcastle Upon Tyne, UK: Cambridge Scholars Publishing.

Magill, Gerard, Kia Aramesh, eds. 2017. *The Urgency of Climate Change*. Newcastle Upon Tyne, UK: Cambridge Scholars Publishing.

Magnusson, Emilie, Anna Karin Axelsson, and Malin Lindroth. 2020. "'We Try' – How Nurses Work with Patient Participation in Forensic Psychiatric Care." *Scandinavian Journal of Caring Sciences* 34, no. 3: 690–97. https://doi.org/10.1111/scs.12773.

Mahajan, Rajiv. 2019. "Onasemnogene Abeparvovec for Spinal Muscular Atrophy: The Costlier Drug Ever." *International Journal of Applied and Basic Medical Research* 9, no. 3: 127–28. https://doi.org/10.4103/ijabmr.IJABMR_190_19.

Maier-Lorentz, Madeline M. 2008. "Transcultural Nursing: Its Importance in Nursing Practice." *Journal of Cultural Diversity* 15, no. 1: 37-43.

Malchiodi, Ms Cathy A. *The Handbook of Art Therapy and Digital Technology*. Jessica Kingsley Publishers, 2018.

Mallikarjuna, Mallana Gowdra, et al,, ed. *2022. Next-Generation Plant Breeding Approaches for Stress Resilience in Cereal Crops*. Springer.

"Management of Common Adverse Effects of Antipsychotic Medications." 2022. Accessed July 26.
https://onlinelibrary.wiley.com/ doi/epdf/10.1002/wps.20567.

Mandal, Jharna, Dinoop Korol Ponnambath, and Subhash Chandra Parija. 2016. "Utilitarian and Deontological Ethics in Medicine." Tropical Parasitology 6, no. 1: 5–7. https://doi.org/10.4103/2229-5070 .175024.

Manqi, Shu. 2020. "The Human Health Impacts of Global Climate Change." *E3S Web of Conferences* 204 (January 1):01006.
https://doi.org /10.1051/e3sconf/202020401005.

Marshall, Deborah A., Karen V. MacDonald, Sebastian Heidenreich, Taila Hartley, Francois P. Bernier, Meredith K. Gillespie, Brenda McInnes, A. Micheil Innes, Christine M. Armour, and Kym M. Boycott. 2019. "The Value of Diagnostic Testing for Parents of Children with Rare Genetic Diseases." Genetics in Medicine 21, no. 12 (December 1): 2798–2806. https://doi.org/10.1038/s41436-019-0583-1.

Martinez-Escobar, Daniel F., and Jennie Mallela. 2019. "Assessing the Impacts of Phosphate Mining on Coral Reef Communities and Reef Development." *Science of The Total Environment* 692: 1257-66.

Masson-Delmotte, V., P. Zhai, H.-O. Pörtner, D. Roberts, J. Skea, P. R. Shukla, A. Pirani, W. Moufouma-Okia, C. Péan, R. Pidcock, S. Connors, J. B. R. Matthews, Y. Chen, X. Zhou, M.I. Gomis, E. Lonnoy, T. Maycock, M. Tignor, and T. Waterfield (eds.). 2018. "Summary for Policymakers." In, Global Warming of 1.5°C. An IPCC Special Report on the impacts of global warming of 1.5°C above pre-industrial levels and related global greenhouse gas emission pathways, in the context of strengthening the global response to the threat of climate change, sustainable development, and efforts to eradicate poverty. Cambridge and New York: Cambridge University Press.

Matas, A. J., J. M. Smith, M. A. Skeans, B. Thompson, S. K. Gustafson, D. E. Stewart, W. S. Cherikh, J. L. Wainright, G. Boyle, and J. J. Snyder. 2015. "OPTN/SRTR 2013 Annual Data Report: Kidney." *American Journal of Transplantation* 15, no. S2: 1–34.

Mattick, Karen, and John Bligh. 2006. "Teaching and Assessing Medical Ethics: Where are We Now?." *Journal of Medical Ethics* 32, no. 3: 181-185.

Maust, Donovan T., Lewei A. Lin, and Frederic C. Blow. 2019. "Benzodiazepine Use and Misuse Among Adults in the United States." *Psychiatric Services* 70, no. 2 (February): 97–106.
https://doi.org/ 10.1176/appi.ps.201800321.

Mavilio, Fulvio, and Giuliana Ferrari. 2008. "Genetic Modification of Somatic Stem Cells." *EMBO Reports* 9, no. Suppl 1 (July): S64–69.

Mawlong, Banshaikupar Lyngdoh. 2018. *Climate Change and Developing* Countries. Cambridge Scholars Publishing. https://search-ebscohost-com.authenticate.library.duq.edu/login.aspx?direct=true&db =cat03556a&AN=duebr.EBC5549071&site=eds-live&scope=site.

Maxwell, Melody J., Mary H. Freire de Carvalho, Armando E. Hoet, Marco AN Vigilato, Julio C. Pompei, Ottorino Cosivi, and Victor J. del Rio Vilas. 2017. "Building the Road to a Regional Zoonoses Strategy: A Survey of Zoonoses Programmes in the Americas." *PLoS One* 12, no. 3: e0174175.

Mayo Clinic, *Dilation and curettage*. 2022. Accessed December 7. https://www.mayoclinic.org/tests-procedures/dilation-andcurettage/about/pac20384910#:~:text=Dilation%20and%20curettage%20(D%26C)%20is,after%20a%20miscarriage%20or%20abortion.

Mayo Clinic. 2022. "The Role of Genes in Your Alzheimer's Risk." Accessed April 26. https://www.mayoclinic.org/diseases-conditions/alzheimers-disease/in-depth/alzheimers-genes/art-20046552.

Mayzell, George. 2020. *The Resilient Healthcare Organization*. Productivity Press.

Mbugua, Karori. 2012. "Respect for Cultural Diversity and the Empirical Turn in Bioethics: A Plea for Caution." *Journal of Medical Ethics and History of Medicine* 5.

McAfee, Andrew, and Erik Brynjolfsson. 2012. "Big Data: The Management Revolution; Exploiting Vast New Flows of Information Can Radically Improve Your Company's Performance. But First You'll Have to Change Your Decision-Making Culture." *Harvard Business Review* 90, no. 10: 60.

McBride, Sebastian D., and A. Jennifer Morton. 2018. "Indices of Comparative Cognition: Assessing Animal Models of Human Brain Function." *Experimental Brain Research* 236, no. 12: 3379–90.

McCormick, Thomas R. 2013. *Principles of Bioethics*, School of Medicine, University of Washington. https://depts.washington.edu/bhdept/ethics-medicine/bioethics-topics/articles/principles-bioethics

McCrudden, Christopher, and King, Jeff. 2017. "The Dark Side of Nudging: The Ethics, Political Economy, and Law of Libertarian Paternalism." In *Choice Architecture in Democracies, Exploring the Legitimacy of Nudging,* edited by Alexandra Kemmerer, Christoph Mollers, Maximilian Steinbeis and Gerhard Wagner. 75-140.

McKee, Jerry, and Nancy Brahm. 2016. "Medical Mimics: Differential Diagnostic Considerations for Psychiatric Symptoms." *The Mental Health Clinician* 6, no. 6 (November 3): 289–96. https://doi.org/ 10.9740/mhc.2016.11.289.

McLean, S. 2009. *Autonomy, consent, and the law*. Oxfordshire, UK: Routledge-Cavendish.

McMurrin, Sterling M. 2011. *The Tanner Lectures on Human Values*. Vol. 8. Cambridge University Press.

Meadows, Donella. 1996-2022. Leverage Points: Places to Intervene in a System. (Academy for Systems Change, 1996) The Donella Meadows Project. https://donellameadows.org/archives/leverage-points-places-to-intervene-in-a-system/.

Med, J. Gene, Michael L. Edelstein, Mohammad R. Abedi, and Jo Wixon. 2007. "Gene Therapy Clinical Trials Worldwide to 2007 – An Update.".

Medlin, Marianne. 2010. Catholic Health Association Defies Phoenix Bishop Over Abortion Case, (Catholic News Agency). Accessed December 1, 2022. https://www.catholicnewsagency.com/news/21663/catholic-health-association-defies-phoenix-bishop-over-abortion-case

Meehan, Danielle. 2020. "Epistemic Vices and Epistemic Nudges: A Solution?" *Epistemic Paternalism: Conceptions, Justifications and Implications.* Amiel Bernal and Guy Axtell, eds. 249-261.

Melieres, Marie-Antoinette and Chloe Marechal. 2015. *Climate Change: Past, Present, and Future.* Chichester, West Sussex, UK: Wiley-Blackwell. https://search-ebscohost-com.authenticate.library.duq.edu/login.aspx?direct=true&db=nlebk&AN=948372&site=eds-live&scope=site.

Mensch, James. 2021. "Subjectivity in the Age of Pandemics." *Open Theology* 7, no. 1: 83–90.

Merriam-Webster.com Dictionary. 2022. "egalitarianism," accessed October 13. https://www.merriam-webster.com/dictionary/egalitarianism

Merriam-Webster.com Dictionary. 2022. "Pandemic," accessed October 7. https://www.merriam-webster.com/dictionary/pandemic.

Mesch, Gustavo and Ilan Talmud. 2006. "The Quality of Online and Offline Relationships: The Role of Multiplexity and Duration of Social Relationships." *The Information Society*, 22, no. 3: 137-148.

DOI:10.1080/01972240600677805 https://www.dhi.ac.uk/san/waysofbeing/ data/communication-zangana-mesch-2006b.pdf. Accessed 27 April 2022.

Mexico, University of New. 2011. "Jail Based Substance Abuse Treatment Literature Review ". *Institute for Social Research*: 1-9.

Midwestern Baptist Theological Seminary. 2022. "MDiv (Master of Divinity) – Standard Program | MBTS." Midwestern Baptist Theological Seminary. Accessed October 19. https://www.mbts.edu/degrees/masters-studies/mdiv/.

Mill, John Stuart. 2001. *Utilitarianism*. 2nd Edition. Ed. Sher. George. Indianapolis/Cambridge: Hackett Publishing Company, Inc.

Miller-Karas, Elaine. 2023. *Building Resilience to Trauma*. Routledge.

"Miscarriage - Prenatal Genetics." 2022. https://www.gbmc.org. Accessed February 6. https://www.gbmc.org/miscarriage.

MIT Technology Review. 2022. "2017 Was the Year Consumer DNA Testing Blew Up." Accessed March 13.

https://www.technologyreview.com2018/02/12/145676/2017-was-the-year-consumer-dna-testing-blew-up/.

Mitchell, Chace, Alan Lipps, Luz Padilla, Zoie Werkheiser, David KC Cooper, and Wayne Paris. 2020. "Meta-Analysis of Public Perception toward Xenotransplantation." *Xenotransplantation* 27, no. 4: e12583.

Mittelstadt, Brent D., and Luciano Floridi. 2016. "The Ethics of Big Data: Current and Foreseeable Issues in Biomedical Contexts." *Science and Engineering Ethics* 22, no. 2 (Apr): 303-341.

Mojtabavi, Helia, Nasirudin Javidi, AnneFrederique Naviaux, et al. 2021. "Exploration of the Epidemiological and Emotional Impact of Quarantine and Isolation During the COVID-19 Pandemic." *Covid-19 and Its Global Economic Impact*. Vol. 1318. Cham: Springer International Publishing. doi: 10.1007/978-3-030-63761-3_46

Möller, Hans-Jürgen. 2005. "Antipsychotic and Antidepressive Effects of Second Generation Antipsychotics: Two Different Pharmacological Mechanisms?" *European Archives of Psychiatry and Clinical Neuroscience* 255, no. 3 (June): 190–201. https://doi.org/10.1007/s00406-005-0587-5.

Moniruzzaman, Monir. 2019. "'The Heavier Selves': Embodied and Subjective Suffering of Organ Sellers in Bangladesh." *Ethos* 47, no. 2: 233–53.

Morens, David M., and Anthony S. Fauci. 2007. "The 1918 Influenza Pandemic: Insights for the 21st Century." *The Journal of Infectious Diseases* 195, no. 7: 1018–28.

Morgan, Nirvana, and Giada Del Fabbro. 2018. "Factors Associated with Recidivism at a South African Forensic Psychiatric Hospital." *South African Journal of Psychiatry* 24 (May 31). https://www.ajol.info/index.php/sajpsyc/article/view/172414.

Morita, Hiroyuki, Jonathan Seidman, and Christine E. Seidman. 2005. "Genetic Causes of Human Heart Failure." *The Journal of Clinical Investigation* 115, no. 3: 518–26.

Morris, Emma, Martin Gore, Andrew Baker, and Adrian J. Thrasher. 2013. "Moving Forward on Shifting Sands: Ethical Regulation of Gene Therapy Clinical Trials in the United Kingdom." *Molecular Therapy* 21, no. 4 (April): 715–16. https://doi.org/10.1038/mt.2013.43.

Mottaghipour, Yasaman, and Maryam Tabatabaee. 2019. "Family and Patient Psychoeducation for Severe Mental Disorder in Iran: A Review." *Iranian Journal of Psychiatry* 14, no. 1 (January): 84–108.

Mulatu, Mesfin Samuel and John W. Berry. 2001. "Health Care Practice in a Multicultural Context: Western and Non-Western Assumptions." In *Handbook of Cultural Health Psychology*. Eds. Shané S. Kazarian and David R. Evans, pp. 45-61. Academic Press.

Mullen, Paul E. "Forensic Mental Health." *The British Journal of Psychiatry* 176, no. 4 (April 2000): 307–11. https://doi.org/ 10.1192 bjp.176.4.307.

Munder, T., C. Flückiger, F. Leichsenring, A. A. Abbass, M. J. Hilsenroth, P. Luyten, S. Rabung, C. Steinert, and B. E. Wampold. 2019. "Is Psychotherapy Effective? A Re-Analysis of Treatments for Depression." *Epidemiology and Psychiatric Sciences* 28, no. 03 (June): 268–74. https://doi.org/10.1017/S2045796018000355.

Muvuka, Baraka, Ryan M. Combs, Suur D. Ayangeakaa, Nida M. Ali, Monica L. Wendel, and Trinidad Jackson. 2020. "Health literacy in African-American Communities: Barriers and Strategies." *HLRP: Health Literacy Research and Practice* 4, no. 3: e138-e143.

"My Garden of a Thousand Bees." 2021. Filmed and narrated by Martin Dohrn. PBS Television Series *Nature*. Directed by David Allen, produced by Gaby Bastyra. Thirteen Productions LLC, Passion Planet LTD and Ammonite LTD. Season 40, Episode 1.

Naess, Arne. 1973. "The Shallow and The Deep, Long-Range Ecology Movement". *Inquiry* 16: 95-100.

Naess, Arne. 1973. "The Shallow and the Deep, Long-Range Ecology Movement." *Inquiry 16.* University of Oslo: 95-100.

Naldini, Luigi. 2011. "Ex Vivo Gene Transfer and Correction for Cell-Based Therapies." *Nature Reviews Genetics* 12, no. 5 (May): 301–15.

Nam, Yoonsoo. 2015. "Do Democratic Governments Improve Environmental Quality?" *Graduate Student Theses, Dissertations, & Professional Papers*, 4500. https://scholarworks.umt.edu/cgi/viewcontent.cgi?article=5506&context=etd.

National Catholic Bioethics Center. 2010. "NCBC Commentary on the 'Phoenix Case.' Philadelphia, PA.

Neenan, Michael. 2017. *Developing Resilience: A Cognitive-Behavioural Approach*. Routledge.

Neergaard, Lauran. 2010. "Ethics Debate over Blood from Newborn Safety Tests." Boston.Com, February 8. http://archive.boston.com/news/health/articles/2010/02/08/ethics_debate_over_blood_from_newborn_safety_tests/.

Nelkin, Dana Kay, and Samuel C. Rickless. 2015. "So Close, yet so Far: Why Solutions to the Closeness Problem for the Doctrine of Double Effect Fall Short." *Noûs* 49, no. 2: 376–409.

Nemeth, Christopher, Robert Wears, David Woods, Erik Hollnagel, and Richard Cook. 2011. "Minding the Gaps: Creating Resilience in Health Care." In *Advances in Patient Safety: New Directions and Alternative Approaches (Vol. 3: Performance and Tools)*. Rockville: Agency for Healthcare Research and Quality.

"New Georgetown Value Highlights University Commitment to Planet." 2021. University News, 21 April. Georgetown University. www.georgetown.edu/news/new-georgetown-value-highlights-university-commitment-to-planet/.

New Orleans Baptist Theological Seminary. 2022. "Chaplaincy and Pastoral Care MDIV | Catalog." New Orleans Baptist Theological Seminary. Accessed October 20. https://catalog.nobts.edu/graduate/masters/degrees/divinity/chaplaincy-pastoral-care.

Newell, Peter. 2012. *Globalization and the Environment: Capitalism, Ecology and Power*. Cambridge, UK: Polity.

News 12. 2021. "Arizona Prisons Not Providing Treatment Programs to Most Inmatess, Report Finds " 12 News (Arizona), 1.

Nietzsche, Friedrich. 2006. *Thus Spoke Zarathustra*. Cambridge University Press.

Nill, Alexander, and Gene Laczniak. 2022. "Direct-to-Consumer Genetic Testing and Its Marketing: Emergent Ethical and Public Policy Implications." Journal of Business Ethics 175, no. 4 (February 1): 669–88. https://doi.org/10.1007/s10551-020-04632-z.

Norra, Stefan. 2014. “The Bisophere in Times of Global Urbanization.” *Journal of Geochemical Exploration* 147, no. Part A (December 1): 52-57. https://doi.org/10.1016/j.gexplo.2014.06.004.

Nunez, Christina. 2022. “Deforestation and Its Effect on the Planet.” *Environment National Geographic*. https://www.nationalgeographic.com/environment/article/deforestation#:~:text=About%2017%20percent%20of%20the,have%20been%20on%20the%20rise

Nunez, K. 2019. “What You Need to Know About Informed Consent.” *Healthline*. Retrieved December 15, 2022. https://www.healthline.com /health/informed-consent

NYU Langone Health. 2022. *Types of antibiotic-resistant infections*. Types of Antibiotic-Resistant Infections. Retrieved December 1. https://nyulangone.org/conditions/antibiotic-resistant-infections/types#:~:text=Types%20of%20Antibiotic-Resistant%20Infections%201%20Methicillin-Resistant%20Staphylococcus%20Aureus,a%20lung%20infection.%20...%203%20Carbapenem-Resistant%20Enterobacteriaceae%20

O’Hagan, Andrew & Hardwick, Rachel. 2017 "Behind Bars: The Truth About Drugs in Prison." *Forensic Research & Criminology International Journal* 5, no. 3: 309-20.

Ogundari, Kolawole, and Raphael Onyeaghala. 2021. “The effects of Climate Change on African Agricultural Productivity Growth Revisited.” *Environmental Science and Pollution Research International* 28, no. 23 (June), 30035-30045. https://doi.org/10.1007/s11356-021-12684-5.

Okamoto, Annika, Frank M. Dattilio, Keith S. Dobson, and Nikolaos Kazantzis. 2019. “The Therapeutic Relationship in Cognitive–Behavioral Therapy: Essential Features and Common Challenges.” *Practice Innovations* 4, no. 2 (June): 112–23. https://doi.org/10.1037/ pri0000088.

Organization for Economic Co-operation and Development (OECD). 2023. *Primary Health Care for Resilient Health Systems in Latin America*. OECD Health Policy Studies.

Ossege, Jennifer M., Richard W. Sears. 2017. *The Resilient Mental Health Practice*. Routledge.

Paerregaard, Karsten. 2020. “Communicating the Inevitable: Climate Awareness, Climate Discord, and Climate Research in Peru’s Highland Communities.” *Environmental Communication* 14, no. 1 (February): 112-125. https://doi.org/10.1080/17524032.2019.1626754.

Page, Scott E. 2017. The Diversity Bonus: How great teams pay off in the knowledge economy. Princeton and Oxford: Princeton University Press.

Parigi, Paolo, and Warner Henson. 2014. "Social Isolation in America." *Annual Review of Sociology*, 40: 153–71. http://www.jstor.org/stable/ 43049530.

Park, Yon Soo, and Benjamin Valentino. 2019. "Animals Are People Too: Explaining Variation in Respect for Animal Rights." *Hum. Rts. Q.* 41: 39.

Parker, Kelly A, Heather E. Keith. 2019. *Pragmatist and American Philosophical Perspectives on Resilience*. Lexington Books.

Parsapoor, Alireza, Mohammad-Bagher Parsapoor, Nima Rezaei, and Fariba Asghari. 2014. "Autonomy of Children and Adolescents in Consent to Treatment: Ethical, Jurisprudential and Legal Considerations." *Iranian Journal of Pediatrics* 24, no. 3 (June): 241–48.

Partin, A. W. 2021. "Ethics and Informed consent." *ScienceDirect Topics*. Retrieved December 14, 2022. https://www.sciencedirect.com/topics /medicine-and-dentistry/informed-consent

Paszti, Peter Gergely. 2020. "Measuring Resilience in Prisoners." no. 2: 73-86. DOI: 10.38146/BSZ.SPEC.2020.2.5.

Patel, Mitesh S, Day, Susan C., et. al. 2016. "Generic Medication Prescription Rates After Health System-wide redesign of default options within the electronic health record." *JAMA Internal Med.* 176, no. 6 847-848.

Patel, Mitesh, Volpp, Kevin G., Asch, David A. 2018. "Nudge Units to Improve the Delivery of Health Care." *N Engl J Med.* 378(3):. 214-216.

Patel, Mitesh, Volpp, Kevin G., Small, Dylan S., et. al. 2017. "Using Active Choice Within the Electronic Health Record to Increase Influenza Vaccination Rates." *Journal of General Internal Medicine.* 32, no. 7: 790-795.

Pathak, Bhawana, Rama Shanker Dubey, eds. 2023. *Climate Change and Urban Environment Sustainability*. Springer.

Patra, Jayanta Kumar, et al, eds. 2020. *Advances in Pharmaceutical Biotechnology: Recent Progress and Future Applications*. Springer.

Patterson, David W, Richard Harvey, Vlatka Matkovic, et al. 2022. "Post COP26: Legal Action Now Part of Public Health's Environment and Climate Change Toolbox," *European Journal of Public Health* 32, no.4 : 519-520.

Paul, Norbert W., Arthur Caplan, Michael E. Shapiro, Charl Els, Kirk C. Allison, and Huige Li. 2017. "Human Rights Violations in Organ Procurement Practice in China." *BMC Medical Ethics* 18, no. 1: 1–9.

Pellegrino, Edmund D. 2010. "Codes, Virtue, and Professionalism." In *Methods in Medical Ethics*, 2nd ed. Edited by Jeremy Sugarman and Daniel P. Sulmasy, 91-108. Washington, DC: Georgetown University Press.

Pellegrino, Edmund D. and David C. Thomasma. 1988. *For the Patient's Good: The Restoration of Beneficence in Healthcare.* New York: Oxford University Press.

Pellegrino, Edmund. 2001. "Philosophy of Medicine: Should it be Telelogically or Socially Constructed?" *Kennedy Institute of Ethics Journal.* 11(2). 2001, 169-180.

Penna, Suprasanna, et al, eds. 2023. *Mutation Breeding for Sustainable Food Production and Climate Resilience*. Springer.

Perakslis, Erik D., et al. 2021. *Digital Health: Understanding the Benefit-Risk Patient-Provider Framework*. Oxford University Press.

Pereno, Amina and Daniel Eriksson. 2020. "A Multi-stakeholder Perspective on Sustainable Healthcare: From 2030 Onwards." *Futures*, 122: 1-21.

Pergolizzi Jr, Joseph V., Jo Ann LeQuang, and Robert B. Raffa. 2021. "Benzodiazepines: Thinking Outside the Black Box." *Journal of Clinical Pharmacy and Therapeutics* 46, no. 3: 554–59. https://doi.org/10.1111/jcpt.13333.

Persad, Govind. 2019. "Will More Organs Save More Lives? Cost-Effectiveness and the Ethics of Expanding Organ Procurement." *Bioethics* 33, no. 6: 684–90.

Persson, Ingmar and Savulescu, Julian. 2010. "Moral Transhumanism." *Journal of Medicine and Philosophy*, 35: 656-669.

"Pesticides Active." N. d. *Pesticides | U.S. Geological Survey*, www.usgs.gov/centers/ohio-kentucky-indiana-water-sciencecenter/science/pesticides.

PETA. 2022. *People for the Ethical Treatment of Animals (PETA)*. Published April 24. Retrieved April 24, 2022, https://www.peta.org/

Petrini, Carlo. 2017. "Preemptive Kidney Transplantation: An Ethical Challenge for Organ Allocation Policies." *Clin Ter* 168, no. 3: e192-3.

"PGD and IVF Costs – What Is the Cost for Preimplantation Genetic Diagnosis–PGD and PGS, Preimplantation Genetic Screening?–Advanced Fertility Center of Chicago." 2022. Accessed April 25. https://advancedfertility.com/fertility-treatment/affording-care/pgd-cost/.

Pierce, Matthias, Hayhurst, Karen, Bird, Sheila M., Hickman, Matthew, Seddon, Toby, Dunn, Graham, & Millar, Tim. 2017. "Insights into the Link between Drug Use and Criminality: Lifetime Offending of Criminally-Active Opiate Users." *Drug and Alcohol ependence* 179: 309-16.

Pierre-Louis, Kendra, and John Schwartz. 2020. "Why Does California Have So Many Wildfires?" New York Times (Online). New York, United States: New York Times Company, August 21. https://www.proquest.com/docview/2435775530/citation/876FF3A7B2C14D31PQ/1.

Pitt, James J. 2010. "Newborn Screening." The Clinical Biochemist Reviews 31, no. 2 (May): 57–68.

Planted, One Tree. 2022. "1.2 Million Trees Planted in California for Forest Fire Restoration in 2021." One Tree Planted. Accessed April 25, 2022.https://onetreeplanted.org/blogs/stories/california-forest-fire-restoration-2021.

Plato. "Phaedo." 1961. In *The Collected Dialogues of Plato Including the Letters.* Eds. Hamilton, Edith. Cairns, Huntington. Bollingen Series LXXI. Princeton, NJ: Princeton University Pres, 40-98.

Plough, Alonzo L. 2021. *Community Resilience: Equitable Practices for an Uncertain Future*. Oxford University Press.

"Practice Guidelines for the Psychiatric Evaluation of Adults, Third Edition." 2015. In *The American Psychiatric Association Practice Guidelines for the Psychiatric Evaluation of Adults*, Third. Practice Guidelines. American Psychiatric Association. https://doi.org/10.1176/appi. books.9780890426760.pe02.

"Preimplantation Genetic Testing." 2022. Accessed February 6. https://nyulangone.org/locations/fertility-center/preimplantation-genetic-testing.

Pope Francis. 2015. *Laudato Si': Praise Be To You. On Care for Our Common Home*. Vatican City: Libreria Editrice Vatican.

Pope Francis. 2015. Encyclical Letter "Laudato Si': On Care for Our Common Home." Washington: United States Conference of Catholic Bishops.

Popovsky, M. 2010. "Exaggerated Benefts and Underestimated Harms: The Direct-to-Consumer Genetic Test Market and How to Manage It Going Forward." Dartmouth Law Journal 8: 65–87.

Pormeister, Kärt. 2019. "Genetic Research and Consent: On the Crossroads of Human and Data Research." Bioethics 33, no. 3 (March): 347–56. https://doi.org/10.1111/bioe.12475.

Porrett, Paige M., Babak J. Orandi, Vineeta Kumar, Julie Houp, Douglas Anderson, A. Cozette Killian, Vera Hauptfeld-Dolejsek, et al. 2022. "First Clinical-Grade Porcine Kidney Xenotransplant Using a Human Decedent Model." *American Journal of Transplantation.*

Portney, Kent E. 2015. *Sustainability.* MIT Press Essential Knowledge Series. The MIT Press. https://search-ebscohost-com.authenticate.library.duq.edu/login.aspx?direct=true&db=cat03556a&AN=duebr.EBC4397950&site=eds-live&scope=site.

Post, Eric. 2013. "Ecosystem Function and Dynamics." *In Ecology of Climate Change: The Importance of Biotic Interactions.* Monographs in Population Biology. Princeton University Press. https://search-ebscohost-com.authenticate.library.duq.edu/login.aspx?direct=true&db=edspmu&AN=edspmu.MUSE9781400846139.13&site=eds-live&scope=site.

Post, Eric S. 2013. *Ecology of Climate Change.: The Importance of Biotic* Interactions. Princeton University Press, 2013. https://search-ebscohost-com.authenticate.library.duq.edu/login.aspx?direct=true&db=cat03556a&AN=duebr.EBC1160068&site=eds-live&scope=site.

Power, Emmet, Niamh McCarthy, Ina Kelly, et al. 2022. "Climate Change and Mental Health: Time for Action and Advocacy." *Irish Journal of Psychological Medicine*, 1–3. doi:10.1017/ipm.2021.70

Prainsack, Barbara. 2020. "The value of healthcare data: to nudge, or not?," *Policy Studies* 41, no. 5. https://doi.org/10.1080/01442872.2020 .1723517.

Pratchett, M. S., S. K. Wilson, and A. H. Baird. 2006. "Declines in the Abundance of Chaetodon Butterflyfishes Following Extensive Coral Depletion." *Journal of Fish Biology* 69, no. 5: 1269-80.

Prevention, Centers for Disease Control. 2010. "Distinguishing Public Health Research and Public Health Nonresearch." Edited by Department of Health & Human Services.

Prince, James H., Jagdish Khubchandani, and Fern J. Webb. 2018. "Poverty and Health Disparities: What Can Public Health Professionals Do?" *Health Promotion Practice* 19, no 2: 170-74. doi:10.1177/ 1524839918755143.

Protti, Michele, Roberto Mandrioli, Camilla Marasca, Andrea Cavalli, Alessandro Serretti, and Laura Mercolini. 2020. "New-Generation, Non-SSRI Antidepressants: Drug-Drug Interactions and Therapeutic Drug Monitoring. Part 2: NaSSAs, NRIs, SNDRIs, MASSAs, NDRIs, and Others." *Medicinal Research Reviews* 40, no. 5: 1794–1832. https://doi.org/10.1002/med.21671.

Quane, James M., and William Julius Wilson. 2012. "Critical Commentary. Making the Connection between the Socialisation and the Social Isolation of the Inner-City Poor." *Urban Studies* 49, no. 14: 2977–87. Accessed 27 Apr. 2022. http://www.jstor.org/stable/26144125..

Rabin, Roni Caryn. 2022. "Kidneys From a Genetically Altered Pig Are Implanted in a Brain-Dead Patient." *The New York Times*, January 20, 2022, sec. Health. https://www.nytimes.com/2022/01/20/health/transplants-pig-human-kidney.html.

Raghupathi, Wullianallur, and Viju Raghupathi. 2014. "Big Data Analytics in Healthcare: Promise and Potential." *Health Information Science and Systems* 2, no. 3: 1-10.

Ranisch, Robert. 2020. "Germline Genome Editing versus Preimplantation Genetic Diagnosis: Is There a Case in Favour of Germline Interventions?" Bioethics 34, no. 1 (January): 60–69. https://doi.org/10. 1111/bioe. 12635.

Rastandeh, Amin, Meghann Jarchow, eds. 2022. *Creating Resilient Landscapes in an Era of Climate Change: Global Case Studies and Real-World Solutions*. Routledge.

"Reasonable or Random: 72-Hour Limits to Psychiatric Holds." 2022. Accessed July 29. https://doi.org/10.1176/appi.ps.202000284.

Regan, Tom. 1987. "The Case for Animal Rights." In *Advances in Animal Welfare Science 1986/87*, 179–89. Springer.

Regan, Tom. 2003. "The Case for Animal Rights." In, Donald VanDeVeer and Christine Pierce, *The Environmental Ethics and Policy Book.* Wadsworth, Cengage Learning: 143-149.

Regent University. 2022. "Master of Divinity (M.Div.)–Regent University." Regent University. Accessed October 20. https://www.regent.edu/program/mdiv-chaplain-ministry/#other-concentrations.

Regis College. 2021. "What are examples of social justice in Health Care?" *Regis College Online*. Retrieved December 15, 2022, https:// online.regiscollege.edu/blog/social-justice-in-health-care/.

Reich, Warren T., ed. 1995. *Encyclopedia of Bioethics*. Washington, DC: Simon & Schuster.

Reid Finlayson, Alistair J., Jane Macoubrie, Christy Huff, D.E. Foster, and Peter R. Martin 2022. "Experiences with Benzodiazepine Use, Tapering, and Discontinuation: An Internet Survey." *Therapeutic Advances in Psychopharmacology* 12 (January 1). https://doi.org/10.1177/20451253221082386.

Rendtorff, Jacob D. 2008. "The Limitations and Accomplishments of Autonomy as a Basic Principle in Bioethics and Biolaw." Chap. 6 In *Autonomy and Human Rights in Health Care: An International Perspective*, edited by David N. Weisstub and Guillermo Diaz Pintos, Dordrecht: Springer.

Renken, Elena. 2020. "Most Americans Are Lonely, and Our Workplace Culture May Not Be Helping." *NPR*, NPR, 23 Jan. https://www.npr.org/sections/health-shots/2020/01/23/798676465/most-americans-are-lonely-and-our-workplace-culture-may-not-be-helping.

Revel, Michel. 2009. "Article 12: Respect for Cultural Diversity and Pluralism." In *The UNESCO Universal Declaration on Bioethics and Human Rights: Background, Principles and Application*, edited by Henk ten Have and M.S. Jean, 199-209. Paris: UNESCO Publishing.

Review, The Regulatory. 2021. "Privacy Problems in the Genetic Testing Industry | The Regulatory Review," January 23. https://www.theregreview.org/2021/01/23/saturday-seminar-privacy-problems-genetic-testing/.

Roberts, David. 2019. "The Green New Deal, Explained." *Vox*.

Robinson, Mary. 2019. *Climate Justice: Hope, Resilience, and the Fight for a Sustainable Future*. Bloomsbury Publishing.

Roche, Patricia A., and George J. Annas. 2006. "DNA Testing, Banking, and Genetic Privacy." The New England Journal of Medicine 355, no. 6 (August 10): 545–46. https://doi.org/10.1056/NEJMp068136.

Roche, Ronan C., Chloe V. Harvey, James J. Harvey, Alan P. Kavanagh, Meaghan McDonald, Vivienne R. Stein-Rostaing, and John R. Turner. 2016. "Recreational Diving Impacts on Coral Reefs and the Adoption of Environmentally Responsible Practices within the Scuba Diving Industry." *Environmental Management* 58, no. 1: 107-16.

Rodd, Scott. 2022. "Budget Package Guarantees $1B In Fire Prevention Funding This Year, At Least $200M Annually Moving Forward." Accessed April 24, 2022. https://www.capradio.org/169616.

Rogers, Martha F., Mary Glenn Fowler, and Mary Lou Lindegren. 2001. "Revised Recommendations for Hiv Screening of Pregnant Women." *Morbidity and Mortality Weekly Report: Recommendations and Reports* 50, no. RR-19: 59-85.

Rohr, Jason. 2022. "Is Biodiversity Loss Increasing Diseases of Wildlife and Humans, And if so, What Can We Do?", Environmental Justice and Sustainability Lecture Series, Duquesne University. March 16.

Rooney, Phyllis. 1992. "On Values in Science: Is the Epistemic/Non-Epistemic Distinction Useful" *PSA: Proceedings of the Biennial Meeting of the Philosophy of Science Association* 1992: 13–22.

Root, Terry L., Jeff T. Price, Kimberly R. Hall, Stephen H. Schneider, Cynthia Rosenzweig, and J. Alan Pounds. 2003. "Fingerprints of Global Warming on Wild Animals and Plants." *Nature* 421, no. 6918: 57-60. https://doi.org/10.1038/nature01333.

Roski, J., G. W. Bo-Linn, and T. A. Andrews. 2014. "Creating Value in Health Care through Big Data: Opportunities and Policy Implications." *Health Affairs* 33, no. 7 (Jul): 1115-1122.

Ross, W. David. 1930. *The Right and the Good*. Oxford: Clarendon.

Rowe, Bertrand R., Andre Canosa, Amina Meslem, and Frantz Rowe. 2022. "Increased Airborne Transmission of COVID-19 with New Variants, Implications for Health Policies." *Building and Environment* 219. DOI: 10.1016/j.buildenv.2022.109132.

Rushton, Cynda. 2018. *Moral Resilience: Transforming Moral Suffering in Healthcare*. Oxford University Press.

Sahagun, Louis and Joseph Serna. 2020. "Coronavirus Fears Complicate Apple Fire Evacuation Plans." Los Angeles Times, August 5. https://www.latimes.com/california/story/2020-08-05/la-me-apple-fire-evacuations-coronavirus.

Sahoo, Amiya Kumar, Hari Charan Behera, and Ajit Kumar Behura. 2022. "Philosophy of Sustainable Development: Understanding Public Health," *Environment, Development & Sustainability* 24, no. 10: 12248-62. doi:10.1007/s10668-021-01945-5.

Saito, Osamu, ed. 2020. *Sharing Ecosystem Services: Building More Sustainable and Resilient Society*. Springer.

Sanabria, Edilma, Ronald Edgardo Cuenca, Miguel Ángel Esteso, and Mauricio Maldonado. 2021. "Benzodiazepines: Their Use Either as Essential Medicines or as Toxics Substances." *Toxics* 9, no. 2 (February): 25. https://doi.org/10.3390/toxics9020025.

Sandberg, Anders. 2013. *The Transhumanist Reader.* Max More and Natasha Vita-More, eds. Wiley-Blackwell. 56-64.

Sansone, Randy A., and Lori A. Sansone. 2014. "Serotonin Norepinephrine Reuptake Inhibitors: A Pharmacological Comparison." *Innovations in Clinical Neuroscience* 11, no. 3–4: 37–42.

Santarsieri, Daniel, and Thomas L Schwartz. 2015. "Antidepressant Efficacy and Side-Effect Burden: A Quick Guide for Clinicians." *Drugs in Context* 4 (October 8): 212290. https://doi.org/10.7573/dic.212290.

Sarkhel, Sujit, O. P. Singh, and Manu Arora. 2020. "Clinical Practice Guidelines for Psychoeducation in Psychiatric Disorders General Principles of Psychoeducation." *Indian Journal of Psychiatry* 62, no. Suppl 2 (January): S319–23. https://doi.org/10.4103/psychiatry.IndianJ Psychiatry_780_19.

Saul, Jack. 2022. *Collective Trauma, Collective Healing: Promoting Community Resilience in the Aftermath of Disaster*. Routledge.

Savulescu, Julian, and Guy Kahane. 2009. "The Moral Obligation to Create Children with the Best Chance of the Best Life." Bioethics 23, no. 5: 274–90. https://doi.org/10.1111/j.1467-8519.2008.00687.x.

Savulescu, Julian. 2001. "Procreative Beneficence: Why We Should Select The Best Children", *Bioethics.* 15 (5/6), 413-426

Savulescu, Julian. 2001. "Why Genetic Testing for genes for criminally is morally required." *Princeton Journal of Bioethics*. 4: 79-97.

Scarr, S. 1981. *Race, Social Class, and Individual Differences in I.q.* 1st edition. Hillsdale, N.J: Psychology Press.

Schaefer, Jame. 2022. Respondent to the Lecture by Rohr. Jason. "Is Biodiversity Loss Increasing Diseases of Wildlife and Humans, And if so, What Can We Do?", Environmental Justice and Sustainability Lecture Series, Duquesne University. March 16.

Schaftenaar, Petra, Ivo van Outheusden, Geert-Jan Stams, and Andries Baart. 2018. "Relational Caring and Contact after Treatment. An Evaluation Study on Criminal Recidivism." *International Journal of Law and Psychiatry* 60 (September 1): 45–50. https://doi.org/10. 1016/j.ijlp.2018.07.011.

Schmidt, Catrin. 2021. *Landscape Resilience: Basics, Case Studies, Practical Recommendations*. Springer.

Schoene-Seifert, Bettina. 2004. "Harm." In *Encyclopedia of Bioethics.* 3rd ed. Edited by Stephen G. Post. 1033-38. Volume 2 of 5. New York: Macmillan Reference USA.

Schumer, Charles E. 2021. "S.1260 - 117th Congress (2021-2022): United States Innovation and Competition Act of 2021." Legislation, June 8, 2021. 2021/2022. https://www.congress.gov/bill/117th-congress/senate -bill/1260.

Schwartzapfel, Beth & Jenkins, Jimmy. 2021. "Inside the Nation's Overdose Crisis in Prisons and Jails " *The Marshall Project.*

Schwartzapfel, Beth & Jenkins, Jimmy 2021. "Overdose Death in State Prisons Have Jumped Dramatically since 2001." *NPR.*

Science News. 2018. "Special Report: Genetic Testing Goes Mainstream," May 22. https://www.sciencenews.org/article/consumer-genetic-testing -ancestry-dna.

Sciorio, Romualdo, Raffaele Aiello, and Alfonso Maria Irollo. 2021. "Review: Preimplantation Genetic Diagnosis (PGD) as a Reproductive Option in Patients with Neurodegenerative Disorders." Reproductive Biology 21, no. 1 (March): 100468. https://doi.org/10.1016/j.repbio. 2020. 100468.

Sealey, L.A., B.W. Hughes, A. N. Sriskanda, J. R. Guest, A.D. Gibson, L. Johnson-Williams, D. G. Pace, O. Bagasra. 2016. "Environmental Factors in the Development of Autism Spectrum Disorders." *Environment International* 88 (March 1): 288-298. https://doi.org/10.1016/j. envint.2015.12.021

Senge, Peter. 2006. The Fifth Discipline: The art and practice of the learning organization. New York: Random House Books.

Senut, Marie-Claude, and Fred H. Gage. 1999. "Prenatal Gene Therapy: Can the Technical Hurdles Be Overcome?" *Molecular Medicine Today* 5, no. 4 (April 1): 152–56. https://doi.org/10.1016/S1357-4310(99) 014 41-0.

Seppänen, Allan, Iida Törmänen, Christopher Shaw, and Harry Kennedy. 2018. "Modern Forensic Psychiatric Hospital Design: Clinical, Legal and Structural Aspects." *International Journal of Mental Health Systems* 12 (October 20): 58. https://doi.org/10.1186/s13033-018-0238-7.

Seymour, Leonard W. 2006. "The Future of Gene Therapy in the UK." *Trends in Biotechnology* 24, no. 8 (August 1): 347–49. https://doi.org/10.1016/j.tibtech.2006.06.002.

SHARE, Michael. 2021. "COVID-19: The Pandemic That Swept the World: Its Political and Economic Effects on Human Security." *Intercultural Communication Studies* 30, 1: 49-66.

Shatner, William. Interviewed by Bolduan, Kate. 2021. CNN "At This Hour With Kate Bolduan." December 9. https://archive.org/details/CNNW_20211209_160000_At_This_Hour_With_Kate_Bolduan/start/3540/end/3600-3660

Shiner, Roger A. 1994. "Aristotle's Theory of Equity." *Loyola of Los Angeles Law Review* 27, no. 4: 1245-1264. https://digitalcommons .lmu.edu/llr/vol27/iss4/1.

Shoniregun, Charles A., Kudakwashe Dube, and Fredrick Mtenzi. 2010. *Electronic Healthcare Information Security.* New York, NY: Springer.

Shroff, Jai. 2016. "The Significant Impact of Animal Testing on the Environment." Jai Shroff's Blog. Published November 11. Retrieved May 6, 2022. https://jaishroff.wordpress.com/2016/11/15/the-signifi cant-impact-of-animal-testing-on-the-environment/

Silvernail, Carrie M., and Steven L. Wright 2022. “Surviving Benzodiazepines: A Patient’s and Clinician’s Perspectives.” *Advances in Therapy* 39, no. 5 (May): 1871–80. https://doi.org/10.1007/s12325-022-02055-y.

Simon, Matt. 2022. “California’s Wildfire and Covid-19 Disasters Just Collided.” *Wired*. Accessed February 14. https://www.wired.com/story/californias-wildfire-and-covid-19-disasters-just-collided/.

Sin, Julie. 2021. “Public Health: Three Key Domains.” In *Commissioning and a Population Approach to Health Services Decision-Making.* Oxford, 2020; online edn, Oxford Academic. https://doi-org.authenticate.library.duq.edu/10.1093/oso/9780198840732.003.0005,

Singer, Peter. 1996. "Animal Liberation." In *Animal Rights: The Changing Debate*, edited by Robert Garner. London, Palgrave Macmillan UK.: 135-142. https://doi.org/10.1007/978-1-349-25176-6_1

Singer, Peter. 2003. “Animal Liberation.” In, Donald VanDeVeer and Christine Pierce, *The Environmental Ethics and Policy Book.* Wadsworth, Cengage Learning: 135-142.

Singer, Peter. 2009. “Speciesism and Moral Status.” *Metaphilosophy* 40, no. 3–4: 567–81.

Singh, Rakesh Kumar, and Suprakash Chaudhury 2009. “Hyponatremia-Induced Psychosis in an Industrial Setting.” *Industrial Psychiatry Journal* 18, no. 2: 137–38. https://doi.org/10.4103/0972-6748. 62277.

Sklarew, Jennifer F. 2022. *Building Resilient Energy Systems*. Routledge.

Skinner, Michael K. 2018. *Encyclopedia of Reproduction*. Academic Press.

Smedley, Brian D., Adrienne Y. Stith, and Alan R. Nelson, eds 2002. *Unequal Treatment: Confronting Racial and Ethnic Disparities in Health Care*. Washington: The National Academies Press.

Smigrodzki, Rafal, Ben Goertzel, Cassio Pennachin, Lucio Coelho, Francisco Prosdocimi, and W. Davis Parker. 2005. “Genetic Algorithm for Analysis of Mutations in Parkinson’s Disease.” Artificial Intelligence in Medicine 35, no. 3 (November): 227–41. https://doi.org /10 .1016/j.artmed.2004.11. 006.

Smith, Cedric M. 2005. “Origin and Uses of Primum Non Nocere--Above All, Do No Harm!” *Journal of Clinical Pharmacology* 45: 371-77. doi: 10.1177/0091270004273680.

Smith, Sean. 2017. "Drug Treatment Programs of the Federal Bureau of Prisons Exist, but Need More Availability." *American Legislative Exchange Council*, 1.

Smith, Thomas W. 1999. "Aristotle on the Conditions for and Limits of the Common Good." *The American Political Science Review* 93, no. 3: 625-636.

Soini, S. 2007. "Preimplantation Genetic Diagnosis (PGD) in Europe: Diversity of Legislation a Challenge to the Community and Its Citizens." Medicine and Law 26, no. 2 (June): 309–23.

Solter, Davor, D. Beyleveld, M. B. Friele, J. Holówka, H. Lilie, R. Lovell-Badge, C. Mandla, U. Martin, and R. Pardo Avellaneda. 2003. Embryo Research in Pluralistic Europe. Springer Science & Business Media.

Soucie, J. M., B. Evatt, and D. Jackson. 1998. "Occurrence of Hemophilia in the United States. The Hemophilia Surveillance System Project Investigators." American Journal of Hematology 59, no. 4 (December): 288–94. https://doi.org/10.1002/(sici)1096-8652(199812)59:4<288::aid-ajh4>3.0.co;2-i.

Soucie, John Michael, Connie H. Miller, Brandi Dupervil, Binh Le, and Tyler W. Buckner. 2020. "Occurrence Rates of Haemophilia among Males in the United States Based on Surveillance Conducted in Specialized Haemophilia Treatment Centres." Haemophilia: The Official Journal of the World Federation of Hemophilia 26, no. 3 (May): 487–93. https://doi.org/10.1111/hae.13998.

South, Emily, Edward Cox, Nick Meader, Nerys Woolacott, and Susan Griffin. 2019. "Strimvelis® for Treating Severe Combined Immunodeficiency Caused by Adenosine Deaminase Deficiency: An Evidence Review Group Perspective of a NICE Highly Specialised Technology Evaluation." *PharmacoEconomics - Open* 3, no. 2 (June): 151–61. https://doi.org/10.1007/s41669-018-0102-3.

Southeastern Baptist Theological Seminary. 2022. "Program: Christian Ethics Track MDiv–Southeastern Baptist Theological Seminary–acalog ACMS." Southeastern Baptist Theological Seminary. Accessed October 20. https://catalog.sebts.edu/preview_program.php?catoid=22&poid=4315

Southern Baptist Convention. 2000. "The Baptist Faith and Message 2000." June 14th. https://bfm.sbc.net/bfm2000/.

Southern Baptist Theological Seminary. 2022. "Master of Divinity (MDiv) in Ethics and Philosophy." Southern Baptist Theological Seminary. Accessed October 19. https://www.sbts.edu/theology/degree-programs/mdiv/ethics-public-theology/.

Southwestern Baptist Theological Seminary. 2022. "Master of Divinity Courses and Concentrations, MDiv Courses." Southwestern Baptist Theological Seminary. Accessed October 20. https://swbts.edu/degree/master-of-divinity-concentrations/

Southwick, Stephen M. 2023. *Resilience*. Cambridge University Press.

Spadaro, Giuliana, Caroline Graf, Shuxian Jin, Sakura Arai, et al. 2022. "Cross-Cultural Variation in Cooperation: A Meta-Analysis." In *Journal of Personality and Social Psychology* 123, 5: 1024-88. doi:10.1037/pspi0000389.supp

St, World Bible Publishing. 1987. *The New American Bible*. New American Bible Revised ed. edition. Iowa Falls, Iowa: Catholic World Press/World Bible Publishers.

STAT. 2021. "Doctors Pledge to Do No Harm. The Entire Health Care Sector Should Do the Same by Battling Climate Change," August 11. https://www.statnews.com/2021/08/11/doctors-health-care-sector-battling-climate-change/.

Statistics, Bureau of Justice. 2021. "Drug Treatment under Correctional Supervision ". *Bureau of Justice Statistics*, 1.

Stegenga, Jacob. 2015. "Effectiveness of medical interventions." *Studies in history and philosophy of biological and biomedical sciences,* 54: 34-44.

Stegenga, Jacob. 2018. *Care & Cure: An Introduction to Philosophy of Medicine.* University of Chicago Press.

Stegenga, Jacob. 2018. *Medical Nihilism.* Oxford University Press.

Stenmark, Mikael. 2016. *Environmental Ethics and Policy-Making*. Florence: Taylor& Francis Group. ProQuest Ebook Central. Accessed May 18, 2022.

Stern, Harvey J. 2014. "Preimplantation Genetic Diagnosis: Prenatal Testing for Embryos Finally Achieving Its Potential." *Journal of Clinical Medicine* 3, no. 1 (March): 280. https://doi.org/10.3390/jcm 3010280.

Stern, Harvey J. 2014. "Preimplantation Genetic Diagnosis: Prenatal Testing for Embryos Finally Achieving Its Potential." Journal of Clinical Medicine 3, no. 1 (March 17): 280–309. https://doi.org/ 10.3390 /jcm3010280.

Stirnadel-Farrant, Heide, Mahesh Kudari, Nadia Garman, Jessica Imrie, Bikramjit Chopra, Stefania Giannelli, Michela Gabaldo, et al. 2018. "Gene Therapy in Rare Diseases: The Benefits and Challenges of Developing a Patient-Centric Registry for Strimvelis in ADA-SCID." *Orphanet Journal of Rare Diseases* 13, no. 1 (April 6): 49. https://doi.org/10.1186/s13023-018-0791-9.

Stone, Christopher D. 2003. "Should Trees Have Standing? – Toward Legal Rights for Natural Objects." (1974). In, Donald VanDeVeer and Christine Pierce, *The Environmental Ethics and Policy Book.* Wadsworth, Cengage Learning: 189-200.

Storrow, Richard F. 2011. "Assisted Reproduction on Treacherous Terrain: The Legal Hazards of Cross-Border Reproductive Travel." Reproductive BioMedicine Online, Symposium: Cross-Border Reproductive Care–Ethical, Legal, and Socio-Cultural Perspectives, 23, no. 5 (November 1): 538–45. https://doi.org/10.1016/j.rbmo.2011 .07.008.

Stout, Mike. 2021. *Building Community and Family Resilience: Research, Policy, and Programs*. Springer.

Strange, H., and R. Chadwick. 2010. "The Ethics of Nonmedical Sex Selection." Health Care Analysis: An International Journal of Health Care Philosophy and Policy 18, no. 3 (September 1): 252–66. https:// doi.org/10.1007/s10728-009-0135-y.

Streiffer, Robert. 2019. "Human/Non-Human Chimeras." In *The Stanford Encyclopedia of Philosophy*, edited by Edward N. Zalta, Summer. Metaphysics Research Lab, Stanford University, 2019.

Strong, Alan, and C. Barrientos. 2020. "Improved Satellite Techniques for Monitoring Coral Reef Bleaching." https://coralreefwatch.noaa.gov/ satellite/publications/crbpub_8icrs97txt.pdf

Strong, Carson. 2011. "Regulatory and Ethical Issues for Phase I In Utero Gene Transfer Studies." *Human Gene Therapy* 22, no. 11 (November): 1323–30. https://doi.org/10.1089/hum.2011.062.

Stucki, Saskia. 2020. "Towards a Theory of Legal Animal Rights: Simple and Fundamental Rights." *Oxford Journal of Legal Studies* 40, no. 3: 533–60.

Suggett, David J., and David J. Smith. 2020. "Coral Bleaching Patterns Are the Outcome of Complex Biological and Environmental Networking." Article. *Global Change Biology* 26, no. 1: 68-79. https:// doi.org/10.1111/gcb.14871.

Sullivan, Winnifred Fallers. 2014. *A Ministry of Presence: Chaplaincy, Spiritual Care, and the Law.* Chicago: University of Chicago Press. Proquest Ebook Central.

Sulmasy, Daniel P. 2014. "Ethical Principles for Spiritual Care." In *Oxford Textbook of Spirituality in Healthcare*. Edited by Mark Cobb, Christina M. Puchalski, and Bruce Rumbold, 465-70. New York: Oxford University Press.

Sunstein, Cass R. 2014. *Why Nudge? The Politics of Libertarian Paternalism.* Yale University Press.

Sutrop, Margit. 2011. "Changing Ethical Frameworks: From Individual Rights to the Common Good?". *Cambridge Quarterly of Healthcare Ethics* 20, no. 4: 533-545.

Sutthacheep, Makamas, Mathinee Yucharoen, Wanlaya Klinthong, Sittiporn Pengsakun, Kanwara Sangmanee, and Thamasak Yeemin. 2013. "Impacts of the 1998 and 2010 Mass Coral Bleaching Events on the Western Gulf of Thailand." *Deep Sea Research Part II: Topical Studies in Oceanography* 96: 25-31. https://doi.org/10.1016/j.dsr2. 2013.04.018.

Swift, Joshua K., Jennifer L. Callahan, Mick Cooper, and Susannah R. Parkin. 2018. "The Impact of Accommodating Client Preference in Psychotherapy: A Meta-Analysis." *Journal of Clinical Psychology* 74, no. 11: 1924–37. https://doi.org/10.1002/jclp.22680.

Takala, Tuija. 2007. "Setting a Dangerous Precedent: Ethical Issues in Human Genetic Database Research." *Medical Law International* 8, no. 2: 105-138.

Tan, Fei, Hongqiang Yang, Xiangrong Xu, Zhou Fang, Huilong Xu, Qi Shi, Xiyang Zhang, et al. 2020. "Microplastic Pollution around Remote Uninhabited Coral Reefs of Nansha Islands, South China Sea." *Science of The Total Environment* 725: 138383. https://doi.org/10.1016/j. scitotenv.2020.138383.

Tang, Ran, and Zhigang Xu. 2020. "Gene Therapy: A Double-Edged Sword with Great Powers." *Molecular and Cellular Biochemistry* 474, no. 1–2 (November): 73–81. https://doi.org/10.1007/s11010-020-0383 4-3.

Tarazona, Ariel M., Maria C. Ceballos, and Donald M. Broom. 2019. "Human Relationships with Domestic and Other Animals: One Health, One Welfare, One Biology." *Animals* 10, no. 1: 43.

Taryn, Luna. 2020. "Temperature Checks and Hotels: Newsom Details New Precautions for Wildfire Evacuations." *Los Angeles Times*, July 9. https://www.latimes.com/california/story/2020-07-09/newsom-coronavirus-california-fire-prevention-evacuations-covid-19.

Tauber, Alfred I. 2005. *Patient Autonomy and the Ethics of Responsibility.* Basic Bioethics. Cambridge, MA: MIT Press.

Taylor, S. 2020. "Anxiety Disorders, Climate Change, and the Challenges Ahead: Introduction to the Special Issue." *Journal of Anxiety Disorders* 74. doi: 10.1016/j.janxdis.2020.102313

Temin, Howard M. 1990. "Safety Considerations in Somatic Gene Therapy of Human Disease with Retrovirus Vectors." *Human Gene Therapy* 1, no. 2 (July 1): 111–23. https://doi.org/10.1089/hum.1990.1 .2-111

Tene, Omer, and Jules Polonetsky. 2011. "Privacy in the Age of Big Data: A Time for Big Decisions." *Stanford Law Review Online* 64: 63-69.

Terbeck, Sylvia, Kahane, Guy, McTavish, Sarah, et. al. 2012. "Propranolol reduces implicit negative racial bias" *Psychopharmacology.* 222: 419-424.

Testa, Megan, and Sara G. West. 2010. "Civil Commitment in the United States." *Psychiatry (Edgmont)* 7, no. 10 (October): 30–40.

Thacker, Stephen B. "Historical Development." 2010. In *Principles and Practice of Public Health Surveillance*, edited by Lisa M. Lee, Steven M. Teutsch and Michael E. St. Louis, New York: Oxford University Press, 1-17.

Thaler, Richard H., Sunstein, Cass R. 2021. *Nudge: The Final Edition.* Penguin Books.

The Association of Theological Schools (ATS) Commission on Accrediting. 2022. "Annual Data Tables." ATS Data Tables 2021-2022. https://www.ats.edu/files/galleries/2021-2022_Annual_Data_Tables.pdf.

The Association of Theological Schools (ATS) Commission on Accrediting. 2020 "ATS Commission Standards of Accreditation: June 2020." Approved June. The Association of Theological Schools (ATS). Pittsburgh: The Association of Theological Schools, 2020. https://www.ats.edu/files/galleries/standards-of-accreditation.pdf.

The Forest Foundation. N. d. "WILDFIRE." Accessed April 25, 2022. http://www.calforestfoundation.org/wildfire.

The Henry J. Kaiser Family Foundation. 2022. "Focus on Health Care Disparities Key Facts." Accessed November 9. https://kaiserfamilyfoundation.files.wordpress.com/2012/11/8396-disparities-in-health-and-health-care-five-key-questions-and-answers.pdf.

The National Law Review. 2022. "New California Law SB 1159 Creates Workers' Compensation Presumption for Certain Eligible Employees." Accessed April 2. https://www.natlawreview.com/article/new-california-law-sb-1159-creates-workers-compensation-presumption-certain-eligible.

The Sacramento Bee. 2022. "California Wildfire Victims Sent to Motels Instead of Shelters, to Lessen COVID-19 Spread." Accessed April 25. https://www.sacbee.com/news/california/fires/article245092745.html.

Theodosiou, Anastasia A., and Martin H. Johnson. 2011. "The Politics of Human Embryo Research and the Motivation to Achieve PGD." Reproductive BioMedicine Online 22, no. 5 (May 1): 457–71.

https:// doi.org/10.1016/j.rbmo.2011.01.008.

Therrell, Bradford L. 2003. "Ethical, Legal and Social Issues in Newborn Screening in the United States." The Southeast Asian Journal of Tropical Medicine and Public Health 34 Suppl 3: 52–58.

Thomas, Vinod. 2023. *Risk and Resilience in the Era of Climate Change*. Palgrave MacMillen.

Thomasma, David C. 2008. "Evolving Bioethics and International Human Rights." Chap. 1 In *Autonomy and Human Rights in Health Care: An International Perspective*, edited by David N. Weisstub and Guillermo Diaz Pintos, Dordrecht: Springer.

Thompson, Phillip. 2004. "Seeking Common Ground in a World of Ethical Pluralism: A Review Essay of Moral Acquaintances: Methodology in Bioethcs by Kevin wm. Wildes, Sj." *HEC* F. 16: 114.

Tilzey, Mark. 2017. *Political Ecology, Food Regimes, and Food Sovereignty: Crisis, Resistance, and Resilience*. Palgrave MacMillen.

Tokar, Brian. 2020. *Climate Justice and Community Renewal*. Routledge.

Torrey, Fuller, Doris Fuller, Jeffrey Geller, Carla Jacobs, and Kristina Ragosta. 2012. "No Room at the Inn- Trends and Consequences of Closing Public Psychiatric Hospitals-2005-2010." The Treatment Advocacy Center, July 19.

Trasande, Leonardo. 2012. Children's Environmental Health: Mercury and Lead, in William N. Rom, *Environmental Policy and Public Health: Air Pollution, Global Climate Change, and Wilderness*. San Francisco: Jossey-Bass.

Trejo-Gabriel-Galán, J. M. 2021. "Euthanasia and Assisted Suicide in Neurological Diseases: A Systematic Review." Neurologia (Barcelona, Spain), June 2, S0213-4853(21)00090-6. https://doi. org/ 10.1016/j.nrl. 2021.04.016.

Tsagkaris, Christos, and Konstantinos Kalachanis. 2020. "The Hippocratic Account of Mental Health: Humors and Human Temperament." *Mental Health: Global Challenges Journal* 3, no. 1 (September 20): 33–37. https://doi.org/10.32437/mhgcj.v3i1.83.

Tse, Wai S., Bond, Alyson J. 2002. "Serotonergic intervention affects both social dominance and affiliative behaviour". *Psychopharmacology.* 161: 324-330.

Tubbs, Nigel. 2018. "Green metaphysics: A sustainable and renewable liberal arts education." *Educational Philosophy and Theory* 50, no. 11: 1469-5812.

Tubbs, Nigel. 2008. *Education in Hegel*. London and New York: Continuum International Publication Group.

Tur-Kaspa, I., G. Aljadeff, S. Rechitsky, H. E. Grotjan, and Y. Verlinsky. 2010. “PGD for All Cystic Fibrosis Carrier Couples: Novel Strategy for Preventive Medicine and Cost Analysis.” Reproductive Biomedicine Online 21, no. 2 (August): 186–95. https://doi.org/10. 1016/j.rbmo.2010.04.031.

Turner, Leigh. 2003. "Bioethics in a Multicultural World: Medicine and Morality in Pluralistic Settings." *Health Care Analysis* 11: 99-117.

Turner, Leigh. 2005. "From the Local to the Global: Bioethics and the Concept of Culture." *The Journal of Medicine and Philosophy* 30, no. 3: 305-320.

UK Bioindustry Association. 2018. “A Guide to Cell and Gene Therapy and UK Excellence in the Field,” October.

Union of Concerned Scientists. 2003. “Prologue: World Scientists' Warning to Humanity.” In D. Vandeveer, & C. Pierce, *The Environmental Ethics and Policy Book.*

United Nations. 2019. *Climate Change*, at https://unfccc.int/event/cop-25.

United Nations Development Programme. 2015. *Sustainable Development Goals*. https://www.undp.org/content/undp/en/home/sustainable-development-goals.html.

United Nations Educational, Scientific and Cultural Organization. 2005. “Universal Declaration on Bioethics and Human Rights.” Paris, France: United Nations Educational, Scientific and Cultural Organization. http://portal.unesco.org/en/ev.phpURL_ID=31058& URL_DO=DO_TOPIC&URL_SECTION=201.html.

United Nations Environment Programme. 2020. UNEP Copenhagen Climate Centre. Emissions Gap Report 2020. Nairobi: UNEP.

United Nations General Assembly. 1948. “Universal Declaration of Human Rights” 217 (III) A. Paris: United Nations. https://www.un.org/en/about-us/universal-declaration-of-human-rights.

“United Nations Millennium Declaration.” 2000. *United Nations Chronicle* 37, no 3 (January 1): 38-49. https://search-ebscohost-com.authenticate.library.duq.edu/login.aspx?direct=true&db=edshol&AN=edshol.hein.unl.unchron0037.98&site=eds-live&scope=site.

United Nations. 2022. “Transforming our world: the 2030 Agenda for Sustainable Development,” under the *Department of Economic and Social Affairs*. Retrieved, September 18. https://sdgs.un.org/2030 agenda

United Nations. 2019. “Climate Change.” *United Nations*. https:// sustainable development.un.org/milestones/unced.

United Nations. 2022. Department of Economic and Social Affairs: Sustainable Development. Retrieved, September 15. https://sdgs.un.org/goals /goal3

University of Texas. 2022. *Ethics Unwrapped.* Accessed November 1. https://ethicsunwrapped.utexas.edu/glossary/consequentialism

UNOS. 2022. "Organ Transplant Trends | More Transplants than Ever." Accessed February 1. https://unos.org/data/transplant-trends/

U.S. Conference of Catholic Bishops. 2013. *Ethical and Religious Directives for Catholic Health Care Services*. Fourth printing.

U.S. Department of Health and Human Services Office of Minority Health. 2013. *National Standards for Culturally and Linguistically Appropriate Services in Health and Health Care: A Blueprint for Advancing and Sustaining Clas Policy and Practice*. Washington.

U.S. Department of the Interior. 2015. "Office of Wildland Fire: Budget," July 17. https://www.doi.gov/wildlandfire/budget.

U.S. Department of Veterans Affairs (VA). 2022. "VA Chaplain Services–Ecclesiastical Endorsement–Patient Care Services," October 11. https://www.patientcare.va.gov/chaplain/Ecclesiastical_Endorsement.asp.

U.S. Department of Veterans Affairs (VA). 2020. "VA Handbook 5005/135, Part II, Appendix G67." Updated September 24. https://www.va.gov/vapubs/viewPublication.asp?Pub_ID=1358&FType=2.

U.S. Environmental Protection Agency. 2021. Climate Change and Social Vulnerability in the United States: A Focus on Six Impacts. Washington: U.S. Environmental Protection Agency.

US EPA, OAR. 2016. "Greenhouse Gas Emissions from a Typical Passenger Vehicle." Overviews and Factsheets, January 12. https://www.epa.gov/greenvehicles/greenhouse-gas-emissions-typical-passenger-vehicle.

Van Ryn, Michelle, and Jane Burke. 2000. "The effect of Patient Race and Socio-Economic Status on Physicians' Perceptions of Patients." *Social Science & Medicine* 50, no. 6: 813-828.

VanDeVeer, Donald. 1986. *Paternalistic Interventions: The Moral Bounds of Benevolence.* Princeton University Press.

VanDeVeer, Donald and Christine Pierce. Eds. 2003. *The Environmental Ethics & Policy Book.* Third Edition. Belmont, CA: Wadsworth, 114-310.

Vanezis, Andrew Peter, and Dudley Manns. 2010. "Physical Examinations of Mental Health Service Users." *Progress in Neurology and Psychiatry* 14, no. 4: 19–23. https://doi.org/10.1002/pnp.168.

Veatch, Robert M. 2012. *Hippocratic, Religious, and Secular Medical Ethics: The Points of Conflict*. Washington D.C.: Georgetown University Press.

Vehmas, Simo, and Tom Shakespeare. 2014. "Disability, Harm, and the Origins of Limited Opportunities." Cambridge Quarterly of Healthcare Ethics 23, no. 1 (January): 41–47. https://doi.org/10.1017/ S096318011300042X.

Ventriglio, Antonio, Massimo Mari, Antonello Bellomo, and Dinesh Bhugra. 2015. "Homelessness and Mental Health: A Challenge." *International Journal of Social Psychiatry* 61, no. 7 (November 1): 621–22. https://doi.org/10.1177/0020764015585680.

Vergunst, F., & Berry, H. L. 2021. Climate Change and Children's Mental Health: A Developmental Perspective. *Clinical Psychological Science*. https://doi.org/10.1177/21677026211040787

Veterans Health Administration (VHA). 2021. "VHA Directive 1111 Spiritual Care." July 1. https://www.va.gov/vhapublications/View Publication.asp?pub_ID=9350.

Viator, J. A., Hazur, M., Sajewski, A., Tarhini, A., Sanders, M. E., & Edgar, R. H. 2020. "Photoacoustic detection of circulating melanoma cells in late stage patients." *Journal of innovative optical health sciences*. Retrieved October 10, 2022. https://www.ncbi .nlm.nih.gov/ pmc/articles/PMC8218985/

Vittozzi, Luciano, Georg F Hoffmann, Martina Cornel, and Gerard Loeber. 2010. "Evaluation of Population Newborn Screening Practices for Rare Disorders in Member States of the European Union." Orphanet Journal of Rare Diseases 5, no. S1 (October): P26, 1750-1172-5-S1-P26. https://doi.org/10.1186/1750-1172-5-S1-P26.

Walker, Rae, Wendy mason, and Karen De Heer. 2015. *Climate Change Adaptation: For Health and Social Services* (CSIRO Publishing), xix. https://search-ebscohost-com.authenticate.library.duq.edu/login.aspx?direct=true&db=cat03556a&AN=duebr.EBC4010517&site=eds-live&scope=site.

Wallis, Jacqueline Mae. 2020. "Is It Ever Morally Permissible to Select for Deafness in One's Child?" Medicine, Health Care, and Philosophy 23, no. 1 (March): 3–15. https://doi.org/10.1007/s11019-019-09922-6.

Walter, Fiona M., and Jon D. Emery. 2012. "Genetic Advances in Medicine: Has the Promise Been Fulfilled in General Practice?" The British Journal of General Practice 62, no. 596 (March): 120–21.

Walters, LeRoy and Julie G. Palmer. 1997.*The Ethics of Human Gene Therapy*. Oxford University Press.

Waltz, Emily. 2017. “When Pig Organs Will Fly.” *Nature Biotechnology* 35, no. 12: 1133-38

Wang, Leah & Sawyer, Wendy. 2021. "New Data: State Prisons Are Increasingly Deadly Places " In *Prison Policy Initiative*, 1.

Washington, Donna L., Jacqueline Bowles, Somnath Saha, Carol R. Horowitz, Sandra Moody-Ayers, Arleen F. Brown, Valerie E. Stone, Lisa A. Cooper. 2008. Writing Group for the Society of General Internal Medicine, Disparities in Health Task Force. "Transforming Clinical Practice to Eliminate Racial–Ethnic Disparities in Healthcare." *Journal of General Internal Medicine* 23: 685-691.

Weissman, Joel S., Joseph Betancourt, Eric G. Campbell, Elyse R. Park, Minah Kim, Brian Clarridge, David Blumenthal, Karen C. Lee, and Angela W. Maina. 2005. "Resident Physicians’ Preparedness to Provide Cross-Cultural Care." *JAMA* 294, no. 9: 1058-1067.

“What Are the Risks and Limitations of Genetic Testing? MedlinePlus Genetics.” 2022. Accessed March 13. https://medlineplus.gov/genetics/ understanding/testing/riskslimitations/.

“What Is Assisted Reproductive Technology? 2020. “Reproductive Health,” CDC. https://www.cdc.gov/art/whatis.html.

White, Alan T., Helge P. Vogt, and Tijen Arin. 2000. "Philippine Coral Reefs under Threat: The Economic Losses Caused by Reef Destruction." *Marine Pollution Bulletin* 40, no. 7: 598-605. https://doi.org/10.1016/S0025-326X(00)00022-9.

WHO. 2021. *2021 Annual Global Public Health Intelligence Report*, under the Creative Commons AttributionNonCommercial-ShareAlike 3.0 IGO licence (CC BY-NC-SA 3.0 IGO). https://creativecommons.org/ licen ses/by-nc-sa/3.0/igo

WHO. 2022. “10 Global Health Issues to Track in 2021.” Accessed December 1. https://www.who.int/news-room/spotlight/10-global-health-issues-to-track-in-2021.

WHO. 2022. “Quality of Care.” Accessed December 23. https://www.who .int/health- topics/quality-of-care#tab=tab_1.

“Why Treat Gene Editing Differently in Two Types of Human Cells? 2015. SciTech Connect.” Accessed February 6, 2022. https://scitechconnect.elsevier.com/treat-gene-editing-differently-human-cells/;

Wilson, Edward O. 2016. *Half-Earth: Our Planet’s Fight for Life*. New York/London: Liveright Publishing Corporation.

Wilson, M. Gabriela, Ball, J. Marion, Szczesny, Peter, et al. 2021. "Health Intelligence Atlas: A Core Tool for Public Health Intelligence," *Applied Clinical Informatics*, 12 (4): 944-53. Doi:10.1055/s-0041-1735973

Wirth, Thomas, Nigel Parker, and Seppo Ylä-Herttuala. 2013. "History of Gene Therapy." *Gene* 525, no. 2 (August): 162–69. https://doi.org/10.1016/j.gene.2013.03.137.

Wismer, Sharon, Sterling B. Tebbett, Robert P. Streit, and David R. Bellwood. 2019. "Spatial Mismatch in Fish and Coral Loss Following 2016 Mass Coral Bleaching." *Science of The Total Environment* 650 : 1487-98. https://doi.org/10.1016/j.scitotenv.2018.09.114.

Wong, Kaufui Vincent. 2016. *Climate Change.* Environmental Engineering Collection. Momentum Press, 41. https://search-ebscohost-com.authenticate.library.duq.edu/login.aspx?direct=true&db=cat03556a&AN=duebr.EBC4307182&site=eds-live&scope=site.

World Bank. 2022. "World Bank Group Supports Ghana's Covid-19 Response," Press Release. Assessed, November 12. https://www.worldbank.org/en/news/press-release/2020/04/02/world-bank-group-supports-ghanas-covid-19-response

Wray, Britt. 2022. Generation Dread: Finding purpose in an age of climate crisis. Toronto: Penguin Random House.

Wright, Jason D., Jacob C. Cogan, Yongmei Huang, Ana I. Tergas, Caryn M. St. Clair, June Y. Hou, Fady Khoury-Collado, et al. 2021. "Association of New Perioperative Benzodiazepine Use With Persistent Benzodiazepine Use." *JAMA Network Open* 4, no. 6 (June 3): e2112478. https://doi.org/10.1001/jamanetworkopen.2021.12478.

Wubeshet, Yirgalem Shewakena, Oumer Sada Mohammed, and Tigestu Alemu Desse. 2019. "Prevalence and Management Practice of First Generation Antipsychotics Induced Side Effects among Schizophrenic Patients at Amanuel Mental Specialized Hospital, Central Ethiopia: Cross-Sectional Study." *BMC Psychiatry* 19, no. 1 (December): 32. https://doi.org/10.1186/s12888-018-1999-x.

Xiao, Maggie Z. X., Syed Ali Akbar Abbass, Lisa Bahrey, Edward Rubinstein, and Vincent W. S. Chan. 2021. "A Roadmap for Environmental Sustainability of Plastic Use in Anesthesia and the Perioperative Arena." Anesthesiology 135, no. 4 (October 1): 729–37.

Yale E360. 2022. "Does a Fire-Ravaged Forest Need Human Help to Recover?" Accessed April 25. https://e360.yale.edu/features/does-a-fire-ravaged-forest-need-human-help-to-recover.

Yamashita, Ayako, Yoshioka, Shin-ichi, & Yajima, Yuki. 2021. "Resilience and Related Factors as Predictors of Relapse Risk in Patients with Substance Use Disorder: A Cross-Sectional Study ". *Substance Abuse Treatment, Prevention, and Policy* 16, no. 4: 1-9.

Yasny, Jeffrey S., and Jennifer White. 2012. "Environmental Implications of Anesthetic Gases." Anesthesia Progress 59, no. 4: 154–58.

Yohanna, Daniel 2013. "Deinstitutionalization of People with Mental Illness: Causes and Consequences." *AMA Journal of Ethics* 15, no. 10 (October 1): 886–91. https://doi.org/10.1001/virtualmentor.2013. 15.10.mhst1-1310.

Zhang, R., Tang, X., Liu, J. et al. 2022. "From concept to action: a united, holistic and One Health approach to responding to the climate change crisis." *Infect Dis Poverty* 11, 17. https://doi.org/ 10.1186/ s40249-022-00941-9

Zhang, Xu, and Xiao-Ying Li. 2020. "Prevalence of Hyponatremia among Older Inpatients in a General Hospital." *European Geriatric Medicine* 11, no. 4 (August): 685–92. https://doi.org/10.1007/s41999-020-00320-3.

Zimmerman, Mark. 2003. "What Should the Standard of Care for Psychiatric Diagnostic Evaluations Be?" *The Journal of Nervous and Mental Disease* 191, no. 5 (May): 281–86. https://doi.org/10.1097/01.NMD.0000066149.40946.FA.

Ziska, Lewis H. 2022. "Rising Carbon Dioxide and Global Nutrition: Evidence and Action Needed." *Plants (Basel, Switzerland)* 11, no. 7 (April 6):1-11. https://doi.org/10.3390/plants11071000.

Zola, Irving Kenneth. 1976. "Medicine as an institution of social control" *Ekistics.* 41, no. 245: 210-214.

Zuckerman, Edward L. 2019. *Clinician's Thesaurus, 8th Edition: The Guide to Conducting Interviews and Writing Psychological Reports*. Guilford Publications.

Zwicky, Jan. 2018. "A Ship From Delos." In Robert Bringhurst and Jan Zwicky, *Learning to Die: Wisdom in the Age of Climate Crisis*. Saskatchewan: University of Regina Press, 41-71.

Zwitter, Andrej. 2014. "Big Data Ethics." *Big Data & Society* 1, no. 2: 1-6.

CONTRIBUTORS

GABRIELLA AGOSTARO, Ph.D. received her Masters of Business Administration from California University of Pennsylvania, where she focused on effective management techniques and incentive programs for retail employees. She completed her doctoral degree at Duquesne University studying Healthcare Ethics, where she hopes to utilize her skills to delve into the field of clinical ethics. Her research focuses include end-of-life care, pain management, and the equitable treatment of individuals with psychiatric conditions.

ERCAN AVCI, Ph.D. is currently a Post-Doctoral Fellow at the Grefenstette Center for Science, Technology, and Law and the *Integrity of Creation* Conference Coordinator at Duquesne University. He has a MA in Political Sciences from Hacettepe University, an MS in Healthcare Administration from the University of New Haven, and a Ph.D. in Healthcare Ethics from Duquesne University. Dr. Avci worked for the Ministry of Health in Turkey as a Health Inspector for over 20 years. As an adjunct faculty, he has also taught ethics courses at different universities, including Duquesne University and Chatham University. Dr. Avci's research interests lie in several areas of bioethics, including ethics education, clinical ethics, end-of-life ethics, religious ethics, and AI ethics.

JENNIFER ANN BATES, Ph.D. is Full Professor of Philosophy at Duquesne University. She specializes in Hegel and 19th Century German Philosophy; she also publishes and teaches Environmental Philosophy. She is the author of *Hegel's Theory of Imagination*, and *Hegel and Shakespeare on Moral Imagination*, and co-editor (with Richard Wilson) of *Shakespeare and Continental Philosophy*. Alongside articles and book chapters on German Philosophy, Bates has published "An Inquiry Into the Nature Of Environmentally Sound Thinking" (*Environmental Ethics*), "Hegel and the Concept of Extinction" (*Philosophy Compass*), and "Sustaining the Trivial: Why the Liberal Arts and the Environment Need to be Sustained Together" (in *The Global Sustainability Challenge,* 2020). She is General Editor of *Idealistic Studies: an Interdisciplinary Journal of Philosophy.*

JAMES L. BENEDICT, Ph.D. is a Scholar-in-Residence in the Center for Global Health Ethics at Duquesne University. He graduated with his Ph.D. degree from Duquesne University in 2015. He has also earned a Doctor of Ministry degree from the Lutheran Seminary at Philadelphia in 2004, a Master of Sacred Theology from the Lutheran Seminary at Gettysburg in 1995, and a Master of Divinity degree from Bethany Theological Seminary in 1986. He has served over 30 years in parish ministry in the Church of the Brethren and is currently a member of the Ethics Committee of the York, Pennsylvania VNA, a division of WellSpan. He is the author of *A Revised Consent Model for the Transplantation of Face and Upper Limbs: Covenant Consent* (International Library of Ethics, Law, and the New Medicine), numerous study guides, and several scholarly and popular articles in ethics and religion. In addition, over recent years he was an invited scholar to a Brocher Conference on Bioethics in Reconstructive Transplantation (Geneva, Switzerland), and to the Symposium on the Evolving Issues of Vascularized Composite Allotransplantation at New York University and Johns Hopkins University, and he was an invited presenter at the 13th Congress of the International Society for Vascularized Composite Transplantation in Salzburg, Austria. He has served on the editorial board and as a book review editor for Brethren Life and Thought and received a post-doctoral study grant from the Lilly Foundation for the study of Early Christian art in Rome, Italy.

MEGHAN CHAPPLE is the Vice President for Sustainability at Georgetown University in Washington D.C. She oversees a unique set of operational, academic, and policy responsibilities to position Georgetown as a prominent contributor both nationally and globally on sustainability. She brings expertise in innovation, entrepreneurship, renewable energy, urban farming, biodiversity, inclusive sustainability, environmental justice, circularity, and green buildings. She has worked as a change agent for sustainability in the corporate and nonprofit sectors. Most recently she led the George Washington University (GW) through its comprehensive sustainability strategy process, and established GW's reputation as a leader in sustainability. She advised F50 corporations at SustainAbility and Odwaga, managed the global business education network at World Resources Institute, led youth education for The Nature Conservancy on the southside of Chicago, served in Americorps with Public Allies Chicago, and has supported many social innovation and sustainable enterprises. She has an MBA in Corporate Strategy and a MS in

Environmental Policy from the University of Michigan, and a BA in Environmental Science from Northwestern University.

SAMUEL KWAME DANSO is an ordained Catholic Priest incardinated into the Catholic Diocese of Konongo-Mampong, Ghana. Since December 2018, he has been living in Chariton, Iowa (USA), as a pastor in three rural parishes in Corydon, Leon, and Chariton. He had his seminary formation in the Pontifical Beda College in Rome, where he graduated with a Bachelor of Divinity and a License in Theology from the University of Wales, Lampeter, (UK) in 2008. Before his seminary formation, he had a bachelor's degree in Business Education from the University of Cape Coast, Ghana, and has taught at the high school level for thirteen years. He has also done some pastoral and administrative work in his home diocese. He also holds a Master of Business Administration (MBA) degree from the University of the People, California. Currently, he is pursuing a Ph.D. program in Healthcare Ethics at Duquesne University, Pittsburg (USA). He is researching and focusing his final dissertation on Public Health Ethics with associated issues of the environment.

IAN M. DOHERTY is a Ph.D. student in Healthcare Ethics at Duquesne University. He earned a Bachelor of Arts degree from Ohio State University in Journalism and Political Science. He has a Master of Arts degree in Ethics and Society from Fordham University. Ian hopes to become a clinical ethicist.

SCOTT DYER is from the small New England state of New Hampshire. There he received his undergraduate degree at Keene State College before earning his Master's in Philosophy of Aesthetics from Stony Brook University in New York, graduating Suma Cum Laude in the philosophy program. After a five year stay in New York City and a brief stint in journalism, he moved to Pittsburgh, Pennsylvania to attend Duquesne University in pursuit of a Ph.D. in Healthcare Ethics. At Duquesne, he focuses on clinical ethicist skills as well as bioethical policymaking regarding euthanasia and assisted suicide.

ANDREW HARRINGTON, Ph.D. is a graduate of Duquesne University in Pittsburgh, PA where he earned his Doctoral degree and Master's degree in Healthcare Ethics. For the past five years, he has worked at Duquesne University's Center for Global Health Ethics, where he initially started as a graduate assistant and then was ßthe Center's Clinical Ethics Fellow. He has successfully supervised healthcare ethics graduate students through their clinical rotations and ethics seminars. He has also assisted and co-facilitated many of the *Integrity of Creation* conferences hosted by Duquesne University. His research interests are on the ethical governance of data analytics in healthcare with specific focus on artificial intelligence, big data and privacy/confidentiality issues related to healthcare data.

BRIANNE HELFRICH is a PhD student studying Healthcare Ethics at Duquesne University with a goal of providing ethical consultations to patients and being an instrumental figure in creating and implementing worldwide healthcare policies. Originally from Alberta, Canada, she first attended the University of Arizona where she completed her undergraduate degree in Physiology, further moving onto her PhD at Duquesne University. She has been recognized for her academic achievements having won the Award for Graduate Research in Ethics and presenting at the 10th International Conference for Education in Ethics in Padua, Italy. She is extremely eager to pursue her education in this field. She is extremely passionate about raising awareness on important topics and in 2021, she started her own clothing brand in which a portion of all proceeds are donated to various charities.

SCOTT N. HELSTAD is a PhD student at Duquesne University, specializing in health care ethics. He has a Master of Arts in Catholic Studies from the University of St. Thomas. Where he studied the Catholic intellectual tradition through the lens of thinkers like Augustine, Thomas Aquinas, John Henry Newman, and Romano Guardini. He has a passion for social teaching and has experience within health care, advocating and implementing both the integration of the systems mission and ethical values into the daily practices of operations. His goal is to communicate the Catholic faith in a captivating way for all people to engage with.

ETHAN HERSTEK completed both his Masters of Science degree in Biomedical Engineering and his Bachelors of Science degree in Biomedical Engineering at Duquesne University in Pittsburgh, PA.

GERARD MAGILL, Ph.D. holds the Vernon F. Gallagher Chair for the Integration of Science, Theology, Philosophy, and Law at Duquesne University in Pittsburgh (appointed in 2007) where he is a tenured Professor in the Center for Global Health Ethics. Also, he holds the position of Senior Research Fellow at the Carl G. Grefenstette Center for Ethics in Science, Technology and Law at Duquesne University. He graduated with his Ph.D. degree from Edinburgh University in Scotland.

In 1987 he started his scholarly career at Saint Louis University, where in 1996 he was appointed as the Department Chair of the Center for Healthcare Ethics. As Executive Director of that Center from 1999, he held multiple appointments including being a member of the Deans' Council for the University's health sciences campus, a Division Director in the Department of Internal Medicine at the University Hospital, a member of the University's Institutional Review Board, a member of the Hospital Ethics Committee, a Professor in the School of Medicine (secondary appointment), and a Professor in the School of Public Health (secondary appointment). He has authored, co-authored, or edited 15 books, including a co-authored textbook on Health Care Ethics (Georgetown University Press) and a co-authored book on Governance Ethics for Healthcare Organizations (Routledge). He has published over 80 scholarly and professional articles, and he has given over 250 scholarly presentations at scholarly conferences. He is a member of 17 Professional Associations. He has extensive experience on Institutional Review Boards, Hospital Ethics Committees, and Ethics Consultation Services. He continues to serve as Chair of the Planning Committee for the annual endowed conference series on the *Integrity of Creation* at Duquesne University.

ASHLEY MAZUR-WILLIAMS is a healthcare provider who transitioned into a career in private health advocacy & the medical device industry after being accepted as one of six individuals from the entire Arizona State University student population to intern for the Mayo Clinic & ASU MedTech Accelerator program. She graduated Cum Laude from ASU with her dual Bachelor of Science degrees in Biology and Political Science. She was a skilled clinician and educator for 10+ years both pre-hospital and in hospital specializing in emergency & critical care. She has now transitioned into the roles of health advocate for patients & commercialization consultant for medical device companies, focusing on quality, regulatory, ethics, and compliance, and is a Certified HIPAA Privacy Security Expert. She has a MA in Catholic Healthcare Ethics and is beginning the dissertation phase of her PhD in healthcare ethics.

FERDINAND OKAFOR is a Catholic priest from Nigeria. He holds bachelor's degrees in philosophy and Sacred Theology; a post graduate Diploma in Education, and a Master of Arts in Religious Education. He also has a certification, *Director of Catechetical Apostolate,* from the Franciscan University of Steubenville, Ohio. And he is currently pursuing his PhD studies in Healthcare Ethics at Duquesne University, with focus in autism spectrum disorder. His interests include research, writing in Catechesis, Psychology, and Clinical Ethics.

ROBERT ALAN PRICE, D. Min., led the Chaplaincy and Ethics programs across three complex VA Healthcare Systems. In the nonprofit sector, he served as Executive Director of a counseling agency and as Senior Pastor of two diverse churches. In 2022, he joined VA's National Center for Ethics in Health Care as a Regional Healthcare Ethicist. He is a Certified Healthcare Ethics Consultant (HEC-C), Board Certified Chaplain with a specialty certification in suicide prevention (BCC-SP), and has earned the Doctor of Ministry degree, three Master's Degrees in Religion, Divinity, and Counseling, and is currently completing his Doctor of Philosophy degree in Healthcare Ethics at Duquesne University.

FRANCIS SARPONG is a Ph.D. student in Healthcare Ethics at Duquesne University. He holds a Master of Business Administration from the University of the People, USA, and a Master of Philosophy in Philosophy from the University of Ghana. In addition, he obtained a Bachelor of Theology from the Pontifical Urban University, Rome, and a Bachelor of Arts in Sociology and Religions from the University of Ghana. He has served as an assistant lecturer at the St. Gregory Seminary, Kumasi, Ghana, and St. Paul's Catholic Seminary, Accra, Ghana, where he taught Ancient Philosophy, Philosophical Ethics, Logic, Metaphysics, and African Philosophy.

JENNIFER SCHINKE completed a Master of Science in Biomedical Engineering at Duquesne University in Pittsburgh. This is also where she also earned a Bachelor of Science in Biomedical Engineering and Applied Mathematics. She is a member of SWE (Society of Women Engineers) and a mentor for Women in STEM. She is currently a verification and validation engineer in the radiology department at Bayer Healthcare in Indianola, Pennsylvania.

JAMES SWINDAL, Ph.D., James Swindal is a Professor of Philosophy at the McAnulty College and Graduate School of Liberal Arts and the Henry Koren, C.S.Sp Endowed Chair in Scholarly Excellence, and Director of Undergraduate Studies. He specializes in Continental Philosophy, Critical Theory, Action Theory, and Catholic Philosophy. He is currently working on a book project, with co-author William Wright, on *An Existentialist Account of Eucharistic Action.* He is also engaged in projects on Critical Theory and Psychoanalysis, Charles Taylor's Catholic Philosophy, and Robert Brandom's Social Pragmatism.

Student Contributors

HAILEY ADAMIK is a Forensic Science and Law student at Duquesne University. She has a keen interest in wildlife conservation and environmental science, seeking to pursue a career that would nurture this life-long love of wildlife. She is especially fascinated with both marine biology and ocean conservation. Amongst the many dire environmental issues our generation faces, she considers matters surrounding our oceans as most troubling. As a result, she pursues this passion by undertaking an internship at the Pittsburgh Zoo and PPG Aquarium as a Water's Edge intern. Her goal is to spread awareness of the disturbing issues surrounding ocean conservation, hoping to begin a career in the realm of wildlife or marine life conservation.

LEI'ASHA BATTLE is enrolled at Duquesne University in the school of Nursing as a first generation student. She is a member of several organizations, including Black Student Union (BSU), serving as the Engagement Coordinator Chair, and the Ebony Women for Social Change, serving as the Event Coordinator. She created the campus's Art Club, focusin on art is a form of free expressing and good for mental stress release. Also, she is a student scholar at the YWCA, where she teaches at Sterrett Classical Academy STEM Stars lessons.

CALLI CAHILL graduated with a Master of Arts in Bioethics from the Ohio State University in 2023. Her interest in all things bioethics, particularly climate change, began in 2016 when she won the National Ethics Bowl Competition with the Snow College Ethics Bowl team. From here, she pursued a Bachelor's degree in biology with a minor in philosophy at Westminster College in Salt Lake City

HANNAH CAWLEY is a student from New Jersey who studies at Duquesne University. She graduated in May of 2022 with a Bachelor of Arts in Biochemistry, and since the fall of 2022 she has been pursuing a Master of Science in Forensic Science and Law. The California wildfires are something she paid close attention to in the 2020 season and decided to do some further research after learning of the environmental impacts. After graduation, she hopes to pursue a career in forensics, focusing on investigations or DNA Analysis.

COURTNEY FEREK is from North Huntingdon, Pennsylvania, and is studying speech-language pathology at Duquesne University, being a Dean's List student. Courtney enjoys travelling and being outdoors with friends and family. Courtney is an animal lover, with a passion for helping others to make a positive impact on peoples' lives.

AMANDA KOLSON is a recent graduate of Duquesne University's McAnulty College and Graduate School of Liberal Arts, with a Bachelor of Arts in Psychology, a minor in Sociology, and a certificate in Pre-Law. Throughout her undergraduate career, Amanda has completed two research projects related to psychology and environmental issues. The first was conducted in 2019 at Mercyhurst University, in which she and a group of fellow students analyzed the Liking-Tactic of Persuasion in order to encourage more recycling on campus. In 2021, she also completed an independent research study on Anxiety Levels in Relation to Climate Change in Undergraduates. Amanda plans to continue her studies by attending the University of Edinburgh, Scotland, in the Master's program in Psychology of Mental Health.

ALEXANDER MOLINARO is a graduate of Duquesne University Class of 2022. While attending Duquesne, he majored in Sociology with a focus in Criminal Justice and a minor in Psychology, and has taken multiple classes regarding culture, religion, sex and gender, and juvenile delinquency. He is a member of AKD International Honor Society of Sociology. He spoke at the *Integrity of Creation* Conference in Fall of 2020, and created a Poster Presentation in Spring 2022, from which his essay in this collection derives. After spending the past four years at Duquesne, Alex is continuing to pursue his education as he is looking to earn a master's degree in sociology or social work.

KATELYNN PIDGORODETSKIY is an Occupational Therapy Student in the Rango's School of Health Sciences at Duquesne University. She has always loved the outdoors and has had an interest in learning about nature and what we can do to preserve it. As a program requirement, she chose to take the elective, Philosophy of the Environment course. This enabled her to learn about the different ways that our environmental struggles and can be thought about.

JOHN SPARBER was born and raised in Cleveland, Ohio. Growing up an only child he attended an all-boys private school in Hunting Valley,

Ohio. After graduating, he pursued his education at Duquesne University where he is studying finance. After finishing studies at Duquesne, he hopes to acquire a master's degree in my field. Once he finishes his education, he looks forward to joining the corporate world and using the knowledge acquired in school.